Acoustics of Speech Production

R. S. McGowan

CReSS
Books

Acoustics of Speech Production

Copyright ©2018, by Richard S. McGowan. All rights reserved. This book or any portion thereof may not be reproduced or used in any manner whatsoever without the express written permission of the publisher except for the use of brief quotations in a book review.

First printing 2018

Second printing with corrections 2023

ISBN: 978-0-9997574-4-4

CReSS Books
1 Seaborn Place
Lexington, MA 02420

www.cressbooks.com

for Winifred and Rebecca

Table of Contents

Preface xi

Chapter 1: Air: The Acoustic Material in the Vocal Tract, and Basic Some Basic Physics

Introduction 1
Properties of air 1
One-dimensional kinematics and dynamics of a point mass 2
Thermodynamic properties of air 8
Geometry 13
Mass-spring systems 15
Conclusion 23
References 24
Appendix to Chapter 1 24

Chapter 2: Acoustic Air Motion in a Straight Tube

Introduction 25
Preliminaries 25
The conservation equations 27
The wave equation 39
A piston boundary 46
Energy in acoustics 50
Conclusion 57
References 58
Appendix to Chapter 2 58

Chapter 3: Acoustic Air Motion in a Tube of Finite Length

Introduction 61
An initial brief piston movement 61

Properties of the circular functions sine and cosine 69
Sinusoidal piston movement 72
Steadiness 78
The time domain and the frequency domain 86
Conclusion 91
References 92

Chapter 4: Mass-Spring Systems

Introduction 93
Linear mass-spring systems 93
Impulse response functions 96
The mass-spring system with a sinusoidal external force 106
The mass-spring system with friction damping 111
Numerical simulations of finite-length tube acoustics with a piston
    and damping 123
The forced mass-spring system with generalized damping 127
Conclusion 129
References 130

Chapter 5: Standing Waves and Normal Modes

Introduction 131
Trigonometric identities 132
Frequency, wavenumber, period, and wavelength 133
Traveling and standing waves 135
Normal modes 138
Independence of normal modes 150
Conclusion 164
References 165

Chapter 6: Applications of Normal Modes: Acoustic Perturbation Theory and Green's Functions

Introduction 167

Acoustic perturbation theory 167
Green's functions for the acoustics of the finite-length tube 177
Conclusion 190
References 191

Chapter 7: Damped Acoustic Motion in a Finite-Length Tube with Sinusoidal Piston Motion

Introduction 193
Energy flow for the sinusoidally forced, damped mass-spring
    system 194
The finite-length tube with damping and a piston source 199
Source location effects 206
About representations with modes 209
Connection with source-filter theory 211
Conclusion 212
References 213

Chapter 8: Introduction to Complex Variables for Acoustics

Introduction 215
Complex numbers as two-space vectors 215
The algebra of complex numbers 219
Physical quantities in complex notation 223
References 229
Appendix to Chapter 8 229

Chapter 9: Two Sub-Tubes of Unequal Cross-Sectional Area

Introduction 233
Low-frequency acoustics 234
Wave propagation considerations 234
Normal modes with two sub-tubes of unequal cross-sectional
    area 239
Steady radiation pressure and acoustic perturbation theory 254
The continuity conditions and their amendment 257

Conclusion 265
References 266

Chapter 10: Multiple Sub-Tubes

Introduction 267
Multiple sub-tubes with pressure and volume velocity continuity conditions 267
Examples of computing normal modes from area functions 275
Accounting for lumped mass elements 286
Conclusion 303
References 303

Chapter 11: Damping and Green's Functions Modifications

Introduction 305
Radiation damping 306
Jetting 312
Wall vibration damping 313
Acoustic momentum boundary layer and acoustic thermal boundary layer damping 319
Damping mechanisms that affect phase speed 324
Damping mechanisms that work locally 326
Generalities and comparisons of mode frequency reductions that accompany damping 327
Modifying Green's functions 329
Conclusion 329
References 330

Chapter 12: Helmholtz Resonators and Side Branches

Introduction 333
Helmholtz resonators 333
Side branches 347
Conclusion...357
References 358

Chapter 13: Fluid Mechanics and Aeroacoustic Sources

Introduction 359
The Euler model 360
Dynamics of rotational air motion 378
Aeroacoustic sources in speech 391
References 410

Chapter 14: Scaling, Curvature, and Speech Development

Introduction 413
Scaling and curvature 414
Hypotheses regarding tongue surface curvature 419
If the curvature hypotheses are true, then young children cannot
    produce strong [ɹ] 420
Sibilant fricatives and tongue curvature 424
Difficulties in velar and alveolar stop releases 426
Conclusion 434
References 435

Chapter 15: A Layered Structure Model for Vocal Fold Vibration: First Results

Introduction 437
The static base configuration 438
Dynamics 441
Simulations 456
Robustness of vibrations 461
Extensions to larger vibration amplitudes with vocal fold
    collision 465
Conclusion 467
References 468

Index 471

Preface

Acoustics is the study of sound. Acoustics, as a part of mathematical physics, is a theory of small disturbances in an otherwise quiescent medium, like air. This means that we take the conservation equations describing air motion, and linearize them. There are other assumptions made in the *acoustic approximation*, but thinking of acoustics in air as linearized fluid mechanics is not far off the mark. While general mechanics of air in the human respiratory system is complex, and often non-acoustic, we can usually make the acoustic approximation for pertinent air motion in most of the supraglottal vocal tract during speech production. When we cannot, the approximation usually breaks down in localized regions, such as at the glottis itself, or in other highly constricted regions. These regions often act as sources of sound for the regions of the vocal tract where pertinent air movement is well described by the acoustic approximation.

We make an important distinction immediately regarding measurable quantities and the various approximations that can be used to describe air motion. This distinction has become clouded when air motion described in the acoustic approximation has been related to the motion described by alternative approximations to the conservation equations, such as the Bernoulli equation. These relations have often been made in an ad hoc manner that seem to make it necessary refer to such things as "acoustic pressure" or "Bernoulli pressure". When considering the physics of air, we understand air has certain physical, measurable properties, such as density, pressure, particle velocity, volume velocity, and so on. These quantities are defined in such a way that they do not depend on the particular theoretical approximation used to describe their changes in space and time. For instance, there is no such thing as "acoustic pressure" or "Bernoulli pressure", but there are instances and regions where the behavior of pressure is best described using the acoustic approximation, and others when its behavior is best described by Bernoulli's equation. The motion of air does not accommodate itself according to the theoretical or mathematical approximations that are intended to describe the air's motion.

There were several motivations that I had to write this book. I believe, that to make progress in understanding sounds that are propagated in the atmosphere as a result of speech, the physics of acoustics must be understood by, at least, some researchers. Further, previous books on speech acoustics, such as those of Fant (1960), Flanagan (1965), and Stevens (1998) were written using electrical analogues. Electrical analogues are fine for many calculations, but they assume that the linearized acoustic approximation is valid. It is important to examine acoustics in the broader context of fluid mechanics to understand the use of acoustics for calculation and its connection to fluid motions in the vocal tract that do not conform to the acoustic approximation.

Other motivation came from questions that I have received from linguists, psychologists, and speech clinicians regarding acoustics. One of these questions regarded filter banks used to synthesize speech. If I remember correctly, the question was how is it that frequencies of a source, $\mathcal{F}$, such as the harmonics of the voice source, drive the filters in a filter bank, but do not follow the relation $\mathcal{F} = c/\lambda$, where c is the speed of sound and λ is the wavelength of the resonance represented by a particular filter in the filter bank? Questions like this made it clear that the way acoustics is traditionally presented to speech researchers is as a computational tool. The physics has often been lost in our understanding when computing outputs from inputs with transfer functions and filter banks. I hold to the idea that "The primary purpose of theory is understanding, not calculation."

This is a book about settled science, as one of my colleagues has said. Almost all the the material in this book comes from nineteenth and early twentieth century physics. The newest topic that is in the book, other than the two research topics in Chapters 14 and 15, is in Chapter 13, where we discuss how air motion that is not described by the acoustic approximation can serve as a source of acoustic energy for acoustic wave motion. This is the study of aeroacoustics, and its modern beginnings came just after the first half of the twentieth century with the publication of Lighthill's paper on jet noise in 1952.

The book is divided into four parts. Chapters 1 through 7 provide a thorough explanation of acoustic motion in a tube of finite length and constant cross-sectional area with a moving piston at one end. Some discussion of another type of source is also included. In the first part of the book, almost all of the results are presented in what we consider to be the time domain. This part lays out the fundamental mathematical physics of the situation, so that further developments in variable-area tubes in Chapters 8 through 12 become more a matter of computation. However, we never switch completely to just computation of physical quantities, but continue to provide physical understanding with the derivation of equations. Much of the second part of the book involves frequency domain considerations. The third part of the book, which is in Chapter 13, is an introduction to fluid mechanics that does not follow the acoustic approximation. This leads to a short study of aeroacoustics and the way that the fluid motions that do not satisfy the acoustic approximation can provide energy for acoustic wave motion in the vocal tract. Finally, Chapters 14 and 15 outline two research projects that have not hitherto been published. The first involves a hypothesized relation between tongue surface curvature constraints applied to young children and their resulting acoustic output during speech. The second work is a proposed model for vocal fold vibration that uses some of the ideas introduced in Chapter 13. It is intended as a replacement for lumped element models, such as the two-mass model. Biomechanical parameters can be more easily related to the proposed model than to lumped element models. This final topic uses advanced mathematical tools.

This book is targeted at people interested in the physical aspects of acoustics that arise during the act of speaking. This includes people involved directly with the science of speech production, but without advanced mathematical knowledge, and those in the physical sciences with some advanced mathematical knowledge. [Here, we consider anything more than a smattering of calculus to be advanced knowledge of mathematics.] It is written to be accessible by those who are not trained in mathematics, while not holding back on presenting important physical ideas that are expressed mathematically. There are many simulations presented in plots and with numerical values to aid understanding. The book is something of a narrative about the physical acoustics encountered during speech production. It is intended to be read in sequence. This is not a text book, or even a reference book, in the traditional sense. For one thing, there are no problems for the student. However, if the reader wishes a more thorough understanding, he or she could derive some of the mathematical expressions presented in the book.

The reader should just forge ahead if he or she finds that some mathematical expressions are too difficult to understand. There is always a discussion or simulation that should still be illuminating. We would also recommend that readers use software to simulate some of the results, and turn to references for more in-depth discussion of topics of interest.

Equations are numbered separately in each chapter. Further, Chapters 1, 2, and 8 have short mathematical appendices with equations numbered as, for example (A1.2), where A1 refers to the Appendix to Chapter 1, and 2 refers to the equation number in that appendix. In some derivations, equation numbers are written in square brackets next to the steps of the derivation, so the reader can justify the step. We do this only for the first few uses of equations in the appendices.

There are numerous people to thank. There are friends, those who I have worked with in speech production research, and those across the United States who have either hosted me or encouraged me on my cross-country treks to teach in the American southwest. The number of people is too great to mention them by name. I want to thank my colleagues at Imperial Valley College, and my friends on both sides of the American-Mexican border in the college's region. All of these people helped me realize that a pedagogical vocation may not be out of reach for me.

Finally, I thank the people who read earlier versions of this book: Michael Howe, Khalil Iskarous, Lynn McGowan, Philip Rubin, and Reiner Wilhelms-Tricarico. They helped to improve this book immensely and are encouraging friends as well.

R. S. McGowan
Lexington, Massachusetts

References

Fant, G. (1960). *Acoustic Theory of Speech Production*. Mouton, The Hague.

Flanagan, J.L. (1965). *Speech Analysis, Synthesis, and Perception.* Springer-Verlag, New York.

Lighthill, M.J. (1952). On sound generated aerodynamically, Part 1: General Theory. *Proc. Roy. Soc. A.* **211**, 564-587.

Stevens, K.N. (1998). *Acoustic Phonetics.* MIT Press, Cambridge, MA.

Chapter 1: Air: The Acoustic Material in the Vocal Tract, and Some Basic Physics

Introduction

Acoustics is the study of small amplitude, unsteady, or time-varying, disturbances to physical materials. These materials provide *restoring forces* to the parts of the material that are moved away from their undisturbed rest position. These restoring forces tend to return the parts that are moved away from rest position to their position of undisturbed rest. A restoring force is the force exerted by a spring after it has been compressed or stretched, and as it returns to its rest length when it is released. *Potential energy* is gained when a spring is compressed or stretched. This potential energy is transformed into *kinetic energy* if it is released from a state of compression or stretch.

The definition of acoustics in terms of restoring forces and potential energy is too broad though, because it applies to vibrations, as well as to what we generally think of as acoustics. Often it is difficult to distinguish vibration from acoustics, so we immediately restrict our considerations to air, as this is the principal medium for vocal tract acoustics. For a classification of the general subject matter covered by the term acoustics, see Pierce (1989).

Properties of air

Air is the gas that surrounds the earth. We have an intuitive idea of a gas like air, but here we characterize air as physicists do. Air is a part of the class of fluids that also includes liquid water. While both gases and liquids are fluids, gases have the property that they can be compressed much more readily than liquids, and gases expand to fill the available space. Fluids gain potential energy when they are compressed, because they possess restoring forces, which means that acoustic motion can occur in these media. However, neither bodies of liquids nor gases retain their shapes when *external forces* are applied (Batchelor 1967).

Let's consider air under so-called *normal atmospheric conditions*. For us, normal atmospheric conditions correspond to the state of atmospheric air at sea level at 15° Celsius, or 15° C. In discussing volumes of air, the cubic centimeter, cm^3, which can be denoted cc, is a common unit. Initially, one may consider this to be a cube measuring 1 cm on each edge. In our examination of air, the smallest length scales of interest are very much larger than the inter-molecular distances, i.e. the mean distance between molecules in air. For example, 10^{-3} cm, or 1 one-thousandth of a cm, is much larger than inter-molecular distances (Batchelor 1967). This corresponds to a volume of 10^{-9} cm^3 or 10^{-9} cc, or 1 billionth of a cc: a very small volume of air on a terrestrial, or, even, human scale. Batchelor (1967) writes that there are about 3×10^{10}, or 30 billion, molecules of air in that apparently small volume at normal *pressure* and *temperature*. Small volumes of air have something called *mass*, m, associated with them; mass is a measure of the amount of "matter" in the small volume. A particular unit of mass is the gram, which is denoted g.

We have been introducing measurement units, such as volume in cm^3, and mass in g. Of course, the volume cm^3 is based on a length measure cm. We are using the system of measurement that has cm as length, g as mass, and seconds, s, for time. We usually work in the metric system, and within that system, we usually work in what is known as *c-g-s units* for centimeters, grams, and seconds.

One-dimensional kinematics and dynamics of a point mass

Before continuing the discussion of air, we review some concepts associated with the movement of small bodies of mass. These small bodies are often idealized as mathematical points that possess mass, called *point masses*. Thus, we consider some of the kinematics and dynamics of masses that are reduced to mathematical points in the next sections. Kinematics is the study of motion without regard to the causes of the motion, while dynamics includes the causes of motion, such as force.

Kinematics

All motions of a point mass considered here are one dimensional along the x-axis, say. The point mass is sometimes referred to as a

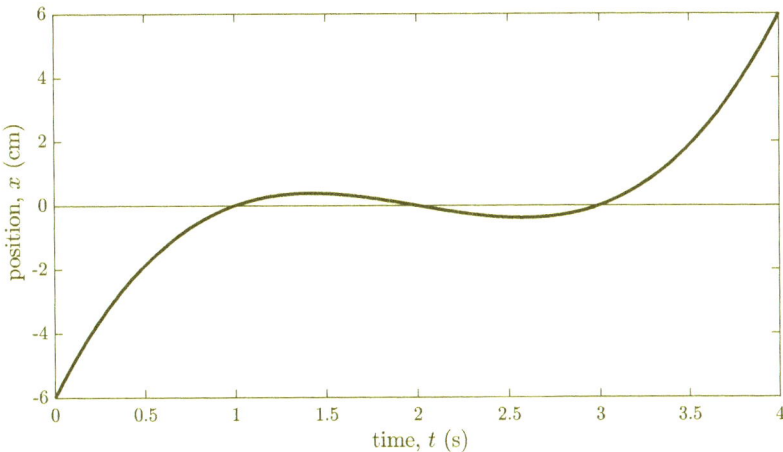

FIGURE 1. Body, or mass, position versus time

body and is given the coordinate x that is a function of the time, t, i.e. $x = x(t)$. This body could represent one of the small volumes of air that was introduced above. The $x(t)$ coordinate, the body's position, provides information on the body's distance from the origin along the x-axis, $|x(t)|$, as well as information on whether it is to the right or left of the origin. That is, the distance of the body from the origin is $+x(t)$ when the body is to the right of the origin, when $x(t) > 0$, and the distance to the origin is $-x(t)$ when the body is to the left of the origin, when $x(t) < 0$. Also, motion can be in the positive x-direction (rightward motion), or the negative x-direction (leftward motion). Figure 1 shows an example plot of the position of a body as a function of time. For $t < 1$ s the body has positions $x(t) < 0$. It moves to the right so that its positions become positive, i.e. $x(t) > 0$ after $t = 1$ s for some time. At some time between $t = 1$ s and $t = 2$ s the body starts to move to the left, so that by $t = 2$ s it is heading back into negative-x territory. Between $t = 2$ s and $t = 3$ s, the body starts to move to the right again, so that after $t = 3$ s, $x(t) > 0$.

If the point mass is moving rightward, its position has a positive time rate-of-change, or positive *velocity*, where velocity $v(t)$ at time t can be defined approximately as the ratio of the change in position over the change in time over an interval $t + \Delta t/2 > t > t - \Delta t/2$, with

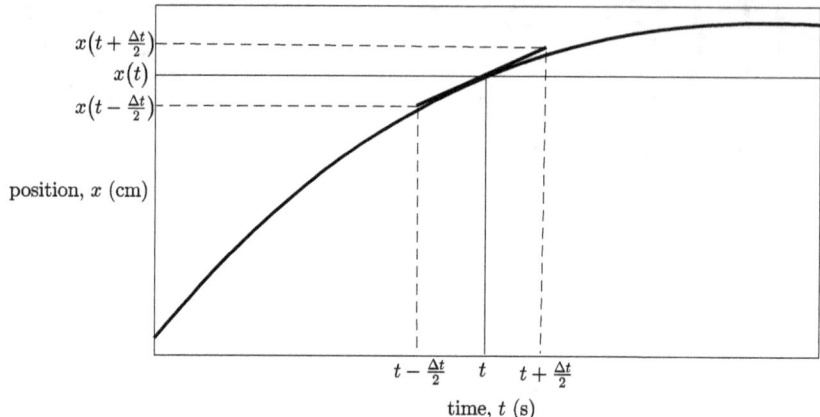

FIGURE 2. The curve $x(t)$ and the secant line at $t = 1$ and $x(t = 1) = 0$

a small $\Delta t > 0$.

$$v(t) \approx \frac{x(t + \Delta t/2) - x(t - \Delta t/2)}{\Delta t} \tag{1}$$
$$\equiv \frac{\Delta_t \, x(t)}{\Delta t} \; .$$

where the approximation improves as the time interval Δt decreases. The sign $\equiv$ means that we are defining Δ_t as

$$\Delta_t f(t) \equiv f(t + \Delta t/2) - f(t - \Delta t/2) \; . \tag{2}$$

This is known as a *centered first difference* of the function $f(t)$. Two properties of the centered first difference are examined in the Appendix to Chapter 1.

Velocity $v(t)$ is approximately the ratio of the centered first difference of position, $x(t)$, to the difference in time, t. [We often drop the approximation symbol, "$\approx$", and simply write equal, "$=$".] This is the *slope* of the line joining points $(t - \Delta t/2, x(t - \Delta t/2))$ and $(t + \Delta t/2, x(t + \Delta t/2))$. This line is called the *secant line*, and Figure 2 shows the secant line for the particular time $t = 1$ and position $x(t = 1) = 0$ of Figure 1.

Slopes at any time t can be calculated. These slopes, are, in turn, a function of time t. $v(t) > 0$ when the body is moving to the right at time t, and $v(t) < 0$ when the body is moving to the left. These

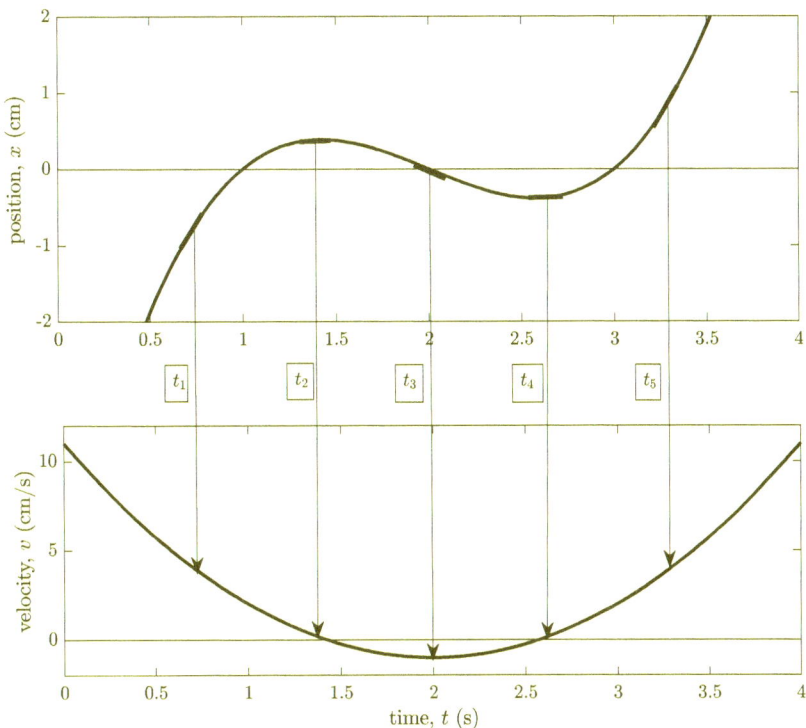

FIGURE 3. Mass position versus time and velocity versus time

ideas are illustrated in Figure 3 for the position $x(t)$ shown in Figure 1. Figure 3 shows the plot of the velocity $v(t)$ below the plot of the position as a function of time $x(t)$. The lower plot is simply a plot of the slopes of the small line segments around each point of the plot of position versus time. Samples of these line segments on the position versus time plot and their corresponding slopes on the velocity versus time graph are indicated by arrows. We see that the velocity is positive through the time t_2, while the position goes from negative to positive before this time. The body stops at the time t_2 with zero velocity, and then starts to move left, as indicated by the fact that the velocity is negative, and the position of the body goes from positive to negative at the time t_3. At the time t_4, the body has stopped and then begins moving to the right again, because the velocity is positive after this time. It continues to move in this direction indefinitely. The reader should note that there can be any combination of positive and negative signs for the pair $x(t)$ and $v(t)$. For example, a body can be to the left of the origin, yet moving rightward, with $x(t) < 0$ and $v(t) > 0$, as in Figure 3. The absolute value of the velocity, or velocity magnitude, is called the *speed*.

Acceleration $a(t)$ is the time rate-of-change of velocity.

$$a(t) \approx \frac{\Delta_t v(t)}{\Delta t} = \frac{v(t + \Delta t/2) - v(t - \Delta t/2)}{\Delta t}, \qquad (3)$$

where the approximation improves as the time interval $\Delta t > 0$ decreases. It follows from Equations (1) and (3) that

$$\begin{aligned}
a(t) &= \frac{\frac{x(t+\Delta t)-x(t)}{\Delta t} - \frac{x(t)-x(t-\Delta t)}{\Delta t}}{\Delta t} \\
&= \frac{\frac{\Delta_t\, x(t+\Delta t/2)}{\Delta t} - \frac{\Delta_t\, x(t-\Delta t/2)}{\Delta t}}{\Delta t} \\
&= \frac{\Delta_t\, x(t + \Delta t/2) - \Delta_t\, x(t - \Delta t/2)}{(\Delta t)^2} \qquad (4) \\
&= \frac{\Delta_t\, \left(x(t + \Delta t/2) - x(t - \Delta t/2)\right)}{(\Delta t)^2} \quad \text{[Equation (A1.4)]} \\
&= \frac{\Delta_t(\Delta_t\, x(t))}{(\Delta t)^2} \quad \text{[Equation (2)]} \\
&= \frac{\Delta_t^2 x(t)}{(\Delta t)^2}.
\end{aligned}$$

Acceleration $a(t)$ is the ratio of the centered first difference of velocity $v(t)$ to the difference in time Δt. Equation (4) shows that acceleration is equivalent to the centered second difference in position $x(t)$ divided by the square of the difference in time $(\Delta t)^2$. The quantities $x(t), v(t)$, and $a(t)$ are related to one another by the ratios of differences shown in Equations (2) and (4).

Figure 4 shows the relationship between the velocity $v(t)$ and acceleration $a(t)$ for the position and velocity of Figures 1 and 3. The acceleration is negative up until the time t_3. From before the time t_1 until the time t_2, the mass' velocity is positive, but it is decreasing in magnitude, or speed. From the time t_2 to the time t_3, the mass is going to the left with an increasing speed, or magnitude of velocity. Both of these intervals provide a negative acceleration. From the time

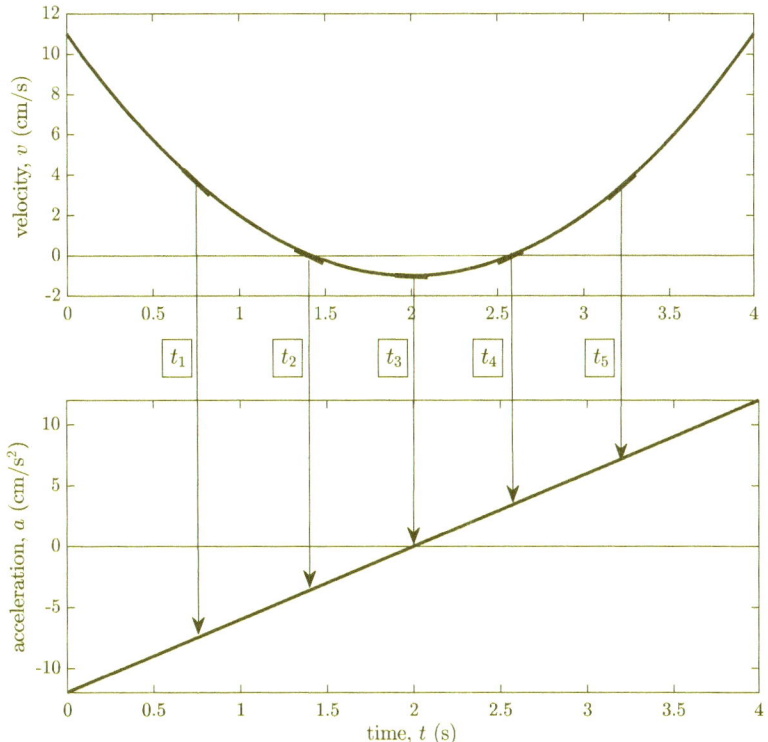

FIGURE 4. Mass velocity versus time and acceleration versus time

t_3, the acceleration is positive. From this time until the time t_4, the velocity of the mass is negative, but its speed is decreasing. From the time t_4 onward, the velocity of the body is positive and increasing in speed. Both of these intervals provide a positive acceleration. Position, velocity, speed, and acceleration are all *kinematic properties* or *kinematic quantities*.

Dynamics

The *momentum* of a body of mass m and velocity v is defined to be mv. *Newton's second law of motion* states that the time rate-of-change of momentum of a body is equal to the force exerted on the body.

$$F = \frac{\Delta_t(mv)}{\Delta t} = m\frac{\Delta_t v}{\Delta t} = ma \ . \qquad (5)$$

Newton's second law relates force on a body of constant mass to acceleration. The constant of proportionality between force F and acceleration a is mass m.

The earth exerts a *gravitational force* on objects with mass. If we were to throw a ball up from the ground, the ball would accelerate toward the center of the earth, that is, down toward the ground, at approximately constant acceleration. The ball would go up initially, but with decreasing speed. It would eventually start to fall back toward the ground with increasing speed. The constant acceleration due to gravity is **g**, which is known as the *acceleration of gravity*. The numerical value of **g** is approximately 980 cm/s^2 near the earth's surface. The force exerted on an object by the earth is given the symbol W, where

$$W = m\mathbf{g} \ . \qquad (6)$$

When an object is at rest on the surface of the earth, the surface exerts a force of magnitude W upward on the object, and the object exerts a force of the same magnitude downward on the earth, as required by *Newton's third law of motion*. W is known as the *weight* of the object.

In daily life we tend to mix mass and weight. In some countries for instance, meat is often labeled according to pounds and grams. Pound (lb) is a force, or weight, unit in the English system, and the gram is a mass unit in the metric system. The weight or force measure in the c-g-s system is a *dyne*, where dyne = g cm/s^2. So, for instance, 2.5 g of meat actually weighs 2450 dyne.

Thermodynamics of air

We return to the discussion of air in terms of thermodynamics. We often refer to things like air pressure and temperature in our everyday life, as in the weather forecast. Pressure and temperature are examples of *thermodynamic properties* or *thermodynamic quantities*. Thermodynamic properties characterize small volumes of air that are composed of many molecules, such as the 10^{-9} cm^3 volume with upwards of 30 billion air molecules. Statistical analyses can be performed with this many molecules, and thermodynamic quantities can be related to averages of dynamic quantities of the molecules, as well as other expected values. For instance, temperature is related to average molecular kinetic energy. One of the triumphs of theoretical physics is statistical mechanics, which takes statistical properties of a large number of molecules in a substance like air and relates the bulk thermodynamic properties to the motions of molecules through space and other degrees-of-freedom of molecular motion.

Density is a thermodynamic quantity that is relatively easy to define. It is simply the amount of mass in a unit of volume, and it is denoted by ρ. Under the normal atmospheric conditions defined above, air has a density of 1.225×10^{-3} g/cm^3 = 0.001225 g/cm^3 (Batchelor 1967). We write $\rho_{atm} = 1.225 \times 10^{-3}$ g/cm^3. This compares to liquid water, which has a density very close to 1 g/cm^3 under the same conditions. It is important to keep in mind that air is very much less dense than water; very roughly, air has one-thousandth the density of water. Because the human body is largely composed of water, this gives us a good approximation of the ratio of the density of air in the vocal tract to the density of surrounding tissue.

Pressure is a thermodynamic quantity that can be characterized by considering the dynamics of air molecules. Pressure is denoted p. The intuitive idea of pressure is that of air pushing against the elastic walls of a balloon as it is being blown up. The agent that is blowing the balloon up is imparting a higher pressure to the air inside the balloon than the pressure of air outside the balloon, *atmospheric pressure*, which is the pressure in normal atmospheric conditions for this book. If one evacuates air from a basketball, it crinkles because the interior pressure is below atmospheric pressure. Air moves from regions of high pressure to regions of low pressure, although this is getting into the relation between pressure and motion of bodies of air, so this is ahead of the discussion.

From a mechanical perspective, pressure is a *stress* and has the units of force per unit area, such as dyne/cm^2. For air at rest, it is the only stress and it is termed *normal stress*, because it is directed in the *normal direction* to any mathematical surface considered to be within the air. Normal direction means perpendicular here; to be directed normal to a surface is to be directed perpendicular to that surface. If we approximate air using a model of gas molecules analogous to small

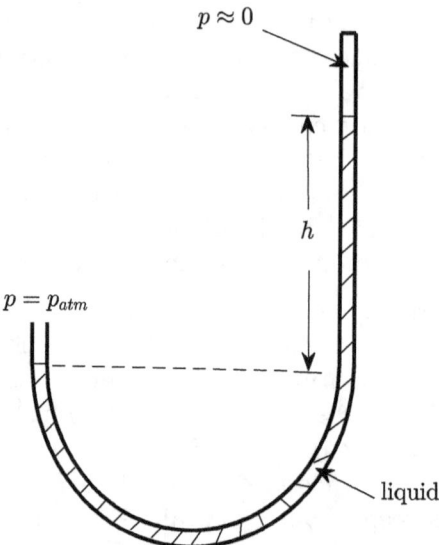

FIGURE 5. U-tube manometer

billiard balls, then the pressure within a small volume is proportional to the average translational kinetic energy per unit volume of the molecules within the small volume (Batchelor 1967).

Often, we consider mathematical surfaces surrounding a small volume of air. For air at rest, it does not matter how the surfaces of the volume are defined: a spherical surface, a cubical surface, or any other surface surrounding a small volume. A non-zero pressure p means that there is the possibility of force in the normal direction relative to surfaces. The magnitude of force across the surface of small area A is pA. If the surface is planar, then force acts in the normal direction relative to the plane. If the surface is spherical then the force acts along the radius of the sphere. However, the force exerted on the surface by the air inside the volume is exactly counterbalanced by the pressure exerted by the air outside the volume when the volume of air is not moving. Thus, the *net force* on the surface is zero.

Atmospheric pressure can be measured with many devices, including water and mercury *U-tube manometers*, as shown in Figure 5. A U-tube manometer can be made from a glass tube of constant cross-section, A_m, that is bent into a U shape. The U-tube is partially filled with a liquid, water or mercury, and closed at one end, which is evacuated, so that the pressure at the surface of the liquid on that side of the U-tube is approximately zero. The other end is left open to the atmosphere so that the liquid surface on this side of the U-tube experiences the atmospheric pressure, p_{atm}. We take the value of p_{atm} to be its value under normal atmospheric conditions, 1.013×10^6 dyne/cm^2 (Batchelor 1967). In order for the liquid to remain at rest the net force on the liquid must be zero, so that the gravitational force on the liquid must balance the force supplied by the pressure of the atmosphere. Let h denote the difference in the liquid elevation between the sides of the U-tube. The upward force on the liquid in the open side of the U-tube is equal to the difference in weight of liquid on the closed side and the open side. The mass of the excess liquid on the closed side is $\rho_l A_m h$, where ρ_l is the density of the liquid. Thus, its weight is $W = \rho_l A_m h \mathbf{g}$, where $\mathbf{g}$ is the acceleration of gravity. The force pushing the liquid down on the open side of the U-tube is $p_{atm} A_m$. In order for these forces to balance one another, their magnitudes must be equal, so that $p_{atm} = \rho_l g h$. Thus, for a given liquid, the atmospheric pressure p_{atm} is proportional to the height difference in the liquid h. This is why pressure is often quoted in units of cm H$_2$O (centimeters of water), or in units of mm Hg (millimeters of mercury).

It is usual to express speech related pressures in terms of centimeters of water, cm H$_2$O. In speech research pressure is in relation to atmospheric pressure, so that, for instance, 10 cm H$_2$O denotes a pressure above atmospheric pressure and -10 cm H$_2$O denotes a pressure below atmospheric pressure. The following is the conversion factor between atmospheres and cm H$_2$O.

$$1 \text{ atmosphere} = 1033 \text{ cm H}_2\text{O} \approx 10^3 \text{ cm H}_2\text{O} . \tag{7}$$

Thus, 10 cm H$_2$O is about 1% of atmospheric pressure. Another useful relation is

$$1 \text{ atmosphere} = 1.013 \times 10^6 \text{ dyne/cm}^2 \approx 10^6 \text{ dyne/cm}^2 . \tag{8}$$

In the discussion of pressure, we have assumed that the small volumes are, as a whole, at rest. Note that the velocity of a small volume of air with its 30 billion molecules is distinct from the velocities of the molecules that compose the volume. Velocities of molecules are generally random in direction and the distribution of their magnitudes, or speeds, is determined by the temperature of the air. If the volume of air has a velocity as a whole, then the velocity of the molecules is the sum of the random velocities and the velocity of the volume. When small volumes of air are in motion, the definition of pressure becomes a little more problematic, because the normal stresses are not the same in all directions. Thus, pressure is simply defined as the mean (over the three Cartesian directions) of the normal stresses (Batchelor 1967).

There are thermodynamic properties of air that we don't talk about in our everyday lives, such as *entropy*. However, there is one fact to remember: we need only two thermodynamic quantities to completely characterize the equilibrium thermodynamics of a given volume of air. That is, we can choose any two quantities, such as temperature and pressure, and calculate the other quantities, such as density and entropy from an *equation of state*. We do not discuss equations of state in general, as this would take us too far from the topic at hand. A particular relation between pressure and density under restrictive assumptions is provided below.

The concept of entropy is difficult, and it is best understood with statistical mechanics. However, it is fortunate that we are able to make a large simplification because of the following assumption. We assume that entropy is uniform through the region of air that is under investigation, and, further, the entropy does not change with time. [For an initially homogeneous medium this would mean that changes in thermodynamic properties of small air masses due to *heat conduction* and *friction* are negligible. There are places where this can be violated in speech acoustics, and we attempt to point those out later in Chapter 11. The regions where there is change in entropy are usually confined to be very close to solid surfaces.] With a completely constant entropy, we can find the equilibrium value for any of temperature, pressure, or density as a function of any one of these quantities. Let's suppose we take pressure p as the independent thermodynamic quantity, so that both temperature and density ρ are considered functions of pressure.

In our discussion of the relation among thermodynamic quantities, we state a result for small thermodynamic disturbances in lieu of a full equation of state. Let p_0 be the pressure of undisturbed air, called *rest pressure*, and ρ_0 the undisturbed density of air, called *rest density*. Further, let $\delta p = p - p_0$ and $\delta \rho = \rho - \rho_0$, be the *perturbation pressure perturbation density*, respectively, of a small disturbance. In other words, we assume that $|\delta p|/p_0$ and $|\delta \rho|/\rho_0$ are both much smaller than one, or that $|\delta p|/p_0 \ll 1$ and $|\delta \rho|/\rho_0 \ll 1$. The disturbances are related by

$$\frac{\delta \rho}{\delta p} = c_0^{-2} \quad \text{or} \quad \frac{\delta p}{\delta \rho} = c_0^2 \ . \tag{9}$$

One can work out that the dimensions of c_0 are those of a velocity, e.g. cm/s. It is called the *adiabatic speed of sound*, and we take the value of $c_0 = 34,100$ cm/s under normal atmospheric conditions, for which $p_0 = p_{atm}$ and $\rho_0 = \rho_{atm}$ (Batchelor 1967). The term adiabatic comes from the constancy of entropy in time. An *adiabatic process* is a physical process where entropy does not change with time. We see later why it is a speed for sound. In order to reference physical quantities pertaining to normal atmospheric conditions, we write them in equation form.

$$\begin{aligned} \rho_{atm} &= 1.225 \times 10^{-3} \text{ g/cm}^3 \\ p_{atm} &= 1.013 \times 10^6 \text{ dyne/cm}^2 = 1033 \text{ cm H}_2\text{O} \\ c_0 &= 34,100 \text{ cm/s} \ . \end{aligned} \tag{10}$$

Geometry

The air continuum

In mathematical theory air is considered to be a *continuum*: it is a mathematical three-dimensional space that is filled with a mathematical "material" that has the physical properties of air. Each point in the mathematical space can be conceived as representing a small volume of air, with its thermodynamic and kinematic properties (Landau & Liftshitz 1959). The molecular level of physical description is completely ignored here.

The continuum framework permits each point to be treated as an *air particle*. Air particles have associated thermodynamic properties, such as pressure, density, temperature, and entropy, as well as the kinematic properties of position, velocity and acceleration. Air particles are analogous to point masses, except that they have density, instead of mass, as well as other thermodynamic properties associated with them. Obviously, this is an abstraction, because a point without extent is now representing a small volume. The distinction between physical air

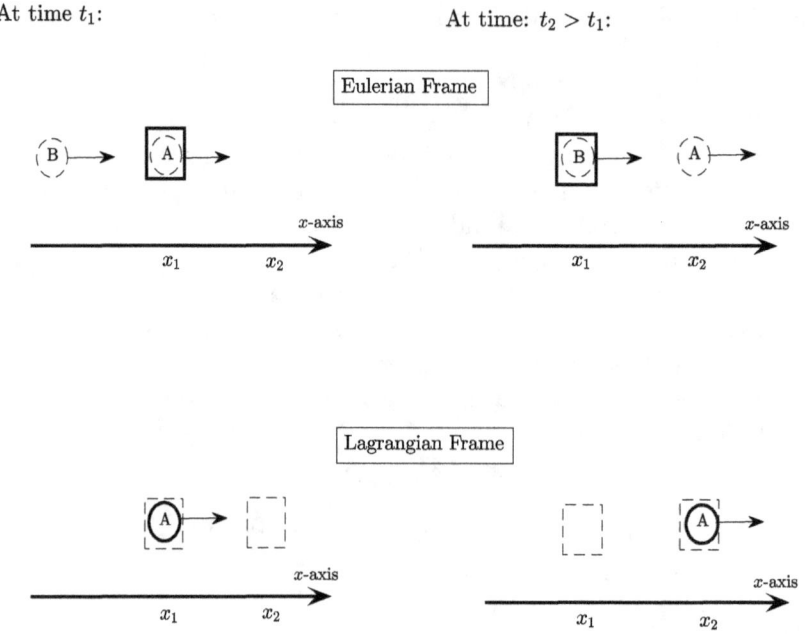

FIGURE 6. Eulerian and Lagrangian frames

and the mathematical space that represents a continuum with each point having the properties of air may seem ponderous. However, we are presenting a physico-mathematical theory, so clarity is important.

Eulerian and Lagrangian frames-of-reference

In our discussion of the kinematics of point masses, we followed an individual point mass through a one-dimensional space. This frame-of-reference is called the *Lagrangian frame-of-reference*. It is the frame-of-reference in which we describe the physics of a finite number of bodies, whether point masses or extended masses. We may be familiar with this frame-of-reference from previous experience with introductory academic physics. On the other hand, our default frame-of-reference for the thermodynamics, kinematics and dynamics of air is called the *Eulerian frame-of-reference* (Landau & Lifshitz 1959), for which we fix points in space and consider what happens as air particles pass through these points. We do not follow the individual air particles. We often shorten Eulerian or Lagrangian frame-of-reference to Eulerian or Lagrangian frame.

Figure 6 illustrates the difference between the two frames of reference. The top two panels show the Eulerian frames at times t_1 and $t_2 > t_1$. The bottom two panels show the Lagrangian frame at the same two times. The box at $x = x_1$ is the object of investigation in the Eulerian frame, with different air particles A and B at that location at the two different times. The object of investigation in the Lagrangian frame is the air particle A, which is in two different boxes, at x_1 and x_2 at the two different times. While the laws of motion are the same in either frame of reference, the way that they are written depends on our frame of reference. Before continuing into the mechanics of air in the Eulerian frame in the next chapter, we examine mass-spring systems in the Lagrangian frame.

Mass-spring systems

A simple mass-spring system

Consider a massless spring that can be compressed or extended in the x-direction. We suppose that the spring's left end is attached to a infinitely massive wall. Supposes that an external agent either stretches or compresses the spring to length L, from a rest length of L_0. The rest length is the spring's equilibrium, or rest, state. [We often use the term agent to denote some person or machine that manipulates the system of interest, but itself is not a part of that system.] We assume that the spring obeys *Hooke's law*.

$$F_{res} = -\kappa(L - L_0) . \tag{11}$$

The $(L - L_0)$ is the difference between the spring length, L, and the undisturbed, or rest length, L_0. F_{res} is the restoring force of the spring, with $\kappa > 0$ known as the *spring constant, spring stiffness*, or just *stiffness*. If the spring is compressed by an agent at the right end of the spring, so that $(L - L_0) < 0$, then the spring exerts a force against the agent in the positive x-direction, that is, toward its equilibrium. If the spring is stretched, then $(L - L_0) > 0$, and the restoring force acts against the agent to pull in the negative x-direction, again, toward the rest length L_0.

Now we attach a mass m to the right end of the spring so that the mass moves along a frictionless surface as the spring compresses and stretches The position of the mass when the spring has length L is denoted x_m, and the particular position of the mass when the spring is at its rest length, L_0, is written x_0. Thus, $x_m - x_0 = L - L_0$. We let $X_m = x_m - x_0$, which is referred to as *displacement position*, or, simply, *displacement* here. When there is no external agent present, the restoring force of the spring, F_{res}, acts on the mass. The entire configuration, a *simple mass-spring system*, is shown in Figure 7.

If there is no external agent acting on the simple mass-spring system, then by Newton's second law, expressed in Equation (5),

$$ma = F_{res} = -\kappa(L - L_0) = -\kappa X_m(t),$$

$$\text{where } a = \frac{\Delta_t^2 x_m(t)}{(\Delta t)^2} \tag{12}$$

$$= \frac{\Delta_t^2 x_m(t)}{(\Delta t)^2} - \frac{\Delta_t^2 x_0}{(\Delta t)^2} \quad \text{[Equation A1.2]}$$

$$= \frac{\Delta_t^2 (x_m(t) - x_0)}{(\Delta t)^2} \quad \text{[Equation A1.4]}$$

$$= \frac{\Delta_t^2 X_m(t)}{(\Delta t)^2}.$$

a is the acceleration of the mass. Rewriting Equation (12), we obtain

$$m \frac{\Delta_t^2 X_m}{(\Delta t)^2} + \kappa X_m(t) = 0. \tag{13}$$

Equation (13) is the *equation of motion* for the simple mass-spring system.

Suppose the simple mass-spring system is left undisturbed until an external agent agent moves the mass to x_m just before time $t = 0$, and then releases the mass from a state of rest at $t = 0$. This means that the spring is initially compressed or stretched by the external agent. That is, $X_m(t = 0) \neq 0$ and $\Delta_t X_m(t = 0)/(\Delta t) = 0$. These are *initial conditions*.

The limiting case of $\Delta t \to 0$ is considered in the following. We happen to know the analytic form of the solution to Equation (13) in

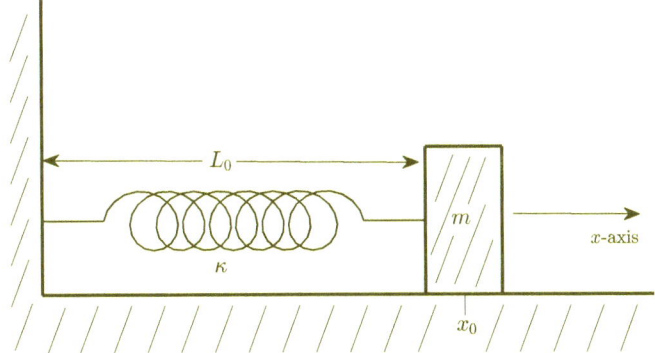

FIGURE 7. A simple mass-spring system

this limit.
$$X_m(t) = x_m(t) - x_0 = \mathcal{A}\cos(\omega_0 t + \theta), \quad \text{for } t > 0. \quad (14)$$
where $\mathcal{A}$ and θ are determined by the initial conditions. In our case $\mathcal{A} = X_m(t=0) = x_m(t=0) - x_0$ and $\theta = 0$. Here we have defined
$$\omega_0 = \sqrt{\frac{\kappa}{m}}. \quad (15)$$
ω_0 has the dimensions of *circular frequency*, s^{-1}, and is the *natural circular frequency* and $\mathcal{F}_0 = \omega_0/(2\pi)$ is the *natural frequency* for the mass-spring system. The dimensions of $\mathcal{F}_0$ are also s^{-1}. The values of $\mathcal{F}_0$ are usually stated in Hz (Hertz). The values of ω_0 are stated in units of radians/s = rad/s because they appear in *circular functions*, like sine and cosine functions. The reason for the term *frequency* is made clear in the following paragraph.

Equation (14) specifies *simple harmonic motion*, because its solution is written as a circular function of time. In this instance the circular function is cos, or cosine. The properties of circular functions cosine and sine are reviewed in Chapter 3. For the present purposes we simply note the properties in Figure 8, where the cos function oscillates between -1 and 1 symmetrically about the value 0. Figure 8 shows the solution in Equation (14) when the initial displacement $X_m(t=0) = -1$ and the natural frequency $\mathcal{F}_0 = 100$ Hz. The displacement position $X_m(t)$, repeats itself every $T = 0.01$ s. T is the *period* of this *periodic motion*, and $1/T$ is termed frequency. It

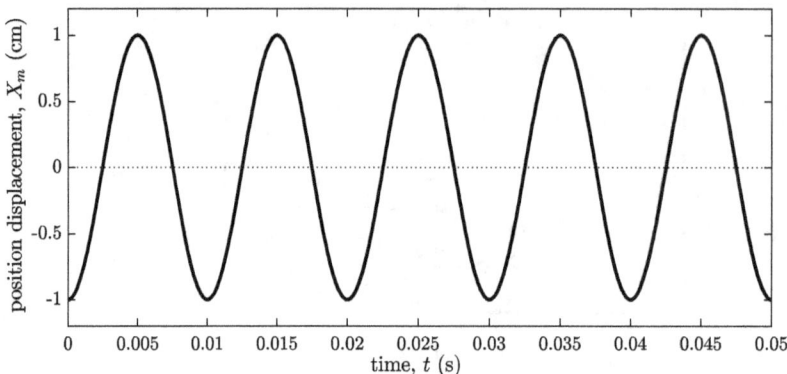

FIGURE 8. Mass displacement position versus time

turns out that simple harmonic motion is a particular kind of periodic motion, and that the concepts of period and frequency play roles in the more general periodic motions.

According to Equation (15), natural circular frequency, and, hence, natural frequency, can be increased by either increasing the spring constant, or stiffness, κ or by decreasing the mass m. The times $t \leq 0$ are not shown in Figure 8, because the system is not described by the equation of motion, Equation (13), for $t \leq 0$.

Figure 9 shows the displacement position $X_m(t)$ as a function of time along with the corresponding velocity $v(t)$. Figure 10 shows the velocity $v(t)$ and corresponding acceleration $a(t)$ as functions of time.

Consider Figures 9 and 10 at the times corresponding to the arrows. At the time t_1, the mass is at its most negative position, and changing its direction of motion from negative to positive, and its velocity is zero. At the same time, the acceleration is at its maximum positive value. This follows from the fact that acceleration is proportional to force, which is in the opposite direction to the displacement of the mass from equilibrium. Note that the change in direction corresponds to maximum acceleration, even as the velocity is zero. At the time t_2, the mass passes through equilibrium in the positive direction, where it has its maximum velocity. The acceleration is zero because the mass is at its equilibrium position at that instant. At time t_3, the situation is that the mass has reached its maximum positive position, so that the acceleration is maximally negative, and the velocity is zero. We leave

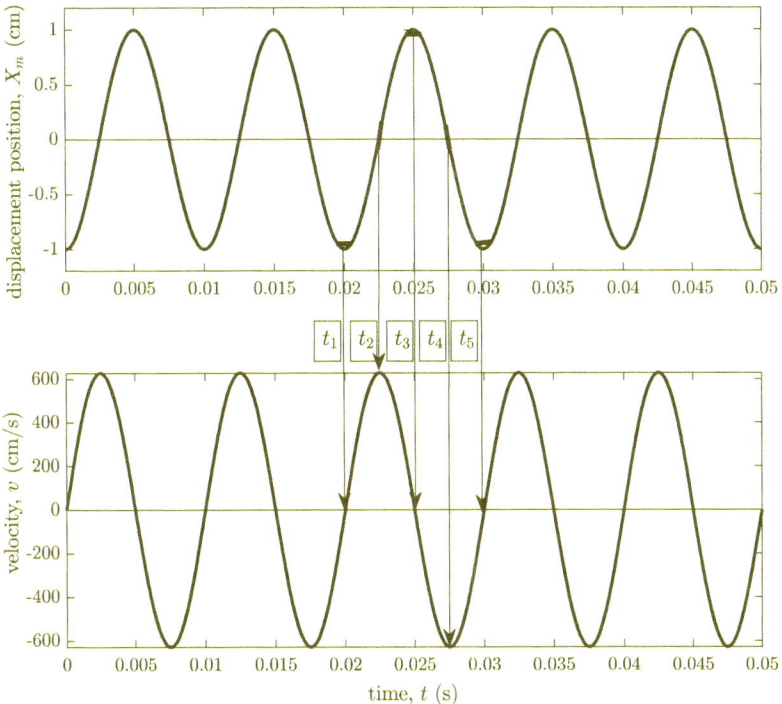

FIGURE 9. Mass displacement position versus time and mass velocity versus time

the reader to discover the properties of motion at times t_4 and t_5. The plots seem to indicate that both the velocity and acceleration are also circular functions, which is indeed the case.

We have seen that when displacement is small that the speed of the mass is large, and when displacement is large that the speed of the mass is small [see Figure 9]. Kinetic energy is $E^{kin} = mv^2/2$. Therefore, the kinetic energy of the mass is inversely relate to the stretch or compression of the spring. We can even think that the mass' kinetic energy "gives" itself up to stretch or compression of the spring, and, conversely, the stretch or compression of the spring "give" themselves up to kinetic energy of the mass. If we define potential energy as $E^{pot} = \kappa(L - L_0)^2/2 = \kappa X_m^2/2$, then it is seen that there is an indefinite *oscillation* between the kinetic energy of the mass and the potential energy of the spring. This is an important characteristic

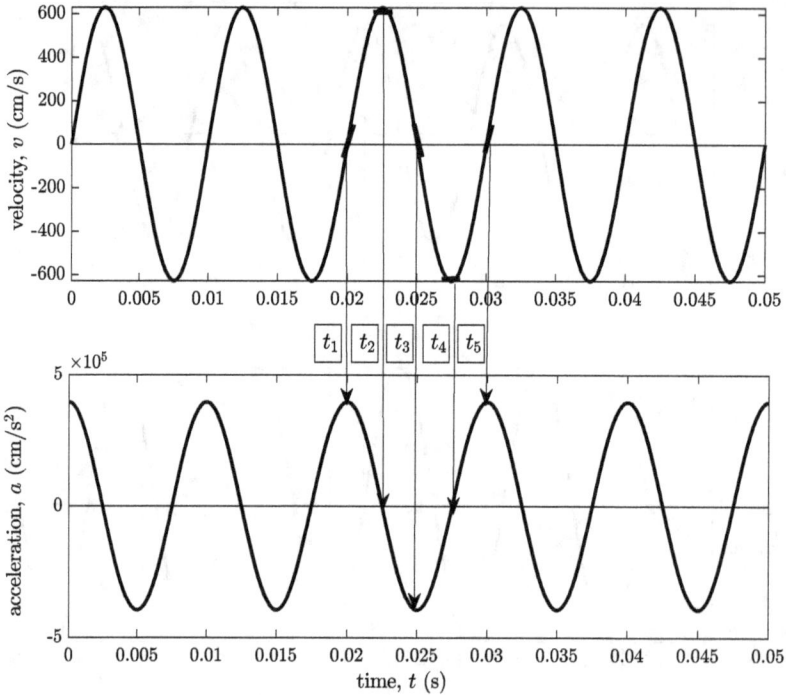

FIGURE 10. Mass velocity versus time and mass acceleration versus time

of oscillating systems, such as this simple mass-spring system.

A mass-spring system involving air

Let's examine a small container of air with a constriction tube that contains a movable solid mass m in one of the sides of the container. The constriction tube has cross-sectional area A_c. Figure 11 shows the configuration. We assume that the mass provides an air-tight seal, and that it is able to move freely without friction in the x-direction. Under normal atmospheric conditions, the mass is in equilibrium at $x_m = x_0$, and the volume of air in the container behind the mass is V_0. In this equilibrium the air both inside and outside the container are under normal atmospheric conditions.

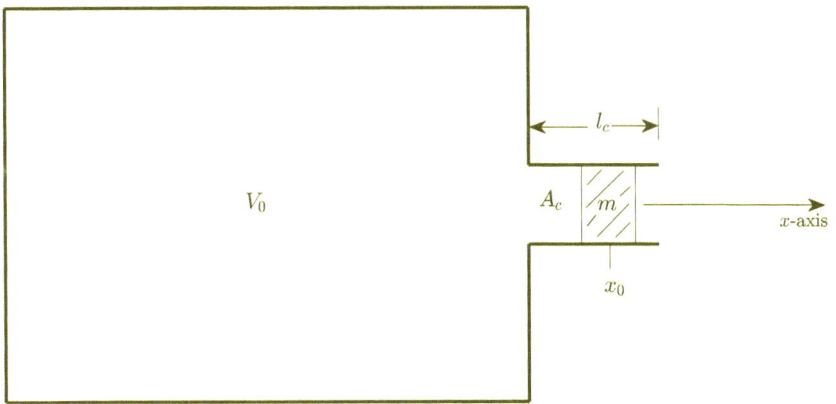

FIGURE 11. A mass-spring system for air

Suppose that the mass is moved to position x_m, so that $X_m = x_m - x_0$. The volume available to the air in the container, V, is changed from V_0 to $V_0 + \Delta V = V_0 + A_c X_m$, where $\Delta V = A_c X_m$. Let ρ and p, respectively, denote the density and pressure of air in the container. Because there is no air taken away or added to the container, the relationship between changes of air density ρ and changes in volume V is an inverse relation

$$\frac{\Delta \rho}{\rho_0} = -\frac{\Delta V}{V_0}. \qquad (16)$$

Under certain conditions that are outlined later in the book, we can assume that changes in density and pressure in the container are approximately uniform and established approximately instantaneously. Thus, while density and pressure can be functions of time, their values at any time are uniform throughout the container. It is best to have a container with small linear dimensions for this assumption to hold true.

If we assume that the relationship between changes in density and changes in pressure given in Equation (9) holds for small, but finite changes in density and pressure, then

$$\Delta p = c_0^2 \Delta \rho$$
$$= -\frac{\rho_0 c_0^2}{V_0} \Delta V \qquad (17)$$
$$= -\frac{\rho_0 c_0^2}{V_0} A_c X_m \ .$$

This follows from Equation (16). The net force on the mass in the x-direction is $F = (p - p_0)A_c = \Delta p A_c$, where p_0 is atmospheric pressure p_{atm}. It follows that

$$F = -\frac{\rho_0 c_0^2 A_c^2}{V_0} X_m \ . \qquad (18)$$

This force is a restoring force, because it acts in a direction opposite to X_m. The constant $\rho_0 c_0^2 A_c^2 / V_0$ is equivalent to the spring constant that appears in Hooke's law, in Equation (11). In fact, the equation of motion for the mass is given by Equation (13) with

$$\kappa = \rho_0 c_0^2 \frac{A_c^2}{V_0} \ , \qquad (19)$$

and the mass executes simple harmonic motion. The air in the large volume, V_0, supplies the restoring force for simple harmonic motion.

It should be of no surprise that the mass in the constriction tube could be supplied by air, instead of a solid mass. Suppose we remove the solid mass so that only air is in the constriction tube. If the constriction tube has length G_c, then the mass of air in the sleeve, m_c, is

$$m_c = \rho_0 l_c A_c \ . \qquad (20)$$

[We see in Chapter 12 that we should use a length slightly larger than the physical length of the tube l_c.] With air as the mass in the constriction tube we obtain the natural frequency from Equations (15), (19), and (20).

$$\mathcal{F}_H = \frac{1}{2\pi} \sqrt{\frac{\kappa}{m_c}} = \frac{1}{2\pi} c_0 \sqrt{\frac{A_c}{l_c V_0}} \ . \qquad (21)$$

This is a rough estimate of the natural frequency, $\mathcal{F}_H$, of a *Helmholtz resonator*. It takes some time to see that this is a valid model for the acoustics of the vocal tract in special geometries. We have derived this only to illustrate that air can behave like a simple harmonic oscillator. Unlike the simple mass-spring system of the previous section, the air provides for both the mass and the spring in the Helmholtz resonator.

Conclusion

The quantity in the expression for spring constant κ in Equation (19) has a factor that is intrinsic to air, namely $\rho_0 c_0^2$. This can be thought of as the inherent stiffness of air, where something that is stiff is not easily compressed. Note that water also has springiness, but it is much stiffer than air: not only is water almost 10^3 times denser than air, but the c_0^2 is also larger in water. In air $c_0 \approx 3.41 \times 10^4$ cm/s and in water $c_0 \approx 1.45 \times 10^5$ cm/s (Batchelor 1967). This means that the ratio of $\rho_0 c_0^2$ in water to the same quantity in air is approximately 1.8×10^4, meaning that water is much stiffer than air. One could expect that acoustic communication in the oceans to differ substantially from that in the atmosphere.

We relate $\rho_0 c_0^2$ to fractional changes in density of air and small change in pressure. From Equation (9)

$$\frac{\delta\rho}{\rho_0} = \frac{1}{\rho_0 c_0^2} \delta p \equiv D\, \delta p . \qquad (22)$$

where $D = 1/(\rho_0 c_0^2)$ is known as the *distensibility* of air (Lighthill 1978). This quantity is the constant of proportionality between small changes in pressure δp and fractional changes in density. The smaller D the more pressure it takes to attain a given fractional density change. Thus, again, we see that $\rho_0 c_0^2 = 1/D$ is something of a spring constant for air. This spring provides the restoring force that allows for what we usually associate with acoustic air motion.

In Chapters 2 and 3, we show that the springiness and mass-like like qualities of air results in *wave motion*. This takes us well beyond the very particular situation of the Helmholtz resonator, where air is configured to act as a simple mass-spring system. On the other hand, the wave motion can be represented by systems that are mathematically like mass-spring systems in the case of wave motion in a tube of finite length, such as the vocal tract. Therefore, after further study of mass-spring systems in Chapter 4, we pursue this mass-spring-like mathematical representation in Chapter 5.

References

Batchelor, G.K. (1967). *An Introduction to Fluid Dynamics.* Cambridge University Press: Cambridge, England. (pp 1-4, 6, 38-40, 141, 594, 596)

Landau, L.D. & Lifshitz, E.M. (1959). *Fluid Mechanics.* Pergamon Press: Oxford, England. (pp. 1-5)

Lighthill, M.J. (1978). *Waves in Fluids.* Cambridge University Press: Cambridge, England. (p 93)

Pierce, A.D. (1989). *Acoustics: An Introduction to its Physical Principles and Applications.* Acoustical Society of America: Woodbury, NY. (pp 3, 28-36)

Appendix to Chapter 1

Let A be a number. This number does not depend on any variable. Therefore, from the definition of centered first difference in Equation (2)

$$\Delta_t A = A - A = 0 \ . \tag{A1.1}$$

Therefore,

$$\frac{\Delta_t A}{\Delta t} = 0 \ . \tag{A1.2}$$

for $\Delta t > 0$.

We now prove that Δ_t and $\Delta_t/\Delta t$ are *linear operators*. Let $f(t)$ and $g(t)$ be functions of t, including the possibility of a constant. Let A and B be numbers.

$$\begin{aligned}\Delta_t\bigl(Af(t) + Bg(t)\bigr) = & \\ \bigl(Af(t + \Delta t/2) &+ Bg(t + \Delta t/2)\bigr) - \bigl(A \cdot f(t - \Delta t/2) + B \cdot g(t - \Delta t/2)\bigr) \\ = \bigl(A \cdot f(t + \Delta t&/2) - A \cdot f(t - \Delta t/2)\bigr) \\ + \bigl(B \cdot g(t + &\Delta t/2) - B \cdot g(t - \Delta t/2)\bigr) \ .\end{aligned} \tag{A1.3}$$

It follows that

$$\frac{\Delta_t\bigl(A \cdot f(t) + B \cdot g(t)\bigr)}{\Delta t} = A \cdot \frac{\Delta_t f(t)}{\Delta t} + B \cdot \frac{\Delta_t g(t)}{\Delta t} \ . \tag{A1.4}$$

Chapter 2: Acoustic Air Motion in a Straight Tube

Introduction

The present chapter focuses on the mechanics of air, which is a very complex subject; consider the difficulty of predicting the weather with accuracy. We are fortunate, however, in that we are considering a very special case of air mechanics: air motion in a tube where the perturbation pressures are small. Also, as discussed in Chapter 1, we do not concern ourselves with entropy variation. Another assumption is made now: there is no *rotational air motion*. That is, we allow only *irrotational air motion*. This means that no air particle is spinning about any line that intersects it. More is said about rotational air motion in Chapter 13. All of the above assumptions mean that we are working under the *acoustic approximation*. The acoustic approximation greatly simplifies the description of air motion.

Preliminaries

A straight tube of constant cross-sectional area, A, is considered. Such a tube is shown in Figure 1, along with Cartesian axes, x, y, and z, as well as the section of a plane orthogonal, or perpendicular, to the x-axis. Cross-sectional planes are orthogonal to the x-axis and parallel to the y-z plane. We often refer to the constant area straight tube as simply a straight tube. The tube is straight because its sides are parallel to the x-axis. We are concerned with the kinematics and thermodynamics of air particles. The air particles could, potentially, be moving in all three Cartesian directions, but we restrict the motion to be parallel to the x-axis. Further, the kinematic and thermodynamic properties are the same in any section of cross-sectional plane of the tube. All of these assumptions mean that we are considering *one-dimensional motion*. [One-dimensional motion excludes the possibility of rotational air motion, which is one of the parts of the acoustic approximation.] The conditions under which these assumptions are valid are discussed later in the book. Figure 1 shows a tube with circular cross-sections, but other shapes are also possible.

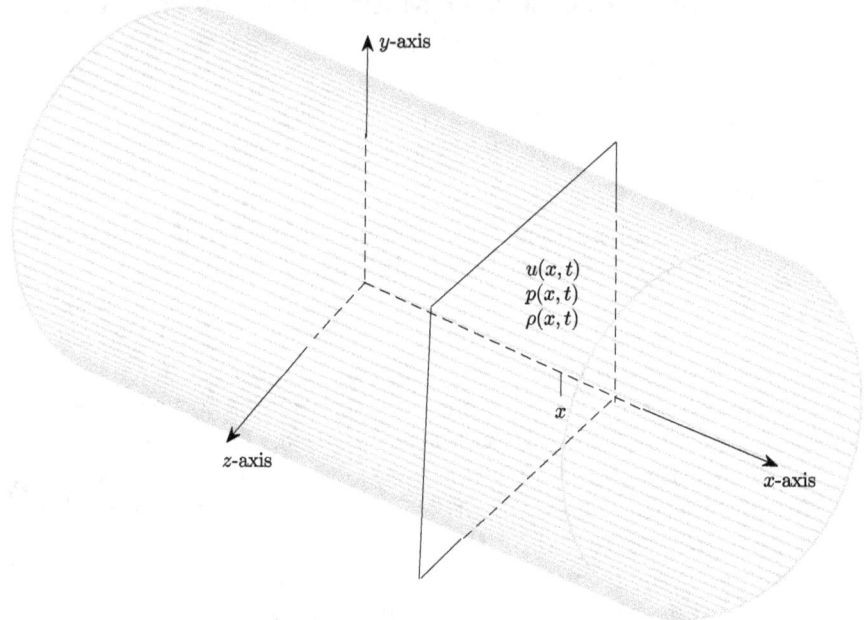

FIGURE 1. Straight tube

In the mathematical continuum, each point in the three-space within the tube coincides with an air particle at any time. However, we consider fixed points in the space in which the tube is at rest, that is, the Eulerian frame, so a fixed point can coincide with different air particles as time evolves. Thus, at each time t we attach to each point in the tube a *particle velocity* u in the x-direction and a thermodynamic property such as pressure p or density ρ. The symbol for velocity has been changed from v to u to emphasize that we are now working in the Eulerian frame. [There is no difference in the definition of velocity between the frames-of-reference, it is that we are not following individual air particles. There is an important difference in how acceleration is calculated in the two frames, and we touch on this briefly below.] Also, at any instant, u, p, and ρ are functions of x and t alone: there is no variation over cross-sections. So we can write

$$u = u(x,t) \quad p = p(x,t) \quad \rho = \rho(x,t) \ . \tag{1}$$

Because all of the air particles in a cross-sectional plane all move in the same way, it is possible to define *volume velocity* Q

$$Q = Q(x,t) \equiv u(x,t)A ,\qquad(2)$$

which has the dimensions of volume-per-unit time, for example cm^3/s. It is important to keep in mind that these quantities are fixed to spatial points and that air particles pass through these points: the quantities do not follow individual air particles.

When the air is undisturbed $u = 0$, $Q = 0$, $p = p_{atm}$, and $\rho = \rho_{atm}$, where p_{atm} and ρ_{atm} are atmospheric pressure and atmospheric density under normal atmospheric conditions, respectively. We often write p_0 for p_{atm} and ρ_0 for ρ_{atm}, which refer to the rest pressure and rest density, respectively. The deviations from rest pressure and rest density are perturbation pressure $\delta p = \delta p(x,t) = p(x,t) - p_0$ and perturbation density $\delta \rho = \delta \rho(x,t) = \rho(x,t) - \rho_0$, respectively. A change in notation in these quantities is made later in this chapter for purposes of simplification. The following discussion depends on the assumption that the amplitude of perturbation pressure $|\delta p|$ is much less than the surrounding atmospheric pressure, or that $|\delta p|/p_{atm} = |\delta p|/p_0 \ll 1$.

The conservation equations

The four equations that govern the motion and thermodynamics of air particles are equations for *mass conservation, momentum conservation,* and *energy conservation,* as well as an equation of state for air. This set of equations is immediately simplified under the acoustic approximation. With the constancy of entropy and irrotational motion, we need not consider energy conservation separately from mass and momentum conservation. Further, the simple Equation (9) of Chapter 1 is used under the acoustic approximation in lieu of a complete equation of state. Finally, the equations for mass and momentum conservation are linearized, or made linear, in a process called *linearization*.

The approximate form of the conservation equations for one-dimensional motion in a straight tube are derived. These derivations refer to Figure 2, which depicts the tube shown in Figure 1 with y-z planes orthogonal to the page. The vertical dashed lines represent cross-sections of the tube, both with area, A. That is, we imagine

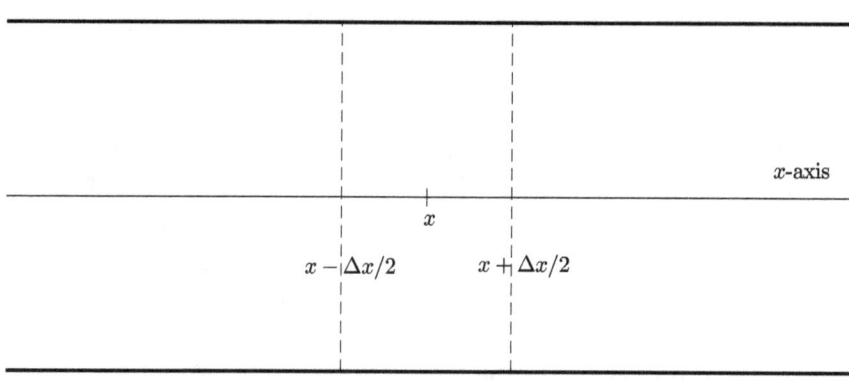

FIGURE 2. Slab of air in the straight tube

a "slab" of air centered at x, and bounded by the cross-sections at $x - \Delta x/2$ and $x + \Delta x/2$, with $\Delta x > 0$. We cannot draw time, so we just state that time is considered at t, and that there is an interval of time from $t - \Delta t/2$ to $t + \Delta t/2$, with $\Delta t > 0$. We imagine that both Δx and Δt are small in some sense.

The conservation equations are examined under the acoustic approximation, which permits linearization of these equations and has the practical effect of allowing us to neglect certain terms. In particular, we can neglect the products of small quantities, such as products of terms δp, $\delta \rho$, u, or Q. The strict mathematical justification for this approximation is beyond the scope of this book. However, it involves non-dimensionalization of physical quantities, so that, for instance, a quantity such as δp is replaced by $\delta p/p_0$ in a consistent way. This is done in order to compare the magnitude of terms in the equations of motion non-dimensionally. It turns out that $|u|/c_0$ is the non-dimensional quantity corresponding to particle velocity u, which later is shown to be small if $|\delta p|/\rho_0 c_0^2$ is small. Products of terms having small magnitudes are said to be negligible. The process where products of small terms are neglected is what we have termed linearization. When products of small terms are neglected in equations, we continue to use an equal sign "=", instead of an approximation sign, "≈". Also, we note that differences in ρ or p values, in either space or time, are actually differences in their deviations $\delta \rho$ and δp from rest condition, because the rest conditions cancel out in the differences.

The mass conservation equation

Mass conservation is a very basic principle. For a fixed volume in space, which can be as small as we like in our mathematical space, mass conservation requires that the difference between the mass flowing into and the mass flowing out of the volume equals the rate at which the mass accumulates in the volume.

For the one-dimensional situation of Figure 2, consider the volume between $x - \Delta x/2$ and $x + \Delta x/2$, which is our slab of air. The volume of the slab is $A\Delta x$, so its mass at time t is approximately $\rho(x,t)A\Delta x$. Thus, the time rate of change of the mass of air in the slab at time t is

$$\text{time rate-of-change of mass} = \frac{\rho(x, t+\Delta t/2) - \rho(x, t-\Delta t/2)}{\Delta t} A\Delta x \qquad (3)$$
$$= \frac{\Delta_t \rho(x,t)}{\Delta t} A\Delta x \;.$$

The rate at which mass flows into the slab through the left boundary at $x - \Delta x/2$ is $\rho_0 Q(x - \Delta x/2, t)$, and the rate at which mass flows out of the right boundary at $x + \Delta x/2$ is $\rho_0 Q(x + \Delta x/2, t)$. The net flow of mass into the slab at time t is

$$\text{net inflow of mass} = \rho_0 \bigl(Q(x - \Delta x/2, t) - Q(x + \Delta x/2, t) \bigr)$$
$$= -\rho_0 \bigl(Q(x + \Delta x/2, t) - Q(x - \Delta x/2, t) \bigr) \qquad (4)$$
$$= -\rho_0 \Delta_x Q(x,t) \;.$$

There is a reason that the rest density ρ_0 is used instead of the actual density $\rho = \rho_0 + \delta\rho$, in Equation (4). Density is multiplied by a small quantity, the volume velocity Q. Thus, the added deviation of density $\delta\rho$ creates a very small term that can be neglected under the acoustic approximation.

If the quantities in Equations (3) and (4) are equated, and we divide through by $A\Delta x$ to obtain

$$\frac{\Delta_t \rho(x,t)}{\Delta t} = -\frac{\rho_0}{A} \frac{\Delta_x Q(x,t)}{\Delta x} \;. \qquad (5)$$

In terms of the particle velocity, $u = Q/A$, Equation (5) can be written

$$\frac{\Delta_t \rho(x,t)}{\Delta t} = -\rho_0 \frac{\Delta_x u(x,t)}{\Delta x} \;. \qquad (6)$$

Equations (5) and (6) are statements of mass conservation. They indicate that the time rate-of-change of the density is proportional to the negative *gradient* in volume velocity or particle velocity. Volume velocity or particle velocity gradients are spatial inhomogeneities, or spatial differences, in these velocities. There is an increase in the density of a slab of air if there is less air flowing out to the right than there is flowing in to the left. There is a decrease in density of a slab if there is more air flowing out to the right than entering to the left. Therefore we encounter local density changes in time, if there are gradients in particle velocity. The approximations in Equations (5) and (6) improve as Δt and Δx get smaller.

The momentum conservation equation

The equation of momentum conservation that best suits terrestrial air motion at a human scale is known as the *Navier-Stokes equation*. This equation states Newton's second law of motion, Equation (5) of Chapter 1, for the air continuum in an Eulerian frame of reference. It actually applies to a large class of fluids, including water. Because the equation is stated in an Eulerian frame, it contains a term of the product of the air particle velocity with its gradient. This is neglected in the acoustic approximation. As with our derivation of Equations (5) and (6) for mass conservation, rest density ρ_0 is used in place of $\rho = \rho_0 + \delta\rho$. The effects of air friction are included in the Navier-Stokes equation, but they are assumed to be negligible under the acoustic approximation.

For the special case of one-dimensional motion in a tube, the momentum conservation equation produces a single equation relating Q with p, or u with p. Referring to Figure 2, the approximate mass of air m in the slab between $x - \Delta x/2$ and $x + \Delta x/2$ with $\Delta x > 0$ is

$$m = \rho_0 A \Delta x \ . \tag{7}$$

Again, the deviation from rest density is neglected here under the acoustic approximation.

The ratio of the centered first difference of u in time to the change in time, can be given a special symbol, a_E.

$$\begin{aligned} a_E(x,t) &= \frac{u(x, t + \Delta t/2) - u(x, t - \Delta t/2)}{\Delta t} \\ &= \frac{\Delta_t u(x,t)}{\Delta t} \\ &= \frac{1}{A} \frac{\Delta_t Q(x,t)}{\Delta t} \ . \end{aligned} \quad (8)$$

Because we are considering a fixed point in space in the Eulerian frame, and not following individual air particles, as in the Lagrangian frame, Equation (8) is not the full expression for acceleration a of an air particle. The exact expression involves a nonlinear term in velocity u that is neglected under the acoustic approximation. [This nonlinear term, the product of the particle velocity and its gradient, is important when sources of sound in the vocal tract are considered in Chapter 13.] However, under the acoustic approximation the *particle acceleration* $a \approx a_E$, and

$$\begin{aligned} F(x,t) &= \bigl(p(x - \Delta x/2, t) - p(x + \Delta x/2, t)\bigr) A \\ &= -\bigl(p(x + \Delta x/2, t) - p(x - \Delta x/2, t)\bigr) A \\ &= -\Delta_x p(x,t) A \ . \end{aligned} \quad (9)$$

This expression neglects the friction between the air and the solid tube. a can be related to force using Newton's second law, $F = ma$. The net force F on the slab of air in the positive x-direction at time t is, from Equations (7), (8) and (9)

$$-\Delta_x p(x,t) A = (\rho_0 A \Delta x) \frac{1}{A} \frac{\Delta_t Q(x,t)}{\Delta t} \ ,$$

or, rewriting, $\quad(10)$

$$\frac{\rho_0}{A} \frac{\Delta_t Q(x,t)}{\Delta t} = -\frac{\Delta_x p(x,t)}{\Delta x} \ .$$

In terms of particle velocity u

$$\rho_0 \frac{\Delta_t u(x,t)}{\Delta t} = -\frac{\Delta_x p(x,t)}{\Delta x} \ . \quad (11)$$

Equations (10) and (11) are simply Newton's second law for air under the acoustic approximation: the acceleration of an air particle at x and t is proportional to the negative of the pressure gradient at the same place and time. Air accelerates as it moves from regions of relatively high pressure toward regions of lower pressure; it decelerates in moving from regions of relatively low pressure to regions of higher pressure. This is as we would expect. Thus, local changes in particle velocity occur when there are pressure gradients, or spatial inhomogeneities in pressure.

With acceleration of air given in Equation (11), we can expect that gradients in particle velocity also are present. In turn, these gradients determine the time rate-of-change of density $\rho(x,t)$ according to Equation (6), and, hence, the time rate-of-change of pressure $p(x,t)$. We could expect gradients in pressure accompany their time rate-of-change. If this connection between Equations (6) and (11) does, in fact, occur, then we can expect some interesting physical phenomena.

Summary of conservation equations

The section on conservation equations under the acoustic approximation is now complete. Let's write the mass and momentum conservation equations in terms of Q and p, Equations (5) and (10), together. Before doing that, however, we express ρ in terms p in Equation (5) using Equation (9) of Chapter 1.

$$\frac{\Delta_t p(x,t))}{\Delta t} = -\frac{\rho_0 c_0^2}{A} \frac{\Delta_x Q(x,t)}{\Delta x} . \tag{12}$$

Similarly,

$$\frac{\Delta_t Q(x,t)}{\Delta t} = -\frac{A}{\rho_0} \frac{\Delta_x p(x,t)}{\Delta x} . \tag{13}$$

Recall that differences in p values, in either space or time, are actually differences in their perturbation pressures δp, because the rest conditions cancel out in the differences. For instance, $\Delta_t p(x.t) = \Delta_t \delta p(x,t)$, by Equations (A1.2) and (A1.4). Using the relation between u and Q, Equations (12) and (13) can be written

$$\frac{\Delta_t p(x,t)}{\Delta t} = -\rho_0 c_0^2 \frac{\Delta_x u(x,t)}{\Delta x} , \tag{14}$$

and

$$\frac{\Delta_t u(x,t)}{\Delta t} = -\frac{1}{\rho_0} \frac{\Delta_x p(x,t)}{\Delta x} . \tag{15}$$

Apart from the constants $\rho_0 c_0^2/A$ or $\rho_0 c_0^2$, and A/ρ_0 or $1/\rho_0$ there is a very nice symmetry in both of these sets of equations: they relate the time variation in one variable to the spatial gradient of the other. $\rho_0 c_0^2$ is an effective spring constant for air, from Chapter 1, and ρ_0 denotes mass per unit volume.

Notational changes and the meaning of signs

We now change the meaning of the symbols $p = p(x,t)$ and $\rho = \rho(x,t)$. This is mostly for convenience in speaking about pressure and density in an acoustic context, and it leaves the equations of motion, Equations (12) through (15) unchanged. The symbols p and ρ from here on refer to the perturbation pressure δp and perturbation density $\delta \rho$, respectively. This leaves the equations of motion unchanged because we refer only to changes in pressure and density in these equations. We continue to denote the rest pressure as p_0 and rest density as ρ_0, or as atmospheric quantities, p_{atm} and ρ_{atm}, respectively.

The meaning of plus and minus signs can be confusing when dealing with kinematic and thermodynamics properties. Because p and ρ are deviations from rest values, a positive value for either quantity means an elevation of p or ρ above the rest condition; a negative value means a decrease below rest condition. For the kinematic properties of particle velocity u and particle acceleration a, a positive value means in-the-positive-x-direction, and a negative value means in-the-negative-x-direction. Thus, the mathematical sign denotes direction for the kinematic properties. Further, decreasing speed, or absolute value of velocity, in the positive x-direction is the same as increasing speed in the negative x-direction, and both are denoted by a negative value of particle acceleration a.

A simulation

We now examine the behavior of a particular disturbance governed by Equations (14) and (15). The particular situation to be considered is that of a triangularly shaped *pressure pulse*, with zero particle velocity everywhere at time $t = 0$, as shown in Figure 3. The peak of the pressure pulse is given the value of 10 cm $H_2O \approx 10^4$ dyne/cm^2.

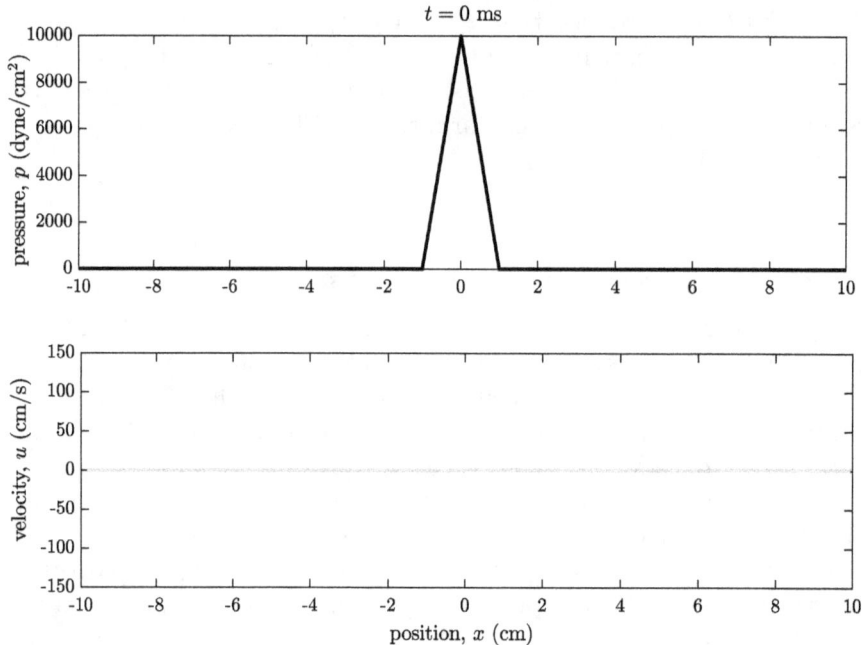

FIGURE 3. Initial pressure and particle velocity

Let's see if we can understand, at least qualitatively, what Equations (14) and (15) predict right after $t = 0$. The gradient, or slope, of the pressure pulse for $x < 0$ is positive, which means that air particles in this region of the pulse have negative acceleration by Equation (15). This, in turn, means that these air particles attain a negative particle velocity u in the disturbance for $x < 0$. On the other hand, air particles attain a positive u near the pulse for $x > 0$. As the air accelerates to either side of the pressure pulse, the pressure and density increase to either side of the pulse, which means that new regions of high pressure are created. This follows from Equation (14), because the gradient in particle velocity is negative on either side of the pressure pulse. Simultaneously, the pressure in the region near $x = 0$ diminishes. The process continues on to the right and to the left, or in the positive x-direction and in the negative x-direction.

In order to quantify this description, we simulate the physical processes with a numerical procedure. In the following, the limit of

Equations (14) and (15) as $\Delta t \to 0$ is taken to produce *differential equations* in the time variable. *Time-stepping*, or *extrapolation in time*, is performed using the *fourth-order Runge-Kutta algorithm*. The parameters of the numerical solution are as follows. The time increments for the time-step are 10^{-9} s, and the spatial increments are $\Delta x = 10^{-2}$ cm. The constants employed in the simulation are $\rho_0 = 0.001225$ g/cm^3 and $c_0 = 34,100$ cm/s.

An alternative to the above procedure is to numerically solve the centered first difference equations, Equations (14) and (15), directly. If a direct procedure is employed, then a relation between Δx and Δt should be observed in order to ensure stability of the calculation (McGowan 1987). This relation is known as the *Courant condition*.

$$\frac{\Delta t}{\Delta x} \leq c_0^{-1} \:. \qquad (16)$$

Figure 4 shows the evolution of the pressure pulse, where each frame represents a different time in milliseconds, ms. Figure 5 shows the evolution of the *particle velocity pulses* associated with this pressure disturbance. As predicted qualitatively above, there is a pressure pulse traveling to the left and one traveling to the right. Also, the left-going particle velocity pulse is negative and the right-going is positive. However, with this simulation, we can understand more fully what happens after time $t = 0$ ms than we could in our qualitative observations. Once the pressure pulse separates into two pulses, these pulses travel to the left and to the right without changing shape. In the case of the pressure pulse, the two separated pulses are one-half the amplitude of the initial pressure pulse, and they also have a triangular shape. On the other hand, the particle velocity pulses are the same shape, but of opposite polarity.

We can estimate the speed at which, say, the right-going pressure pulse is moving by computing the difference in the location of the pressure peak at time $t = 0$ ($x = 0$ cm) and the location of the pressure peak at time $t = 0.1$ ms $= 0.0001$ s (about $x = 3.5$ cm), and then dividing by 0.0001 s. This provides an estimate of 35,000 cm/s for both the speed at which the pressure pulses are moving and the speed at which the particle velocity pulses are moving. Keep in mind that we are in the Eulerian frame, and that we are describing the propagation of disturbances to thermodynamic and kinematic properties along the x-axis of the tube. That is, the thermodynamic disturbance p and the

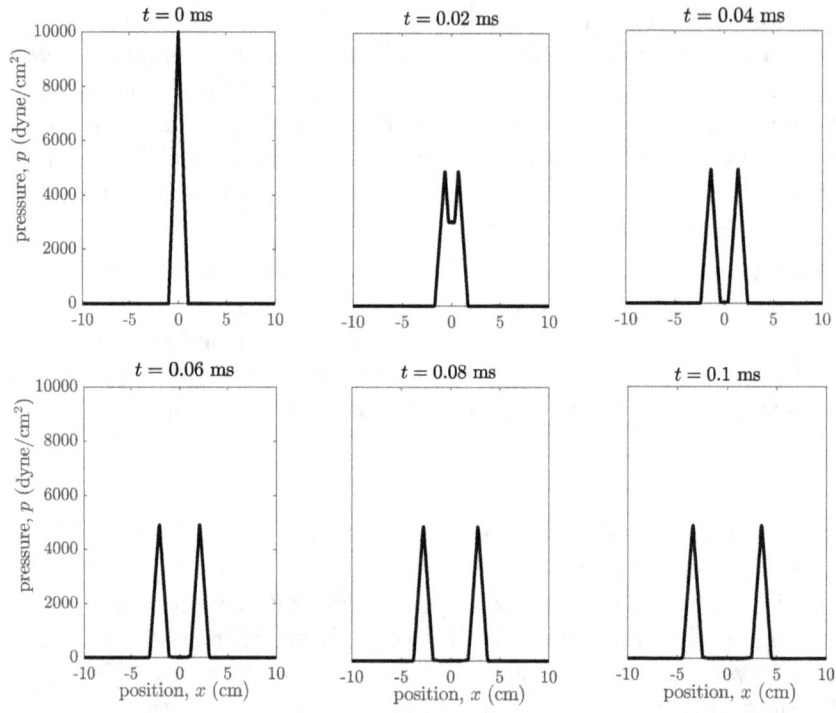

FIGURE 4. Evolution of pressure

kinematic disturbance u are moving along the tube axis at the rate of about $35,000$ cm/s.

The difference between propagation of a disturbance versus the movement of air particles can be seen if we consider the Lagrangian frame. For instance, consider the Lagrangian frame of reference for an air particle initially at $x = 0.5$ cm, which is inside the initial pressure pulse. [Recall that under the one-dimensional motion assumption, this air particle could be anywhere in the cross-section at $x = 0.5$ cm.] Figure 6 shows the kinematics of such an air particle in the Lagrangian frame. The top panel of Figure 6 shows its position in the x-direction, the middle panel its particle velocity, and the bottom panel its particle acceleration. The notation LF on the vertical axes refer to the Lagrangian frame. The air particle accelerates immediately to attain a maximum velocity of just over 200 cm/s at about $t = 0.015$ ms. Thus, the magnitude of particle velocity is much less than the

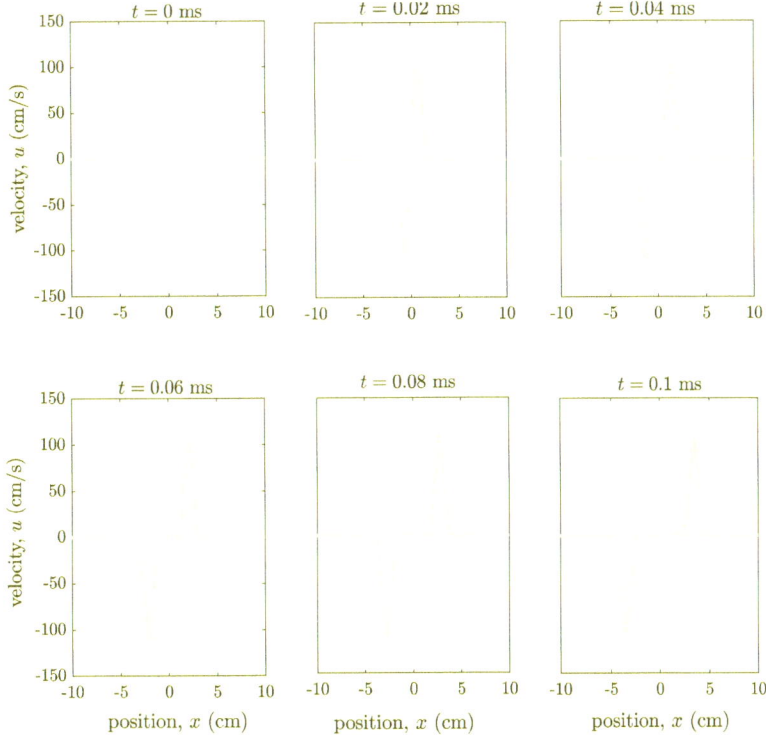

FIGURE 5. Evolution of particle velocity

speed at which the disturbance itself moves. There is a longer duration of deceleration from about $t = 0.015$ ms to about $t = 0.045$ ms. The Lagrangian frame particle velocity does have a triangular shape in time, but it is not symmetric (isosceles), as is the particle velocity pulse in the Eulerian frame. The top panel shows the movement that this particular air particle to be small and brief: it moves only 0.0025 cm and for only 0.04 ms. So while the disturbance of air propagate at a rapid rate along the tube forever at about 35,000 cm/s, the air particles themselves experience only a small, brief movement as the disturbance passes. Further, the maximum particle velocities possess magnitudes much less than 35,000 cm/s.

There is a phenomenon that we ignore here, and that is the *nonlinear steepening* of the pressure pulse. We are using an approximation to the theory of fluid mechanics, and the terms that are neglected

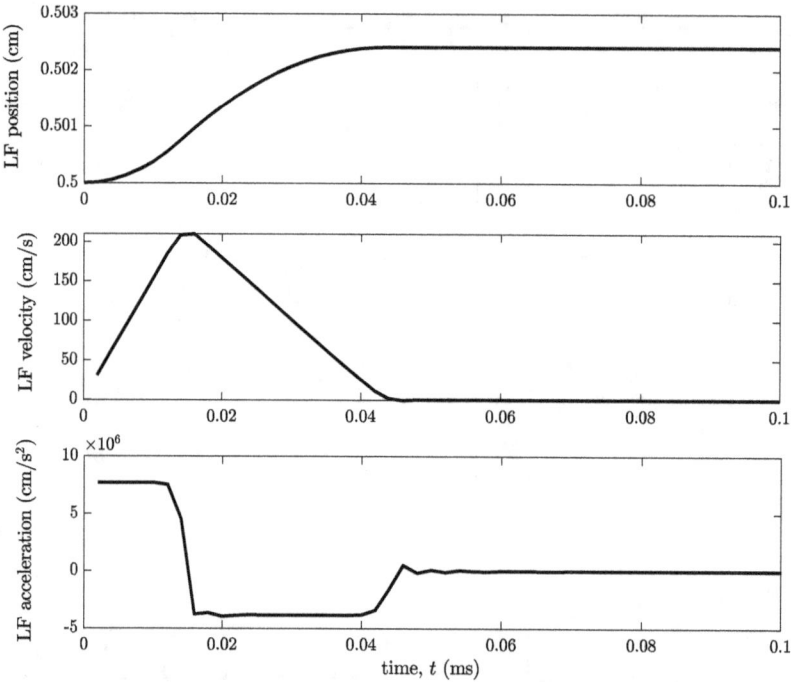

FIGURE 6. Kinematics in the Lagrangian frame of reference

in the acoustic approximation can have an additive effect over time and distance to make them non-negligible over long times and distances. The larger the pressure pulse, the sooner that these effects become non-negligible. With an initial peak pressure of about 10 cm $H_2O \approx 10^4$ dyne/cm^2, we have, in acoustic terms, a very large disturbance. [We formally introduce the idea of *decibel*, or dB, later in this chapter. However, the peak pressure in this situation is on the order of 156 dB, which makes a loud bang.]

When terms of the fluid mechanics equations neglected under the acoustic approximation are retained instead, we find that the peak of the right-going pressure pulse travels slightly faster in the rightward direction than the parts of the pressure pulse near zero. Similarly, the peak of the left-going pressure pulse travels slightly faster in the leftward direction than the parts of the pressure pulse at lower pressures. The magnitude of the speed differences peak and the bottom of the pressure pulse depends on the peak amplitudes. For the peak amplitude of 5×10^3 dyne/cm^2, which is the peak amplitude after the single pressure pulse separates into two pulses traveling in opposite directions, this speed difference for each pulse is about 150 cm/s (Pierce 1989). This is small compared to the speed at which the disturbances as a whole travel, which we estimated to be 34,000 cm/s. However, effects accumulate over time and space as the pulses travel, and result in the steepening of the pulses in the travel directions. We estimate that the pulses actually possess vertical lines with infinite slope on the travel direction side after about 7 ms when the pulses have traveled 240 cm. This means that the pulses have become shock waves at about these times and distances. Thus, we do not consider distances of travel much beyond tens of cm here. In speech, there are mechanisms that tend to mitigate this nonlinear steepening, even as pulses are reflected between the lips and the glottis multiple times. Also, it is seen in Chapter 13 that a subglottal pressure of 8 to 10 cm H$_2$O does not produce pressure pulses with this peak amplitude during voiced speech.

With a little more mathematical manipulation applied to Equations (14) and (15), it can be seen that numerical simulation is unnecessary when modeling the situation with a triangular pressure perturbation at $t = 0$. In fact, the following mathematical manipulation leads to much greater insight into the phenomenon that we are encountering: wave motion.

The wave equation

We describe some mathematical manipulations on the equations of motion, Equations (14) and (15). Apply $\Delta_t/\Delta t$ to both sides of Equation (14) and apply $\Delta_x/\Delta x$ to both sides of the Equation (15). Now divide the first expression by $-c_0^2$ and multiply the second expression by ρ_0. Because $\Delta_t\Delta_x u(x,t) = \Delta_x\Delta_t u(x,t)$, the first and second expressions can be combined in such a way that the terms involving $u(x,t)$ are eliminated. The result is

$$\frac{1}{c_0^2}\frac{\Delta_t^2 p(x,t)}{(\Delta t)^2} - \frac{\Delta_x^2 p(x,t)}{(\Delta x)^2} = 0 \ . \tag{17}$$

Equation (17) is the *wave equation* in one dimension. [Strictly speaking, the wave equation results when $\Delta t \to 0$ and $\Delta x \to 0$ in Equation (17).

Here, we allow Equation (17) to be called the wave equation.] It can be shown that both u and Q also satisfy the wave equation.

Finally, we write the wave equation in factored form

$$\frac{1}{c_0^2}\frac{\Delta_t^2 p(x,t)}{(\Delta t)^2} - \frac{\Delta_x^2 p(x,t)}{(\Delta x)^2} = \left(\frac{1}{c_0}\frac{\Delta_t}{\Delta t} - \frac{\Delta_x}{\Delta x}\right)\left(\frac{1}{c_0}\frac{\Delta_t}{\Delta t} + \frac{\Delta_x}{\Delta x}\right)p(x,t)$$

$$= \left(\frac{1}{c_0}\frac{\Delta_t}{\Delta t} + \frac{\Delta_x}{\Delta x}\right)\left(\frac{1}{c_0}\frac{\Delta_t}{\Delta t} - \frac{\Delta_x}{\Delta x}\right)p(x,t)$$

(18)

$$= 0\ .$$

Solutions to the wave equation

One of the most important properties that a quantity has when it is governed by the wave equation, is that time and space are necessarily related. All of the quantities named above are not best understood as functions of the two variables t and x, but of two new variables: $\zeta = t - x/c_0$ and $\xi = t + x/c_0$. [Recall that $c_0 > 0$ has the dimension of velocity, so that x/c_0 has the dimension of time.]

Consider the factors of the wave equation in Equation (18). If $p(x,t)$ satisfies either factor, that is, if

$$\frac{1}{c_0}\frac{\Delta_t p(x,t)}{\Delta t} = -\frac{\Delta_x p(x,t)}{\Delta x}\ ,\ \text{or}$$

$$\frac{1}{c_0}\frac{\Delta_t p(x,t)}{\Delta t} = \frac{\Delta_x p(x,t)}{\Delta x}\ .$$

(19)

then $p(x,t)$ satisfies the wave equation. In words, if the time rate-of-change of $p(x,t)$ divided by c_0 is $\pm$ the gradient of $p(x,t)$, then $p(x,t)$ satisfies the wave equation. This means that functions $p(x,t) = \rho_0 c_0 f(t - x/c_0) = \rho_0 c_0 f(\zeta)$ and $p(x,t) = \rho_0 c_0 g(t + x/c_0) = \rho_0 c_0 g(\xi)$ are solutions to the wave equation, where f and g are arbitrary functions, each of a single variable. The mathematical details of this reasoning are in the Appendix to Chapter 2. [The meaning of the constant $\rho_0 c_0$ should become clear shortly.] So the most general solution to the wave

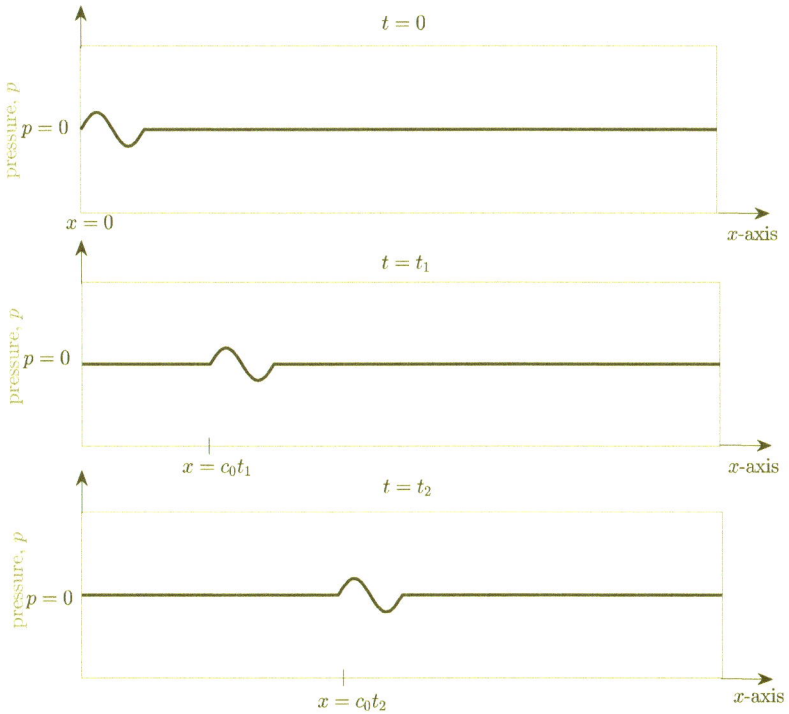

FIGURE 7. Right-going pressure disturbance

equation is
$$p(x,t) = \rho_0 c_0 f(t - x/c_0) + \rho_0 c_0 g(t + x/c_0) = \rho_0 c_0 f(\zeta) + \rho_0 c_0 g(\xi) \ . \quad (20)$$

What does it mean to be a function of $\zeta = t - x/c_0$ or of $\xi = t + x/c_0$? This is best understood using an illustration. Figure 7 shows a pressure disturbance, or perturbation, as a function of the spatial variable x at three different times, $t = 0$, $t = t_1$, and $t = t_2$, with $0 < t_1 < t_2$. Here it is supposed that $p = \rho_0 c_0 f(t - x/c_0)$ for simplicity. Note that the left end of the disturbance at $x = 0$ is translated to $x_n = c_0 t_n$ at $t = t_n$, because this amount of translation leaves $\zeta = t - x/c_0$ equal to zero. In fact the pressure disturbance just translates without changing shape, because all the points translate at the same speed, c_0. The other bits of the pressure disturbance are represented by $\zeta = t - x/c_0 = K$, for some constant K. Because the pressure perturbation is to the right of

$x = 0$ at $t = 0$, the constants K are negative, and the largest value of $|K|$ for the disturbance is the value of x/c_0 at the right edge of the disturbance at $t = 0$. c_0 is the adiabatic speed of sound that we introduced in Equation (9) of Chapter 1. Recall that we quoted a value $c_0 = 34,100$ cm/s in Equation (10) of Chapter 1 for this quantity, which is very close to our estimate, $35,000$ cm/s of the speed of the moving disturbances in the simulations shown in Figures 4 and 5. The function $\rho_0 c_0 f(t - x/c_0)$ represents what we call a right-going wave. It is easy to see that $\rho_0 c_0 g(t + x/c_0)$ represents a left-going wave.

It is not surprising that once the solution for p is written, the solution for u is also specified, because Equations (14) and (15) make it clear that these quantities are closely related. In fact, given Equation (20), u can be written as

$$u(x,t) = f(t - x/c_0) - g(t + x/c_0) \,. \tag{21}$$

so that u is also the sum of right-going and a left-going functions. [Equation (21) can be derived with Equation (15) applied to $\rho_0 c_0 f(t - x/c_0) + \rho_0 c_0 g(t + x/c_0)$, and in conjunction with Equations (A2.4), (A2.2), (A2.1), and (A1.4).] Note, however, that the ratio of the right-going part of the pressure disturbance to that of the particle velocity disturbance is $\rho_0 c_0$; and for the left-going counterparts, this ratio is $-\rho_0 c_0$. Recall that our simulations in Figures 4 and 5 had positive right-going pressure and velocity disturbances, but the positive left-going pressure disturbance was coupled with a negative left-going velocity disturbance. The quantity $\rho_0 c_0$ is referred to as the *characteristic impedance* (Pierce 1989). With $\rho_0 = \rho_{atm}$, from Chapter 1, Equation (10) the value of the characteristic impedance of air is

$$\rho_0 c_0 = 41.8 \text{ g}/(\text{cm}^2 \text{ s}) \,. \tag{22}$$

It seems that *acoustic wave motion* is a bunch of "lumps" of thermodynamic and kinematic properties of air moving to the right and left. Once again, this does not describe the motion of the air particles themselves, because we are considering the Eulerian frame.

Another very important property of the wave equation is that it is a *linear equation*. This is the result of the linearization of the conservation equations. The wave equation inherits this linearity from the conservation equations derived under the acoustic approximation, Equations (14) and (15). Thus, if p_1 and p_2 are solutions to the one-dimensional wave equation,

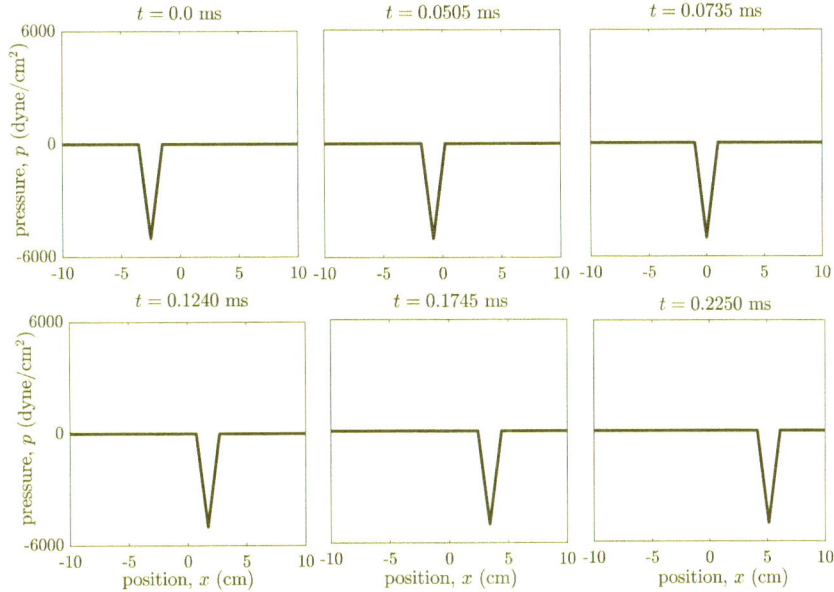

FIGURE 8. Pressure rarefaction

then so is $p = p_1 + p_2$. This is easily seen from the fact that $p_n = \rho_0 c_0 f_n(t-x/c_0) + \rho_0 c_0 g_n(t+x/c_0)$, which means that $p = p_1 + p_2 = \bigl(\rho_0 c_0 f_1(t-x/c_0) + \rho_0 c_0 f_2(t-x/c_0)\bigr) + \bigl(\rho_0 c_0 g_1(t+x/c_0) + \rho_0 c_0 g_2(t+x/c_0)\bigr)$, so that $p = \rho_0 c_0 F(t - x/c_0) + \rho_0 c_0 G(t + x/c_0)$, where $F(t - x/c_0) = f_1(t - x/c_0) + f_2(t - x/c_0)$ and $G(t + x/c_0) = g_1(t + x/c_0) + g_2(t + x/c_0)$. This is the principle of *linear superposition*, or *superposition*. For instance, in the simulation above, we could have added any number of pressure pulses to the initial configuration. Linear superposition also applies to u and Q, which are also governed by the wave equation.

Figures 8 through 11 demonstrate superposition in a somewhat dramatic way. Figure 8 shows a negative pressure disturbance, or *rarefaction*, traveling to the right. Figure 9 shows a positive pressure disturbance, or *compression*, traveling to the left. Because these both are solutions to the one-dimensional wave equation, the same is true of their sum, by superposition. Their sum is shown in Figure 10, where at $t = 0.0735$ ms there is almost no pressure disturbance. In fact, if we had more precision in our calculations we could show the precise instant in which there is absolutely no pressure disturbance. After this

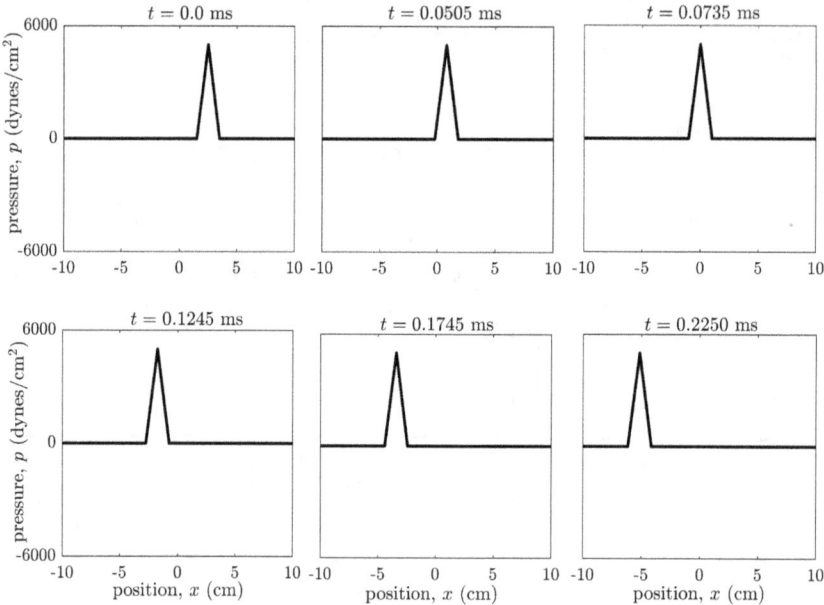

FIGURE 9. Pressure compression

time, the pressure pulses reappear and travel to the right and to the left as before. This may seem as though we are getting something from nothing. How could this be? The answer is that not all disturbances are brought to zero near $t = 0.0735$ ms. Consider the particle velocity pulses associated with the combined pressure pulses, which are shown in Figure 11 using the same principle of superposition. In fact, these particle velocity pulses combine at $t = 0.0735\ ms$ to produce a negative particle velocity pulse that is twice the amplitude of the original two separate pulses.

Small quantities in acoustics

We stated that a disturbance is small if the perturbation pressure p is small compared to the rest pressure p_0, i.e. if $|p|/p_0 \ll 1$. For instance, the pressure disturbance of $p = \pm 10$ cm H_2O is small compared to atmospheric air at normal atmospheric conditions, because it is about 1% of atmospheric pressure. We used the smallness of acoustic

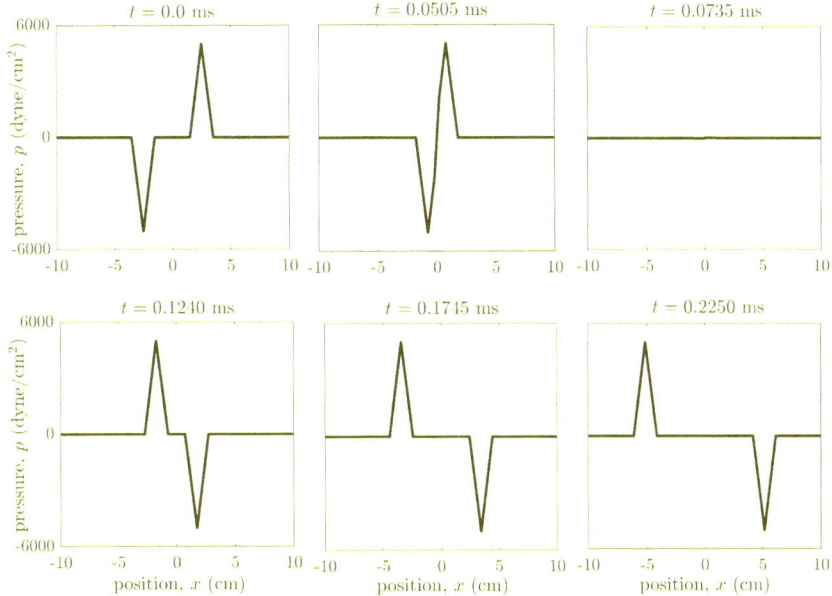

FIGURE 10. Sum of pressure rarefaction and compression

disturbances in the linearization of the conservation equations, which resulted in the wave equation applying to these quantities. [However, it has been seen that if we wait long enough, that a situation that is well described by the acoustic approximation can become a situation where the neglected effects have accumulated to become important through nonlinear steepening.]

In order to claim that a physical quantity is small, it is necessary to compare it to another known quantity with the same dimensions. So what does it mean to say the particle velocity is small for acoustic disturbances? We have seen that the peak particle velocity u is a little over 200 cm/s, when the pressure disturbance is 1% of atmospheric pressure. It turns out that when the perturbation pressure is small that the particle velocity u is small compared to the adiabatic speed of sound c_0. That is, what we call the *acoustic Mach number*, $M_a = |u|/c_0$, is small. Note also that for a disturbance traveling in one direction, this requirement of small acoustic Mach number is equivalent to the requirement that $|p|/\rho_0 c_0^2$ be small, because $|p| = \rho_0 c_0 |u|$ for such a disturbance. This is in addition to requiring that $|p|/p_0$ be small.

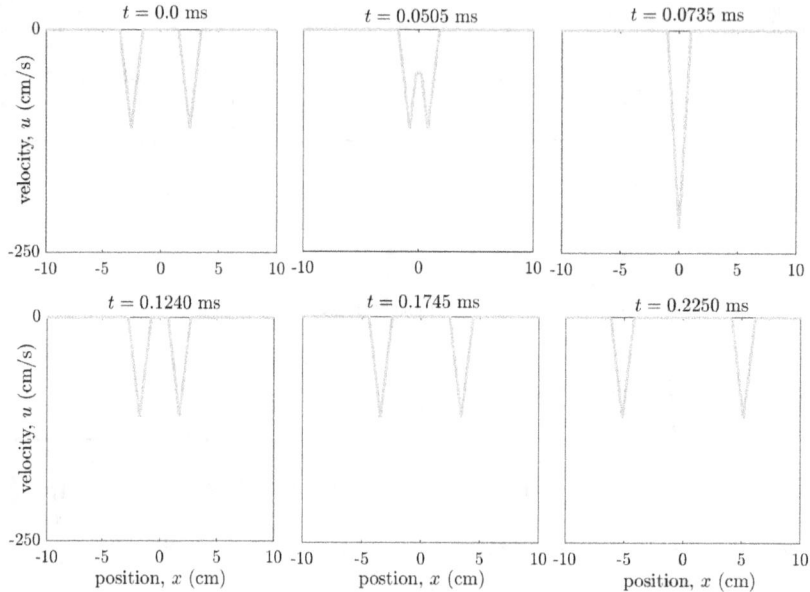

FIGURE 11. Particle velocities

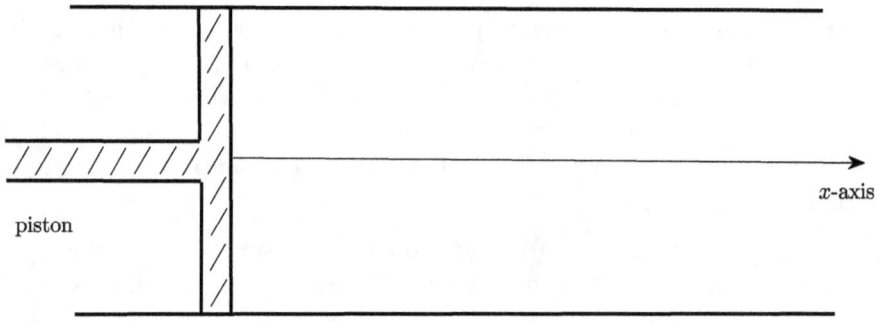

FIGURE 12. Semi-infinite tube with piston

A piston boundary

We introduce a boundary orthogonal to the axis of the infinitely long straight tube. The boundary is a movable piston at the left end of the tube, as shown in Figure 12. The tube is changed from one having infinite extent in both the positive x-direction and negative x-direction, to one having infinite extent in the positive x-direction only, and this is known as a semi-infinite tube. Until now, acoustic disturbances have been assumed to exist at some initial time, and then we followed their time evolution. In the following, the acoustic disturbances, or waves, are created as the piston is moved by an external agent.

What happens to the air in the tube as a result of the piston motion? The *boundary condition* is that the air particles at the piston face follow that surface. Thus, the velocity in the Lagrangian frame is

$$v(x = \text{piston face}, t) = U_{pst}(t) , \qquad (23)$$

where the piston velocity $U_{pst}(t)$ is assumed to be small, or $\max\left(|U_{pst}(t)|\right)/c_0 \ll 1$. The assumption of small velocity amplitude piston movement permits us to replace the boundary condition from the moving piston face to the air particle velocity u at $x = 0$ in the Eulerian frame.

$$v(x = \text{piston face}, t) = U_{pst}(t) \approx u(x = 0, t) . \qquad (24)$$

The result is an acoustic wave propagating to the right, because the piston face blocks any possibility of a left-going wave. This means that we can write

$$u(x,t) = f(t - x/c_0) \quad \text{and} \quad p(x,t) = \rho_0 c_0 f(t - x/c_0) . \qquad (25)$$

for some function of a single variable, $\zeta = t - x/c_0$. It follows from Equations (24) and (25) that

$$f(t) = U_{pst}(t) . \qquad (26)$$

The acoustic disturbance carries the shape of the piston velocity to the right in the semi-infinite tube at the adiabatic speed of sound, c_0.

The piston can create two types of simple disturbances: compressions and rarefactions. Rarefactions are also known as *expansions*. In creating a compression, the piston accelerates and then, eventually, decelerates in the positive x-direction. In creating an expansion, or rarefaction, the piston accelerates and, eventually, decelerates in the negative x-direction. An example of each scenario is shown in Figure 13: the top panels show the piston velocities $U_{pst}(t)$ as functions of time for a particular compression (left panels) and a particular rarefaction (right panels). Below each top panel is the corresponding pressure pulse according to the relation $p = \rho_0 c_0 u$ for right-going waves. For

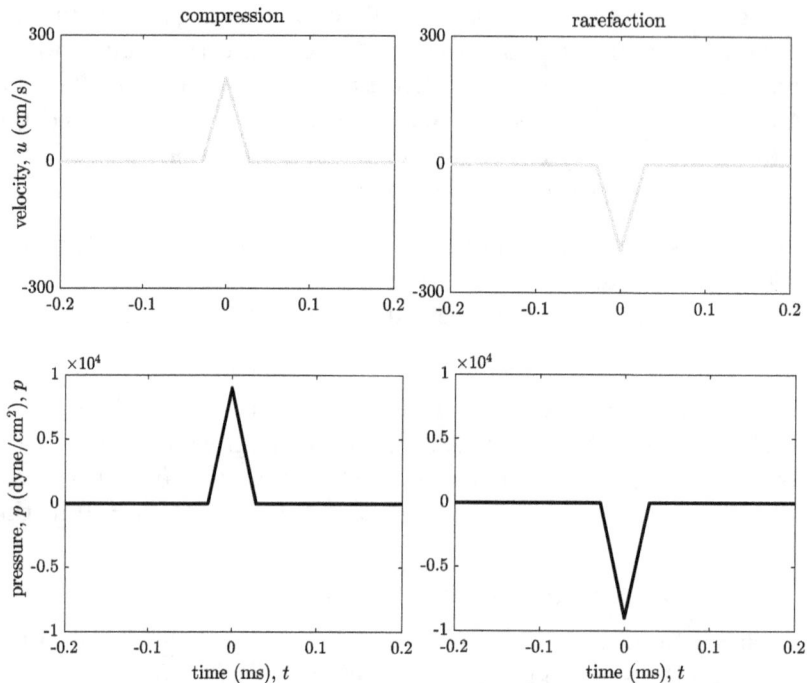

FIGURE 13. Velocities and pressures at the piston face

the compression, there is acceleration before $t = 0$, for about 0.026 ms, and then a deceleration for the same duration. The value of maximum velocity attained by the piston at $t = 0$, which is about 200 cm/s, is small compared to the adiabatic speed of sound, c_0. The peak pressure that corresponds to the peak $u = 200$ cm/s is $p = 8360$ dyne/cm$^2 \approx 8.5$ cm H$_2$O. Figure 13 shows that the piston creates pressure pulses of the same shape and duration as the particle velocity pulses. Figure 14 shows the position of the piston face for the compression in Figure 13. This figure shows that the movement of the piston for the compression is, indeed, brief and small. The piston moves less than 0.006 cm in about 0.03 ms.

Equations (25) and (26) completely determine the acoustic wave motion at all positions in the tube for all times. The disturbance can be considered to have been created at $x = 0$ for -0.026 ms $< t < 0.026$ ms. Thus, $f(\zeta)$ is zero, except for -0.026 ms $< \zeta < 0.026$ ms, or

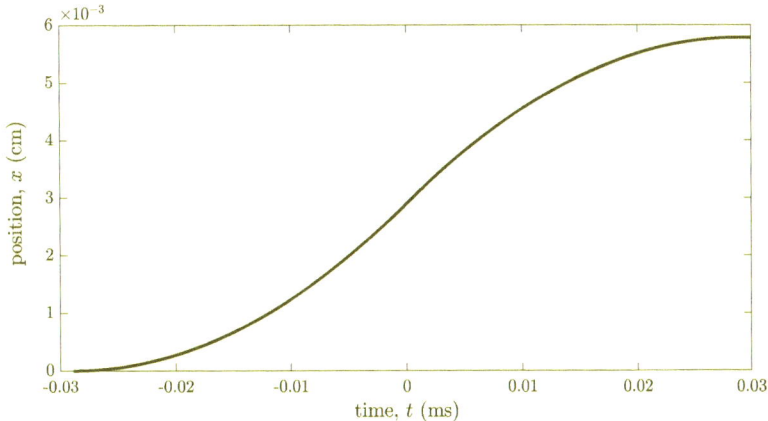

FIGURE 14. Piston face position versus time

-0.000026 s $< \zeta < 0.000026$ s. This determines how the air behaves at all positions and times, because the value of ζ determines the relation between time t and position x. For instance, the peak of the compression is at $x = 0$ for $t = 0$, which means that the peak corresponds to $\zeta = 0$. To follow the progress of the peak of the compression, we set $\zeta = 0 = t_1 - x_1/c_0$ and solve for x_1 at any chosen time t_1. That is, the peak of the compression is at $x_1 = c_0 t_1$. Or we could find when the peak reaches a chosen x_1, by $t_1 = x_1/c_0$. Another piece of the triangular pulse can be chosen by choosing another ζ. The reader can track the piece of disturbance denoted by $\zeta = 0.013$ ms $= 0.000013$ s, for instance.

The quantity x/c_0 is the *acoustic delay time* for a disturbance traveling from $x = 0$ to another position, $x > 0$. $\zeta = t - x/c_0$ is called the *retarded time*. If we were concerned with left going waves from $x = 0$ to $x < 0$, the acoustic delay time would be $-x/c_0$, and the retarded time would be $\xi = t + x/c_0$.

The present scenario of the piston in a semi-infinite tube is the simplest situation with an *acoustic source*. In this case, movement of a solid (i.e., the piston) driven by an external agent (e.g., human, motor) causes an acoustic disturbance. Because this is a physical process, we can expect that energy can be transferred between the piston and the air. The piston can act as an acoustic source or *acoustic sink*. An acoustic source is something that provides energy to acoustic wave motion, and an acoustic sink is something that takes energy away from acoustic wave motion. The next section defines quantities that permits us to relate energy to the movement of the piston.

Energy in acoustics

A body with mass m moving in translational motion with velocity v possesses a kinetic energy E^{kin}.

$$E^{kin} = \frac{m}{2}v^2 \ . \tag{27}$$

This is the kinetic energy of the mass in the mass-spring system discussed in Chapter 1. The unit of energy in the c-g-s system is the *erg*, where erg = g-cm^2/s^2 = dyne-cm.

Energy in acoustics is usually defined on a per volume basis, because we are working with a continuum. Under the acoustic approximation the kinetic energy of air particles per unit volume is

$$e^{kin} = \frac{\rho_0}{2}u^2 \ . \tag{28}$$

Note that this *kinetic energy density* is proportional to the square of a small quantity u and the rest density ρ_0. The reason that rest density is used is subtle, but the definition is consistent with the acoustic approximation. The interested reader is directed to Lighthill (1978) for a somewhat advanced discussion of these matters. The unit of energy density in the c-g-s system is erg/cm^3. If we use the fact that erg = dyne-cm, then it is seen that energy density has the same units as pressure, erg/cm^3 = dyne/cm^2.

As with all physical systems with spring-like restoring forces, there is potential energy as well as kinetic energy. For the simple mass-spring system of Chapter 1, the energy needed to compress or extend the spring so that the mass has position x_m that is different from its rest position x_0 is its potential energy, given by

$$E^{pot} = \frac{\kappa}{2}X_m^2 \ , \tag{29}$$

where $X_m = x_m - x_0$.

In the case of air, the potential energy is the energy needed to compress a volume of air from rest density ρ_0 to a state of density $\rho_0 + \rho$ with $\rho > 0$ (Lighthill 1978). Let e^{pot} denote *potential energy density*. It turns out that for small ρ/ρ_0, which holds under the acoustic approximation, that

$$e^{pot} = \frac{c_0^2}{2\rho_0}\rho^2 = \frac{1}{2\rho_0 c_0^2}p^2 , \qquad (30)$$

where the second equality follows from Equation (9) of Chapter 1. It turns out that this is the potential energy density for expansions ($\rho < 0$) as well.

The *total energy density e* is

$$e = e^{kin} + e^{pot} = \frac{\rho_0}{2}u^2 + \frac{1}{2\rho_0 c_0^2}p^2 . \qquad (31)$$

For the special case of a one-dimensional wave traveling in one direction, $|p| = \rho_0 c_0 |u|$, we obtain

$$e^{pot} = e^{kin} = \frac{\rho_0}{2}u^2 = \frac{1}{2\rho_0 c_0^2}p^2 , \qquad (32)$$

so that, in this special case

$$e = \rho_0 u^2 = \frac{1}{\rho_0 c_0^2}p^2 . \qquad (33)$$

While energy conservation applies to any physical system, there is no equivalent law for energy density conservation. In order to examine energy conservation in an acoustical system, one needs the energy density times the volume for each small volume element of air, and add up all these products for a measure of total energy. Further, one always needs to account for conversion between types of energy. For instance, energy occurs in different forms in acoustic transducers, such as microphones. For such cases there is an exchange between mechanical and electromagnetic energies. Acoustic wave motion possesses mechanical energy.

Figure 15 shows a triangular particle velocity pulse whose base extends over 2 cm. The total energy density at the peak of the wave shown in Figure 15 is, by Equation (33), $e = 0.001225 \times 200^2$ erg/cm^3 = 49 erg/cm^3. If we assume that the tube has a cross-sectional area, A, of 2 cm^2, then for the triangular pulses shown in Figure 15, the total energy is approximately 98 erg. This is the energy attained by a 1 gram mass that is dropped near the earth's surface after 14.3 ms of free fall, when it has traveled 0.1 cm = 1 mm. In terms of *heat energy*, this corresponds to 2.34×10^{-6} calories, which is the energy required

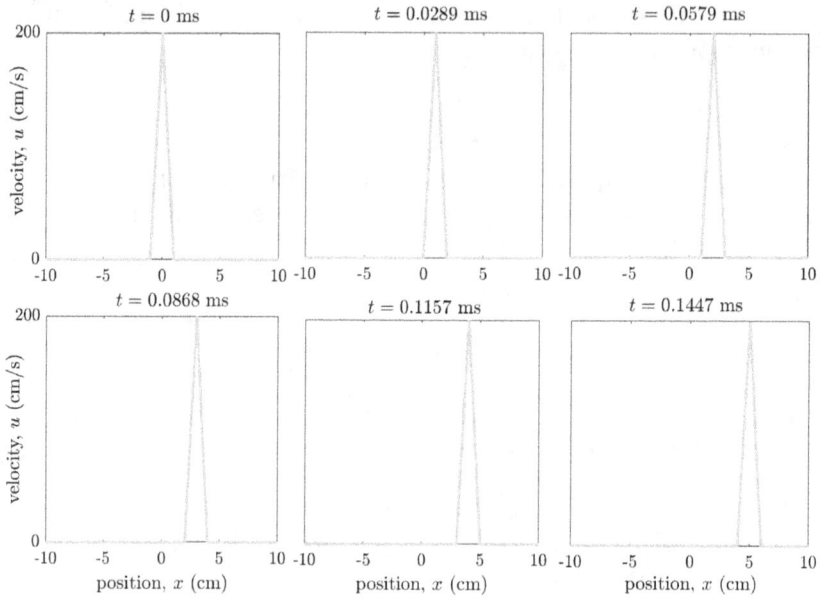

FIGURE 15. Right-going wave

to raise the temperature of 1 gram of water from 14.5° C to about 14.50000234° C (Resnick & Halliday 1966). In terms of household electrical use, 98 erg is 2.72×10^{-12} kWh (Resnick & Halliday 1966). 98 erg is not much energy in terms of any of these measures. We see from the simple example of the triangular compression in Figure 15, that energy propagates with the disturbance at the speed c_0.

The energy density and total energy of an acoustic wave are unfamiliar quantities, even in much of applied acoustics. Where issues of energy conservation and energy flow are concerned, it is often useful to work with the quantities of *power* and *intensity*. A discussion of these quantities relates the movement of the piston to the energy in acoustic wave motion.

Suppose a force F is applied to a body in its direction of movement, then the body accelerates. The force can either cause the body's speed to increase or to decrease, depending on the signs of F and of the body's velocity v. Thus, the body either gains or loses kinetic energy because of the force. The force can be a function of time, denoted

$F = F(t)$, and the body moves with a velocity that is a function of time $v = v(t)$. A result of classical physics is that the rate at which the body attains energy, called power, is

$$\mathcal{P}(t) = F(t)v(t) \,. \tag{34}$$

If $\mathcal{P}(t)$ is positive, the body is gaining energy at time t. If $\mathcal{P}(t)$ is negative, the body is losing energy at time t.

We go back to consider the piston-in-a-tube situation shown in Figure 12. Suppose that air on the piston face has pressure p_0 when the piston is not moving. This may require that an external agent apply a force $p_0 A$ in the positive x-direction in order that the piston remain at rest, depending on whether there is air to the left of the piston. In any case, it is necessary to apply additional force to the piston in order to get it to accelerate to the right or left. The additional force would be positive in the former case and negative in the latter case. There are two parts to the total force: the force due to the mass of the piston that would need to be applied even if there were no air to the right of the piston face, and force pA, where p is any deviation from pressure p_0. It is this latter force that imparts energy to the air. The piston imparts energy to the particles of air at its face at the rate of

$$\mathcal{P}_{pst}(t) \approx \big(p(x=0,t)A\big)U_{pst}(t) \approx \big(p(x=0,t)u(x=0,t)\big)A \,, \tag{35}$$

where $u(x=0,t)$ is the velocity of the air particles in the Eulerian frame at $x = 0$. Equation (25) tells us that a right-going wave is created so that $p(x=0,t) = u(x=0,t)/\rho_0 c_0$. Therefore, the power input by piston movement is

$$\mathcal{P}_{pst}(t) \approx \frac{u^2(x=0,t)}{\rho_0 c_0}A \approx \frac{U_{pst}^2(t)}{\rho_0 c_0}A \,. \tag{36}$$

If we were to multiply an average $\mathcal{P}_{pst}(t)$ by the small increment of time Δt by the duration of time that the piston is moving, we would approximate the amount of energy imparted by the piston to the air in that brief time. The acoustic disturbance created by the piston carries the energy imparted by the piston rightward.

While we often work in c-g-s units here, much of physical acoustics is done in *m-k-s units*, or the meters-kilograms-seconds system. For instance, the preferred unit for power is often the *Watt*, or W, which is an m-k-s unit. The following is the factor for conversion from c-g-s

erg/s to W.
$$1\ W = 10^7\ \text{erg/s} . \tag{37}$$
Electric bills may quote energy usage in terms of kiloWatt-hours, or kWh, which is a *power* unit multiplied by a time unit, which results in an energy unit.

The product pu that appears in Equation (35) can be defined for any x and t. This quantity is called intensity $\mathcal{I}$.
$$\mathcal{I}(x,t) \equiv p(x,t)u(x,t) . \tag{38}$$
From Equations (35) and (38)
$$\mathcal{P}_{pst}(t) \approx \mathcal{I}(x=0,t)A . \tag{39}$$
The rate of which energy crosses any cross-section of the tube is equal to intensity times the cross-sectional area.
$$\mathcal{P}(x,t) = \mathcal{I}(x,t)A . \tag{40}$$
The piston provides power, which is carried by the acoustic wave through the air, and each part of that wave has something called intensity associated with it.

Now we briefly leave the piston-in-a-tube scenario to discuss intensity in right-going and left-going waves in one dimension. In general, intensity has a plus or minus sign associated with it, and this sign denotes direction, just as in the case of particle velocity u. Consider a compression or a rarefaction pulse traveling to the right (positive x-direction) and to the left (negative x-direction). [It should be clear by now that traveling pulses are a kind of acoustic wave motion, therefore we use the term wave for pulse sometimes.] Figure 16 shows the two possible scenarios for right-going waves. Recall that u and p must possess the same sign for right-going pulses, because $p = \rho_0 c_0 u$ for these pulses. Figure 17 shows the two possibilities for the left-going waves, with p and u having opposite signs, because $p = -\rho_0 c_0 u$ for these acoustic waves. It is seen that the intensity has the same sign as the direction of travel for all these possibilities, so that intensity travels in the same direction as the pulse, whether it is a compression or a rarefaction. Our treatment of intensity and its relation to power applies only to the special case of one-dimensional acoustic waves in a tube. More sophisticated treatments involve vectors, and these treatments can be found in most books on acoustics, including Lighthill (1978) and Pierce (1989).

Acoustics of Speech Production

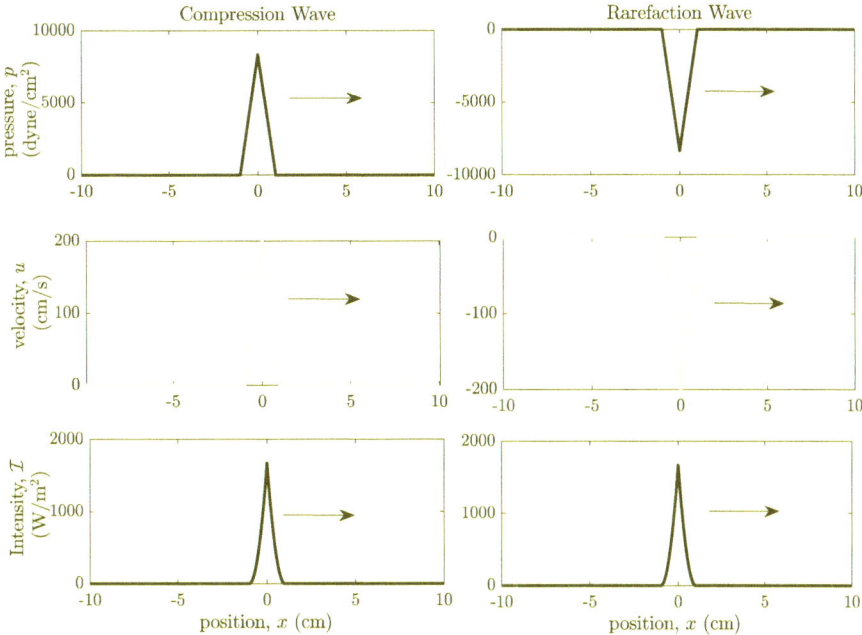

FIGURE 16. Intensity of right-going waves

The reader is probably familiar with the decibel, or dB, measure of power or intensity. This can be defined to be,

$$\text{power in dB} = 10 \log_{10} \left(\frac{|\mathcal{P}|}{\mathcal{P}_{ref}} \right),$$

for some fixed $\mathcal{P}_{ref}$,

(41)

$$\text{intensity in dB} = 10 \log_{10} \left(\frac{|\mathcal{I}|}{\mathcal{I}_{ref}} \right),$$

for some fixed $\mathcal{I}_{ref}$.

We have been discussing one-dimensional acoustic wave motion, which can be called *plane wave motion*, which is said to be composed of *plane waves*. In speech it is typical to assume plane wave motion, where intensity in dB is calculated from a measurement of pressure. Thus, $|\mathcal{I}| = |pu| = p^2/(\rho_0 c_0)$. It follows that Equation (41) can be

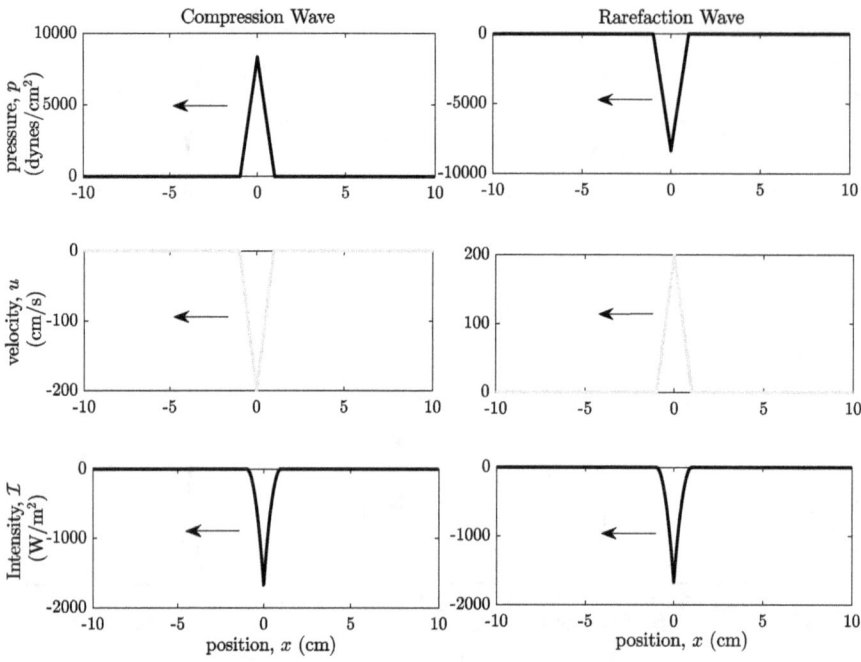

FIGURE 17. Intensity of left-going waves

expressed for a reference pressure, $p_{ref} = 2 \times 10^{-4}$ dyne/cm^2, as

$$\text{intensity in dB} = 20 \log_{10}\left(\frac{|p|}{p_{ref}}\right). \tag{42}$$

The compression in Figure 15 with a peak particle velocity of 200 cm/s, which gives a pressure peak of 8360 dyne/cm^2 = 8.5 cm H$_2$O. From Equation (41) this corresponds a peak intensity of about 152.2 dB, which is a very loud bang. As explained earlier in the chapter, a pressure pulse with a peak of 10 cm H$_2$O, or even 8 cm H$_2$O, eventually exhibits behavior that is not predicted under the acoustic approximation. These pressures are normally associated with waves that are quite intense, exhibit nonlinear steepening, and need special mathematical treatment.

It may appear that intensity obeys the principle of linear superposition, but this is not the case. First, intensity is not governed by the wave equation, so we cannot prove linear superposition based on this property. In fact, we can prove that intensity cannot obey superposition in general. It is because pressure p and particle velocity u are governed by linear superposition that intensity $\mathcal{I}$ cannot be. Consider two acoustic waves with pressure and particle velocity combinations p_1, u_1 and p_2, u_2. Individually, they possess intensities $\mathcal{I}_1 = p_1 u_1$ and $\mathcal{I}_2 = p_2 u_2$. If the waves are superposed, we obtain a wave with pressure $p = p_1 + p_2$ and particle velocity $u = u_1 + u_2$, and its intensity is $\mathcal{I} = pu = p_1 u_1 + p_1 u_2 + p_2 u_1 + p_2 u_2 = \mathcal{I}_1 + \mathcal{I}_2 + p_1 u_2 + p_2 u_1 \neq \mathcal{I}_1 + \mathcal{I}_2$, in general.

Conclusion

Quantities governed by the wave equation have simple and elegant physical behavior. However, we have yet to discuss the details of cross-sectional area variation and aspects of the physics of air not covered by the acoustic approximation, but which impinge on the motion of air in the vocal tract. The addition of more complicated geometries and greater physical reality is what makes the field of acoustics interesting and useful.

Readers interested in knowing more about the one-dimensional wave equation are directed to most first-year college texts in physics such as Resnick & Halliday (1966), or Feynman, Leighton, & Sands (1963). We can also recommend the book by Morse (1976). In these books, the wave equation is often derived for a string with an elastic restoring force. A non-mathematical development of acoustics is found in a book by Benade (1976).

Our discussion of one-dimensional acoustic waves sometimes known, or plane waves, has been fairly general so far. We start to make this ongoing discussion more relevant to speech in Chapter 3 by turning from the semi-infinite tube to a tube of finite length by adding another boundary orthogonal to the tube axis.

References

Benade, A.H. (1976). *Fundamentals of Musical Acoustics*. Oxford University Press: New York.

Feynman, R.P., Leighton, R.B., & Sands, M. (1963). *The Feynman Lectures on Physics, Volume 1*. Addison-Wesley Publishing Company: Reading, Massachusetts.

Morse, P.M. (1976). *Vibration and Sound*. The Acoustical Society of America: Woodbury, New York.

Pierce, A.D. (1989). *Acoustics: An introduction to its physical principles and applications*. The Acoustical Society of America: Woodbury, New York. (p 107, pp 566-71).

Lighthill, J. (1978). *Waves in Fluids*. Cambridge University Press: Cambridge, England. (pp 11-17).

McGowan, R.S. (1987). Articulatory synthesis: numerical solution of a hyperbolic differential equation. *Haskins Laboratories Status Report on Speech Research*, **SR-89/90**, p 69.

Resnick, R. & Halliday, D. (1966). *Physics, Part I*. John Wiley & Sons, Inc.: New York (p. 547, Appendix G p 31).

Appendix to Chapter 2

The following Equations are intended to show that $f(t - x/c_0)$ and $g(t + x/c_0)$ are solutions to the wave equation, Equation (16).

Equations (17) and (18) are equivalent. Suppose a is a number.

$$\begin{aligned}
\frac{\Delta_t f(a \cdot t)}{\Delta t} &= \frac{f\big(a \cdot (t + \Delta t/2)\big) - f\big(a \cdot (t - \Delta t/2)\big)}{\Delta t} \\
&= \frac{f\big(a \cdot t + \Delta a \cdot t/2\big) - f\big(a \cdot t - \Delta a \cdot t/2\big)}{\Delta t} \\
&= a \cdot \frac{f\big(a \cdot t + \Delta(a \cdot t)/2\big) - f\big(a \cdot t - \Delta(a \cdot t)/2\big)}{\Delta a \cdot t} \\
&= a \cdot \frac{f(z + \Delta z/2) - f(z - \Delta z/2)}{\Delta z}, \text{ where } z = a \cdot t \\
&= a \cdot \frac{\Delta_z f(z)}{\Delta z} \\
&= a \cdot \frac{\Delta_t f(t)}{\Delta t} . \quad (A2.1)
\end{aligned}$$

Thus, we write

$$\frac{\Delta_t f(at)}{\Delta t} = a \cdot \frac{\Delta_t f(t)}{\Delta t} . \quad (A2.2)$$

Also,

$$\frac{\Delta_t t}{\Delta t} = \frac{(t + \Delta t/2) - (t - \Delta t/2)}{\Delta t} = \frac{\Delta t}{\Delta t} = 1 . \quad (A2.3)$$

We state something called the *chain rule* without proof. Suppose f is a function of a variable z, and that z is a function of a variable t. That is, $f = f(z)$ and $z = z(t)$, or, in other words, $f = f(z(t))$. The following is the chain rule.

$$\frac{\Delta_t f(z(t))}{\Delta t} = \frac{\Delta_z f(z)}{\Delta z} \cdot \frac{\Delta_t z(t)}{\Delta t} . \quad (A2.4)$$

The chain rule in Equation (A2.4) looks reasonable if we can replace Δz with $\Delta_t z(t)$. In fact, it is valid to do just that.

Now we show that $f(t - x/c_0)$ is a solution to the wave equation by showing that it is a solution to one of the expressions in Equation (18). Let $z = t - x/c_0$.

$$\begin{aligned}
\frac{1}{c_0} \frac{\Delta_t f(t - x/c_0)}{\Delta t} &= \frac{1}{c_0} \frac{\Delta_z f(z)}{\Delta z} \cdot \frac{\Delta_t z}{\Delta t} \quad \text{[Equation (A2.4)]} \\
&= \frac{1}{c_0} \frac{\Delta_z f(z)}{\Delta z} . \quad \text{[Equations (A1.4) and (A2.2)]}
\end{aligned}$$
$$(A2.5)$$

Also,
$$\frac{\Delta_x f(t - x/c_0)}{\Delta x} = \frac{\Delta_z f(z)}{\Delta z} \cdot \frac{\Delta_x z}{\Delta x} \quad \text{[Equation (A2.4)]}$$
$$= \frac{\Delta_z f(z)}{\Delta z} \cdot \left(\frac{-1}{c_0}\right), \quad \text{[Equations (A1.4), (A2.2), and (A2.1)]}$$
$$\tag{A2.6}$$

It follows that $f(t - x/c_0)$ satisfies the first expression in Equation (18). Similarly, $g(t + x/c_0)$ satisfies the second expression in Equation (18). Thus, both $f(t - x/c_0)$ and $g(t + x/c_0)$ satisfy the wave equation.

This is a good approximation when $\delta p \approx \pm 10$ cm H_2O at standard atmospheric conditions, because for this value $|\delta p|/p_{atm} \approx 0.01$ according to Equation (7) of Chapter 1.

Chapter 3: Acoustic Air Motion in a Tube of Finite Length

Introduction

We now consider acoustic wave motion in a tube of finite length. The situation is shown in Figure 1, with a piston at the left end of the tube and the right end of the tube opening into the atmosphere. The new phenomenon that needs to be considered is the air motion at the opening under the acoustic approximation. We depict the opening as possessing adjoining walls perpendicular to the x-axis; the opening is a *flanged opening*. Figure 1 shows the piston at $x = 0$ and the tube opening at $x = L$.

An initial brief piston movement

We again examine the behavior of a compression created by a small, brief rightward movement of the piston. The compression travels to the right, until it reaches the open end of the tube. We rewrite Equations (25) and (26) of Chapter 2, which are valid before the pressure pulse reaches $x = L$, that is for $0 < t < L/c_0$.

$$p(x,t) = \rho_0 c_0 f(t - x/c_0)$$
$$u(x,t) = f(t - x/c_0) \qquad (1)$$
$$f(t) = U_{pst}(t) \ .$$

The shape of the pulse is known completely if the movement of the piston is known.

What can we expect to happen as the compression approaches the open end of the tube? The compression reaches the open end at about $t = L/c_0$. We set $L = 17.1$ cm in order that $t = L/c_0 = 17.1/34,100$ s $= 0.0005$ s $= 0.5$ ms. Before this pressure pulse arrives at the open end of the tube, the air just outside the tube is at rest and the pressure is the same as atmospheric, i.e. $p(x = L, t) = 0$ for $t < L/c_0$. This condition should, approximately, continue to hold after the compression arrives at the open end. The reason for this involves an argument based on terms that we have yet to define, so we do not reproduce that argument

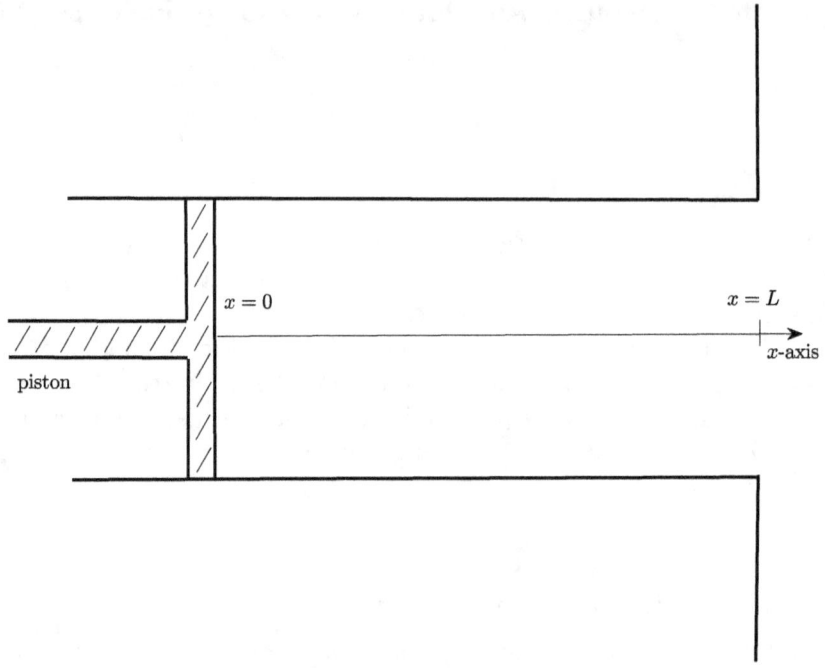

FIGURE 1. Finite-length straight tube with a flanged opening

here (Howe 2015). We write a boundary condition at the opening

$$p(x = L, t) = 0 \, . \tag{2}$$

This says that the pressure disturbance at the opening is zero for all time t. We see later in Chapters 10 and 11 that the boundary condition is approximate, and that there actually is some pressure fluctuation at $x = L$.

How is it possible for Equation (2) to hold with the compression present in the tube? The way to permit both the existence of the right-going compression and satisfaction of the boundary condition expressed in Equation (2) is to create a left-going rarefaction. Thus, a *reflected wave* $g(t + x/c_0)$ appears at the open end during a brief interval around the time $t = L/c_0$.

$$\begin{aligned} p(x, t) &= \rho_0 c_0 \big[f(t - x/c_0) + g(t + x/c_0) \big] \\ u(x, t) &= f(t - x/c_0) - g(t + x/c_0) \, . \end{aligned} \tag{3}$$

The left-going rarefaction $g(t + x/c_0)$ permits us to enforce the boundary condition that cancels the over-pressure in the compression when it arrives at $x = L$. We find the form of $g(t+x/c_0)$ by substituting Equation (3) into the boundary condition, Equation (2), that

$$0 = p(x = L, t) = \rho_0 c_0 \left[f(t - L/c_0) + g(t + L/c_0) \right] . \tag{4}$$

Equation (4) is solved,

$$g(t) = -f(t - 2L/c_0) \quad \text{or} \quad g(t + x/c_0) = -f\left(t + (x - 2L)/c_0\right) . \tag{5}$$

Thus,

$$\begin{aligned} p(x, t) &= \rho_0 c_0 \left[f(t - x/c_0) - f\left(t + (x - 2L)/c_0\right) \right] \\ u(x, t) &= f(t - x/c_0) + f\left(t + (x - 2L)/c_0\right) . \end{aligned} \tag{6}$$

These expressions can be interpreted as follows. Take the compression f traveling to the right, which is located at $x = 0$ when $t = 0$, and add a rarefaction of the same shape g traveling to the left, which is located at $x = 2L$ at $t = 0$. Figures 2 shows this scenario for the pressure pulse and Figure 3 shows this for the corresponding particle velocity pulse. [The other pulse in the figure labeled h is discussed below.] At around $t = 0.5$ ms, the compression f and rarefaction g combine to give zero pressure disturbance at $x = L$, as shown in Figure 2. The particle velocities of the two waves add, however, so that that peak particle velocity at $x = L$ around $t = 0.5$ ms is twice that of the right-going compression, as shown in Figure 3.

From $t \approx L/c_0 = 0.5$ ms to $t \approx 2L/c_0 = 1.0$ ms, the rarefaction is traveling to the left in the tube from $x = L$ to $x = 0$. The vertical dashed lines denote the ends of the finite-length tube. The waves outside the tube are simply mathematically convenient, and the waves in the tube are what we would expect to actually observe.

What occurs when this rarefaction reaches the piston at $x = 0$? Equation (24) of Chapter 2 gives us the boundary condition at the piston. By the time the reflected wave g reaches the piston face, the piston has stopped moving. After the piston has stopped, the boundary condition at $x = 0$ becomes

$$u(x = 0, t) = 0 . \tag{7}$$

In this example, the piston does not act as an acoustic source after the initial brief movement; it becomes just a boundary. Later in this chapter we will consider more general situations in which the piston is

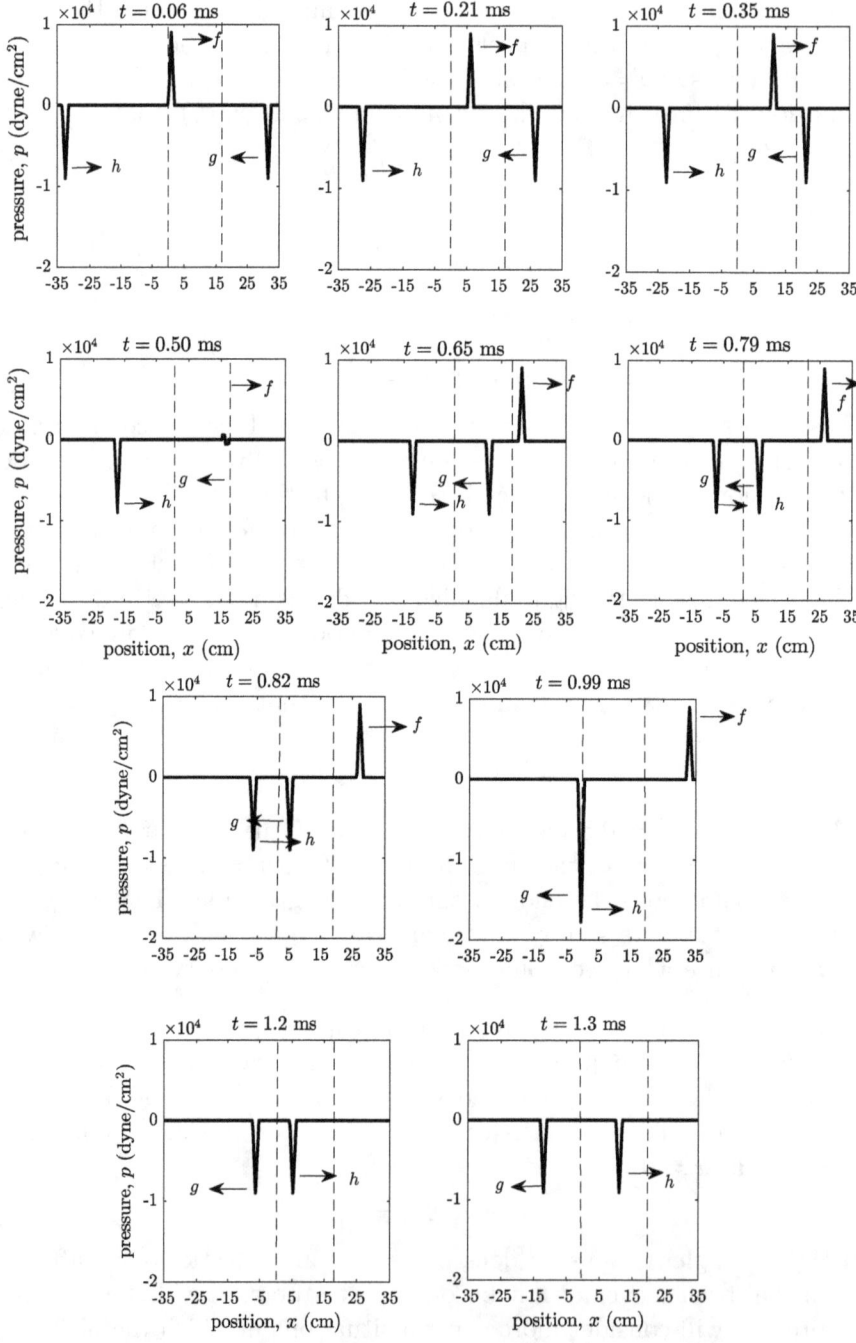

FIGURE 2. Evolution of the pressure pulse

Acoustics of Speech Production 65

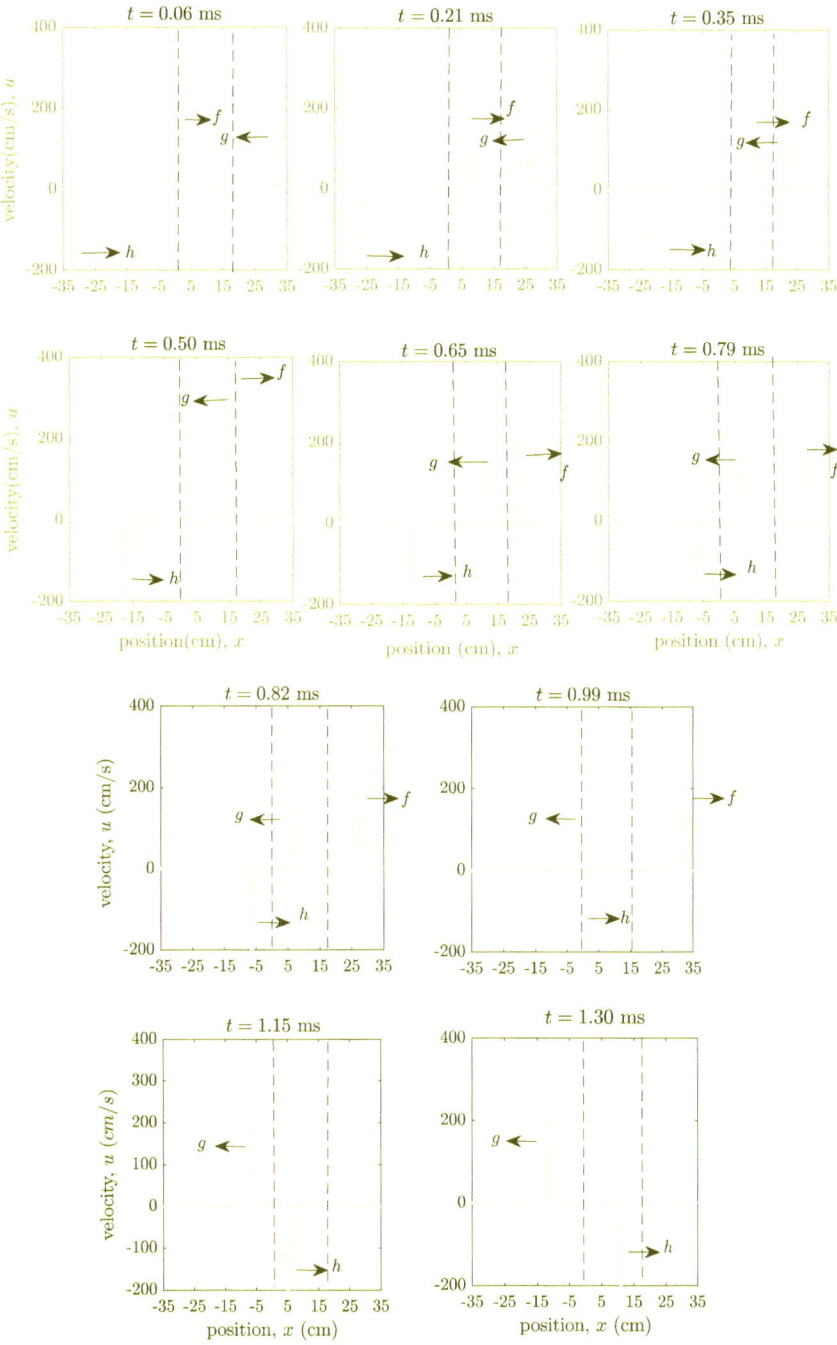

FIGURE 3. Evolution of the particle velocity pulse

both an acoustic source and/or acoustic sink, as well as a boundary at times $t > 0$.

In order that there be zero particle velocity at $x = 0$, a right-going wave, $h(t - x/c_0)$, must be superposed around $t = 2L/c_0 = 1.0\ ms$, in order to cancel the left-going wave at $x = 0$. This means that for $2L/c_0 < t < 3L/c_0$

$$p(x,t) = \rho_0 c_0 \big[f(t - x/c_0) - f\big(t + (x - 2L)/c_0\big) + h(t - x/c_0)\big] \quad (8)$$

$$u(x,t) = f(t - x/c_0) + f\big(t + (x - 2L)/c_0\big) + h(t - x/c_0)\ .$$

Using the boundary condition, Equation (7), the function h is found in terms of function, f, so that,

$$p(x,t) = \rho_0 c_0 \big[f(t - x/c_0) - f\big(t + (x - 2L)/c_0\big) -$$
$$f\big(t - (x + 2L)/c_0\big)\big] \quad (9)$$
$$u(x,t) = f(t - x/c_0) + f\big(t + (x - 2L)/c_0\big) - f\big(t - (x + 2L)/c_0\big)\ .$$

This equation says that we should add a right-going rarefaction originating at $x = -2L/c_0$ at $t = 0$ to the left-going rarefaction that is approaching $x = 0$. This creates a situation where the particle velocity is zero when the left-going wave reaches $x = 0$ at $t \approx 2L/c_0 = 1.0$ ms. The rarefactions reinforce each other when these waves meet at $x = 0$, so that the maximum amplitude of pressure at the time of reflection at $x = 0$ is twice that of the pressure pulse at other times. All of the discussion above is illustrated in the later frames of Figure 2 for pressure and Figure 3 for particle velocity.

What is actually seen in the tube is shown in Figure 4 for pressure and Figure 5 for particle velocity. Except for the time intervals when the pulse is being reflected from one of the two boundaries, there is only one pulse within the tube. Thus, while we keep adding pulses to one another in time intervals of L/c_0, in the case of the brief pulse only one (or two pulses at intervals of reflection) are actually necessary to represent the physical situation.

As time proceeds, the pulse continues to be reflected at the ends of the tube. The right-going rarefaction continues toward the opening at $x = L$, and reaches it at $t \approx 3L/c_0 = 1.5$ ms, when a left-going compression is created. When this left-going compression reaches $x = 0$ at $t \approx 4L/c_0 = 2.0$ ms, a right-going compression is created, and the cycle is complete because this is the same situation as at $t = 0$. The

Acoustics of Speech Production

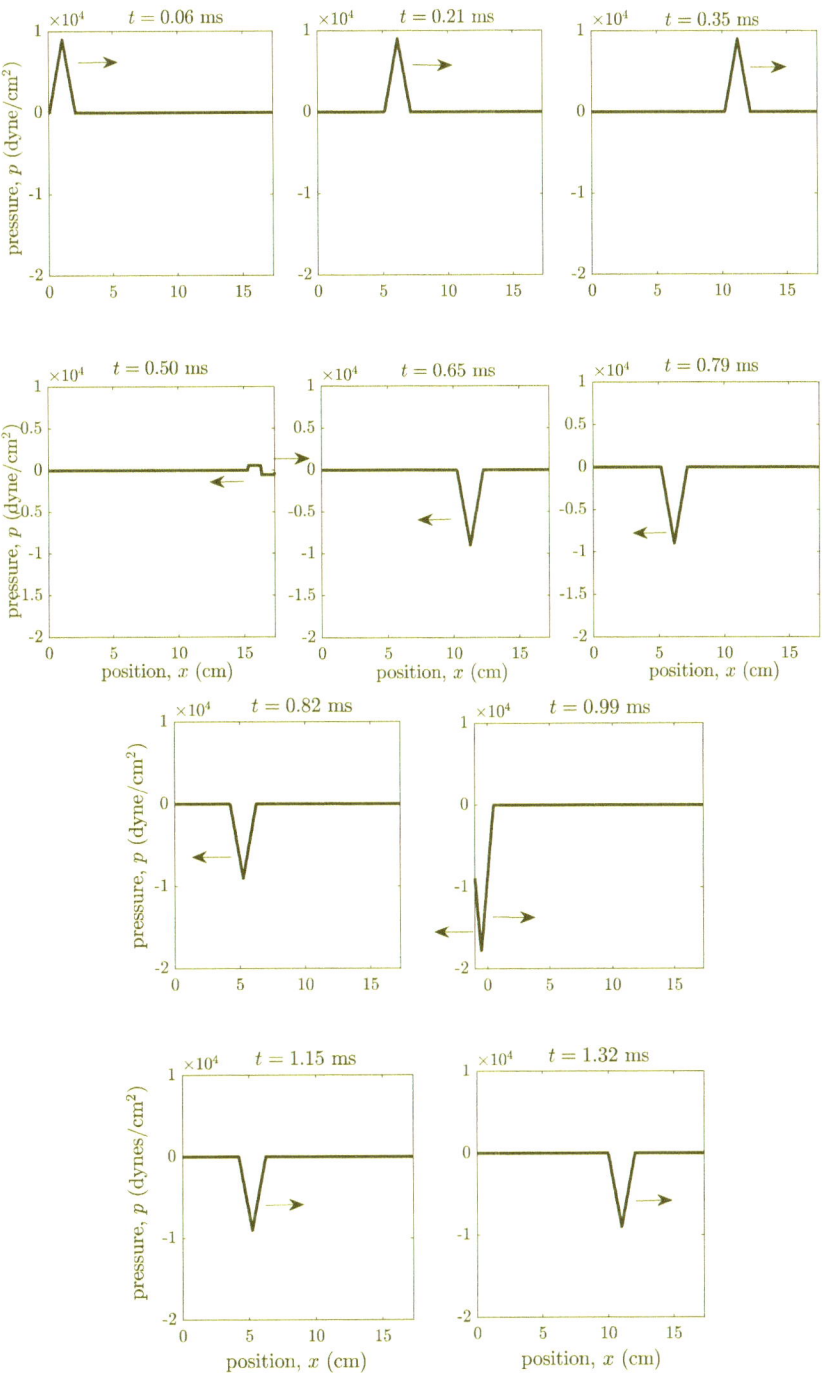

FIGURE 4. Evolution of a pressure pulse in the tube

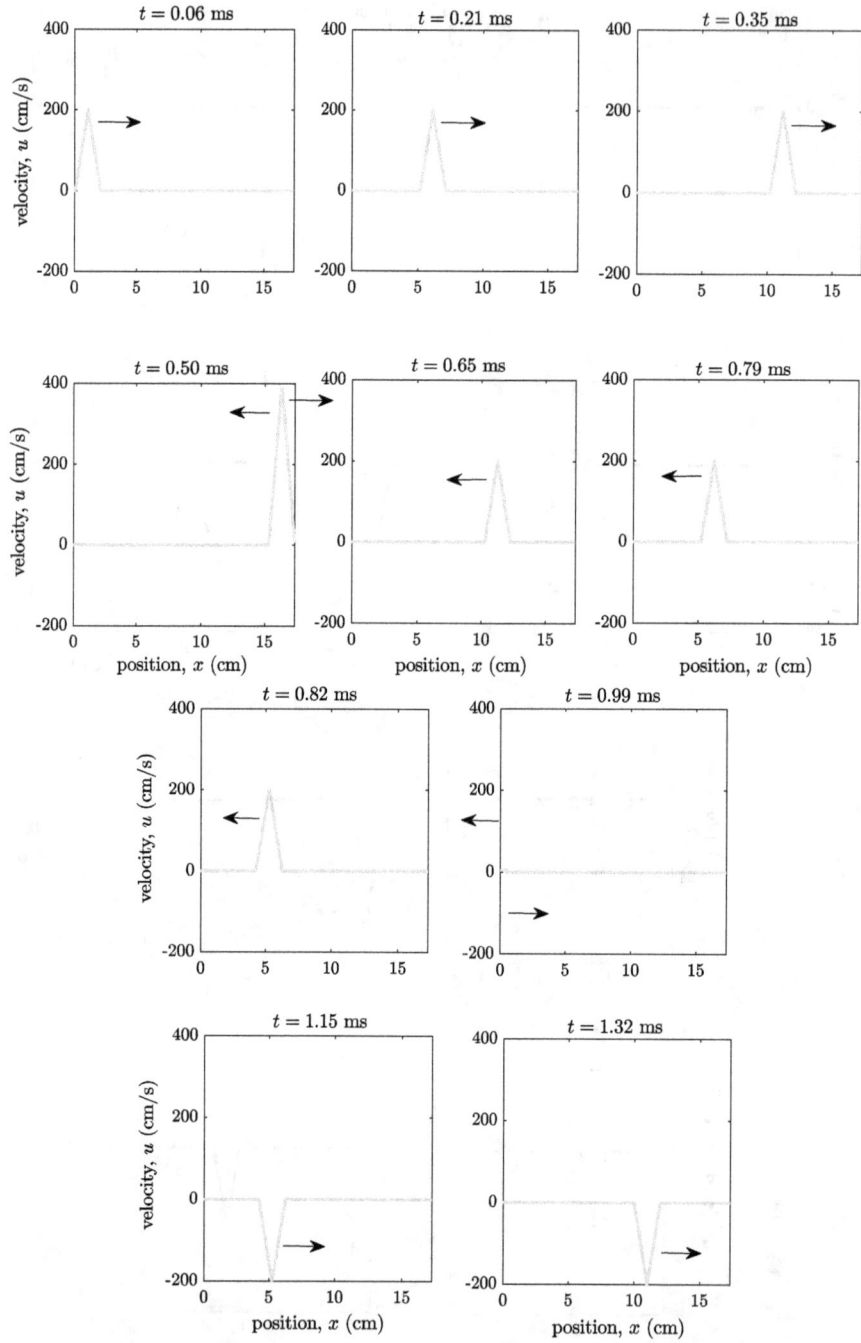

FIGURE 5. Evolution of the particle velocity pulse in the tube

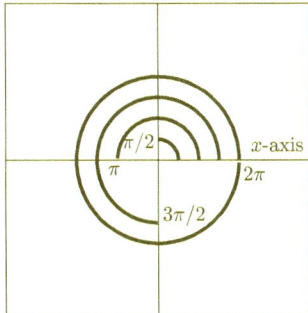 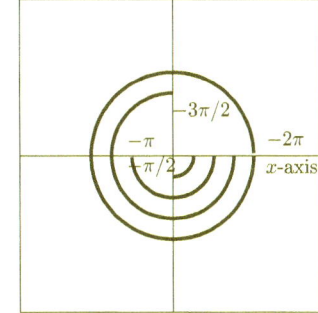

FIGURE 6. Angles in the plane

motion is periodic motion with period of 2.0 ms. It is a good exercise for the reader to work this out graphically, if not algebraically.

It can be noted that the acoustic motion is known for all time, once the piston motion is specified. Once the piston stops, there is no new energy that is provided to the air in the tube. That energy is carried indefinitely by the pulse moving back and forth through the tube as a compression or a rarefaction. There are no mechanisms for the loss of energy from the tube, because we have neglected physical mechanisms for *energy dissipation* so far, such as friction. Also, the boundary conditions expressed in Equations (2) and (7) mean that there is always zero intensity at the ends of the tube: energy cannot leave the tube by "leaking" out the ends of the tube. Thus, there are important physical aspects that are missing in this picture of acoustic wave motion.

Properties of the circular functions sine and cosine

We now turn to a review of some trigonometry in the two-dimensional plane. All angles in the two-dimensional plane are expressed in radians, or rad, so that 2π is the angle inscribed by a rotation of a ray originating at the origin in the two-dimensional plane in the counter-clockwise direction that brings it back to the same location. -2π is inscribed by a similar rotation in the clockwise direction, as in Figure 6.

Recall the definitions of sine and cosine in terms of a ray of unit length as it rotates about its fixed end at the origin of the plane to inscribe a unit circle at its unfixed end. The cosine is the the projection

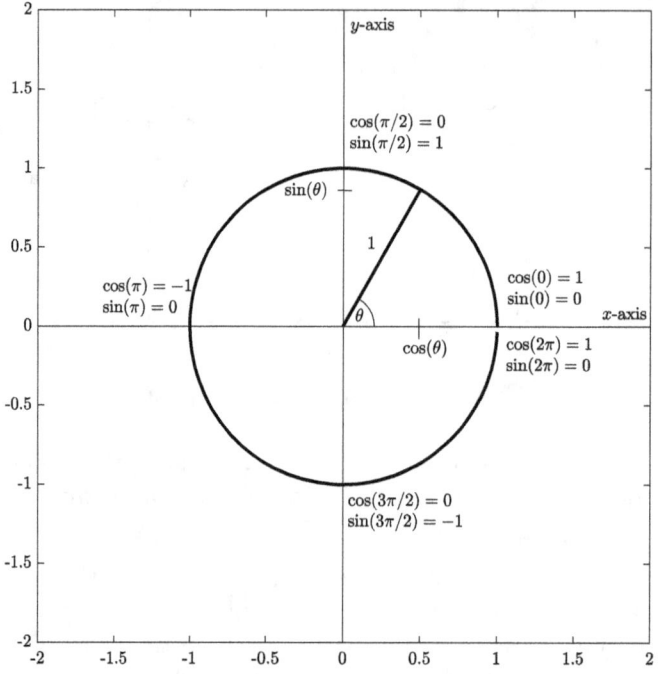

FIGURE 7. Geometry of the cosine and sine functions

of the ray onto the horizontal axis, which is usually denoted as the x-axis, and the sine is the projection of the ray onto the vertical axis, which is usually denoted as the y-axis. This is shown in Figure 7.

It follows that

$\cos(0) = 1$, $\cos(\pi) = \cos(-\pi) = -1$, $\cos(\pi/2) = \cos(-3\pi/2) = 0$, $\cos(-\pi/2) = \cos(3\pi/2) = 0$,

and (10)

$\sin(0) = 0$, $\sin(\pi) = \sin(-\pi) = 0$, $\sin(\pi/2) = \sin(-3\pi/2) = 1$, $\sin(-\pi/2) = \sin(3\pi/2) = -1$.

From the considerations of the previous two paragraphs that, and for any angle θ

$$\cos(\theta \pm 2\pi) = \cos(\theta) \text{ , and}$$
$$\sin(\theta \pm 2\pi) = \sin(\theta) \text{ .} \qquad (11)$$

The cosine and sine functions have period 2π.

Consider circular or sinusoidal functions of time t of the form,
$$\sin(\omega t + \phi) \quad \text{and} \quad \cos(\omega t + \phi) \tag{12}$$
Define $\theta^{time} = \theta^{time}(t) \equiv \omega t + \phi$ in Equation (12). θ^{time} is known is the *argument* of the circular functions, and for circular functions, the argument is called the *phase*. Phase is a linear function of time here, and it is stated in a dimensionless unit of radians, or rad. The constant ϕ can be called *initial phase*.

We slightly generalize some terminology that was introduced in Chapter 1 in the discussion of simple mass-spring systems. In Chapter 1, we introduced the notions of natural circular frequency and natural frequency, which are properties of the simple mass-spring system parameters. Here we do not make reference to a simple mass-spring system. ω is the circular frequency, and it is related to what we usually call frequency $\mathcal{F}$ by $\omega = 2\pi\mathcal{F}$. ω is written with units rad/s. The frequency, $\mathcal{F}$ is written in units of Hertz = Hz = s^{-1}. How long does it take to complete one cycle, where a cycle is defined to be such that $\theta^{time} = \omega t + \phi$ to change by 2π? This is the period, T, and it is given by $2\pi = \omega T$, or $T = (2\pi)/\omega$. With $\mathcal{F} = \omega/(2\pi)$ or $\omega = 2\pi\mathcal{F}$, the period is given by $T = 1/\mathcal{F}$.

We often need to refer to rates-of-change in either time or space for the circular functions. [We refer to rates-of-change in space as gradients. Also, circular functions are sometimes called *sinusoids*.] Thus, we quote mathematical results on the rates-of-change of circular functions, sine and cosine. Let z be the independent variable, which can be either time t or position x. Let b and ϕ be constants, and $\Delta z > 0$, then
$$\frac{\Delta_z \sin(bz + \phi)}{\Delta z} \approx b\cos(bz + \phi)$$
and $\tag{13}$
$$\frac{\Delta_z \cos(bz + \phi)}{\Delta z} \approx -b\sin(bz + \phi) \ .$$
The approximations improve as Δz gets smaller. In fact, the approximations become equalities as $\Delta z \to 0$. Recall that the rates-of-change can be viewed as slopes of secant lines. Figures 8, 9, and 10 illustrate the expressions in Equation (13).

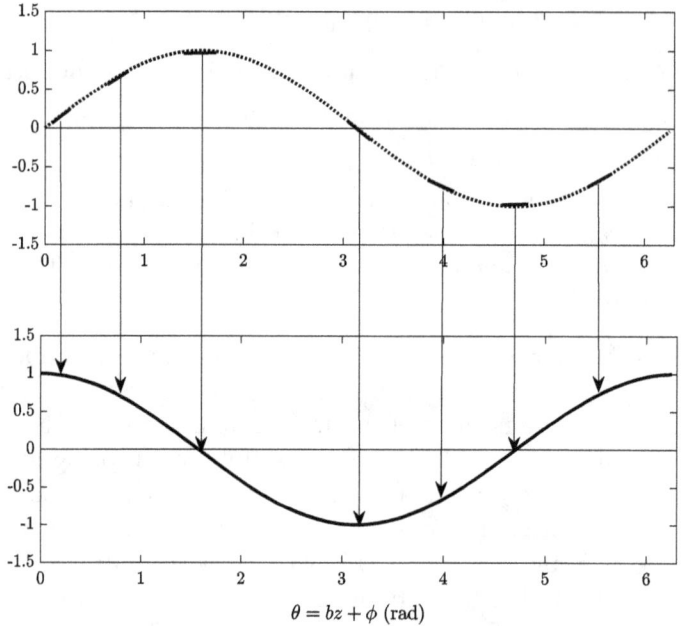

FIGURE 8. Slopes of sinusoids, I

Figure 8 illustrates the first expression in Equation (13) with $b = 1$, with the bottom of Figure 8 plotting the slopes of the secant lines of the plot at the top of Figure 8. In other words, the bottom of Figure 8 shows $\Delta_z \sin(z)/\Delta z \approx \cos(z)$. Figure 9 shows the first expression in Equation (13) with $b = 2$, so that $\Delta_z \sin(2z)/\Delta z \approx 2\cos(2z)$. Figure 10 illustrates the second expression in Equation (13) for $b = 1$, $\Delta_z \cos(bz)/\Delta z \approx -b\sin(bz)$. Figures 9 and 10 of Chapter 1 are also special cases illustrating Equation (13).

Sinusoidal piston movement

Here we consider a piston whose position is a circular function of time. This kind of motion is a specific type of periodic motion that we have called simple harmonic motion. Sometimes, it is referred to as

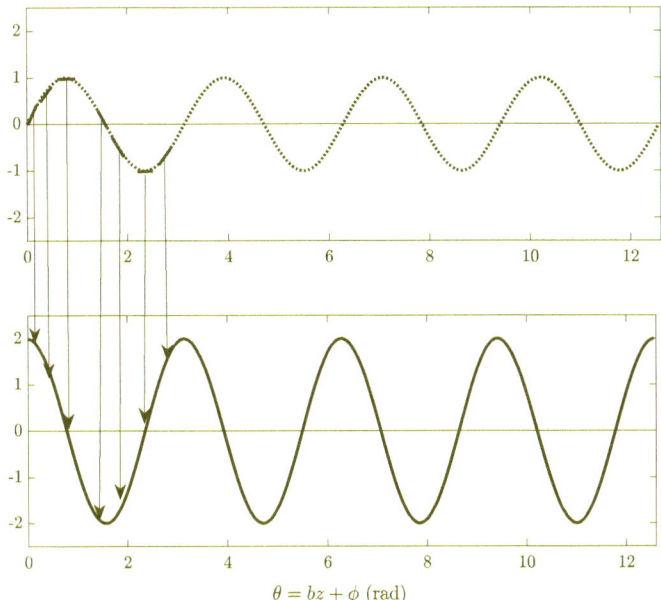

FIGURE 9. Slopes of sinusoids, II

oscillation.

$$U_{pst}(t) = \begin{cases} \omega_{pst}\mathcal{A}_{pst}\sin(\omega_{pst}t) & \text{if } t > 0 \\ 0 & \text{if } t \leq 0 \end{cases}. \qquad (14)$$

$\mathcal{A}_{pst}$ is the *amplitude* of the piston oscillation. $\omega_{pst}\mathcal{A}_{pst}$ is the amplitude of the velocity from Equation (13). Let $\mathcal{F}_{pst} = \omega_{pst}/(2\pi)$ denote the piston frequency of oscillation.

The oscillating piston creates a series of compressions and rarefactions continuously following one another. We do not use the graphical method that we used for the brief pulse above. We apply the boundary condition in Equations (2), and instead of Equation (7), we apply Equation (24) of Chapter 2 at the piston, to find the reflected waves at the ends of the tube. Equations (1) through (6), as well as Equations (8) and (9), remain valid for the continuously moving piston situation here, within their stated time intervals. Unlike the case of the brief pulse, this disturbance is created by the continuous piston motion. This means that we need to retain all the terms from reflections for all

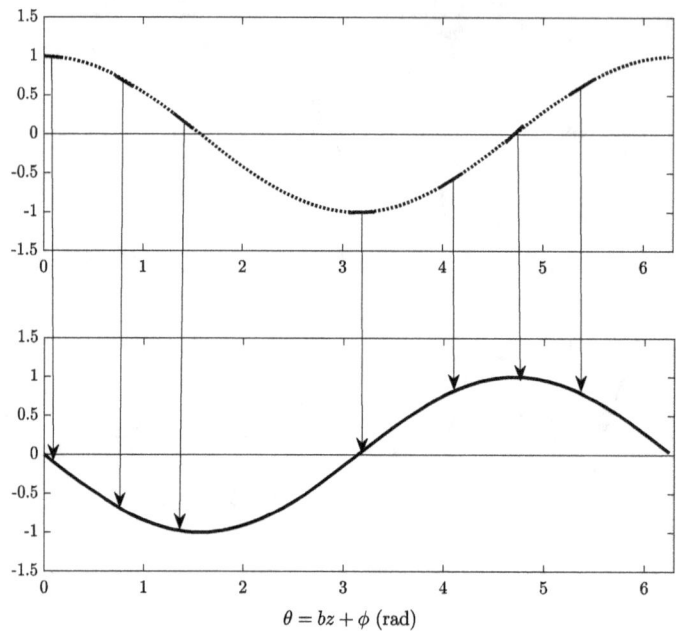

FIGURE 10. Slopes of sinusoids, III

time intervals: this does not simplify to only one or two pulses present in the tube at any time.

In the example examined here, the tube is again assumed to be of length $L = 17.1$ cm and the frequency of piston oscillation is $\mathcal{F}_{pst} = 200$ Hz. The resulting acoustic wave motion is pictured in Figure 11 for pressure, and in Figure 12 for particle velocity. The light lines denote component waves, and the heavy lines are the superposition of the component waves. A component wave is created whenever a wave front created either by the piston or a previous component wave begins to create a reflected wave itself. The component waves are denoted f, g, and h in Figures 11 and 12. The labeled arrows denote the direction of travel of the component waves, as well as the approximate location of the front of the wave. When an arrow is drawn so that part is outside the tube, the front of the corresponding component wave has reached the opposite boundary to where it was created.

We discuss the progress of the acoustic wave created by the piston in time intervals of L/c_0, which is the time interval for a component

Acoustics of Speech Production

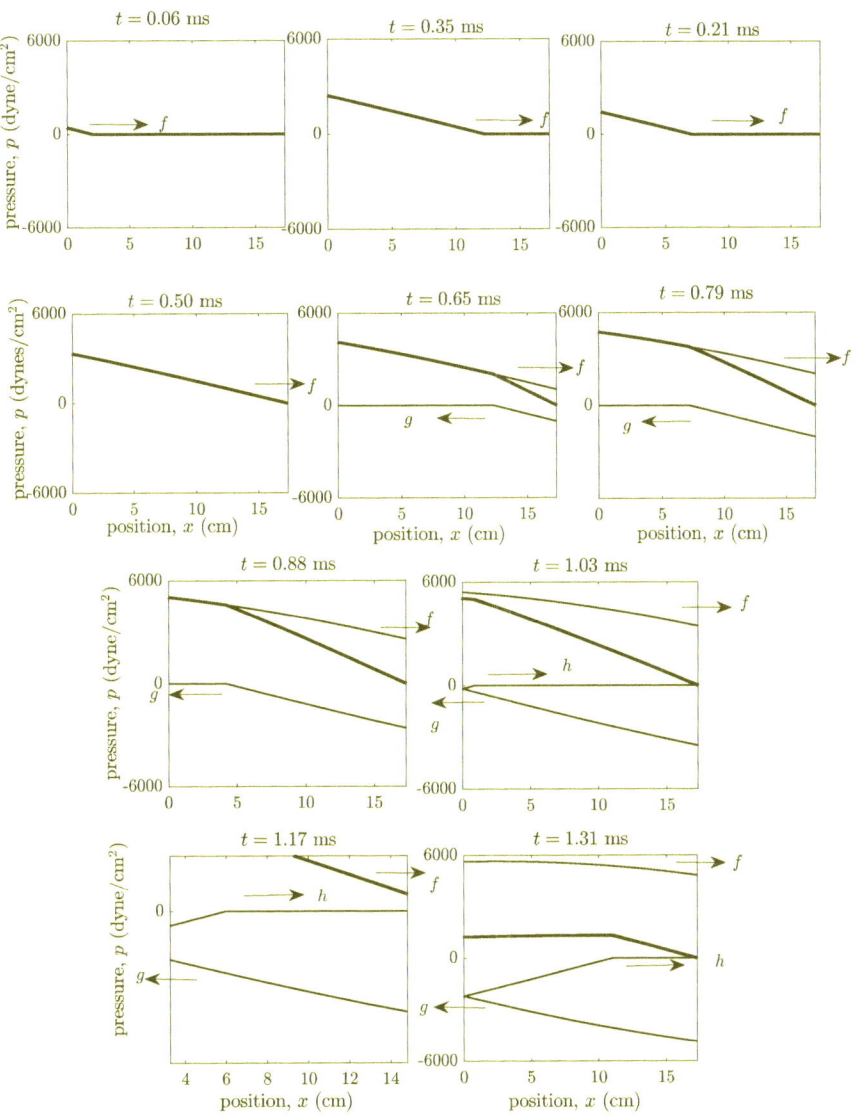

FIGURE 11. Evolution of pressure in the tube, $\mathcal{F}_{pst} = 200$ Hz

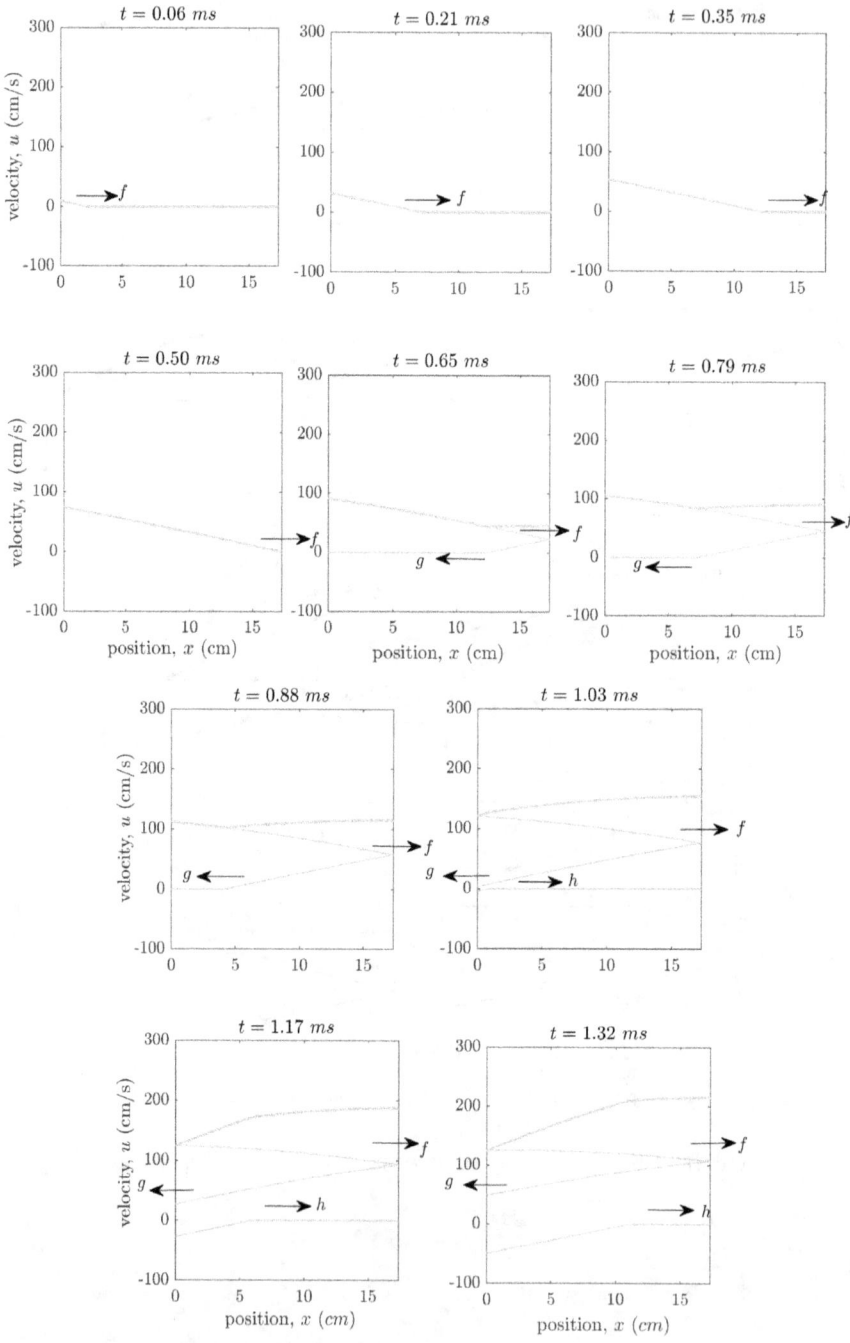

FIGURE 12. Evolution of particle velocity in the tube, $\mathcal{F}_{pst} = 200$ Hz

wave to travel the length of the tube. This time interval corresponds to the first four panels in Figure 11 and Figure 12. Equation (1), with $f(t) = U_{pst}(t)$, describes the wave motion for $0 < t < L/c_0$. The right-going wave, which is the same as the total wave for this initial interval, just reaches the tube opening, $x = L$, at $t = 0.5$ ms.

In the second time interval under consideration, $L/c_0 < t < 2L/c_0$, a left-going reflected wave, g, starts to be created at $x = L$. The second time interval is described by Equation (6) and pictured in Figures 11 and 12 in the panels titled $t = 0.65$ ms, $t = 0.79$ ms, and $t = 0.88$ ms. Note that the sum of the two waves f and g are such that the pressure disturbance is zero at $x = L$, while the particle velocities reinforce one another.

The third time interval $2L/c_0 < t < 3L/c_0$, shown in panels titled $t = 1.03$ ms, $t = 1.17$ ms, and $t = 1.31$ ms, is when things start to differ from the scenario with the brief pulse; the piston is still in motion when the left-going wave reaches it. However, Equations (8) and (9) remain valid for this time interval. We can think of a wave created at the piston boundary, $h(t - x/c_0)$ in Equation (8), as the sum of two waves: one that would result if the piston were at rest and a second due to the piston motion. This is true because the solutions to the wave equation can be superposed. Representing the wave at the piston as being a sum of two waves emphasizes an important property of the piston: the piston acts as both a boundary and an acoustic source or sink. The decomposition of the h wave referred to above is as follows.

$$h(t - x/c_0) = -f(t - (x + 2L)/c_0)$$
$$= -\big[f(t - x/c_0) + f(t - (x + 2L)/c_0)\big] + f(t - x/c_0) \ . \tag{15}$$

$-\big[f(t - x/c_0) + f(t - (x + 2L)/c_0)\big]$ is the reflected wave component of $h(t - x/c_0)$ at $x = 0$ that provides for the boundary condition $u(x = 0, t) = 0$. The other component of $h(t - x/c_0)$, $f(t - x/c_0)$, is the wave due to the piston motion. That is, it accounts for $u(x = 0, t) = U_{pst}(t)$.

The process of accounting for reflected waves could continue indefinitely. The continuously moving piston has both commonalities and differences with the brief pulse example. Just as in the case of the brief pulse, $f(t) = U_{pst}(t)$. What differs here from the case of brief piston movement is that more terms are added as more time intervals of duration L/c_0 are added, because the piston continues to move. The sum of terms involving the f functions at different retarded times, $t - nL/c_0$, for integer n, provides the solution to the acoustic situation in the tube.

Steadiness

There is an issue that is beginning to show itself in our considerations of the acoustics of the finite-length tube: the issue of *steadiness*. We cannot mean static when we speak of steadiness in acoustics, because acoustics is the study of a type of motion. What we often mean when we say steady in acoustics is that periodic motion is occurring, which means that the patterns of motion repeat themselves in equal intervals of time. [In the case where conditions are statistically random, there is a different meaning to steadiness, which is often termed *statistical stationarity*. We do not discuss this important situation here.]

In the considerations above, there seems to be a repeated pattern for the brief pulse. After the brief pulse travels the length of the tube four times, the pattern repeats itself. The time it takes for this to happen is close to 2 ms, which corresponds to a frequency of $\mathcal{F} = 500$ Hz. However, there is no indication of periodic behavior for the case when the piston moves sinusoidally at a frequency of $\mathcal{F}_{pst} = 200$ Hz. We explore the steadiness of acoustic wave motion in the case of a piston moving according to circular functions in more depth, as this leads to important physical considerations.

Another example of the sinusoidal piston in a finite-length tube

We cannot expect the situation with the piston executing periodic motion at $\mathcal{F}_{pst} = 200$ Hz, which has a period of 5 ms, to exhibit periodic motion in just over one millisecond, or even, two milliseconds. Let's consider the case when the piston is oscillating at $\mathcal{F}_{pst} = 1200$ Hz, which corresponds to a period of about 0.83 ms, with all the other parameters the same as the $\mathcal{F}_{pst} = 200$ Hz piston frequency example.

The plots of pressure p and particle velocity u with the component waves, are shown for particular times as a function of tube position in Figures 13 and 14. As before, the component waves are indicated by labeled arrows that denote the direction of travel and the approximate position of the front of the wave. If an arrow is partly outside the plot, this indicates that the wave front has reached a boundary, either at $x = L = 17.1$ cm or $x = 0$ cm. The thicker curves represent the sum, or superposition, of the component waves.

There is no indication that the system is beginning to repeat itself after one millisecond, even at this higher piston frequency, $\mathcal{F}_{pst} = 1200$ Hz. It is possible that we have missed something by not examining the total energy flow, or intensity. Intensity, as a function of tube position, is shown in Figure 15. Recall that positive intensity means that energy is flowing to the right, and negative intensity means that energy is flowing to the left. Also, consider a thin slab, or disc, of air centered around position x_0 in the tube. If the intensity is positive and its curve has a positive slope at x_0, then energy is flowing out of the disc. This is because there is more energy flow out of the disc (higher absolute value of intensity), than is flowing into the disc (lower absolute value of intensity). There is flow of energy into the disc if intensity at x_0 is positive and the slope of the intensity versus space curve is negative there. The opposite flow of energy occurs according to slope of intensity if the intensity is negative around the disc at x_0.

It can be seen that the energy flow directions and which discs of air that are gaining or losing acoustic energy are constantly changing as the various component waves move with speed c_0. Thus, if we were to plot energy density as a function of tube position, energy density would change at the same rate as slopes of intensity change. Figure 15 indicates that we need to consider durations longer than ones of milliseconds to find steadiness in energy flow.

Finally, in this example the intensities are enormous in terms of acoustics. They are greater than 194 dB according to Equation (42) of Chapter 2 at certain times and places in the tube. The reason for this is that we retained the amplitude of piston oscillation at $\mathcal{A}_{pst} = 0.1$ cm while increasing the piston frequency, $\mathcal{F}_{pst}$ by a factor of six from 200 Hz to 1200 Hz. An estimate of maximum acoustic Mach number would be $M_a = \omega_{pst}\mathcal{A}_{pst}/c_0 = (2\pi)(1200)(0.1)/34100 = 0.022$, which is large for the acoustic approximation. It would be appropriate to start to consider some of the nonlinear terms in conservation equations and to leave the realm of strictly linear acoustics, as discussed in Chapter 2. Other than for highly constricted regions of the vocal tract such large

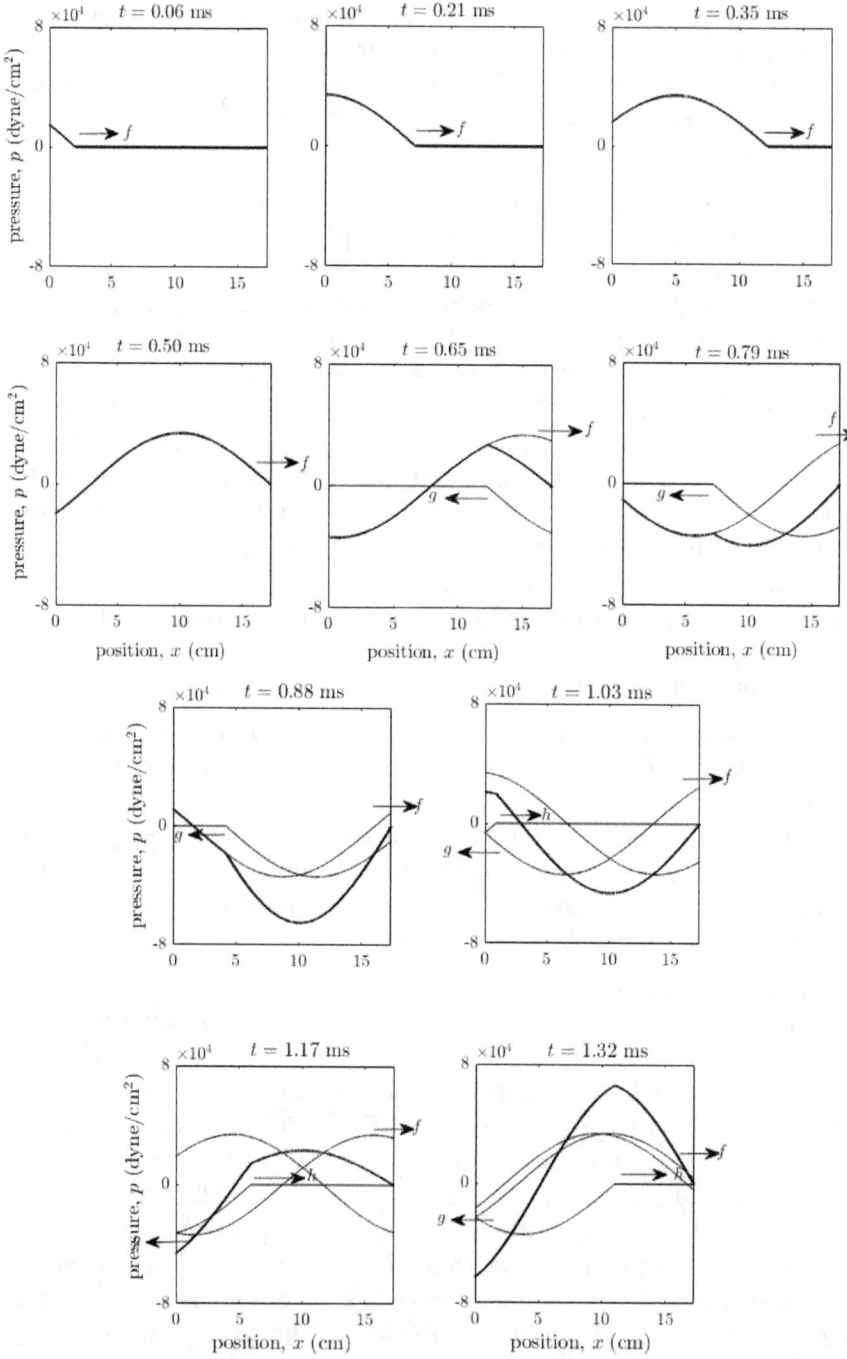

FIGURE 13. Evolution of pressure in the tube, $\mathcal{F}_{pst} = 1200$ Hz

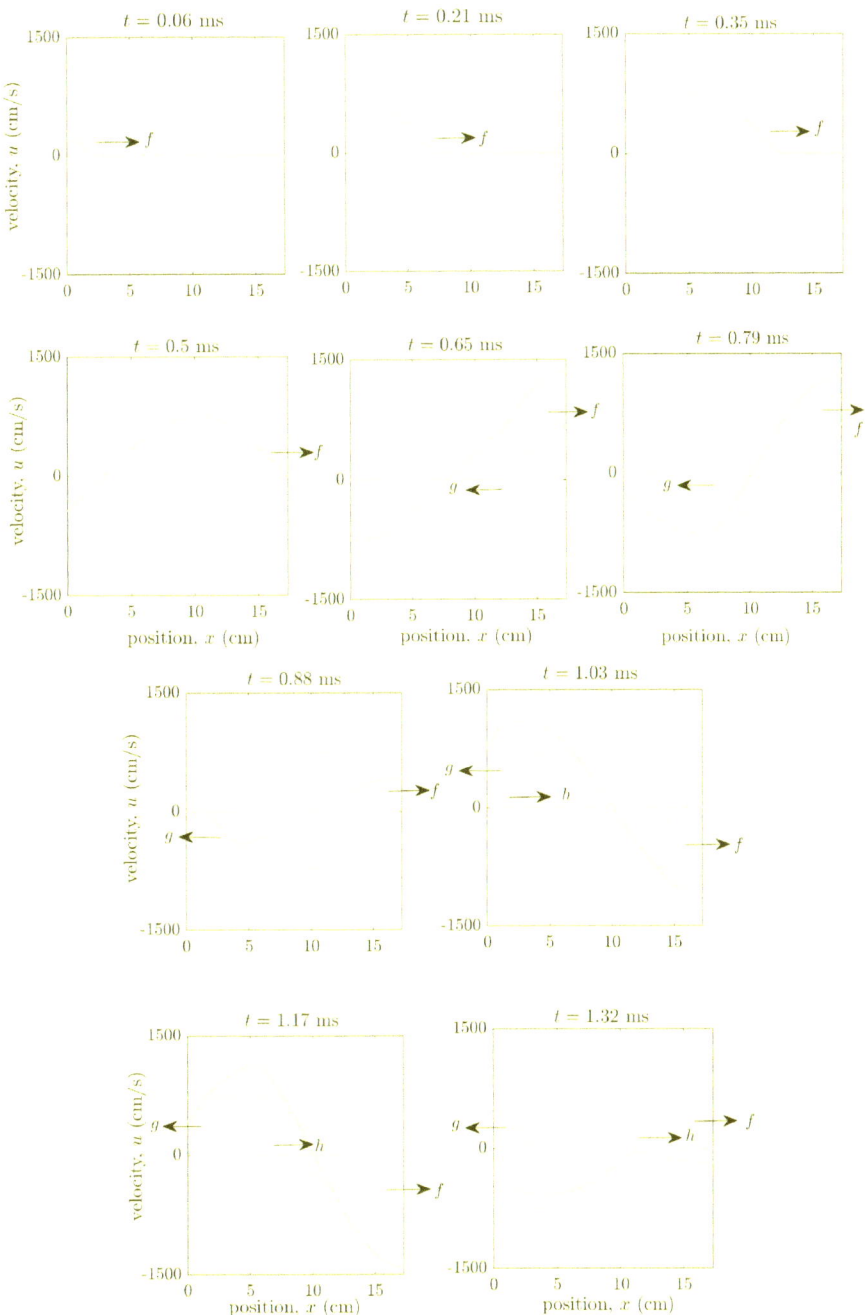

FIGURE 14. Evolution of particle velocity in the tube, $\mathcal{F}_{pst} = 1200$ Hz

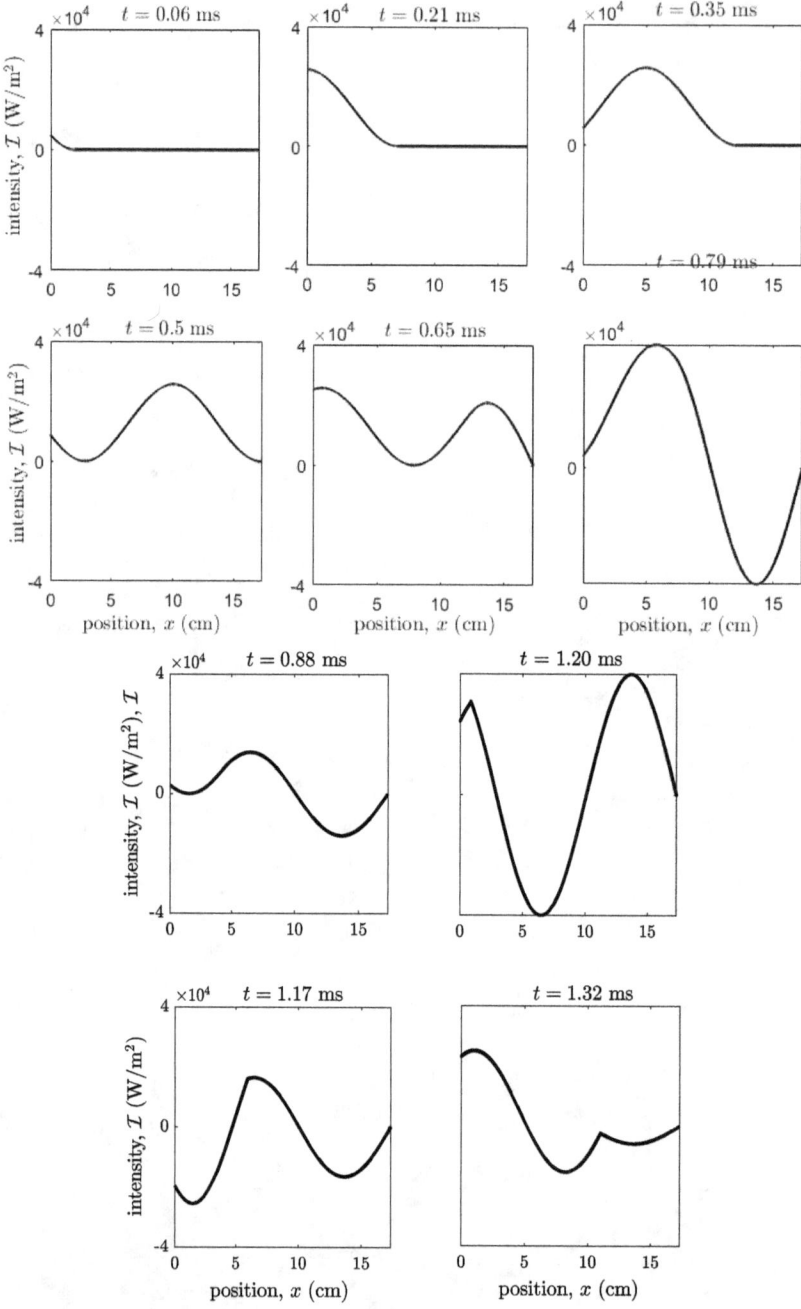

FIGURE 15. Evolution of intensity in the tube, $\mathcal{F}_{pst} = 1200$ Hz

Acoustics of Speech Production

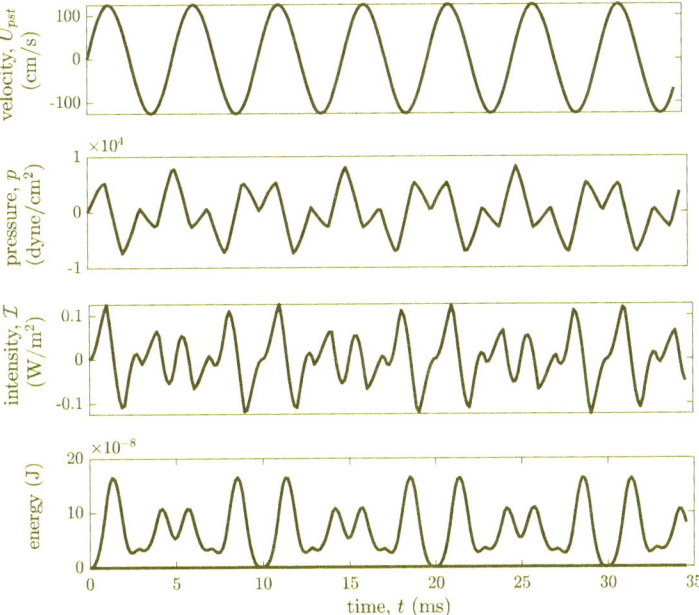

FIGURE 16. Physical quantities at the piston face, $\mathcal{F}_{pst} = 200$ Hz

acoustic Mach numbers are not attained during conversational speech.

Steadiness of physical quantities at the piston face

It seems that we need to examine the acoustics of the finite-length tube for a long duration to discover *steady conditions*. The present approach of following component waves in space and time is unwieldy. Instead, we now examine quantities at one position in space, the piston face at $x \approx 0$, as a function of time to detect steadiness, that is, repeating patterns.

Figures 16 through 20 show the quantities, piston velocity, pressure, and intensity at the piston, as well as the energy input to the acoustics of the tube from time $t = 0$, from top to bottom of each figure, for various piston oscillation frequencies $\mathcal{F}_{pst}$. Again, the amplitude of oscillation is $\mathcal{A}_{pst} = 0.1$ cm and the tube length is $L = 17.1$ cm. Here it is assumed that the tube has a cross-sectional area of 2.0 cm^2.

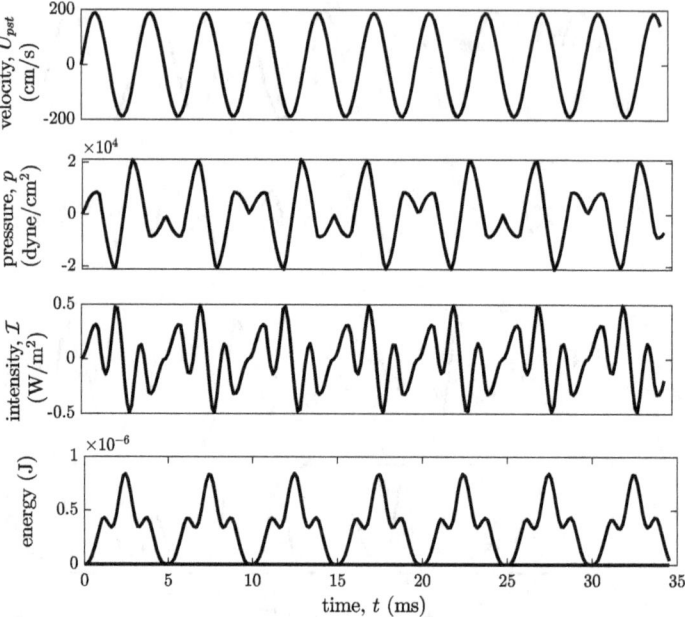

FIGURE 17. Physical quantities at the piston face, $\mathcal{F}_{pst} = 300$ Hz

Figure 16 shows the case when $\mathcal{F}_{pst} = 200$ Hz. The third panel from the top shows that the intensity at the piston face is both positive and negative, which means that the piston both puts energy into and extracts energy from the acoustics of the tube. The fact that the intensity goes from positive to negative in time means that the imposed piston velocity goes continuously from agreeing with the sign of the pressure wave at the piston to having the opposite sign from the pressure at the piston. Note that the total energy input by the piston is either positive or zero. Becasue intensity indicates that energy is both injected and extracted by the the piston, it is not surprising that overall energy input falls to zero periodically. That is, the piston periodically starts from scratch inputting energy. The patterns of pressure, intensity, and energy input repeat themselves every 10 ms, thus establishing periodic motion at a frequency of 100 Hz. Other frequencies appear to be present as well.

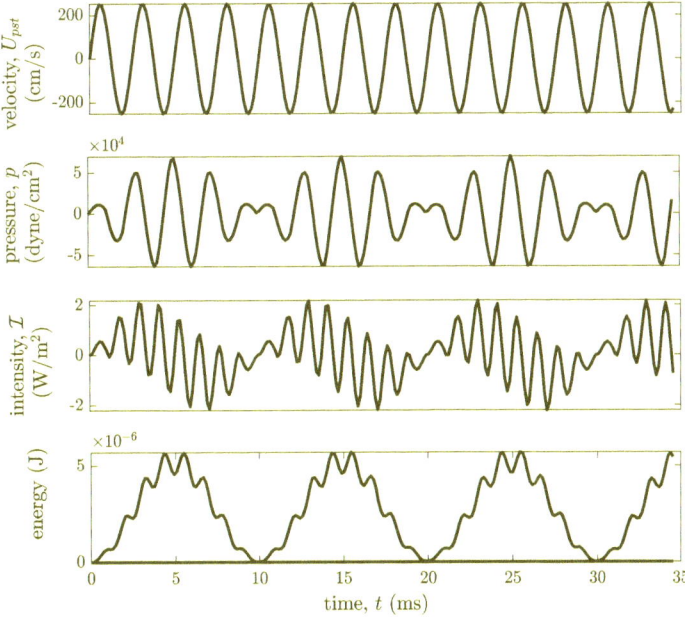

FIGURE 18. Physical quantities at the piston face, $\mathcal{F}_{pst} = 400\ Hz$

Figure 17 shows the case when $\mathcal{F}_{pst} = 300$ Hz. While the pressure repeats itself every 10 ms (100 Hz), the intensity and energy input patterns repeat themselves at a rate of 200 Hz. Figure 18 shows the case when $\mathcal{F}_{pst} = 400$ Hz. Here the frequency for repeating pressure, intensity, and energy input is again 100 Hz. Also, the modulation pattern is simpler than the case when $\mathcal{F}_{pst} = 200$ Hz. Figure 19 shows that case when $\mathcal{F}_{pst} = 470$ Hz. Here the intensity and energy input repeat at a rate of about 30 Hz, corresponding to a period of 33.3 ms. This is a rate that is quite slow in terms of the repetition rate of the brief pulse scenario, which was 500 Hz, corresponding to a period of 2 ms. Also, note that the intensity and energy input excursions become larger as the piston frequency $\mathcal{F}_{pst}$ increases from 200 Hz to 470 Hz.

We return to the case with $\mathcal{F}_{pst} = 1200$ Hz. Again, a pressure, intensity, and energy repetition frequency of 100 Hz is established. Note also, despite the fact that the piston velocity amplitudes are higher for

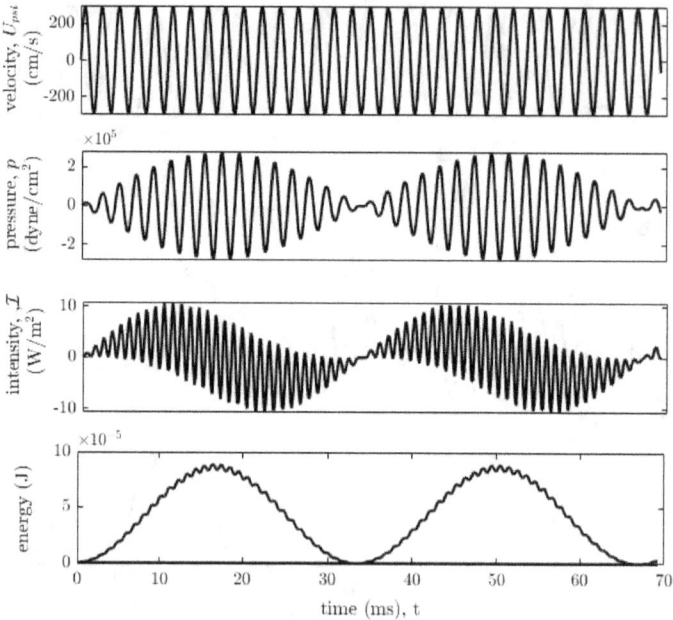

FIGURE 19. Physical quantities at the piston face, $\mathcal{F}_{pst} = 470$ Hz

this case than for the case of $\mathcal{F}_{pst} = 470$ Hz, that the intensity and energy peaks are less for $\mathcal{F}_{pst} = 1200$ Hz than for $\mathcal{F}_{pst} = 470$ Hz.

In sum, it appears that repetition patterns for pressure and energy input with an oscillating piston possess frequencies less than those of either the piston oscillation or the repetition frequency of the brief pulse scenario, 500 Hz. In the case when the piston frequency $\mathcal{F}_{pst}$ is close to the brief pulse repetition frequency, as when $\mathcal{F}_{pst} = 470$ Hz, the system repetition frequency of about 30 Hz is particularly low. Thus, steadiness is not established as quickly as one may have expected from either the piston frequency or the brief pulse repetition frequency.

The time domain and the frequency domain

We reviewed circular functions earlier in this chapter. We can write any circular function as $A\cos(\omega t + \phi)$, with $A > 0$. Circular functions and sums of circular functions play a central role in the theory of

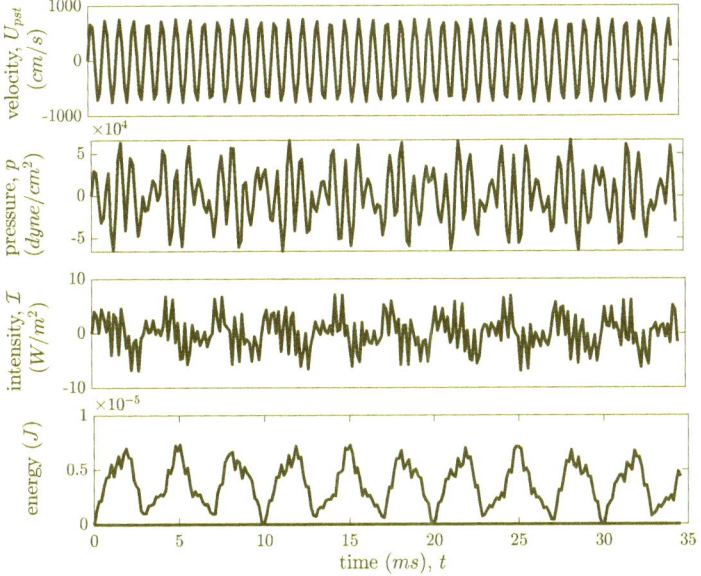

FIGURE 20. Physical quantities at the piston face, $\mathcal{F}_{pst} = 1200$ Hz

acoustics. This fact leads us to describe functions that have both a *time domain* representation and a *frequency domain* representation.

We concentrate on functions in the time domain that describe steady situations. One of he simplest ones is the circular function, for $\omega_1 > 0$, is
$$f(t) = A_1 \cos(\omega_1 t + \phi_1) \ . \tag{16}$$
$A_1 > 0$ is the amplitude of this circular function, and ϕ_1 is its initial phase.

Many functions have a representation in the frequency domain, which we denote $\tilde{f}(\omega)$. This function can be thought of as having two pieces of information: amplitude, or magnitude, as a function of circular frequency ω, and initial phase, also as a function of ω. We denote magnitude, or amplitude, as $|\tilde{f}| = |\tilde{f}(\omega)| \geq 0$, and we denote initial phase, where amplitude is not zero, as $\arg(\tilde{f}(\omega))$. For the function in Equation (16)

$$|\tilde{f}(\omega)| = \begin{cases} A_1/2 & \text{for } \omega = \pm\omega_1 \\ 0 & \text{otherwise}, \end{cases}$$

and (17)

$$\arg(\tilde{f}(\omega)) = \pm\phi_1 \quad \text{for } \omega = \pm\omega_1,$$

where arg stands for argument. [We use both positive and negative frequencies in the frequency domain. This is because of the mathematical properties of the transformation between the time and frequency domains.]

The figures in the previous section showing various quantities when there is sinusoidal piston motion can be represented with sums of functions of the form shown in Equation (16). That is, we can write a superposition as

$$f(t) = A_1\cos(\omega_1 t + \phi_1) + A_2\cos(\omega_2 t + \phi_2) + \cdots + A_n\cos(\omega_n t + \phi_n), \quad (18)$$

where $A_1, A_2, \ldots, A_n > 0$ and $\omega_1, \omega_2, \ldots, \omega_n > 0$. In the frequency domain

$$|\tilde{f}(\omega)| = \begin{cases} A_1/2 & \text{for } \omega = \pm\omega_1 \\ A_2/2 & \text{for } \omega = \pm\omega_2 \\ \cdot \\ \cdot \\ \cdot \\ A_n/2 & \text{for } \omega = \pm\omega_n, \end{cases}$$

and

$$\arg(\tilde{f}(\omega)) = \begin{cases} \pm\phi_1 & \text{for } \omega = \pm\omega_1 \\ \pm\phi_2 & \text{for } \omega = \pm\omega_2 \\ \cdot \\ \cdot \\ \cdot \\ \pm\phi_n & \text{for } \omega = \pm\omega_n. \end{cases} \quad (19)$$

The magnitude $|\tilde{f}(\omega)|$ is often represented graphically in an *amplitude spectrum*, or *magnitude spectrum*. An example of such a magnitude spectrum is shown for $n = 4$ in Figure 21. The symmetry

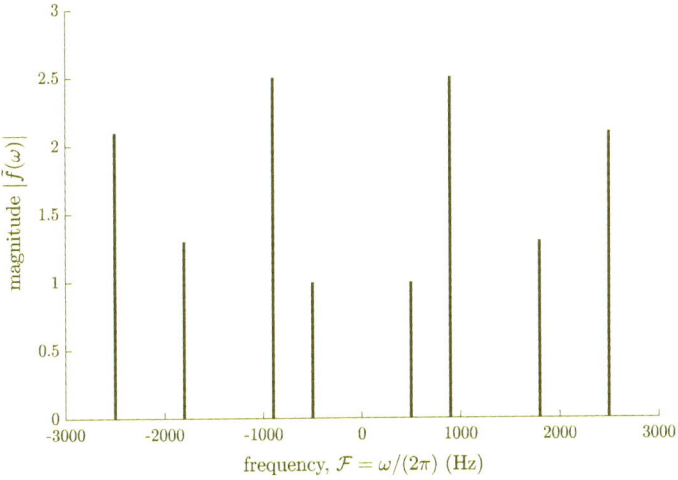

FIGURE 21. Example of a discrete spectrum

of the magnitude spectrum about $\mathcal{F} = 0$, or $\omega = 0$, is a property of the fact that the time functions $f(t)$ take on real number values. $|\tilde{f}(\omega)|$ has the the same dimensions as $f(t)$. Thus, if $f(t)$ describes pressure fluctuation as a function of time, then $|\tilde{f}(\omega)|$ has the dimensions of pressure, perhaps dyne/cm^2. The $\arg(\tilde{f}(\omega))$ is a radian measure.

The functions that we have discussed so far have *discrete spectra*, meaning that the frequencies that have non-zero amplitudes are represented by lines in the magnitude spectrum. There is also the possibility of functions that possess *continuous spectra*, for which the magnitude spectrum is a continuous curve. We have yet to encounter these situations, but they appear later in the book. In this case, the interpretation of $|\tilde{f}(\omega)|$ changes from amplitude to something called amplitude density. For a continuous spectrum, the units of $|\tilde{f}(\omega)|$ are the units of $f(t)$ per radian per second. We mention this for the reader who goes further to explore the relation between the time domain and frequency domain. Also, for those readers, the relation between $f(t)$ and $\tilde{f}(\omega)$ is found with the mathematical operations known as the *Fourier transform* and the inverse Fourier transform. We do not present these operations here, because they involve relatively advanced mathematics.

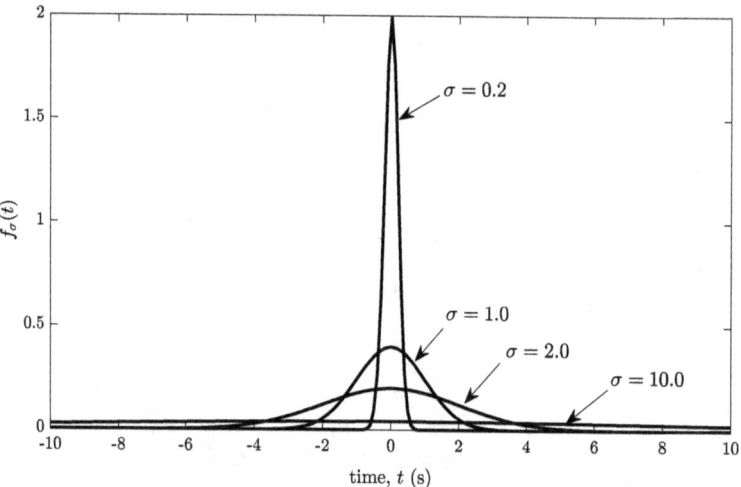

FIGURE 22. Family of normalized Gaussians in the time domain

We conclude this section with an example of a family of functions $f_\sigma(t)$ that have continuous spectra $\tilde{f}_\sigma(\omega)$. Here we use the example of the *normalized Gaussian* for the function of time, because it illustrates the nature of time domain functions that have continuous spectra, particularly for the magnitude spectra. Also, they are familiar to anyone who has had a course in probability. The normalized Gaussian with with its peak centered about $t = 0$ is

$$f_\sigma(t) = \frac{1}{\sigma\sqrt{2\pi}} e^{-t^2/(2\sigma^2)} \:. \tag{20}$$

These functions are plotted in Figure 22 for various values of σ. We can imagine this function representing the speed of an object that goes from near 0 for very large negative time t, accelerates to its maximum value at $t = 0$, and then decelerates to become near 0 again as time increases to large positive values. σ controls the exact speeds and accelerations.

In the frequency domain, it can be shown that,

$$\tilde{f}_\sigma(\omega) = \frac{1}{\sqrt{2\pi}} e^{-\sigma^2\omega^2/2} \tag{21}$$

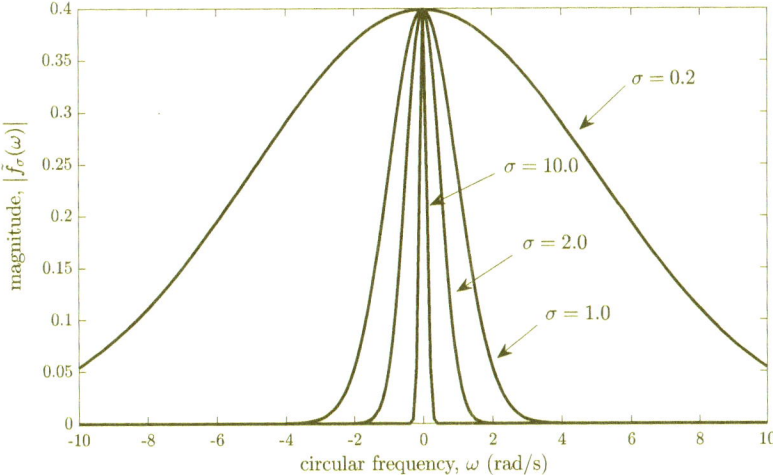

FIGURE 23. Magnitude spectra of the family of normalized Gaussians in Figure 22

The resulting magnitude spectra for various values of σ are shown in Figure 23. All real frequencies go into making up $\tilde{f}_\sigma(\omega)$, and, hence, $f_\sigma(t)$. Further, the magnitude spectra are symmetric about $\omega = 0$.

The important thing to note is the inverse relation between the "fatness", or expanse, of $f(t)$ and that of $|\tilde{f}(\omega)|$. This is a general relationship between the time domain and frequency domain, when there is a continuous spectrum. [There is a similar result for discrete spectra, but we do not pursue this for discrete spectra in this book.] In fact, if the time domain function is a infinitely high spike at a single time, then the magnitude spectrum is a constant for all real frequencies. As an aside, *Heisenberg's uncertainty principle* in quantum mechanics is related to this inverse relation in the breadth of time domain and frequency domain representations.

Conclusion

In this chapter the scenario of a brief piston movement was examined to find that the pattern of acoustic wave motion has a repetition frequency of 500 in a finite-length tube with length $L = 17.1$ cm. That is, we can consider the system to reach steady conditions after

2 ms. However, this is no longer the case when the piston moves in a sinusoidal manner. We found that the system could have repetition frequencies such as 100 Hz, or even as low as 30 Hz, and, perhaps, lower. This could be problematic when analyzing speech signals, where steadiness is often a major assumption in the analysis, such as in *Fourier analysis*, which takes a time domain function and produces its frequency domain representation.

We explore steadiness in the simple mass-spring system in the next chapter. We continue on to discuss the physical factors that can drastically reduce the time for the onset of steady conditions. That is, we will get a glimpse as to how physics enables steadiness to be attained in the finite-length tube within a few milliseconds.

References

Howe, M.S. (2015). *Acoustics and Aerodynamic Sound*. Cambridge University Press, Cambridge, England. (pp 10-12)

Chapter 4: Mass-Spring Systems

Introduction

Many ideas applicable to the acoustics of finite-length tubes can be readily understood when discussed in relation to a *linear mass-spring system*. Here we present the theory of linear mass-spring systems an example of which was introduced in Chapter 1 as a simple mass-spring system. The concepts of *impulse response function* and *damping* are examined in this chapter. The effects of damping on the steadiness of linear mass-spring systems are explained. The effects of damping on the acoustics of the finite-length tube are observed using numerical simulation, and these effects on the steadiness of acoustic wave motion in a finite-length tube are found to be substantial.

Linear mass-spring systems

An example of a linear mass-spring system was introduced in Chapter 1, and the reader may want to review the section in that chapter entitled "A mass-spring system". Here we repeat the equation of motion for the simple mass-spring system in Equation (13) of Chapter 1.

$$m\frac{\Delta_t^2(X_m(t))}{(\Delta t)^2} + \kappa X_m(t) = 0 , \qquad (1)$$

where m is the mass of the body at the end of the spring, and κ is the spring constant for the spring, which is assumed to obey Hooke's law, Equation (11) of Chapter 1. With x_0 the the position of the mass, when the spring is neither compressed or expanded, and x_m the position of the mass, $X_m = x_m - x_0$ is the displacement, or displacement position. Equation (1) is a statement of Newton's second law of motion that mass times acceleration equals force.

The natural circular frequency ω_0 has the dimensions of circular frequency, s^{-1}, but it is stated in units of rad/s, where rad is dimensionless. Recall that in Equation (15) of Chapter 1 that

$$\omega_0 = \sqrt{\frac{\kappa}{m}} \ . \tag{2}$$

$\mathcal{F}_0 = \omega_0/(2\pi)$, stated in units of Hz, is the natural frequency for the simple mass-spring system. With equation (2), Equation (1) can be written

$$\frac{\Delta_t^2(X_m(t))}{(\Delta t)^2} + \omega_0^2 X_m(t) = 0 \ . \tag{3}$$

In this chapter we go into more depth than we did in Chapter 1 regarding the mathematical solutions of Equation (3), as well as discussing properties of other linear mass-spring systems.

Before moving farther into the discussion, we need to emphasize the property of linearity of linear mass-spring systems. The linear mass-spring system shares this property with the wave equation, Equation (17) of Chapter 2. Thus, if $f(t)$ and $g(t)$ are solutions to the equation of motion for a linear mass-spring system, then so are $af(t) + bg(t)$ for any real numbers a and b. This provides the property of superposition for solutions to Equation (3). The adjective "simple" for the linear mass-spring systems is dropped, and, often the word linear is also dropped because all mass-spring systems discussed in the present book are linear.

Initial value problems

In real physical systems we do not usually assume that objects have been moving forever, but that there is some initial time when the object starts to move. This means that we need to consider how objects start to move. In technical terms, we consider *initial value problems* now.

We often choose $t = 0$ as the time for which things start moving. For mass-spring systems, initial values are given for displacement position and velocity at $t = 0$.

$$X_m(t = 0) = a, \text{ and}$$
$$v(t = 0) = \frac{\Delta_t X_m(t = 0)}{\Delta t} = b \ , \tag{4}$$

where a and b are real numbers. These are termed the initial conditions. Equations (3) and (4) constitute an initial value problem. Initial value problems can be solved numerically, but it is more instructive to state the general solution of this particular initial value problem symbolically. We write the solution to the initial value problem represented in Equations (3) and (4) as $\Delta t \to 0$.

$$X_m(t) = \mathcal{A}\sin(\omega_0 t + \phi) \text{ with } \mathcal{A} > 0, \text{ for } t > 0 . \tag{5}$$

The amplitude $\mathcal{A}$ and initial phase ϕ are determined by initial conditions $X_m(t=0) = a$ and $v(t=0) = b$ in Equation (4). First, the expression for velocity $v(t) = \Delta_t X_m(t)/\Delta t$ as $\Delta t \to 0$ is found from Equation (13) of Chapter 3.

$$v(t) = \frac{\Delta_t X_m(t)}{\Delta t} = \omega_0 \mathcal{A} \cos(\omega_0 t + \phi) \text{ for } t > 0 ,$$

or $\tag{6}$

$$v(t) = \omega_0 \mathcal{A} \cos(\omega_0 t + \phi) \text{ for } t > 0 .$$

We now incorporate the initial values of $X_m(t=0)$ and $v(t=0)$ into the constants $\mathcal{A}$ and ϕ. Equations (5) and (6) evaluated as t approaches zero from the right, or as $t \to 0^+$, give

$$\begin{aligned}\lim_{t\to 0^+} \mathcal{A}\sin(\omega_0 t + \phi) &= \mathcal{A}\sin(\phi) = X_m(t=0) , \\ \lim_{t\to 0^+} \omega_0 \mathcal{A}\cos(\omega_0 t + \phi) &= \omega_0 \mathcal{A}\cos(\phi) = v(t=0) .\end{aligned} \tag{7}$$

The expressions in Equation (7) can be solved for $\mathcal{A}$ and ϕ.

$$\begin{aligned}\phi &= \arctan\left[\frac{\omega_0 X_m(t=0)}{v(t=0)}\right] , \\ \mathcal{A} &= \frac{X_m(t=0)}{\sin(\phi)} \text{ or } \mathcal{A} = \frac{v(t=0)}{\omega_0 \cos(\phi)} .\end{aligned} \tag{8}$$

We consider oscillation when $m = 1$ g and $\kappa = 1,000,000 \cdot \pi^2$ dyne/cm. The natural frequency for this configuration is $\mathcal{F}_0 = (1/2\pi)\sqrt{\kappa/m} = 500\ Hz$. Figure 1 shows the resulting motion for different numerical values of the initial conditions. In the top panel, $X_m(t=0) = 0.0$ cm and $v(t=0) = 0.0$ cm/s, for which $\mathcal{A} = 0$ cm and there is no motion. In the next panel down, $X_m(t=0) = 0.0$ cm and $v(t=0) = 100 \cdot \pi$ cm/s, and Equation (8) is used to obtain $\phi = 0.0$ rad and $\mathcal{A} = 0.1$ cm. In the following panel, $X_m(t=0) = 0.1$ cm and $v(t=0) = 0.0$ cm/s, for which $\phi = \pi/2$ rad and $\mathcal{A} = 0.1$ cm by Equation (8). In the bottom panel, $X_m(t=0) = 0.1$ cm and $v(t=0) = 100 \cdot \pi$ cm/s, for which $\phi = \pi/4$ rad and $\mathcal{A} = 0.14$

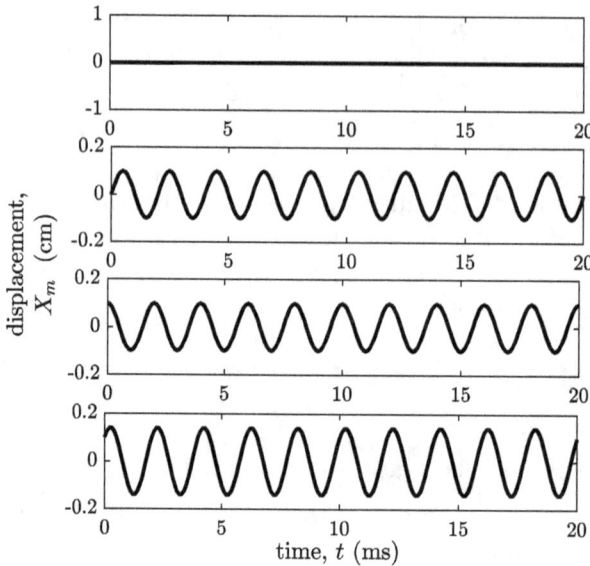

FIGURE 1. Initial value problem solutions

cm. The latter three motions are identical except for amplitude and the "place" in the sine wave where it starts at $t = 0$, that is, the initial phase.

Impulse response functions

We examine an important initial value problem and its solution that takes us well beyond calculating solutions for various solutions for simple mass-spring motion with differing initial conditions. In doing so, we explore impulse response functions, which provide extra insight, as well as a calculation tool, for any linear system with constant parameters. By parameter we mean things such as mass m, spring constant κ, or $\omega_0 = \sqrt{\kappa/m}$.

Suppose that the mass-spring system is forced by an external force in the x-direction for a brief duration. Symbolically, force has amplitude $F_{\tau=0}$, but is of very short duration, say $\Delta\tau > 0$ around time $t = \tau = 0$. The quantity $\mathbb{I} \equiv F_{\tau=0}\Delta\tau$ is known as *impulse*, and it has the dimensions of momentum, dyne s = g cm/s. Suppose that we stipulate that impulse at $t = \tau = 0$ has unit magnitude, i.e. $\mathbb{I} = \mathbb{I}^{unit}_{\tau=0} = 1$ in

Acoustics of Speech Production

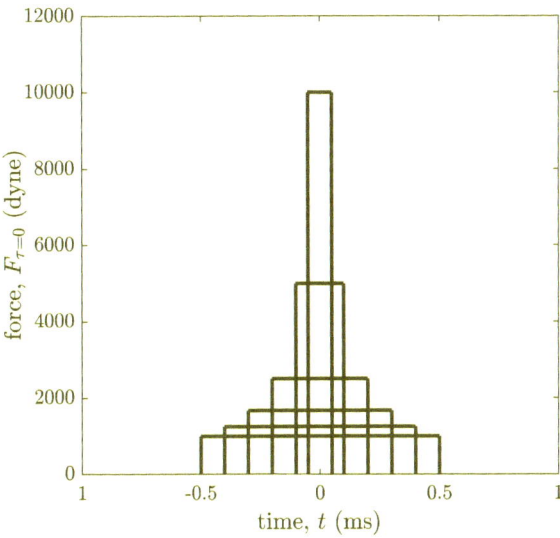

FIGURE 2. Impulsive forces

whatever system of units one is using. $\mathbb{I}^{unit}$ is called *unit impulse*. For instance, in the c-g-s system $\mathbb{I}^{unit}_{\tau=0} = 1$ dyne s $= 1$ g cm/s. Figure 2 shows that once $\Delta\tau$ is chosen, then $F_{\tau=0}$ is determined by the constraint that $F_{\tau=0}\Delta\tau = 1$ dyne s.

The brief force applied to the linear mass-spring system around $t = 0$ provides an impulse of unit magnitude, and this is simply the amount of momentum mv gained by the simple mass-spring system around the time $t = \tau = 0$. For unit impulse, $\mathbb{I}^{unit}_{\tau=0} = mv^{ui}(t = 0)$. The superscript ^{ui} is used to denote the motion due to unit impulse. The unit impulse is a source of momentum, and energy, for the simple mass-spring system. Because impulse imparts momentum to the mass at $t = 0$, the unit impulsive force problem can be written as an initial value problem, incorporating the effect of the impulsive force into the initial conditions.

$$X^{ui}_m(t=0) = 0 \quad \text{and} \quad v^{ui}(t=0) = \frac{\mathbb{I}^{unit}_{\tau=0}}{m}. \tag{9}$$

The solutions are given by Equation (5), in conjunction with Equation (9). Thus, $\phi = 0$ and $\mathcal{A} = \mathbb{I}^{unit}_{\tau=0}/(\omega_0 m)$, and the solution to the initial

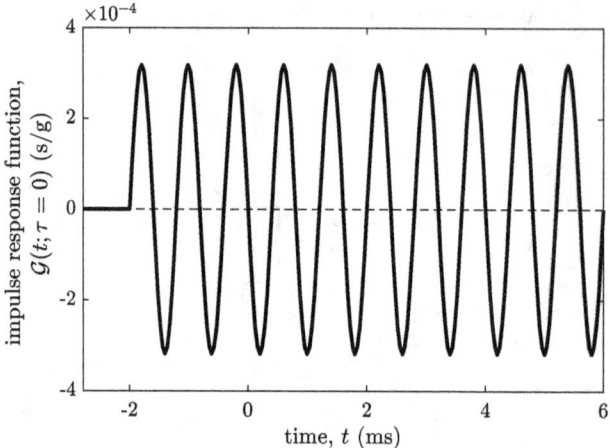

FIGURE 3. Impulse response function of the simple mass-spring system

value problem of Equations (3) and (9) is

$$X_m^{ui}(t) = \frac{\mathbb{I}_{\tau=0}^{unit}}{\omega_0 m} \sin(\omega_0 t) \text{ for } t > 0 . \tag{10}$$

We define the impulse response function as

$$\mathcal{G}(t; \tau = 0) \equiv \begin{cases} \frac{1}{\omega_0 m} \sin(\omega_0 t) & \text{for } t > 0 \\ 0 & \text{for } t < 0 . \end{cases} \tag{11}$$

From equations (10) and (11)

$$X_m^{ui}(t) = \mathbb{I}_{\tau=0}^{unit}\mathcal{G}(t; \tau = 0) = \mathcal{G}(t; \tau = 0) \text{ for } t > \tau = 0 . \tag{12}$$

The final equality in Equation (12) assumes that we have multiplied by the number one with diemnsions of momentum g-cm/s. $\mathcal{G}(t; \tau = 0)$ is illustrated in Figure 3.

Setting the time of impulse τ to 0 is arbitrary. One of the underlying assumptions of classical physics is that time is homogeneous: properties of physical systems depend on relative times and not on absolute time. This means that the response to a unit impulse $\mathbb{I}_\tau^{unit}$ at any τ is

$$X_m^{ui}(t) = \frac{\mathbb{I}_\tau^{unit}}{\omega_0 m} \sin\left(\omega_0(t - \tau)\right) \text{ for } t > \tau . \tag{13}$$

This means that the impulse response function of Equation (11) generalizes to

$$\mathcal{G}(t;\tau) = \mathcal{G}(t-\tau) = \begin{cases} \frac{1}{\omega_0 m} \sin\left(\omega_0(t-\tau)\right) & \text{for } t > \tau \\ 0 & \text{for } t < \tau \,. \end{cases} \quad (14)$$

The impulse response function in Equation (14) is really not a function of two variables: the time of impulse τ and the time of observation t, but it is a function of a single variable $t - \tau$. Of course, this assumes that parameters, such as mass m and spring constant κ do not change with time.

It is not necessary to use a unit impulse in order for the impulse response function to be useful. For any real number a, let $F_\tau \Delta\tau = \mathbb{I}_\tau = a \cdot \mathbb{I}_\tau^{unit}$. From the property of linearity

$$\begin{aligned} X_m(t) &= a \cdot X_m^{ui}(t) \\ &= a \cdot \mathbb{I}_\tau^{unit} \mathcal{G}(t-\tau) \\ &= \mathbb{I}_\tau \mathcal{G}(t-\tau) \\ &= F_\tau \Delta\tau \, \mathcal{G}(t-\tau) \\ &= \frac{F_\tau}{\omega_0 m} \sin\left(\omega_0(t-\tau)\right) \Delta\tau \,, \end{aligned} \quad (15)$$

for $t > \tau$.

Let's examine the same problem of impulsive forcing from a different perspective, and derive the same result. Again we consider the more general case where $\mathbb{I}_\tau = F_\tau \Delta\tau$ has any real value. We rewrite the equation of motion, Equation (3), to take account of the impulsive forcing explicitly. Recalling that the term $-\kappa X_m$ is the force on the mass (in the x-direction) due to the restoring force of the spring, the total force on the mass in the presence of the impulsive force is

$$-\kappa X_m(t) + \begin{cases} F_\tau & \text{for } \tau - \frac{\Delta\tau}{2} < t < \tau + \frac{\Delta\tau}{2} \\ 0 & \text{otherwise} \,. \end{cases} \quad (16)$$

By Newton's second law

$$m \frac{\Delta_t^2 X_m(t)}{(\Delta t)^2} = -\kappa X_m(t) + \begin{cases} F_\tau & \text{for } \tau - \frac{\Delta\tau}{2} < t < \tau + \frac{\Delta\tau}{2} \\ 0 & \text{otherwise} \,. \end{cases} \quad (17)$$

This can be written

$$\frac{\Delta_t^2 X_m(t)}{(\Delta t)^2} + \omega_o^2 X_m(t) = \frac{F(t)}{m}, \qquad (18)$$

where $F(t) = \begin{cases} F_\tau & \text{for } \tau - \frac{\Delta \tau}{2} < t < \tau + \frac{\Delta \tau}{2} \\ 0 & \text{otherwise} \end{cases}$

With the force incorporated into the equation of motion, Equation (18), we do not incorporate its effect into the initial conditions. The appropriate initial conditions to go with Equation (18) are

$$X_m(t=0) = 0 \quad \text{and} \quad v(t=0) = 0. \qquad (19)$$

[We ignore the fact that the force starts at $t = -\frac{\Delta \tau}{2}$, and treat this time as approximately zero.] The solution of the initial value problem stated in Equations (18) and (19) is exactly the same as the solution given in Equation (15). We derived that solution using non-zero initial conditions, but with no external forcing on the right-hand-side of the equation. We are stating that the same solution can be obtained with zero initial conditions, but with a brief force for the right-hand-side of the equation.

We cannot go into the details of how to mathematically solve the initial value problem presented in Equations (18) and (19). However, we provide a description on how this works. We go into the frequency domain to solve Equation (18), which means that the brief force on the right-hand-side is represented in the frequency domain. This brief pulse has a continuous spectrum. The fact that the force is very brief means that the spectrum is very broad, as discussed at the end of Chapter 3. In fact, if we let $\Delta \tau \to 0$, then $F_\tau \to \infty$, and we have a infinitely skinny and high spike. This has a completely flat amplitude spectrum. However, the system can only "grab" the energy at its natural circular frequency ω_0 because the force ceases to be non-zero beyond $t = \tau + \Delta \tau/2$. After returning to the time domain we find the solution expressed in Equation (15) for $t > \tau$.

The linear mass-spring system with a general external force

Suppose that external force $F(t) = 0$ for $t < 0$, and that $F(t)$ may be non-zero for $t > 0$. General initial conditions expressed in Equation (4) are also possible. Thus, instead of "kicking" the mass-spring system with a brief impulsive force at τ, we consider what occurs when there is force continuously applied through time, that is, for some continuous interval of time. We write the equation of motion with initial conditions as

$$\frac{\Delta_t^2 X_m(t)}{(\Delta t)^2} + \omega_0^2 X_m(t) = \frac{F(t)}{m} \quad \text{for } t > 0, \tag{20}$$

with $X_m(t=0)$ and $v(t=0)$ specified.

The general solution to Equation (20) is the sum of two functions of time, $f_I(t)$ and $f_H(t)$. The first function of time, $f_I(t)$, satisfies the first relation in Equation (20), the equation of motion, but not necessarily the initial conditions. The second function, $f_H(t)$, satisfies Equation (3). That is, $f_H(t)$ satisfies the first relation in Equation (20), with $F(t) = 0$. We know the form of this second function is given by Equation (5). The constants $\mathcal{A}$ and ϕ are adjusted so that the initial conditions are satisfied by the sum of the two functions $f_I(t) + f_H(t)$. The property of linearity ensures that the resulting function of time satisfies all of Equation (20). We start with finding $f_I(t)$, and we do not concern ourselves with satisfying the initial conditions in Equation (20) in this section.

Before arriving at the general case, we consider two brief impulses. One of the impulses has amplitude F_{τ_0} and is centered at time $\tau_0 > 0$, and the other has amplitude F_{τ_1} and is centered at time $\tau_1 > \tau_0$.

$$F(t) = \begin{cases} F_{\tau_0} & \text{for } \tau_0 - \frac{\Delta \tau}{2} < t < \tau_0 + \frac{\Delta \tau}{2} \\ F_{\tau_1} & \text{for } \tau_1 - \frac{\Delta \tau}{2} < t < \tau_1 + \frac{\Delta \tau}{2} \\ 0 & \text{otherwise} . \end{cases} \tag{21}$$

Individually, the impulses produce two scaled impulse response functions, one beginning at τ_0 and the other starting at τ_1. We invoke the fact that Equation (15) describes a linear system to conclude that the sum of these solutions is the solution to the problem with two impulses.

$$X_m(t) = F_{\tau_0} \mathcal{G}(t - \tau_0) \Delta \tau + F_{\tau_1} \mathcal{G}(t - \tau_1) \Delta \tau . \tag{22}$$

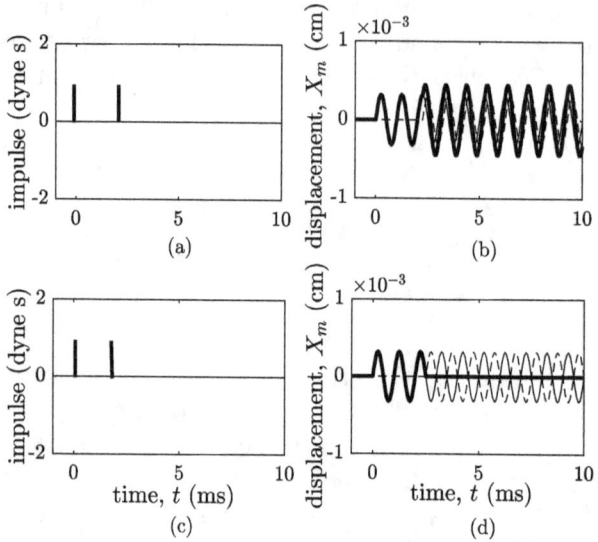

FIGURE 4. Examples of two impulses

Therefore,

$$X_m(t) = \begin{cases} 0 & \text{for } 0 < t \leq \tau_0 \\ \frac{F_{\tau_0}}{\omega_0 m} \sin\left(\omega_0(t - \tau_0)\right) \Delta\tau & \text{for } \tau_0 < t \leq \tau_1 \\ \frac{F_{\tau_0}}{\omega_0 m} \sin\left(\omega_0(t - \tau_0)\right) \Delta\tau + \\ \frac{F_{\tau_1}}{\omega_0 m} \sin\left(\omega_0(t - \tau_1)\right) \Delta\tau & \text{for } t > \tau_1 \, . \end{cases} \qquad (23)$$

In Figures 4 and 5 we show examples of examples of Equations (22) and (23) for $\tau_0 = 0$ and $F_{\tau_0}\Delta\tau = 1$, and for various values of $F_{\tau_1}\Delta\tau$ and $\tau_1 > 0$. The impulses are shown on the left sides and the resulting displacements, $X_m(t)$, on the right sides. The results of the first impulse are shown in thin solid lines, and the result of the second impulse in thin dashed lines. The sum of the two is $X_m(t)$, and is shown in the thick solid line. Figure 4(a) shows two unit impulses separated by 4.5 ms, and Figure 4(b) shows that the resulting two displacements exhibit *constructive interference*. The amplitude of the sum of the impulse response functions is greater than the amplitudes

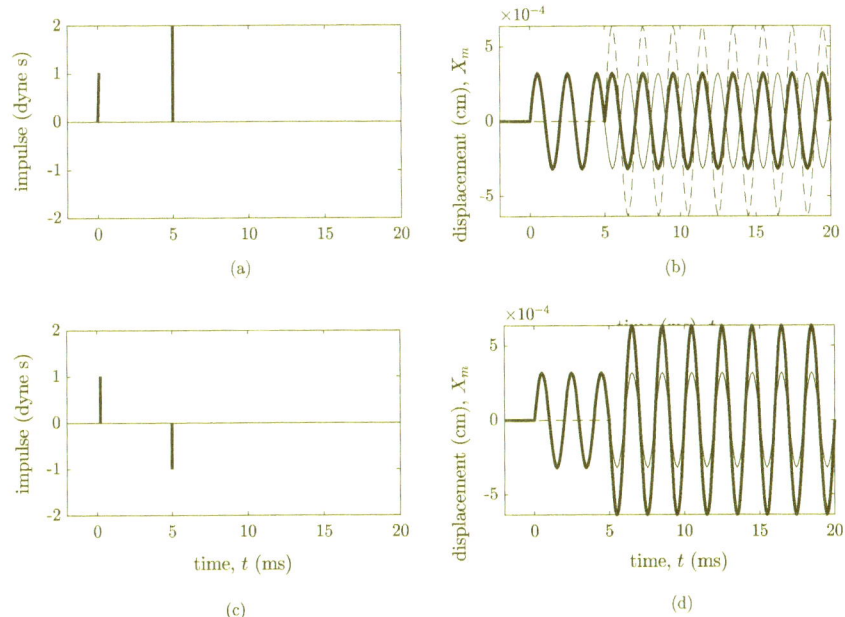

FIGURE 5. More examples of two impulses

of the individual impulse response functions. With the impulses separated by 5.0 ms in Figure 4(c), the displacements show *destructive interference*, so that there is no displacement after the second impulse, as seen in Figure 4(d).

The destructive interference in Figure 4(d) is somewhat diminished if the amplitude of the second impulse in Figure 4(c) is doubled, as shown in Figure 5(a) and 5(b). The resulting output has a constant amplitude from $\tau_0 = 0$, with a phase shift when the second impulse occurs at $\tau_1 = 5$ ms. We obtain complete constructive interference, if the second impulse in Figure 4(c) is given a negative unit amplitude as shown in Figure 5(c) and 5(d).

We now examine continuous external forces using a couple of examples. The first example is with an external force given as

$$F(t) = \begin{cases} Bt & \text{for } 0 \leq t \leq 1, \\ -B(t-2) & \text{for } 1 < t \leq 2, \\ 0 & \text{otherwise}. \end{cases} \quad (24)$$

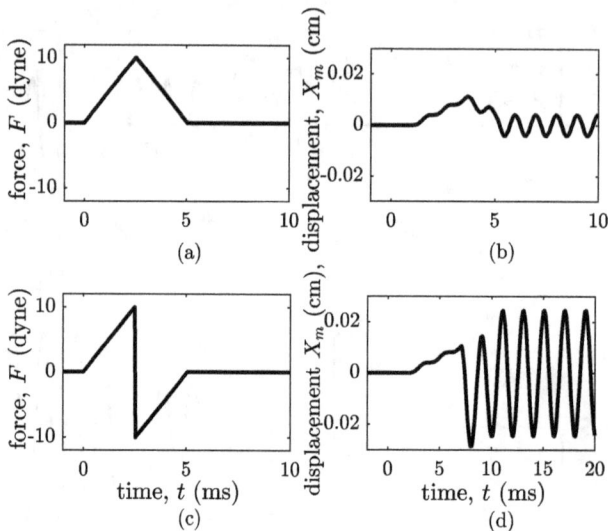

FIGURE 6. Two examples of continuous forcing

where $B > 0$ is a constant with dimensions dyne/s. The second external force is

$$F(t) = \begin{cases} Bt & \text{for } 0 \leq t \leq 1, \\ B(t-2) & \text{for } 1 < t \leq 2, \\ 0 & \text{otherwise}. \end{cases} \quad (25)$$

Figure 6 shows these forces as functions of time in the panels to the left and the resulting displacements to the right. Figure 6(a) and Figure 6(b) correspond to forcing function in Equation (24), and Figure 6(c) and Figure 6(d) correspond to Equation (25). In both cases there is a complicated displacement while the external force is non-zero, but after the forcing has finished there is only oscillation at the natural frequency, $\mathcal{F}_0 = \omega_0/(2\pi) = 500 \; Hz$.

How do we obtain the results shown in Figure 6? We can think of taking a continuous external force function and approximating it as a series of impulses $F(\tau_n)\Delta\tau$, where $\tau_n = \tau_0 + n\Delta\tau$ with n an integer, $n \geq 0$, and where the external force begins at time $\tau_0 - \Delta\tau/2$. Again, by linearity and the time invariance of the system parameters, the resulting displacement is

$$X_m(t) = \sum_{n=0}^{N} F(\tau_n)\mathcal{G}(t - \tau_n)\Delta\tau . \tag{26}$$

The sum is known as a *convolution sum*. Note that we require that $t - \tau_n > 0$ in order for the bit of impulse, $F(\tau_n)\Delta\tau$ be included into the sum. Also, note that $\mathcal{G}$ depends on both t and τ_n, but only in the form $t - \tau_n$. It is a way to obtain the response of the linear mass-spring system for a given external force $F(t)$. The solution in Equation (26), $X_m(t)$, corresponds to the function $f_I(t)$ that was introduced in the discussion after Equation (20). In the next section, we explore how to find $f_H(t)$ so that $f_I(t) + f_H(t)$ satisfies the initial value problem in Equation (20).

In this section we introduced the idea of an impulse response function, and it turns out to be an important tool in solving initial value problems. The impulse response function in Equation (15) characterizes the mass-spring system, as well as helping us compute solutions to the initial value problem of the linear mass-spring system with external forcing, Equation (20). *In fact, we could use the impulse response function in Equation (15) in place of the first equation in Equation (20), the equation of motion, to mathematically characterize the linear mass-spring system.*

The notion of impulse response function can be generalized to go beyond the situation where an impulsive force provides unit momentum to the system. We leave this section with the observation that the mass-spring system can be provided with initial energy if the spring is either compressed or stretched initially, in contrast to the initial condition with an impulsive force providing kinetic energy and momentum. That is, supplying initial potential energy can be a way to start the system, just like supplying impulse can start the system. This means that the idea of impulse response function can be generalized. This is not done for the mass-spring system here. We wait until these ideas are applied to the acoustic system.

The linear mass-spring system with a sinusoidal external force

We consider the initial value problem for the *forced linear mass-spring system* of Equation (20), when the external forcing is sinusoidal forcing with *forcing circular frequency* ω_F

$$F(t) = F_0 \sin(\omega_F t) \quad \text{for } t > 0 \ . \tag{27}$$

F_0 is the force amplitude and it has units of dyne. In the examples of forced linear mass-spring systems so far we have not attended to the initial conditions, but we start to do that in this section. We find a solution for the specified external force and add on the solution for zero external force and adjust the constants $\mathcal{A}$ and ϕ to help satisfy those initial conditions.

We do not derive, but simply quote the solution.

$$X_m(t) = \frac{-F_0}{m(\omega_F^2 - \omega_0^2)} \sin(\omega_F t) + \mathcal{A} \sin(\omega_0 t + \phi) \quad \text{for } t > 0 \ . \tag{28}$$

where $\mathcal{A}$ and ϕ are constants that are determined by the initial conditions. The first term in Equation (28) corresponds to $f_I(t)$ in the discussion after Equation (21). It can be found using the convolution sum, Equation (26), as $\Delta \tau \to 0$, but it is beyond the scope of this book to perform the calculations. The second term is the general form of the solution to Equation (3). It corresponds to the function $f_H(t)$ in the discussion after Equation (21). The values of $\mathcal{A}$ and ϕ can be found in order that $f_I(t) + f_H(t)$ satisfy the initial conditions. This is done below.

[The first term in Equation (28) can be shown to be a solution to Equation (20) with the force $F(t)$ given by Equation (27) as $\Delta t \to 0$. All that is needed is two applications of relations in Equation (13) of Chapter 3. Similarly, the second term in equation (28) solves Equation (3) for any $\mathcal{A}$ and ϕ. Again, two applications of relations in Equation (13) of Chapter 3 is all that is needed to prove this.]

The circular function in the first term in Equation (28) has the *forcing frequency* $\mathcal{F}_F = \omega_F/(2\pi)$, which is consistent with the fact that it satisfies the equation of motion in Equation (20), with the external force specified in Equation (27). The other sinusoidal function of time has the natural frequency of the linear mass-spring system, $\mathcal{F}_0 = \omega_0/(2\pi)$. This is the component that corresponds to zero external force $F(t)$, but must be added to provide for the initial values of $X_m(t=0)$ and $v(t=0)$.

The relations between the constants $\mathcal{A}$ and ϕ and the initial conditions in the case of non-zero sinusoidal external force are

$$\mathcal{A}\sin(\phi) = X_m(t=0) \text{ and} \tag{29}$$
$$\omega_0 \mathcal{A}\cos(\phi) - \frac{\omega_F F_0}{m(\omega_F^2 - \omega_0^2)} = v(t=0).$$

Unlike the case with no external force, where $X_m(t=0)$ and $v(t=0) = 0$ implies $\mathcal{A} = 0$, this is not the case for forced motion. In the case that $X_m(t=0) = 0$ and $v(t=0) = 0$

$$\mathcal{A} = \frac{(\omega_F/\omega_0) F_0}{m(\omega_F^2 - \omega_0^2)}$$
$$\phi = 0. \tag{30}$$

With Equation (30), Equation (28) becomes

$$X_m(t) = \frac{-F_0}{m(\omega_F^2 - \omega_0^2)} \sin(\omega_F t) + \frac{(\omega_F/\omega_0) F_0}{m(\omega_F^2 - \omega_0^2)} \sin(\omega_0 t) \tag{31}$$
$$= \frac{-F_0}{m(\omega_F^2 - \omega_0^2)} \left[\sin(\omega_F t) - \frac{\omega_F}{\omega_0} \sin(\omega_0 t) \right] \quad \text{for } t > 0.$$

In the following series of figures we plot various quantities related to the solution in Equation (31) for different values of forcing frequency $\mathcal{F}_F$. Velocity is $v(t) = \Delta X_m(t)/\Delta t$, which can be computed from Equation (31) using the relations in Equation (13) of Chapter 3. The power $\mathcal{P}(t)$ is computed as $\mathcal{P}(t) = F(t)v(t)$. The energy input at any time t is simply the sum of power times very small time interval durations Δt from $t = 0$ up to time t. The forcing frequencies $\mathcal{F}_F$ are: 200, 300, 400, 470, and 1200 Hz in Figures 7 through 11, respectively.

Figures 7, 8, 9, and 11, with forcing frequencies, $\mathcal{F}_F = 200$, 300, 400, and 1200 Hz respectively show a 100 Hz periodicity. Because we have an explicit mathematical expression for the linear mass-spring system in Equation (31), we can understand the 100 Hz periodicity, for example, in the case when $\mathcal{F}_F = 300$ Hz and $\mathcal{F}_0 = 500$ Hz shown in Figure 8. According to Equation (31) we have sinusoidal terms at both these frequencies, so that in order for the system to complete one full period, both terms need to complete an integer number of periods. The time it takes the external forcing term to complete an integer n_F cycles is $n_F/\mathcal{F}_F$, and the time it takes for term involving the natural frequency to complete integer n_0 cycles is $n_0/\mathcal{F}_0$. We require

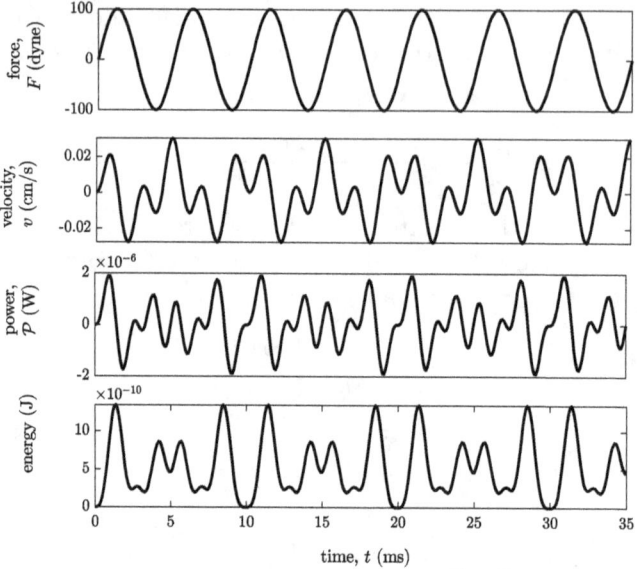

FIGURE 7. $\mathcal{F}_F = 200\ Hz$

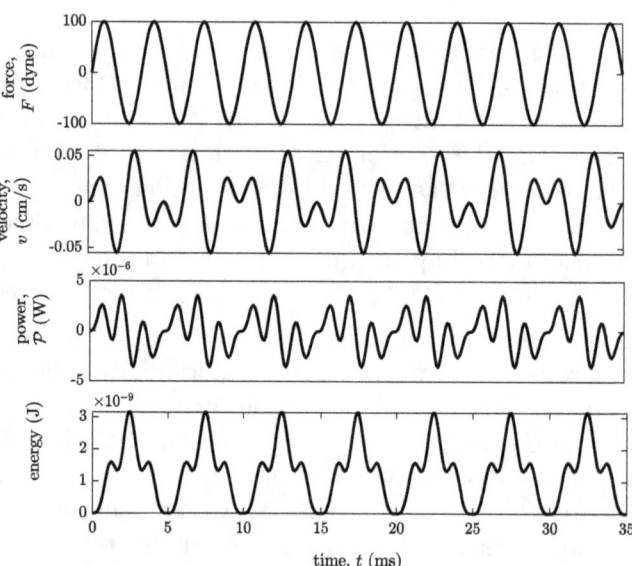

FIGURE 8. $\mathcal{F}_F = 300\ Hz$

$n_F/\mathcal{F}_F = n_0/\mathcal{F}_0$ for an overall period of the system. The implies that $n_F/n_0 = \mathcal{F}_F/\mathcal{F}_0$, or that $n_F/n_0 = 3/5$. The smallest positive integers that satisfy this are $n_F = 3$ and $n_0 = 5$, which means that $n_0/\mathcal{F}_0 = 0.01$ s, or that the overall frequency of the system is 100 Hz.

In the case where the forcing frequency, $\mathcal{F}_F = 470$ Hz, $n_F/n_0 = 47/50$, so it takes 0.94 s for the pattern to repeat itself. However, Figure 10 appears to show that the repetition frequency is 30 Hz, which corresponds to a repetition time of about 0.033 s. The latter is not exact repetition, but it is close enough that we perceive a 30 Hz modulation at this frequency. It can be shown mathematically that when the forcing frequency is close to the natural frequency, there is strong modulation at the difference frequency. In this case this difference frequency is 30 Hz. This frequency modulation phenomenon is called *beating*, and the difference frequency is called the *beat frequency*.

We note that power input to the mass-spring system can be both positive and negative for all of the forcing frequencies in these examples. Thus, the external force both puts energy into and takes energy out of the system. While the net energy input to the mass-spring system is always at least zero, the energy input to the system always returns to zero.

Figure 7 and Figure 16 of Chapter 3 look very similar, as do Figure 8 and Figure 17 of Chapter 3, Figures 9 and Figure 18 of Chapter 3, and Figure 10 and 19 of Chapter 3. Figures 11 and Figure 20 of Chapter 3 differ somewhat in appearance, although both exhibit a 100 Hz repetition rate. There does seem to be something in common between the finite-length tube with a piston source and the forced linear mass-spring system in terms of steadiness. In particular, beating seems to be common to both systems when $\mathcal{F}_F = 470$ Hz. We pursue commonalities between these systems in the following chapters.

We note that the solution given in Equation (31) is undefined when the forcing frequency equals the natural frequency. We need to go further into the physics of these systems in order to understand forcing at the natural frequency. This is done in the next section on the effect of *friction damping*.

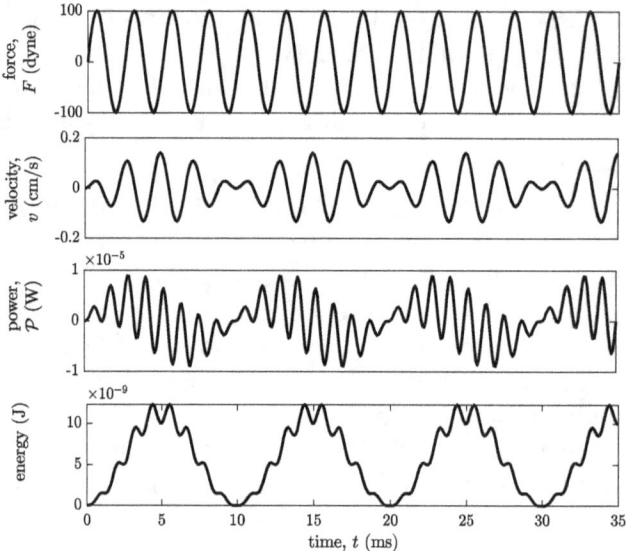

FIGURE 9. $\mathcal{F}_F = 400$ Hz

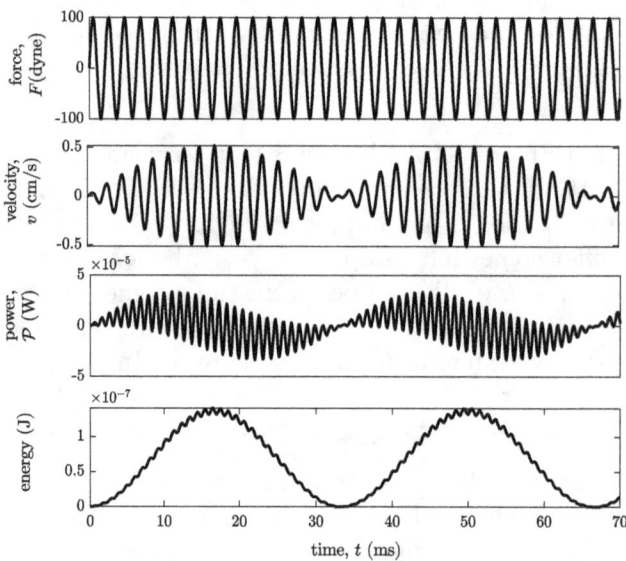

FIGURE 10. $\mathcal{F}_F = 470$ Hz

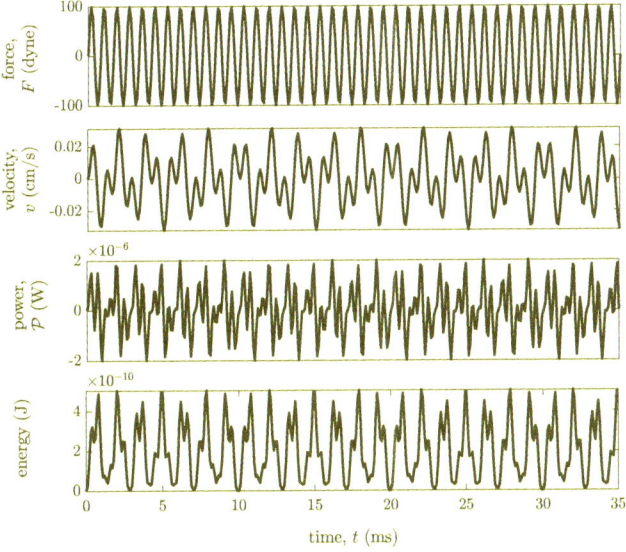

FIGURE 11. $\mathcal{F}_F = 1200\ Hz$

The mass-spring system with friction damping

To understand how a mass-spring system evolves towards steadiness for real oscillating systems it is necessary to consider friction and other phenomena that cause energy of the oscillations to be transferred either to other forms of energy or to places outside the system. The other forms of energy considered here often involve motion on the molecular level, so the energy appears as heat energy. The transfer of energy from the motion of a body to heat energy is called *dissipation*.

Dissipation is not the only way that energy can be lost from the mechanical system. Later we concern ourselves with a number of mechanisms for energy loss in acoustic systems. We refer to all such energy loss mechanisms as damping, whether or not they are dissipative. Presently we concern ourselves with damping of the mass-spring system by friction, which is dissipative.

The *friction force* is most often observed to be proportional to velocity of the mass $v(t) = \Delta_t(X_m(t))/\Delta t$, and acts opposite to the direction of motion. The friction force is

$$F_{friction}(t) = -\mu \frac{\Delta_t X_m(t)}{\Delta t} = -\mu v(t) \ . \tag{32}$$

where the constant of proportionality, $\mu > 0$, is the *coefficient of friction*. This is an approximation usually made in textbooks on physics. Friction force is dissipative, because it causes energy to be transferred from the motion of the mass to heat energy. This is in contrast with the linear mass-spring system, where the the sum of potential energy and kinetic energy remains constant. This can easily be proven by using the definitions of kinetic and potential energies in Equations (27) and (29) of Chapter 2, and the solution in Equation (5).

In the presence of friction and in the unforced case, the equation of motion in Equation (3) becomes, with initial conditions at $t = 0$,

$$m\frac{\Delta_t^2 X_m(t)}{(\Delta t)^2} = -\kappa X_m(t) - \mu \frac{\Delta_t X_m(t)}{\Delta t} \text{ , or}$$

$$\frac{\Delta_t^2 X_m(t)}{(\Delta t)^2} + 2\gamma \frac{\Delta_t X_m(t)}{\Delta t} + \omega_0^2 X_m(t) = 0 \text{ ,} \qquad (33)$$

with $X_m(t=0)$ and $v(t=0)$ specified ,

where $\omega_0 = \sqrt{\kappa/m}$ and $\gamma = \mu/2m$ is the *damping constant*. The general solution to Equation (33) is

$$X_m(t) = \mathcal{A} e^{-\gamma t} \sin(\hat{\omega}_0 t + \phi) \text{ for } t > 0 \text{ ,} \qquad (34)$$

where $\hat{\omega}_0 = \sqrt{\kappa/m - \gamma^2} = \sqrt{\omega_0^2 - \gamma^2}$ is the *reduced natural circular frequency* of the *friction damped mass-spring system* (e.g. Braun 1983). This means that the *reduced natural frequency* is $\hat{\mathcal{F}}_0 = (\hat{\omega}_0/2\pi) = (1/2\pi)\sqrt{\omega_0^2 - \gamma^2}$. We note that friction damping always makes the values of reduced natural frequency less than that of the natural frequency. $\mathcal{A}$ and ϕ are constants that are determined by initial conditions, $X_m(t=0)$ and $v(t=0)$. In order to show that Equation (34) is a solution to Equation (33), we need to generalize the identities in Equation (13) of Chapter 3. Without proof and for constants α, ω, and ϕ:

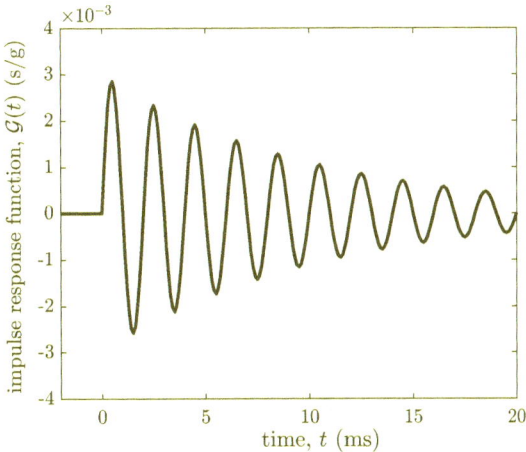

FIGURE 12. The impulse response function for a friction damped mass-spring system

If $f(t) = e^{-\alpha t} \sin(\omega t + \phi)$,

then $\dfrac{\Delta_t f(t)}{\Delta t} \approx e^{-\alpha t}\Big[\omega \cos(\omega t + \phi) - \alpha \sin(\omega t + \phi)\Big]$, and

if $f(t) = e^{-\alpha t} \cos(\omega t + \phi)$, (35)

then $\dfrac{\Delta_t f(t)}{\Delta t} \approx -e^{-\alpha t}\Big[\omega \sin(\omega t + \phi) + \alpha \cos(\omega t + \phi)\Big]$.

The impulse response function, $\mathcal{G}(t - \tau)$ is

$$\mathcal{G}(t - \tau) = \begin{cases} \dfrac{1}{m\hat{\omega}_0} e^{-\gamma(t-\tau)} \sin\left(\hat{\omega}_0(t-\tau)\right) & \text{for } t - \tau > 0, \\ 0 & \text{for } t - \tau < 0, \end{cases} \quad (36)$$

where the approximation improves as $\Delta t \to 0$. Figure 12 shows $\mathcal{G}(t)$ in Equation (36).

The motion of the friction damped mass-spring system is not described by a sinusoidal function, but by an *exponentially damped sinusoid*, or *damped sinusoid*. The amplitude of oscillation is an exponential function of time. The larger γ, for fixed ω_0, the greater the degree of exponential damping. We often refer to the motion described by sinusoidal or circular functions as simple harmonic motion, and when there is exponential damping, it is *damped simple harmonic motion*. Figure 12 can be compared to Figure 3, where there is no damping.

It is often convenient to express the strength of damping in terms of the *damping ratio*, $R = \gamma/\omega_0$. This is a measure of the ratio of the magnitude of the friction force to the magnitude of the restoring force of the spring. We can write the reduced natural frequency in terms of the damping ratio R.

$$\hat{\mathcal{F}}_0 = \frac{\hat{\omega}_0}{2\pi} = \frac{1}{2\pi}\sqrt{\omega_0^2 - \gamma^2}$$
$$= \frac{\omega_0}{2\pi}\sqrt{1 - R^2} \qquad (37)$$
$$= \mathcal{F}_0\sqrt{1 - R^2}\ .$$

In all of our considerations, $R < 1$, and most often $R \ll 1$.

We now turn to the way that the initial conditions $X_m(t=0)$ and $v(t=0)$ determine the constants $\mathcal{A}$ and ϕ. Equations (34) and (35) give

$$\mathcal{A}\sin(\phi) = X_m(t=0) \qquad (38)$$

$$\mathcal{A}\Big[\hat{\omega}_0 \cos(\phi) - \gamma \sin(\phi)\Big] = v(t=0)\ .$$

The constants $\mathcal{A}$ and ϕ can be found in terms of $X_m(t=0)$, $v(t=0)$, γ, and $\hat{\omega}_0$ using the two relations in Equation (38).

$$\phi = \arctan\left(\frac{\hat{\omega}_0 X_m(t=0)}{(v(t=0) + \gamma X_m(t=0))}\right),$$
$$\mathcal{A} = \frac{X_m(t=0)}{\sin(\phi)} \quad \text{or} \quad \mathcal{A} = \frac{v(t=0)}{\hat{\omega}_0 \cos(\phi) - \gamma \sin(\phi)}\ . \qquad (39)$$

We simulate the motion for $m = 1$ g and $\kappa = 1,000,000\ \pi^2$ dyne/cm for four different initial conditions. Note that $\mathcal{F}_0 = (1/2\pi)\sqrt{\kappa/m} = 500$ Hz, as for the undamped case discussed above. Two damping ratios are considered, $R = 0.065$ and $R = 0.35$. Figure 13 shows the resulting motion for two different initial conditions at the two damping ratios. In the top panel $X_m(t=0) = 0.0$ cm and $v(t=0) = 100\ \pi$ cm/s, with a damping ratio $R = 0.065$. In this situation, $\mathcal{A} = 0.1002$ cm, $\phi = 0.0$ rad, and $\hat{\mathcal{F}}_0 = 498.9$ Hz. The following panel shows the motion for the same initial conditions, but with $R = 0.35$. The initial phase remains

Acoustics of Speech Production

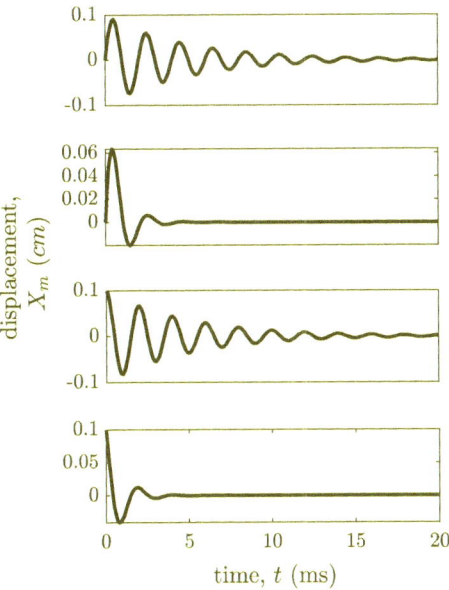

FIGURE 13. Friction damped mass-spring motion

unchanged at $\phi = 0.0$ rad, but now $\mathcal{A} = 0.107$ cm and $\hat{\mathcal{F}}_0 = 468.4$ Hz. In the panel above the bottom panel, $X_m(t=0) = 0.1$ cm and $v(t=0) = 0.0$ cm/s, again, with $R = 0.065$ for which $\mathcal{A} = 0.1002$ cm and $\phi = 0.52$ rad. The bottom panel has $X_m(t=0) = 0.1$ cm, $v(t=0) = 0.0$ cm/s, and $R = 0.35$ for which $\mathcal{A} = 0.107$ cm, $\phi = 0.61\,\pi$ rad, and $\hat{\mathcal{F}}_0 = 468.4$ Hz. The damping is apparent in all of these plots and is particularly strong for $R = 0.35$.

The forced, friction damped mass-spring system

When the friction damped mass-spring system is forced with an external force $F(t)$, the equation of motion and initial conditions at $t = 0$ become

$$\frac{\Delta_t^2 X_m(t)}{(\Delta t)^2} + 2\gamma \frac{\Delta_t X_m(t)}{\Delta t} + \omega_0^2 X_m(t) = \frac{F(t)}{m}\,, \tag{40}$$

with $X_m(t=0)$ and $v(t=0)$ specified. .

The general solution to Equation (40) is the sum of two functions, $f_I(t)$ and $f_H(t)$, just as for the solution of Equation (20). One function, $f_I(t)$, satisfies the first relation in Equation (40). The other function, $f_H(t)$, satisfies the first relation in Equation (3), or Equation (40), with $F(t) = 0$. $f_H(t)$ has the form given in the second expression of Equation (34). The constants $\mathcal{A}$ and ϕ are set when the initial conditions are applied to the sum of the two functions, $f_I(t) + f_H(t)$.

Suppose that a sinusoidal external force is applied to the damped system described in Equation (40), with initial conditions $X_m(t=0) = 0$ and $v(t=0) = 0$.

$$F(t) = F_0 \sin(\omega_F t) \text{ for } t > 0$$
$$\text{with } X_m(t=0) = 0 \text{ and } v(t=0) = 0, \tag{41}$$

where ω_F is the forcing circular frequency of the forcing. The solution can be written

$$X_m(t) = \frac{-F_0}{m\sqrt{(\omega_F^2 - \omega_0^2)^2 + 4(\gamma \omega_F)^2}} \sin(\omega_F t + \theta) + \mathcal{A} e^{-\gamma t} \sin(\hat{\omega}_0 t + \phi), \tag{42}$$

for $t > 0$, where $\theta = \arctan\left[\dfrac{2\gamma \omega_F}{(\omega_F^2 - \omega_0^2)}\right]$, $0 \leq \theta \leq \pi$.

Employing the initial conditions to solve for $\mathcal{A}$ and ϕ gives

$$X_m(t) = \frac{-F_0}{m\sqrt{(\omega_F^2 - \omega_0^2)^2 + 4(\gamma \omega_F)^2}} \times$$
$$\left[\sin(\omega_F t + \theta) - \beta e^{-\gamma t} \sin(\hat{\omega}_0 t + \phi)\right], \text{ for } t > 0, \tag{43}$$

where $\beta = \dfrac{\sin(\theta)}{\sin(\phi)}$, $\phi = \arctan\left[\dfrac{\gamma \tan(\theta)}{\omega_F}\right]$, and

$\theta = \arctan\left[\dfrac{2\gamma \omega_F}{(\omega_F^2 - \omega_0^2)}\right]$, $0 \leq \theta \leq \pi$.

The reader can verify that Equation (43) satisfies Equation (40) by using the identities in Equation (35). Also, the derivation of this solution can be found in almost any book on elementary differential

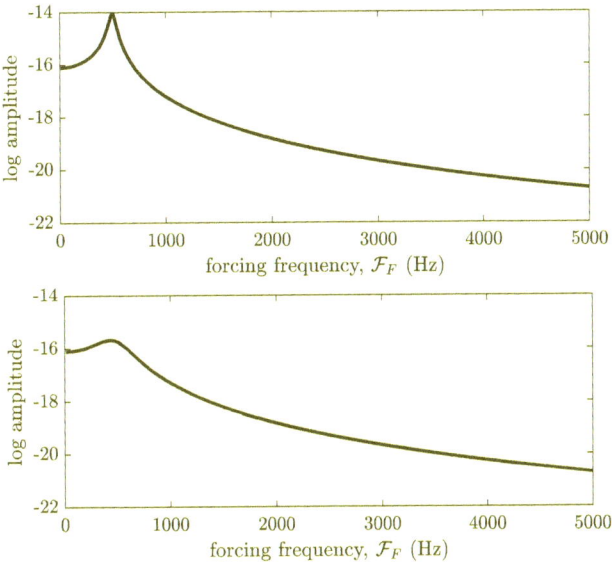

FIGURE 14. log amplitude versus forcing frequency $\mathcal{F}_F$

equations or physics (e.g. Braun 1983). The details of the final two lines in Equation (43) are not important for us now.

The factor $F_0/m\sqrt{(\omega_F^2 - \omega_0^2)^2 + 4(\gamma\omega_F)^2}$ gives the overall amplitude of the solution. Finally we show how the amplitude of vibration depends on forcing frequency $\mathcal{F}_F$. Figure 22 shows the logarithm of amplitude: $\left[\sqrt{(\omega_F^2 - \omega_0^2)^2 + 4(\gamma\omega_F)^2}\right]^{-1}$ as a function of $\mathcal{F}_F$ for two different damping ratios. The top panel is the log amplitude for $R = 0.065$, and the bottom is the log amplitude for $R = 0.35$. In both cases the amplitude is greatest when $\mathcal{F}_F = \mathcal{F}_0 = 500$ Hz. For the smaller damping ratio, $R = 0.065$, the amplitude peak is larger, but thinner than for the larger damping ratio, $R = 0.35$.

Equation (43) is a modification of Equation (31), where damping was not considered. The resulting motion is the sum of a sinusoid, $f_I(t)$, with the forcing frequency $\mathcal{F}_F = \omega_F/(2\pi)$ and the damped sinusoid of unforced motion $f_H(t)$, with reduced natural frequency, $\hat{\mathcal{F}}_0 = \hat{\omega}_0/(2\pi) = (1/2\pi)\sqrt{\kappa/m - \gamma^2} = (1/2\pi)\sqrt{\omega_0^2 - \gamma^2}$. Unlike the undamped solution in Equation (31), the amplitude of $X_m(t)$ remains finite, although relatively large, when $\mathcal{F}_F \approx \mathcal{F}_0$.

The second term in Equation (43), $f_H(t)$, which is associated with the reduced natural frequency, is exponentially damped for $\gamma > 0$. This part of the solution is called the *transient response*, or simply the *transient*, of the system. If there is no damping, the so-called transient does not decay with time. We previously saw that this means that it takes a long time for the system to attain steadiness, because there are low repetition frequencies without damping. With damping, the transient response diminishes exponentially in time. The time interval before the transient substantially dies away is known as the *transient phase*, and the time after that is the *steady-state phase* or *steady phase*. Note that the boundary between the two phases is qualitative and not defined here with a quantitative criterion. When the transient has been substantially damped after a sufficiently long time, steady conditions are attained.

$$X_m(t) \to \frac{-F_0}{m\sqrt{(\omega_F^2 - \omega_0^2)^2 + 4(\gamma\omega_F)^2}} \sin(\omega_F t + \theta)$$

$$v(t) = \frac{\Delta_t X_m(t)}{\Delta t} \to \frac{-\omega_F F_0}{m\sqrt{(\omega_F^2 - \omega_0^2)^2 + 4(\gamma\omega_F)^2}} \cos(\omega_F t + \theta) \quad (44)$$

as $t \to \infty$, and where $\theta = \arctan\left[\dfrac{2\gamma\omega_F}{(\omega_F^2 - \omega_0^2)}\right]$, $0 \leq \theta \leq \pi$.

Next we explore the properties of Equation (43), which includes the transient, using numerical simulations.

Simulations of a forced, damped mass-spring system

In the following we choose $R = 0.065$ and an undamped natural frequency $\mathcal{F}_0 = 500$ Hz. The reduced natural frequency is $\hat{\mathcal{F}}_0 = 498.9$ Hz. The forcing frequency is varied, so that $\mathcal{F}_F = 200, 470, 500,$ and 1200 Hz are considered in Figures 15 through 21. We can compare some of these to some of the simulations without damping shown in Figures 7, 10, and 11 for $\mathcal{F}_F = 200, 470,$ and 1200 Hz, respectively.

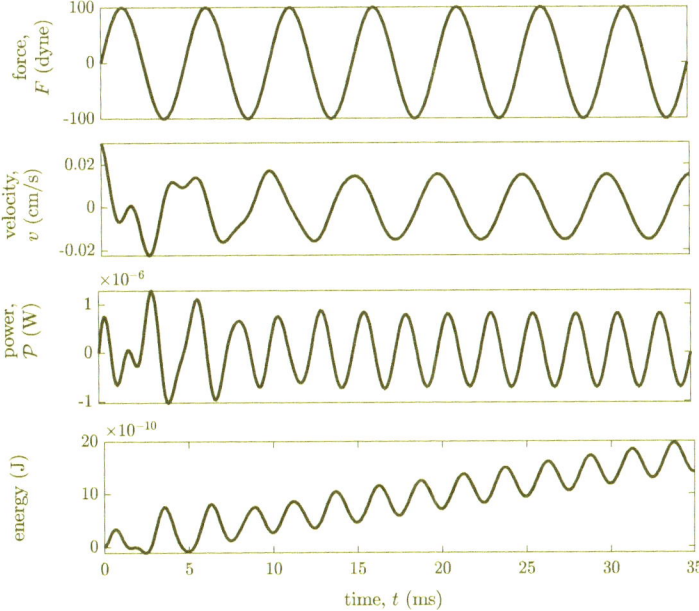

FIGURE 15. Forced, damped mass-spring system, $\mathcal{F}_F = 200$ Hz

Figure 15 shows the results for the sinusoidally forced damped mass-spring system with $\mathcal{F}_F = 200$ Hz. This is to be compared to the undamped case shown in Figure 7. The comparison shows the important effect that even a small amount of damping can have on the time evolution of the forced mass-spring system. The situation in Figure 15 appears to become steady after 10 to 15 ms in the sense that the velocity of the mass in the second panel and the power input by the external force in the third panel exhibit periodic motion. The periodicity of the position and velocity has the same frequency as the forcing frequency $\mathcal{F}_F$. Thus, the transient substantially dies out by 15 ms after the start of forcing, so that only the first term in Equation (43), with a sinusoid at the forcing frequency becoming very dominant.

After the transient dies away, the power input shown in the third panel of Figure 15 is twice the frequency of the the forcing frequency. This is because power $\mathcal{P}(t)$ is the product of two quantities that are oscillating at the forcing frequency: external force $F(t)$ and velocity $v(t)$ shown in the top two panels. Just as in the undamped case

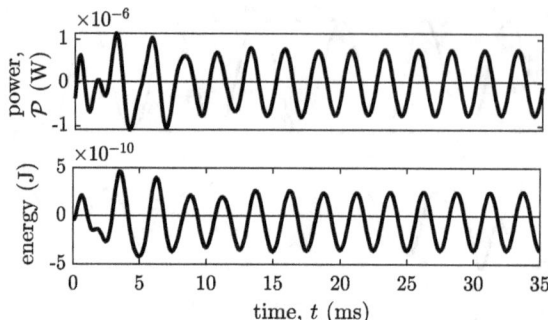

FIGURE 16. Energy flow subtracting the energy lost due to friction damping in Figure 15

in Figure 7, the external force acts both as a source and a sink of energy for the mass-spring system. However, there is a very important difference between the undamped and damped systems that becomes obvious when comparing the bottom panels of Figures 7 and 15. In the damped case, the external force is more of a source than a sink over one oscillation. This means that the mass-spring system gains more energy from the external force than it loses to the external force over a cycle. Thus, the bottom panel of Figure 15 shows a linear trend in time with oscillation, which means that there is a consistent rate of energy input to the system. This is not the case for the undamped system in Figure 7. All of this makes sense, as we know that friction removes energy from the mass-spring system, so in order that the system to continue moving it must receive energy from the external force over time. Indeed, Figure 16 illustrates the consequences of accounting for the power input and the power dissipated as a sum, or total power in the steady phase. The top panel now shows equal amounts of positive and negative total power over the power oscillation period. It is seen in the bottom panel, that accounting for the energy expended in dissipation, removes the linear trend seen for energy input seen in Figure 15. The net energy input oscillates about a negative constant level after the transient has died away. Apparently, there is a net energy loss during the transient phase that is never recovered during the steady phase.

These phenomena are illustrated again for a forcing frequency of $\mathcal{F}_F = 470$ Hz in Figures 17 and 18. There is a net loss of energy during the transient phase that is sustained in the steady phase, as

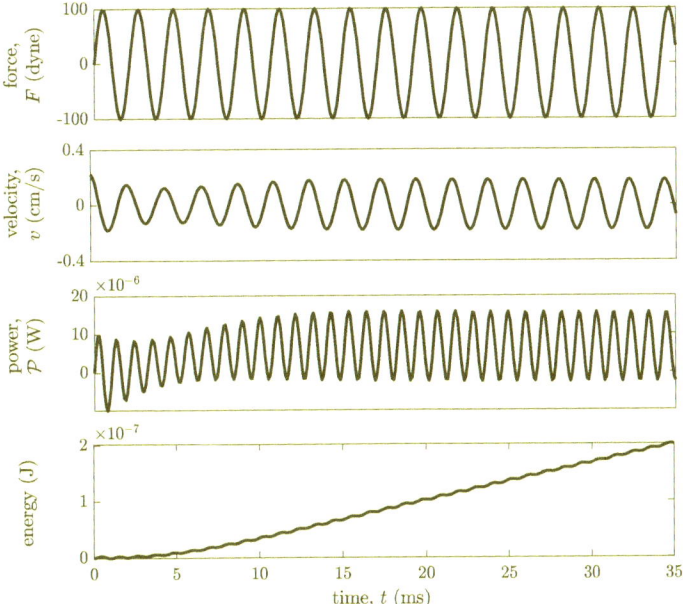

FIGURE 17. Forced, damped mass-spring system, $\mathcal{F}_F = 470$ Hz

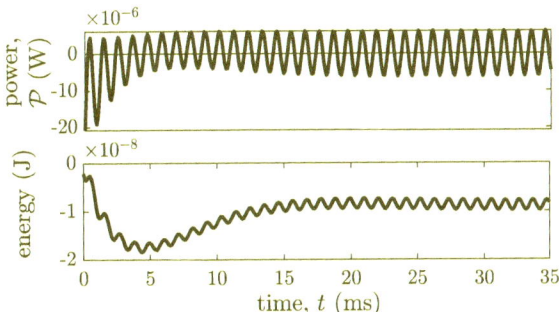

FIGURE 18. Energy flow subtracting the energy lost due friction damping in Figure 17

seen in the bottom panel of Figure 17. Most remarkably though, is the fact that the 30 Hz modulation, or beating, seen for the case with no damping in Figure 10 is almost completely absent, as the transient is significant for only 10 to 15 ms. During the steady phase, the power

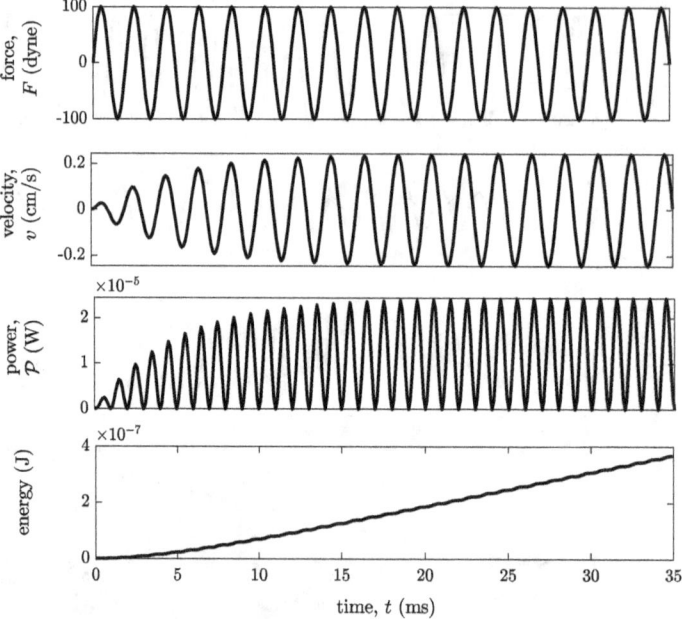

FIGURE 19. Forced, damped mass-spring system, $\mathcal{F}_F = 500$ Hz

input by the external force is mostly positive; more so than for the $\mathcal{F}_F = 200$ Hz case. [Compare the third panels from the top in Figures 15 and 17.] In fact, the energy input by the external force is much greater at the end of 35 ms for $\mathcal{F}_F = 470$ Hz than for $\mathcal{F}_F = 200$ Hz. [Compare the bottom panels of Figures 15 and 17.] This must mean that energy is being dissipated at a faster rate for the $\mathcal{F}_F = 470$ Hz situation than for the $\mathcal{F}_F = 200$ Hz situation. This is plausible because the mass velocity amplitudes for the former system are so much larger than for the latter system, and the friction force is proportional to velocity. [Compare the second panels from the top in Figures 15 and 17.)] The amplitudes are large because the system is being forced at a frequency, $\mathcal{F}_F = 470$ Hz close to its natural frequency $\mathcal{F}_0 = 500$ Hz.

When damping is brought into the system it is possible to have the forcing frequency equal the natural frequency of the system without having the system "blow up". Figures 19 and 20 show the quantities that we have been considering for the case when $\mathcal{F}_F = 500$ Hz, which is just above the reduced natural frequency of $\hat{\mathcal{F}}_0 = 498.9$ Hz and at

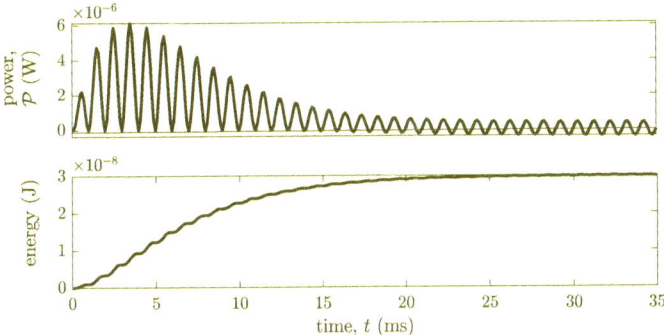

FIGURE 20. Energy flow subtracting the energy lost due friction damping in Figure 19

the natural frequency $\mathcal{F}_0 = 500$ Hz. Figure 19 for $\mathcal{F}_F = 500$ Hz is similar to Figure 17 for $\mathcal{F}_F = 470$ Hz, except that the power input in the third panel is always greater than or equal to zero in the former case. Thus, during the transient phase, the system is gaining energy, and, as the second panel of Figure 20 shows there is a net energy gain by the system at all times. Finally, Figures 21 and 22 show quantities of interest for forcing frequency $\mathcal{F}_F = 1200$ Hz. Power input is positive during the transient phase, but the rate of energy input by the external force is very small during the steady phase compared to forcing at a frequency close to $\mathcal{F}_0$. [Compare the bottom panels of Figures 19 and 21.] The amplitude modulation seen in the third panel of Figure 21 and the first panel of Figure 22 appears to be a numerical artifact.

Numerical simulations of the finite-length tube with sinusoidal piston motion and damping

In this section of the chapter, numerical simulations of the acoustics of a finite-length tube driven by a sinusoidal piston movement with damping are shown. This is done to show the similarities between the mass-spring system and the acoustic system. We do not present the methods used for the simulations here, but only the results. Energy loss occurs at every time step of the numerical simulation with the same damping ratio used for the forced, damped mass-spring system above, namely with $R = 0.065$.

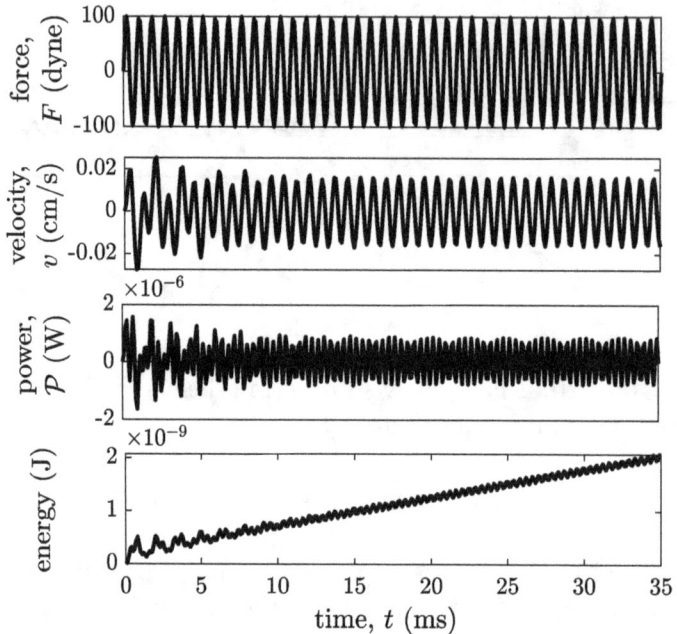

FIGURE 21. Forced, damped mass-spring system, $\mathcal{F}_F = 1200$ Hz

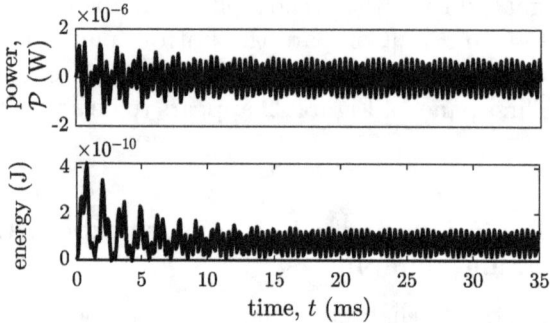

FIGURE 22. Energy flow subtracting the energy lost due friction damping in Figure 21

We first consider the case when $\mathcal{F}_{pst} = 200$ Hz. Figure 23 shows the velocity of the piston, the pressure at the piston, the intensity at the piston, and the net energy input by the piston in four panels, top to

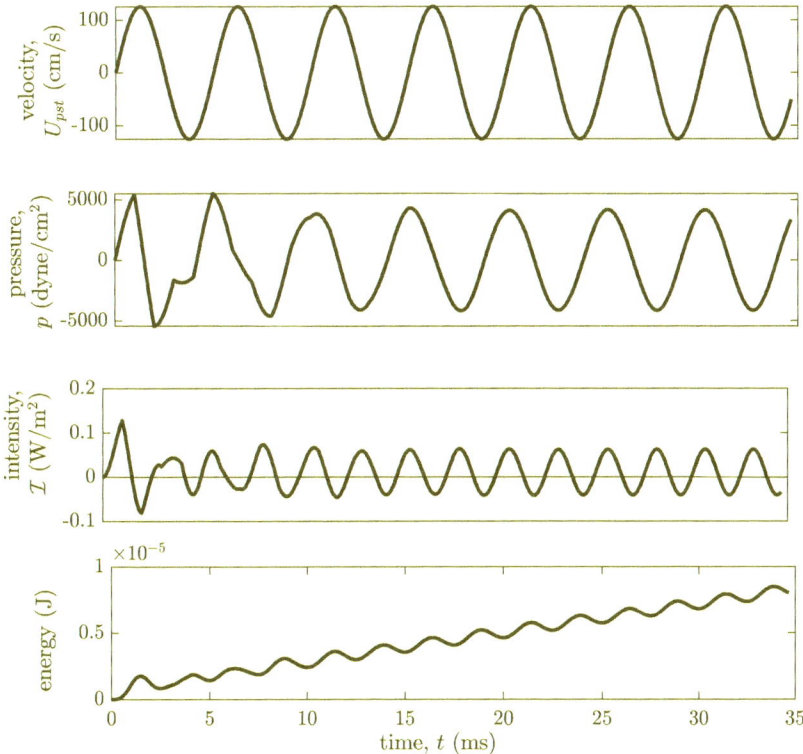

FIGURE 23. Finite-length tube with sinusoidal piston motion and damping, $\mathcal{F}_{pst} = 200$ Hz

bottom. The corresponding panels in Figure 14 for the forced, damped mass-spring system look qualitatively similar, except for the details in the transient phases. The transient phase for the finite-length tube also lasts about 10 to 15 ms.

The same can be said when comparing the responses of the damped mass-spring and damped acoustic systems at the same input frequencies: at $\mathcal{F}_F = \mathcal{F}_{pst} = 470$ Hz (Figures 16 and 24), at $\mathcal{F}_F = \mathcal{F}_{pst} = 500$ Hz (Figures 18 and 25), and at $\mathcal{F}_F = \mathcal{F}_{pst} = 1200$ Hz (Figures 20 and 26). It should be noted that we do not control for the increase in piston velocity input amplitude that is the result of increased piston frequency for the acoustic case, while the external force input for the mass-spring system was maintained at constant amplitude

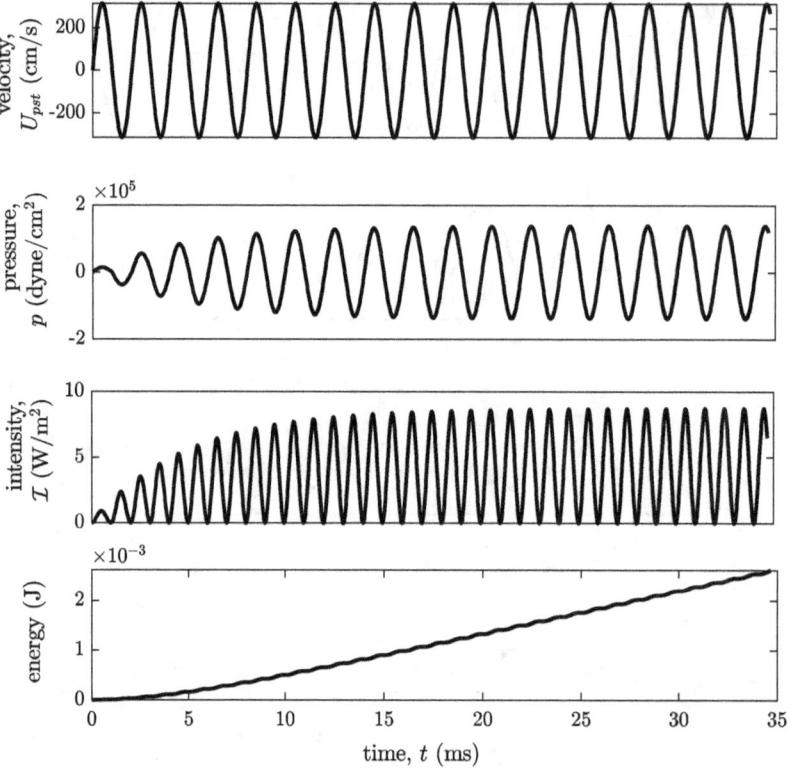

FIGURE 24. Finite-length tube with sinusoidal piston motion and damping, $\mathcal{F}_{pst} = 470$ Hz

for all forcing frequencies. Even with a larger source amplitude in the finite-length tube for $\mathcal{F}_{pst} = 1200$ Hz, than for $\mathcal{F}_{pst} = 500$ Hz (top panels of Figures 25 and 26), the amplitude of the pressure oscillations at the piston are much greater for 500 Hz than for 1200 Hz (second panels in Figures 25 and 26). Further, the piston inputs energy to the acoustic motion at a faster rate at 500 Hz than at 1200 Hz (bottom panels in Figures 25 and 26).

It appears that damping is essential in order for both mass-spring and acoustic systems to attain steadiness in brief durations when they are driven by a source. Before finishing this chapter, in the next section we briefly discuss damping in the mass-spring system in general.

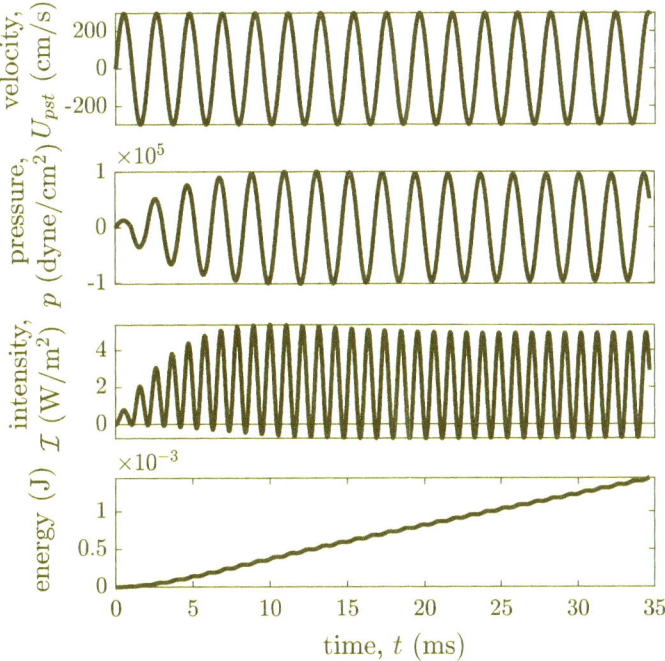

FIGURE 25. Finite-length tube with sinusoidal piston motion and damping, $\mathcal{F}_{pst} = 500$ Hz

The mass-spring system with generalized damping

Equation (40), the equation of motion, is a good mathematical model for the mass-spring system that is lightly damped by friction. Friction results in a damping constant γ that is the same for all natural circular frequencies ω_0. However, not all damping mechanisms are so simple, as we eventually see for the damping of acoustic waves. Also, for the mass-spring system, it is likely that there are damping mechanisms that do depend on the natural frequency, as well as forcing frequency. For example, we could imagine that the internal friction produced by the relative motion of pieces of the spring itself could well depend on the mass-spring system's frequency of motion. To find an equation of motion that includes this effect would be a large enterprise with, perhaps, limited applicability.

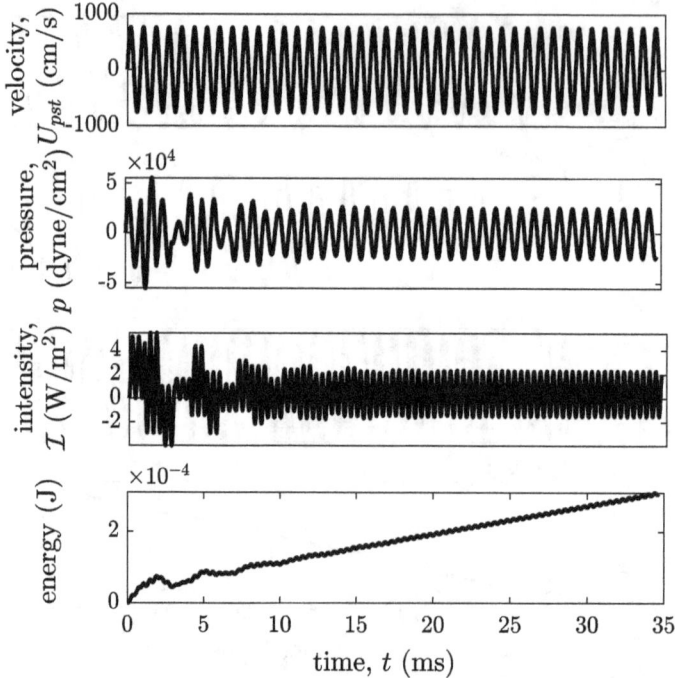

FIGURE 26. Finite-length tube with sinusoidal piston motion and damping, $\mathcal{F}_{pst} = 1200$ Hz

Thus, it is often important to employ the impulse response function to characterize a linear system with constant parameters rather than an equation of motion. We just adjust the damping constant γ that appears in the impulse response function in Equation (36) according to frequency, rather than attempting to find the equation of motion to account for all significant damping mechanisms. Empirical measures of impulse response functions, including the parameter γ, can be made.

Suppose we are able to infer damping as a function of forcing frequencies, i.e. $\gamma = \gamma(\omega_F)$ without knowledge of how the equation of motion should be written. For this more general damping, Equation (36) becomes

$$\mathcal{G}(t) = \frac{1}{m\omega_0'}\mathcal{H}(t)e^{-\gamma t}\sin(\omega_0' t) \ . \tag{45}$$

Equation (45) has the same form as Equation (36) for the particular case of frictional damping. The differences between the two equations is that now $\gamma = \gamma(\omega_F)$, and the *perturbed natural circular frequency*, ω_0', is not necessarily the reduced natural circular frequency $\hat{\omega}_0 = \sqrt{\omega_0^2 - \gamma^2}$. We return to reduced natural frequencies in conjunction with acoustics in Chapters 7 and 11.

Conclusion

We have examined mass-spring systems in detail. Initial value problems, impulse response functions, forced systems, and forced systems with damping were all studied in the present chapter. We found that the sinusoidally forced, undamped mass-spring system possesses similar behavior to the undamped finite-length acoustic tube with a sinusoidally moving piston. In particular, steadiness takes a long duration to establish in both cases. When damping is added, this is no longer the case for either the mass-spring system or the acoustics of the finite-length tube. We saw how this works mathematically for the mass-spring system. After the next three chapters, it will be seen how it works for the acoustics of the finite-length tube. When mass-spring systems are referred to further in the book, it is conjunction with the finite-length tube, and the context should make it clear whether a damped or undamped system is being referred to.

One of the keys to understanding the relation between mass-spring systems and finite-length tube acoustics is a mathematical theory that we introduce in Chapter 5. In this theory, acoustic wave motion is represented by a sum of waves such that each term in the sum has a mathematical representation that is analogous to a mass-spring system. This goes well beyond the Helmholtz resonator introduced in Chapter 1, which is, literally, a mass-spring system made from air, but has little to do with acoustic waves in general.

Along the way to completing the picture of finite-length tube acoustics as a sum of systems that are analogues to mass-spring systems, we discuss the analogue to the impulse response function, which is called the *Green's function* for the acoustical system. This is done in Chapter 6. Loosely, we can think of a Green's function as an impulse response function that takes account of space, as well as time. Importantly, like the fact that an impulse response function characterizes its mass-spring system as well as its equation of motion does, the Green's function characterizes its acoustic system.

With the Green's function results of Chapter 6, we examine, in Chapter 7, the steady behavior of pressure and particle velocity in the finite-length tube when the piston moves sinusoidally and damping is present. We use Green's functions where the damping constants have been determined as a function of frequency of the piston. It is only later, in Chapter 11, that we provide some theoretical derivations for the damping of acoustic wave motion in tubes. It should be clear at the end of the discussion in Chapter 7 that we have described much of fundamental *source-filter theory* in speech acoustics.

References

Braun, M. (1983). *Differential Equations and Their Applications*. Springer-Verlag, New York. (pp 163-9).

Chapter 5: Standing Waves and Normal Modes

Introduction

We return to acoustic wave motion, or *acoustic propagation*, in a tube without damping and without acoustic sources or sinks. In particular, the piston, which was used in Chapters 2 and 3, is assumed to be stationary. Initially, we return to the idea that we follow an acoustic disturbance as it propagates in a tube of infinite length, as in Chapter 2, but we spend most of the present chapter on the tube of finite length introduced in Chapter 3. By the time we have finished Chapter 5, it should be clear that acoustic waves in the finite-length tube can be described using an infinite number of systems that share much in common with mass-spring systems.

In Chapter 2 it was demonstrated that disturbances propagate in both the positive x-direction and the negative x-direction without change of shape under the acoustic approximation. The speed of propagation is c_0, so that we obtain a disturbance of the form $f(t - x/c_0)$ traveling to the right and one of the form $g(t + x/c_0)$ traveling to the left. These are called *traveling waves*. This remains the same in a finite-length tube, except that reflected waves are generated at the boundaries at $x = 0$ and at $x = L$. Examples in Chapter 3 make it clear that it becomes unwieldy to follow all of the component waves that are generated as the number of reflections increases with time t. This is one reason that we now introduce a different representation in terms of something called *standing waves*. Standing waves are not simply a convenient mathematical representation, but they are physical phenomena that can be observed with acoustic waves in tubes of finite length. Standing waves appear as steady patterns of spatial pressure and particle velocity variation that can occur with acoustic propagation. Again, steadiness here means periodic and not static. Standing waves are formed under certain conditions when right-going and left-going waves interfere with one another through linear superposition. The finite-length tube is ripe for standing waves with its reflected waves.

One other condition should be in place before we obtain standing waves in the finite-length tube: the functions describing the traveling waves should be sinusoids, or circular functions. This may appear to be a severe restriction, but mathematics rescues us to show that requiring circular functions is actually not very restrictive. Under very general conditions, functions describing acoustic wave motion in a finite-length tube can be written as a sum of other function involving only circular functions. And this is enough to make the standing wave representation in a finite-length tube very general indeed.

In fact, acoustic waves in the finite-length tube can be written as a sum of standing waves, each behaving in a mass-spring-like way, except for one important property: the kinetic and potential energies of acoustic waves are distributed throughout the tube. This is in contrast to a mass-spring system, where the kinetic energy is a property of the moving mass, and the potential energy is a property of the extension of the spring.

Trigonometric identities

We discuss trigonometric identities, so that we may use them here and in later chapters. The proofs of these identities are best done in the context of *complex variables*, which are introduced in Chapter 8.

The following identities regarding the sum and differences of angles are simply stated.

$$\begin{aligned}
\sin(\theta + \phi) &= \sin(\theta)\cos(\phi) + \cos(\theta)\sin(\phi) \\
\sin(\theta - \phi) &= \sin(\theta)\cos(\phi) - \cos(\theta)\sin(\phi) \\
\cos(\theta + \phi) &= \cos(\theta)\cos(\phi) - \sin(\theta)\sin(\phi) \\
\cos(\theta - \phi) &= \cos(\theta)\cos(\phi) + \sin(\theta)\sin(\phi) \,.
\end{aligned} \quad (1)$$

These identities can be used to derive other identities, such as

$$\begin{aligned}
\sin(\theta)\cos(\phi) &= \frac{1}{2}\bigl(\sin(\theta + \phi) + \sin(\theta - \phi)\bigr) \\
\cos(\theta)\cos(\phi) &= \frac{1}{2}\bigl(\cos(\theta + \phi) + \cos(\theta - \phi)\bigr) \\
\sin(\theta)\sin(\phi) &= \frac{1}{2}\bigl(\cos(\theta - \phi) - \cos(\theta + \phi)\bigr) \,.
\end{aligned} \quad (2)$$

We can also relate the sine and cosine functions using the identities in Equation (1). For example,

$$\sin(\theta + \pi/2) = \sin(\theta)\cos(\pi/2) + \cos(\theta)\sin(\pi/2) \\ = \cos(\theta), \quad (3)$$

because $\cos(\pi/2) = 0$ and $\sin(\pi/2) = 1$. Similarly,

$$\cos(\theta - \pi/2) = \cos(\theta)\cos(\pi/2) + \sin(\theta)\sin(\pi/2) \\ = \sin(\theta). \quad (4)$$

Thus, one can always express the cosine function in terms of the sine function, and vice versa.

Frequency, wavenumber, period, and wavelength

We examine the behavior of traveling sinusoidal pressure disturbances without regard to boundaries for now. Thus, we return to the tube of infinite length in Chapter 2 briefly. Let P have the units of pressure. Right-going and left-going sinusoidal pressure waves can be written

$$p(x,t) = P\sin\left(\omega(t \pm x/c_0) + \phi\right) \\ = P\sin\left(\omega t \pm (\omega/c_0)x + \phi\right) \\ = P\sin\left(\omega t \pm kx + \phi\right), \quad (5)$$

where $k = \omega/c_o$.

The minus sign is for the right-going wave, and the plus sign for the left-going wave. k is known as the *wavenumber*, and it has dimensions of cm^{-1}, but we state it in units of rad/cm. ϕ is a constant.

We fix time t and consider variation in the spatial variable, position, x. There is no change in pressure value if kx changes by 2π. What is the distance that we need to travel in order that kx changes by 2π? That distance, which is referred to as *wavelength*, is denoted λ. Wavelength is illustrated in Figure 1, for a frequency $\mathcal{F} = \omega/(2\pi) = 1000\ Hz$. The wavelength is the distance it takes for the circular function to start to repeat itself for a fixed time. So $2\pi = k\lambda$, or $\lambda = 2\pi/k$. The wavenumber k is the spatial analogue to circular frequency ω, and λ is the spatial analogue to period T. As is shown in Figure 1, it does not matter where in space we start to measure that change in pattern, we always come up with the same value for λ. It is verified in Figure 1 that $\lambda = 2\pi/k = c_0/\mathcal{F} = (34,100/1000)$ cm ≈ 34 cm.

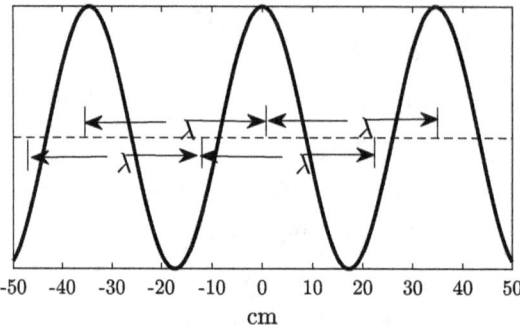

FIGURE 1. Wavelength

We define $\theta^{time} \equiv \omega t + \phi$ and $\theta^{space} \equiv kx$. [We have chosen to include the constant ϕ into θ^{time} somewhat arbitrarily. It turns out to be the most convenient choice for our purposes. With this choice, ϕ is initial phase.] Equation (5) can then be written

$$\begin{aligned} p(x,t) &= P \sin\left(\omega(t \pm x/c_0) + \phi\right) \\ &= P \sin(\omega t \pm kx + \phi) \\ &= P \sin(\theta^{time} \pm \theta^{space}) \\ &= P \sin(\theta^{total}) \,, \end{aligned} \qquad (6)$$

where $\theta^{total} = \theta^{time} \pm \theta^{space}$ is the phase of the traveling wave. Phase has two components: one is the phase that depends on time, θ^{time}, and the other is the phase that depends on spatial position, θ^{space}.

We can also rewrite the expression for the pressure wave in Equation (6) as

$$\begin{aligned} p(x,t) &= P \sin\left(\omega(t \pm x/c_0) + \phi\right) \\ &= P \sin(\omega t \pm kx + \phi) \\ &= P \sin\left(\frac{2\pi}{T} t \pm \frac{2\pi}{\lambda} x + \phi\right). \end{aligned} \qquad (7)$$

The final expression shows that wavelength λ is, indeed, the spatial analogue to period T. Strictly speaking k is acoustic wavenumber and λ is acoustic wavelength. There are other wavenumbers and wavelengths in fluid mechanics, and we run into some in Chapters 13 and 15.

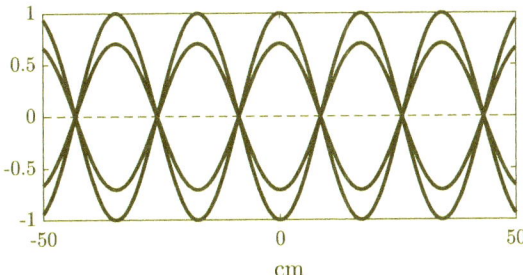

FIGURE 2. Standing waves over time

Traveling and standing waves

Traveling and standing wave representations

We consider a wave of the form $p(x,t) = P\sin(\omega t - kx)$. This wave is of infinite extent in both time and space, and it is traveling to the right. In order for it to satisfy the wave equation, Equation (16) of Chapter 2, it is necessary that $k = \omega/c_0$. We can use the second identity in Equation (1) to write,

$$p(x,t) = P\sin(\omega t - kx + \phi)$$
$$= P[\sin(\omega t + \phi)\cos(kx) - \cos(\omega t + \phi)\sin(kx)],$$

or (8)

$$p(x,t) = P\sin(\theta^{time} - \theta^{space})$$
$$= P[\sin(\theta^{time})\cos(\theta^{space}) - \cos(\theta^{time})\sin(\theta^{space})].$$

$\sin(\omega t + \phi)\cos(kx)$ and $\cos(\omega t + \phi)\sin(kx)$ are examples of standing waves because each describes a pattern in space that is modulated by a function of time. Mathematically, they have the form $f(t)g(x)$. Further, it is shown below that both $\sin(\omega t + \phi)\cos(kx)$ and $\cos(\omega t + \phi)\sin(kx)$ satisfy the wave equation when $k = \omega/c_0$.

Figures 2 and 3 show the first of these standing waves, $\sin(\omega t)\cos(kx)$, at various times, where the times are chosen so that $\omega t = 0$, $\pi/4$, $\pi/2$, $3\pi/4$, π, $5\pi/4$, $3\pi/2$, and $7\pi/4$. Figure 2 shows these time frames all together. It is clear that there are points on the x-axis that correspond to places where there is no variation in time. These are known as *nodes*. The points on the x-axis where there is maximum

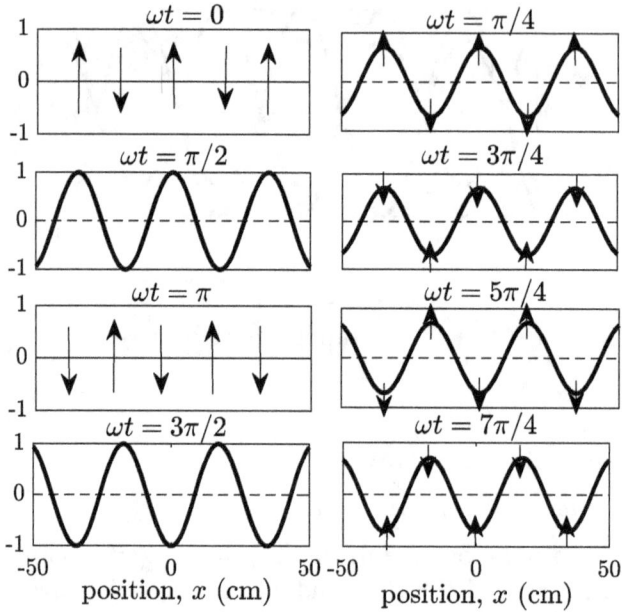

FIGURE 3. Standing waves over time in separate panels

movement, or maximum amplitude, are known as *anti-nodes*. Figure 3 shows the time frames separately. The arrows in each frame illustrate the magnitude and direction of the time rate-of-change of the amplitude of the standing wave at the anti-nodes.

It is clear that standing waves, which are products of circular functions, can be written as sums of traveling waves as well. For instance, it follows from the identities in Equation (2) that

$$\sin(\omega t)\cos(kx) = \frac{1}{2}\big(\sin(\omega t + kx) + \sin(\omega t - kx)\big)$$
$$\cos(\omega t)\sin(kx) = \frac{1}{2}\big(\sin(\omega t + kx) - \sin(\omega t - kx)\big). \quad (9)$$

These standing waves are the result of adding traveling waves of equal amplitude: one traveling to the right and the other traveling to the left. Further, because both standing waves can be written as the sum of two traveling waves, each of which satisfy the wave equation [as long as $k = \omega/c_0$], the standing waves also satisfy the wave equation by the property of superposition.

Equation (9) shows that sinusoidal standing waves result from superposition, or interference, of traveling waves. Equation (8) shows that traveling waves result from interfering standing waves. In sum, the relationships between traveling and standing waves are just a matter of trigonometric identities when it comes to circular functions.

Why choose a standing wave or a traveling wave representation?

Standing waves can be written as a sum of traveling waves, as shown in Equation (9), and traveling waves can be written as a sum of standing waves, as shown in Equation (8). Are there reasons for choosing one representation over another? In the case of the finite-length acoustic tube we generally use the standing wave representation. The special values of circular frequency are examined in the next section.

A standing wave representation is employed in the finite-length tube, because the opening at the right end of the tube causes strong reflection of waves traveling to the right, and the piston causes strong reflection of waves traveling to the left. The wave that is being reflected is called the *incident wave*, which is traveling right in this case. The reflected wave travels left in this case. Strong reflection of waves means a wave of similar amplitude as the incident wave, but traveling in the opposite direction, is created. With the current boundary conditions, we obtain total reflection. In other words, with the current boundary conditions given by Equations (2) and (7) of Chapter 3 total reflection occurs: the boundary conditions do not allow for leakage of acoustic energy out of the tube. This means that the right-going and left-going waves posses the same amplitude. Using pressure transducers along the length of a tube of finite length, we would measure the variation in pressure similar to that shown in Figure 3, again, assuming that the frequencies have special values.

Later in the book, we permit the situation where total refection does not occur. Even with this modification, reflected left-going waves possess nearly the same amplitudes as right-going waves. In the case of non-total reflection we can add a small traveling wave to the standing wave.

Normal modes

Basics of normal modes

We now consider the details of representing the wave propagation in the finite-length tube with sinusoidal standing waves. The boundary conditions at $x = 0$ and $x = L$ select the standing waves that occur. When the piston is not moving, the boundary conditions given in Equations (2) and (7) of Chapter 3 are

$$u(x = 0, t) = 0$$
$$p(x = L, t) = 0 \ . \qquad (10)$$

It is not very convenient to work with both particle velocity u and pressure p when finding a standing wave solution. Here we choose pressure p as the quantity of interest. Once $p(x,t)$ is found, $u(x,t)$ can be found using the conservation equations in Chapter 2. The first boundary condition in Equation (10) can be rewritten in terms of pressure using the momentum conservation equation, Equation (11) of Chapter 2.

$$\frac{\Delta_t u(x,t)}{\Delta t} = -\frac{1}{\rho_0} \frac{\Delta_x p(x,t)}{\Delta x} \ , \qquad (11)$$

where both $\Delta t > 0$ and $\Delta x > 0$ are small in some sense. If we consider $x = 0$, then for any time t, it follows by the first relation in Equation (10) and Equation (11) that

$$0 = \frac{\Delta_t u}{\Delta t}(x = 0, t) = -\frac{1}{\rho_0} \frac{\Delta_x p}{\Delta x}(x = 0, t) \ . \qquad (12)$$

This says that the gradient of pressure at $x = 0$ is zero at all times. Thus, the boundary conditions in Equation (10) can be rewritten

$$\frac{\Delta_x p}{\Delta x}(x = 0, t) = 0$$
$$p(x = L, t) = 0 \ . \qquad (13)$$

We assume that p is a standing wave with $p(x,t) = P \sin(\omega t + \phi) g(x)$, where $g(x)$ is a circular function, such as $\sin(kx + \theta)$, and P is a constant with units of pressure. With this assumption, we explore the properties that the function $g(x)$ must have in order for $p(x,t)$ to satisfy both the wave equation and the boundary conditions in Equation (13).

Consider the possibility that $g(x)$ is the cosine function, $\cos(kx)$. The gradient of $\cos(kx)$ goes from positive to negative as x goes from negative to positive through zero. In fact, the gradient of the cosine function is $-k\sin(kx)$, from Equation (13) of Chapter 3. The gradient of $\cos(kx)$ is zero at $x = 0$. Thus, with $g(x) = \cos(kx)$, the boundary condition at $x = 0$ in Equation (13) is satisfied.

How do we satisfy the second boundary condition in Equation (13)? With $g(x) = \cos(kx)$

$$p(x = L, t) = P\sin(\omega t + \phi)\cos(kL) = 0 \tag{14}$$

$$\Rightarrow \cos(kL) = 0.$$

The solutions to this equation are $kL = \pm\pi/2, \pm 3\pi/2, \pm 5\pi/2, \ldots$ In other words, the values of kL must be restricted to odd integer multiples of $\pi/2$. Thus, the values of k are restricted to

$$k_m = \frac{(2m-1)\pi}{2L}. \tag{15}$$

for integer m. It turns out to be redundant to use negative integers m, and we need only consider the positive integers $m = 1, 2, 3, \ldots$. Therefore, as long as the wavenumber is restricted to be k_m for positive integer m, we have satisfied both boundary conditions with $p(x,t) = P\sin(\omega t + \phi)\cos(k_m x)$. However, in order that $P\cos(k_m x)\sin(\omega t + \phi)$ satisfy the wave equation, particular circular frequencies must be coupled to particular wavenumbers as

$$\omega_m = c_0 k_m. \tag{16}$$

With the values of k and ω restricted by Equations 15 and 16, we have a set of standing waves built from circular functions that satisfy the wave equation and the boundary conditions for the finite-length tube.

$$p_m(x,t) = P_m \sin(\omega_m t + \phi_m)\cos(k_m x)$$
$$\equiv p_m^{time}(t) \cdot p_m^{space}(x), \tag{17}$$

where $p_m^{time}(t) = \sin(\omega_m t + \phi_m)$ and $p_m^{space}(x) = P_m \cos(k_m x)$,

and k_m and ω_m are specified in Equations (15) and (16) for positive integer m. The amplitude P_m and initial phase ϕ_m are constants that are determined by initial conditions. The $p_m(x,t)$, with $P_m = 1$, are *normal modes* for this particular finite length tube of length L. The $p_m(x,t)$ are also referred to as *eigenfunctions*. Each the functions of time, $p_m^{time}(t)$, and the functions of position, $p_m^{space}(x)$, are often referred to as eigenfunctions. The parameters of the normal modes, *normal mode wavenumber* k_m and *normal mode circular frequency* ω_m, are referred to as *eigenvalues* of the eigenfunctions.

What are the wavelengths, that correspond to the eigenvalues k_m? We rewrite Equation (15) in terms of wavelengths λ_m.

$$\lambda_m \equiv \frac{2\pi}{k_m} = \frac{4L}{2m-1},$$

or (18)

$$\lambda_1 = 4L, \quad \lambda_2 = \frac{4L}{3}, \quad \lambda_3 = \frac{4L}{5}, \ldots,$$

for $m = 1, 2, 3, \ldots$. This implies, of course, that

$$L = \frac{2m-1}{4}\lambda_m,$$

or (19)

$$L = \frac{\lambda_1}{4}, \quad L = \frac{3\lambda_2}{4}, \quad L = \frac{5\lambda_3}{4}, \ldots.$$

Equation (19) is the reason that the finite-length tube that we wave been considering is known as a *quarter-wave resonator*: the first normal mode possesses a wavelength equal to four times the length of the tube, or the tube has a length that is one-quarter of the wavelength of the first normal mode. Figure 4 shows the spatial dependence of the amplitude for the first eight normal modes, $\cos(k_m x)$, for $m = 1, 2, \ldots 8$. From Equations (15) and (16)

$$\frac{\mathcal{F}_m}{\mathcal{F}_1} = \frac{\omega_m}{\omega_1} = 2m - 1, \qquad (20)$$

for $m = 1, 2, 3, \ldots$ *Normal mode frequencies* are odd, positive integral multiples of the frequency of the first normal mode. What are the frequencies of the normal modes for the finite-length tube of length, $L = 17.25$ cm?

$$\begin{aligned}\mathcal{F}_m = \frac{\omega_m}{2\pi} &= \frac{c_0 k_m}{2\pi} = \frac{(2m-1)c_0}{4L} \\ &= \frac{(2m-1)34,100 \text{ cm/s}}{(4)(17.05) \text{ cm}} \\ &= (2m-1)500 \text{ Hz},\end{aligned} \qquad (21)$$

for $m = 1, 2, 3, \ldots$. So, the first normal mode frequency is 500 Hz and the higher normal mode frequencies are 1500 Hz, 2500 Hz, 3500 Hz, and so on.

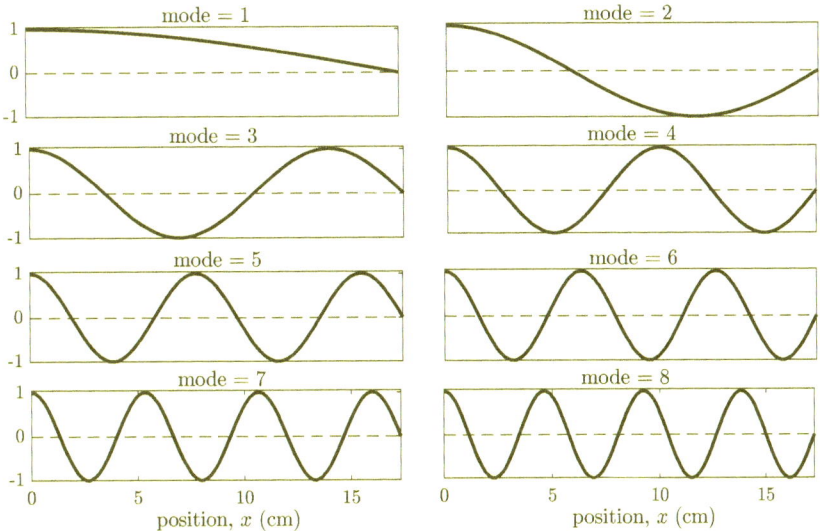

FIGURE 4. Normal modes for pressure

The lowest normal mode frequency, 500 Hz, is the repetition frequency of the brief pulse of Chapter 3. Also, the reader may recognize that the eigenvalues $\mathcal{F}_m$ are related to the so-called *formant frequencies* for the neutral vocal tract in speech production. This is the case, although their values are modified slightly by other physical processes. We emphasize that formant frequencies are associated with particular spatial distributions of pressure and particle velocity within the vocal tract.

An alternative method for finding normal modes

Now we describe a slightly different way to derive the normal modes. This alternative method is useful for future study of normal modes. We assume that p is a standing wave with $p(x,t) = P\sin(\omega t + \phi)g(x)$, where $g(x) = \cos(kx)$, $k = \omega/c_0$. We know that this function satisfies the first boundary condition in Equation (13). With $k = \omega/c_0$, the proposed $p(x,t)$ also satisfies the wave equation.

We examine what occurs to $g(x) = \cos(kx)$ at $x = L$ as $\mathcal{F} = \omega/(2\pi)$ increases from 440 Hz to 580 Hz in a series of panels shown in Figure

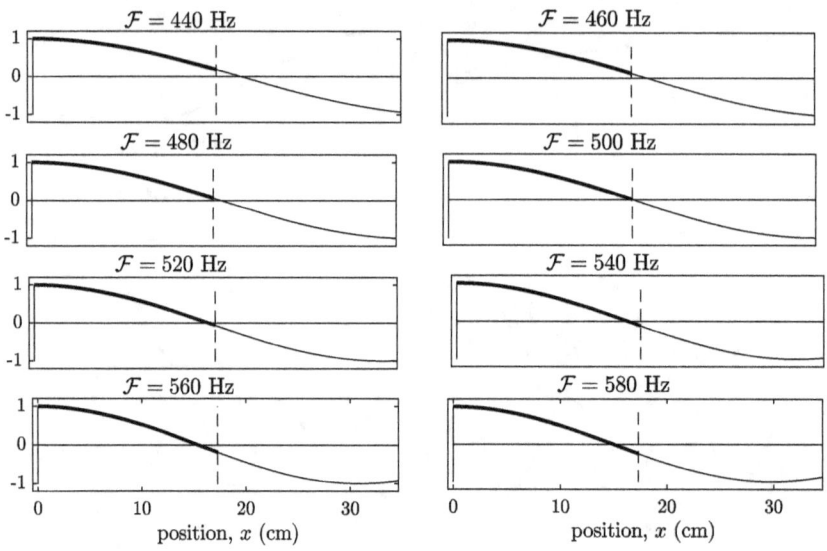

FIGURE 5. Fitting the right boundary, I

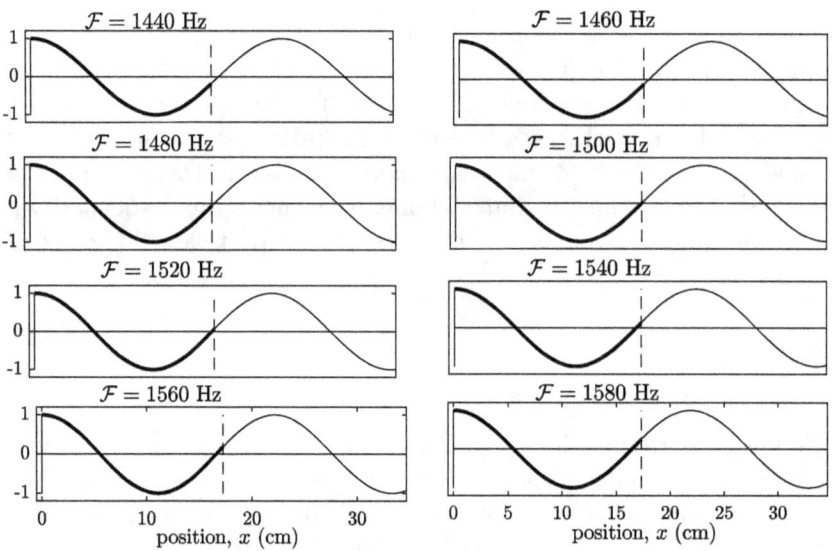

FIGURE 6. Fitting the right boundary, II

5. The function $g(x)$ is plotted for 20 Hz intervals in $\mathcal{F}$, and it can be seen that the boundary condition $g(x = L) = 0$ is satisfied for $\mathcal{F} = 500$ Hz. That is, as ω approaches $\pi c_0/2L$, k approaches $\pi/2L$ and $g(x = L) = \cos(kL)$ approaches 0, which is the second boundary condition expressed in Equation (13). Thus, $\mathcal{F}_1 = \omega_1/(2\pi) = c_0/4L$ is an eigenvalue, or normal mode frequency, and $k_1 = \pi/2L$ is the corresponding wavenumber. Thus, we can start with a circular function that satisfies the boundary condition at $x = 0$ for all frequencies, and adjust the frequencies until the boundary condition at $x = L$ is also satisfied. This is an eigenvalue and the corresponding function, $\cos(k_1 x)$, is an eigenfunction, or normal mode.

If we continue increasing $\mathcal{F}$ we get the set of panels shown in Figure 6. Again, as $\mathcal{F}$ approaches $\mathcal{F}_2 = 1500$ Hz, which means that ω approaches $\omega_2 = 3\pi c_0/2L$, k approaches $k_2 = 3\pi/2L$ and $g(x = L) = \cos(kL)$ approaches 0, which is the second boundary condition expressed in Equation (13). This process can be continued indefinitely.

We can also start with a $g(x)$ that satisfies the boundary condition at $x = L$, and increase $\mathcal{F}$ to find particular $\mathcal{F}_m$'s such that $g(x)$ satisfies the boundary condition at $x = 0$. Consider $g(x) = \sin\bigl(k(x - L)\bigr)$, as this gives $g(x = L) = 0$ for any $k = \omega/c_0$, and therefore, for any $\mathcal{F}$. Equation (13) of Chapter 3 shows the gradient of $g(x)$ can be written

$$\frac{\Delta_x g(x)}{\Delta x} = k\cos\bigl(k(x - L)\bigr),$$

so that (22)

$$\frac{\Delta_x g(x = 0)}{\Delta x} = k\cos(kL).$$

Figure 7 shows what happens as $\mathcal{F}$ increases from $440\, Hz$ to $580\, Hz$. As $\mathcal{F}$ approaches $\mathcal{F}_1 = 500\, Hz$, ω approaches $\pi c_0/2L$, k approaches $\pi/2L$, Equation (22) shows that $g(x)$ has zero gradient at $x = 0$, as required by Equation (13).

The same process can be continued with $\mathcal{F}$ increasing as we did for $g(x) = \cos(kx)$. This is shown in Figure 8 from $1440\, Hz$ to $1580\, Hz$. The zero gradient boundary condition at $x = 0$ is met when $\mathcal{F} = \mathcal{F}_2 = 1500\, Hz$. Thus, we generate the same normal mode frequencies in this process.

We seem to have generated two different sets of eigenfunctions now with $\cos(k_m x)$ and $\sin\bigl(k_m(x - L)\bigr)$. How can we resolve two, apparently

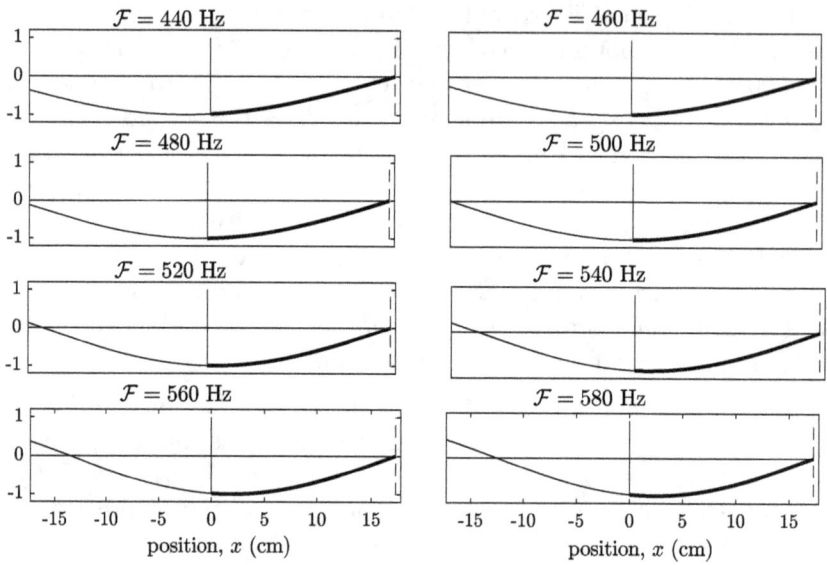

FIGURE 7. Fitting the left boundary, I

different spatial distributions of the normal mode for a particular m? The fact is that these two spatial distributions are not different at the normal mode wavenumbers k_m. Consider $\sin\bigl(k_m(x-L)\bigr)$, with $k_m = (2m-1)\pi/2L$, which are the wavenumbers that correspond to the normal mode frequencies $\mathcal{F}_m$, and invoke the second trigonometric identity in Equation (1) for

$$\begin{aligned}\sin\bigl(k_m(x-L)\bigr) &= \sin(k_m x)\cos(k_m L) - \cos(k_m x)\sin(k_m L) \\ &= (-1)^m \cos(k_m x)\,,\end{aligned} \qquad (23)$$

because $\cos(k_m L) = 0$ and $\sin(k_m L) = (-1)^{(m+1)}$. Therefore, the two sets of eigenfunctions are the same, within a constant multiple, when $k = k_m$.

In the future, we refer to the functions that satisfy the boundary condition at $x = 0$ for all frequencies as the *left semi-eigenfunction*. This is $\cos(kx)$ for the finite-length tube that we are considering. We refer to the function that satisfies the boundary condition at $x = L$ for all frequencies as the *right semi-eigenfunction*. This function is $\sin(k(x-L))$ for the finite-length tube.

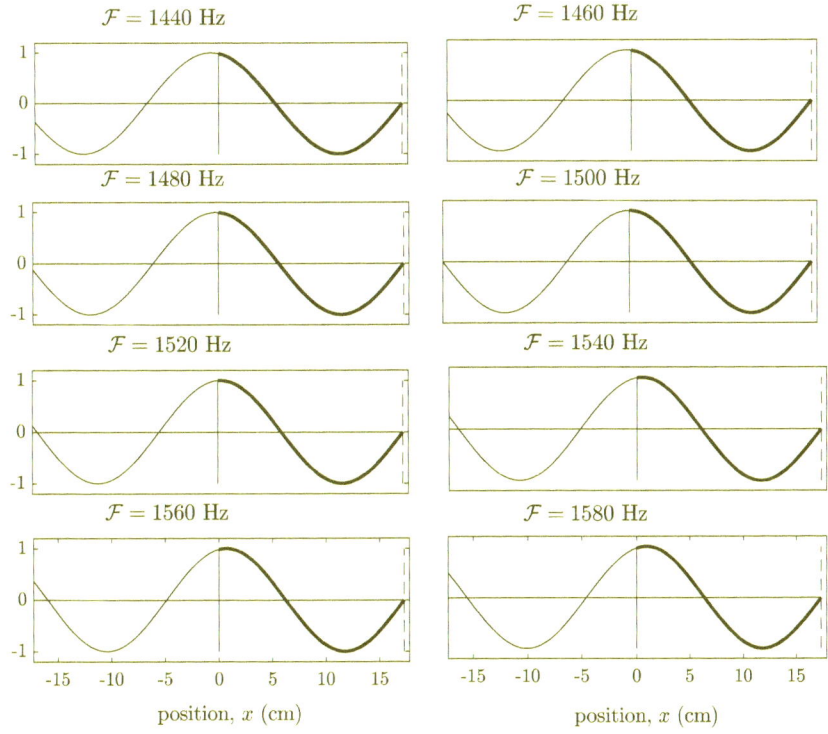

FIGURE 8. Fitting the left boundary, II

Normal modes for particle velocity and volume velocity

As might be expected, particle velocity and volume velocity possess their own normal modes, $u_m(x,t)$ and $Q_m(x,t)$, respectively, that correspond to the pressure normal modes. It is easy to write down the normal modes for particle velocity and volume velocity when the normal modes of pressure $p_m(x,t)$ are known. We refer to Equation (17) for the pressure normal modes, and to the conservation equations in Equation (12) and (14) of Chapter 2, as well as Equation (13) of

Chapter 3, to obtain

$$u_m(x,t) = -\frac{P_m}{\rho_0 c_0} \cos(\omega_m t + \phi_m) \sin(k_m x)$$
$$Q_m(x,t) = -A\frac{P_m}{\rho_0 c_0} \cos(\omega_m t + \phi_m) \sin(k_m x) \ . \qquad (24)$$

That $u_m(x,t)$ and $Q_m(x,t)$ both satisfy the wave equation, is because they both can be written as the sum of two traveling waves that satisfy the wave equation.

Just as we put the boundary conditions $u(x=0,t)=0$ and $p(x=L,t)=0$ in terms of pressure alone by using the conservation equations of Chapter 2, this can also be done for particle velocity and volume velocity

$$u(x=0,t) = Q(x=0,t) = 0$$
$$\frac{\Delta_x u(x=L,t)}{\Delta x} = \frac{\Delta_x Q(x=L,t)}{\Delta x} = 0 \ , \qquad (25)$$

for $\Delta x > 0$ arbitrarily small. It is easy to verify that the purported normal modes given in Equation (24) satisfy the boundary conditions in Equation (25). The spatial dependences of the normal modes given by $\sin(k_m x)$ are shown in Figure 9.

An initial value problem

Can a single mode be realized in a finite-length tube? Mathematically this is easy to do, but it would be difficult to realize physically without an acoustic source. Mathematically, we simply stipulate that the initial distribution of pressure is a spatial eigenfunction.

$$p(x,t=0) = P_m \cos(k_m x) \ ,$$
with $\qquad (26)$
$$\frac{\Delta_t p(x,t=0)}{\Delta t} = 0 \ ,$$

for any m at time $t=0$. For small enough disturbance amplitude P_m this initial distribution evolves according to the wave equation. In fact, for $t > 0$

$$p(x,t) = P_m \cos(k_m x) \cos(\omega_m t) \ , \qquad (27)$$

where ω_m and k_m are related by Equation (16), that is $\omega_m = c_0 k_m$.

Acoustics of Speech Production

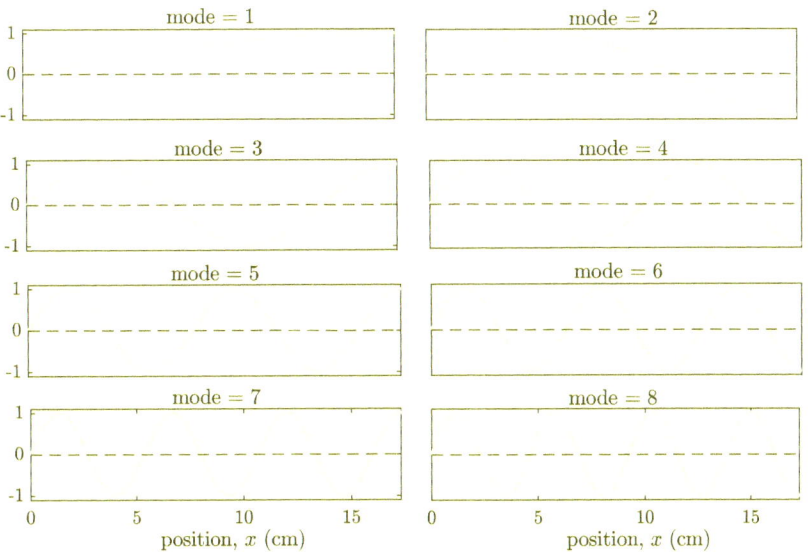

FIGURE 9. Normal modes for particle velocity and volume velocity

What we have done here is analogous to the initial value problems of the mass-spring system without external force in Chapter 3. The difference is that we start with a pressure or particle velocity distribution throughout the tube here, while we start with a single position and/or velocity for the single mass in the case of the mass-spring system.

We reiterate that the standing wave representation is preferred to the traveling wave representation for the acoustics of the finite-length tube. The standing wave in Equation (27) can be written as the sum of right-going and left-going traveling waves according to the identities in Equation (1).

$$p(x,t) = P_m \cos(k_m x) \cos(\omega_m t)$$
$$= \frac{P_m}{2} \cos(\omega_m t - k_m x) + \frac{P_m}{2} \cos(\omega_m t + k_m x) \qquad (28)$$
$$= \frac{P_m}{2} \cos\left(\omega_m(t - x/c_0)\right) + \frac{P_m}{2} \cos\left(\omega_m(t + x/c_0)\right).$$

However, if we were to measure the time evolution of the pressure distribution in such an initial value problem, we would detect the standing wave with pressure transducers along the length of the tube. At any position x in the tube, the amplitude of the pressure oscillating according to $\cos(\omega_m t)$ has an amplitude of $P_m \cos(k_m x)$. The representation of the standing wave in Equation (27) as the sum of two traveling waves in Equation (28) seems to be an abstraction to the physical measurements in this circumstance.

A deep mathematical fact

We now go on to assert an incredible mathematical fact. If $p(x,t)$ is the pressure in the finite-length tube that satisfies the wave equation and the boundary conditions in Equation (13), and there are no acoustic sources or sinks present, then

$$\begin{aligned} p(x,t) &= p_1(x,t) + p_2(x,t) + p_3(x,t) + \ldots \\ &= p_1^{time}(t) p_1^{space}(x) + p_2^{time}(t) p_2^{space}(x) + p_3^{time}(t) p_3^{space}(x) + \ldots \\ &= P_1 \sin(\omega_1 t + \phi_1) \cos(k_1 x) + P_2 \sin(\omega_2 t + \phi_2) \cos(k_2 x) \\ &\quad + P_3 \sin(\omega_3 t + \phi_3) \cos(k_3 x) + \ldots, \end{aligned} \quad (29)$$

for some set of amplitudes with units of pressure, $P_1, P_2, P_3, \ldots > 0$, and initial phases, $\phi_1, \phi_2, \phi_3, \ldots$ (Courant & Hilbert 1953). That is, $p(x,t)$ necessarily is a weighted sum of the normal modes. The ω_m and k_m are given in Equations (15) and (16). The magnitude spectrum for any such pressure is similar to the discrete spectrum shown in Figure 21 of Chapter 3. The magnitude lines are spaced at 1000 Hz symmetrically $\mathcal{F} = 0$ Hz, and they continue indefinitely in both the positive and negative frequency directions. The normal modes specify all the solutions to the wave equation in the tube of length L. With pressure expressed as in Equation (29), we use Equation (24) to express particle velocity $u(x,t)$

$$\begin{aligned} u(x,t) &= u_1(x,t) + u_2(x,t) + u_3(x,t) + \ldots \\ &= -\frac{P_1}{\rho_0 c_0} \cos(\omega_1 t + \phi_1) \sin(k_1 x) - \frac{P_2}{\rho_0 c_0} \cos(\omega_2 t + \phi_2) \sin(k_2 x) \\ &\quad - \frac{P_3}{\rho_0 c_0} \cos(\omega_3 t + \phi_3) \sin(k_3 x) - \ldots. \end{aligned} \quad (30)$$

Instead of a single natural frequency, as for the mass-spring system, we have a countable infinity of normal mode frequencies for the finite-length tube. [A set that is countably infinite has as many members as there are positive integers.] Solutions to the wave equation in the finite-length tube, with the present boundary conditions is a sum of special standing waves. This applies to all waves. In particular, it applies to traveling waves, including the brief pulse that we have been studying. The property that any function of space and time that satisfies the wave equation and the boundary conditions of the finite-length tube can be written as a sum of the form shown in Equation (29) and Equation (30) make the circular functions very important in acoustics. This representation of a function as a weighted sum of normal modes is called a *Fourier sum*, and it is a type of *Fourier synthesis*. Equations (29) and (30) provide the generalization from solutions in terms of simple products of circular functions to any wave that satisfies the wave equation and the boundary conditions. It should be noted that there are no acoustic sources or sinks present, nor any damping that we encountered regarding the mass-spring system in Chapter 4.

Normal modes and the undamped mass-spring system

The motion of the undamped mass-spring system without external forcing and with general initial conditions is given in Equation (5) of Chapter 4. The mass-spring system possesses a single natural frequency, while the finite-length tube possesses an infinity of "natural" frequencies: the normal mode frequencies. These normal mode frequencies correspond to normal modes of acoustic waves with particular spatial dependence for pressure described by the functions $\cos(k_m x)$. Acoustic problems in the finite-length tube can be simplified immensely using normal modes.

Normal modes differ from actual mass-spring systems because the pressures and particle velocities are distributed spatially throughout the tube. This produces a *distributed system*. This implies that kinetic energy and potential energy are distributed throughout the finite-length tube. For the mass-spring system, the spring as it stretches or compresses possesses potential energy, and the mass in movement has kinetic energy. The potential and kinetic energies are associated with different parts of the mass-spring system. The mass-spring system is a *lumped system* The Helmholtz resonator of Chapter 1 is also a lumped system. It is seen in Chapter 12 that acoustic distributed systems sometimes can be approximated by the lumped Helmholtz resonator at sufficiently low frequencies.

Independence of normal modes

What makes normal modes normal?

The term normal is used for both the eigenfunctions $p_m(x,t)$ and their spatial components $p_n^{space}(x)$ because of a particular mathematical property of the set of function $p_m^{space}(x), m = 1, 2, 3....$ It is said that the set of functions $\{p_m^{space}(x), m = 1, 2, 3, ...\}$ are *orthogonal functions*, or *normal functions*, if, for any two functions, $p_m^{space}(x)$ and $p_n^{space}(x)$ with $m \neq n$

$$\sum_{i=0}^{N} p_m^{space}(x_i) \cdot p_n^{space}(x_i) \approx 0 , \qquad (31)$$

for any very fine division of the x-axis from 0 to L, $0 = x_0 < x_1 < x_2 < ... < x_{N-2} < x_{N-1} < x_N = L$. If the spatial functions of different normal modes were a statistical time series, we would say that they are uncorrelated. This is a very important property, because it means that the normal modes are independent of one another in terms of the spatial variable x.

Time averages and energy densities of normal modes

In this section we examine correlation in time between normal modes. This leads to important results regarding the independence of energy densities associated with the normal modes in the time average. Mathematically, time averages of are over all time, $-\infty < t < \infty$. The notation that we use for the time-average of a function of time, $f(t)$, is $\langle f(t) \rangle$. Note that time-averaging is a linear operator, which means that

$$\langle af(t) + bg(t) \rangle = a\langle f(t) \rangle + b\langle g(t) \rangle , \qquad (32)$$

where $f(t)$ and $g(t)$ are functions of time, and a and b are constants.

We are most interested in time averages of circular functions and products of circular functions. For all real ω and ϕ

$$\langle \sin(\omega t + \phi)\rangle = \langle \cos(\omega t + \phi)\rangle = 0 . \tag{33}$$

This should be obvious because the functions are positive and negative by exactly the same amounts for each period, $T = (2\pi)/\omega$, and the time-average is taken from time $t = -\infty$ to $t = +\infty$. Practically, time-averages, like the one in Equation (33), are taken over a time interval of several periods of the longest period involved.

From the identities in Equation (2), and for $m \neq n$

$$\langle \sin(\omega_m t + \phi_m)\sin(\omega_n t + \phi_n)\rangle =$$
$$\frac{1}{2}\langle \cos\left((\omega_m - \omega_n)t + (\phi_m - \phi_n)\right)\rangle - \tag{34}$$
$$\frac{1}{2}\langle \cos\left((\omega_m + \omega_n)t + (\phi_m + \phi_n)\right)\rangle = 0 .$$

because $\omega_m - \omega_n \neq 0$ and $\omega_m + \omega_n \neq 0$ for $m \neq n$. From the same identities, for all m and n, we also have

$$\langle \cos(\omega_m t + \phi_m)\sin(\omega_n t + \phi_n)\rangle =$$
$$\frac{1}{2}\langle \sin\left((\omega_m - \omega_n)t + (\phi_m - \phi_n)\right)\rangle + \tag{35}$$
$$\frac{1}{2}\langle \sin\left((\omega_m + \omega_n)t + (\phi_m + \phi_n)\right)\rangle = 0 .$$

Thus, not only are normal modes "uncorrelated" in space, but they are also uncorrelated in time. Equation (2) provides, for any real ω and ϕ

$$\langle \sin^2(\omega_m t + \phi_m)\rangle = \frac{1}{2}\langle \cos(0)\rangle - \frac{1}{2}\langle \cos(2\omega_m t + 2\phi_m)\rangle = \frac{1}{2} ,$$
and \hfill (36)
$$\langle \cos^2(\omega_m t + \phi_m)\rangle = \frac{1}{2}\langle \cos(0)\rangle + \frac{1}{2}\langle \cos(2\omega_m t + 2\phi_m)\rangle = \frac{1}{2} .$$

From the definition of intensity in Equation (36) of Chapter 2, and from the fact that all acoustic disturbances can be written as a weighted sum of normal modes in Equations (29) and (30), it follows that

$$\langle \mathcal{I}\rangle = \langle pu\rangle$$
$$= \left\langle \left(P_1 \cos(k_1 x)\sin(\omega_1 t + \phi_1) + P_2 \cos(k_2 x)\sin(\omega_2 t + \phi_2) + ...\right)\right.$$

$$\left(\frac{-P_1}{\rho_0 c_0} \sin(k_1 x) \cos(\omega_1 t + \phi_1) - \right. \tag{37}$$

$$\left.\left. \frac{P_1}{\rho_0 c_0} \sin(k_2 x) \cos(\omega_2 t + \phi_2) + ...\right)\right\rangle$$

$$= 0 ,$$

because this reduces to the sum of terms of the form $P_m P_n/(\rho_0 c_0) \cos(k_m x) \sin(k_n x) \langle \sin(\omega_m t + \phi_m) \cos(\omega_n t + \phi_n)\rangle$. Each one of these terms is zero according to Equation (35). Therefore, energy in the tube of finite length does not travel in the time-average when there is no damping or acoustic source or sink.

From the definitions of kinetic energy density and potential energy density in Equations (28) and (30) of Chapter 2, their time-averages can be written

$$\begin{aligned}\langle e^{kin}\rangle &= \left\langle \frac{\rho_0}{2} u^2 \right\rangle = \frac{\rho_0}{2} \langle u^2 \rangle , \\ \langle e^{pot}\rangle &= \left\langle \frac{1}{2\rho_0 c_0^2} p^2 \right\rangle = \frac{1}{2\rho_0 c_0^2} \langle p^2 \rangle .\end{aligned} \tag{38}$$

We again invoke the representations of $p(x,t)$ and $u(x,t)$ written as a sum of normal modes in Equations (29) and (30).

$$\begin{aligned}\langle p^2 \rangle &= \langle p_1^2 + p_2^2 + p_3^2 + ... + cross\ terms \rangle \\ &= \langle p_1^2 \rangle + \langle p_2^2 \rangle + \langle p_3^2 \rangle + ... + \langle cross\ terms \rangle ,\end{aligned} \tag{39}$$

where $\langle cross\ terms \rangle$ is a sum of terms of the form $\langle p_m p_n \rangle$, with $m \neq n$. It follows from Equation (34), that $\langle cross\ terms \rangle = 0$. Therefore

$$\langle p^2 \rangle = \langle p_1^2 \rangle + \langle p_2^2 \rangle + \langle p_3^2 \rangle + ... \tag{40}$$

Similar equations hold for particle velocity by using Equation (35).

$$\langle u^2 \rangle = \langle u_1^2 \rangle + \langle u_2^2 \rangle + \langle u_3^2 \rangle + ... \tag{41}$$

Equations (40) and (41) say that the time-average of the square of pressure and velocity disturbances is equal to the sum of the time-average squares of their respective normal mode decomposition. There is no interaction between the normal mode energy densities in the time-average.

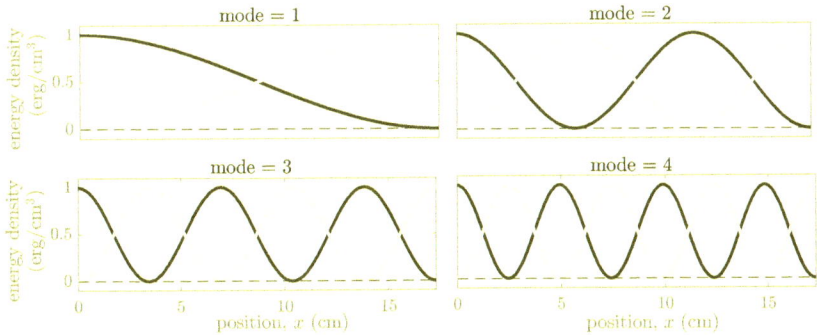

FIGURE 10. Time-average energy densities

We calculate $\langle p_m^2 \rangle$ for our case of a finite-length tube using Equation (36).

$$\langle p_m^2 \rangle = P_m^2 \cos^2(k_m x)\langle \sin^2(\omega_m t + \phi_m) \rangle \qquad (42)$$
$$= P_m^2 \frac{\cos^2(k_m x)}{2} .$$

Similarly,

$$\langle u_m^2 \rangle = \left(\frac{P_m}{\rho_0 c_0}\right)^2 \frac{\sin^2(k_m x)}{2} . \qquad (43)$$

If e_m^{kin} and e_m^{pot} denote the kinetic and potential energy densities associated with the mth normal mode, respectively, then Equations (38), (42) and (43) give

$$\langle e_m^{pot} \rangle = \frac{P_m^2}{4\rho_0 c_0^2} \cos^2(k_m x) ,$$
$$\langle e_m^{kin} \rangle = \frac{P_m^2}{4\rho_0 c_0^2} \sin^2(k_m x) . \qquad (44)$$

$\langle e_m^{kin} \rangle$ and $\langle e_m^{pot} \rangle$ are plotted in Figure 10 for the first four normal modes, with $P_m^2/(4\rho_0 c_0^2) = 1$ erg/cm^3. The black lines are for $\langle e_m^{pot} \rangle$, and the grey lines are for $\langle e_m^{kin} \rangle$. Adding the time-average potential energy and kinetic energy densities for the mth mode, to obtain time-average total energy density $\langle e_m \rangle$

$$\langle e_m \rangle \equiv \langle e_m^{kin} \rangle + \langle e_m^{pot} \rangle = \frac{P_m^2}{4\rho_0 c_0^2} \left(\sin^2(k_m x) + \cos^2(k_m x) \right) = \frac{P_m^2}{4\rho_0 c_0^2}. \quad (45)$$

Thus, the time-average energy density for each mode is uniform in space for the finite-length tube of constant cross-sectional area when there is no damping and no acoustic sources or sinks are present. This is consistent with the fact that energy does not travel in the time-average, which is expressed in Equation (37).

Summarizing, we can write both the time-average kinetic and potential energy densities as sums of the time-average energy densities of each mode.

$$\langle e^{kin} \rangle = \sum_{m=1}^{\infty} \langle e_m^{kin} \rangle,$$

$$\langle e^{pot} \rangle = \sum_{m=1}^{\infty} \langle e_m^{pot} \rangle. \quad (46)$$

This follows from Equation (40) and the definitions of energy densities given in Equations (28) and (30) of Chapter 2. These sums are possible because the normal modes do not interact in the time-average.

Another example of an initial-value problem

Another, more complicated initial-vale problem for acoustic motion in a tube of finite length is presented now. The following example involves some amount of mathematical detail; however, the description of the steps is straight-forward. First, the initial conditions are specified. That is, the pressure and the time rate-of-change of pressure are given throughout the length of the tube at time, $t = 0$. Secondly, we use Equation (29) to write the pressure in the tube $0 < x < L$ at any time t in terms of the sum of weighted normal modes. Thirdly, the amplitudes and initial phases in the sinusoidal functions of time of the normal modes are determined from the initial conditions.

Consider an example with a pressure disturbance at time $t = 0$ specified by

$$p(x, t = 0) = P\left[1 - \left(\frac{x}{L}\right)^2\right],$$

and
$$\frac{\Delta_t p(x, t = 0)}{\Delta t} = 0,$$
(47)

for $0 < x < L$, where Δt is arbitrarily small, and P is a constant with units of pressure. The shape of $p(x, t = 0)$ is shown approximately in the panel in the top left of Figure 16 below. The first relation specifies the initial pressure distribution, $p(x, t = 0)$, to be a parabola. $p(x, t)$ is a solution to the wave equation, satisfies the boundary conditions in Equation (13), and the initial conditions in Equation (47).

First, we use Equation (29) to write $p(x, t)$ in terms of normal modes.

$$\begin{aligned}p(x, t) = &P_1 \sin(\omega_1 t + \phi_1)\cos(k_1 x) + P_2 \sin(\omega_2 t + \phi_2)\cos(k_2 x) + \\ &P_3 \sin(\omega_3 t + \phi_3)\cos(k_3 x) + \ldots,\end{aligned}$$
(48)

where P_m and ϕ_m are determined by the initial conditions in Equation (47). The second expression in Equation (47) helps to determine the initial phases ϕ_m. From the identities in Equation (13) of Chapter 3, we obtain

$$\frac{\Delta_t p(x, t)}{\Delta t} =$$

$$P_1 \frac{\Delta_t \sin(\omega_1 t + \phi_1)}{\Delta t}\cos(k_1 x) + P_2 \frac{\Delta_t \sin(\omega_2 t + \phi_2)}{\Delta t}\cos(k_2 x) + \\ P_3 \frac{\Delta_t \sin(\omega_3 t + \phi_3)}{\Delta t}\cos(k_3 x) + \ldots$$
(49)

$$= P_1 \omega_1 \cos(\omega_1 t + \phi_1)\cos(k_1 x) + P_2 \omega_2 \cos(\omega_2 t + \phi_2)\cos(k_2 x) + \\ P_3 \omega_3 \cos(\omega_3 t + \phi_3)\cos(k_3 x) + \ldots,$$

where Δt is arbitrarily small. For $t=0$ in Equation (49)

$$\frac{\Delta_t p(x,t=0)}{\Delta t} = \\ P_1\omega_1 \cos(\phi_1)\cos(k_1 x) + P_2\omega_2 \cos(\phi_2)\cos(k_2 x) + \\ P_3\omega_3 \cos(\phi_3)\cos(k_3 x) + \ldots \quad (50)$$

If $\phi_m = \pi/2$, for all positive integers m, then the second initial condition in Equation (47) is satisfied. Therefore, we stipulate that $\phi_m = \pi/2$, for all positive integers m. Noting that $\sin(\omega_m t + \pi/2) = \cos(\omega_m t)$ from Equation (3), Equation (48) can be written

$$p(x,t) = P_1 \cos(\omega_1 t)\cos(k_1 x) + P_2 \cos(\omega_2 t)\cos(k_2 x) + \\ P_3 \cos(\omega_3 t)cos(k_3 x) + \ldots \quad (51)$$

Equation (51) can be used to satisfy the first initial condition in Equation (47), namely that $p(x,t=0)$ possesses a parabolic distribution in the region $0 < x < L$. Evaluating Equation (51) at t= 0 gives

$$p(x,t=0) = P\left[1 - \left(\frac{x}{L}\right)^2\right] \\ = P_1 \cos(k_1 x) + P_2 \cos(k_2 x) + P_3 \cos(k_3 x) + \ldots \quad (52)$$

The amplitudes of the normal modes in Equation (36) P_m are determined in the process of Fourier analysis of the parabola. The required mathematical machinery is beyond the scope of this book, but the process relies on the fact that normal modes are normal functions expressed in Equation (31). We simply state the result,

$$P_m = (-1)^{m+1} \frac{32P}{((2m-1)\pi)^3}, \quad \text{for } m = 1,2,3,\ldots \quad (53)$$

The log magnitudes of the P_m s, or more precisely, the $\log|P_m/P|$ s, are plotted in Figure 11. The logarithm function is employed because of the rapid decrease in the amplitudes P_m with m.

Acoustics of Speech Production

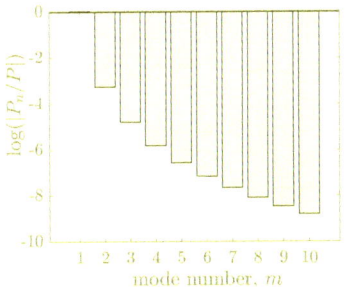

FIGURE 11. Log magnitude of amplitudes

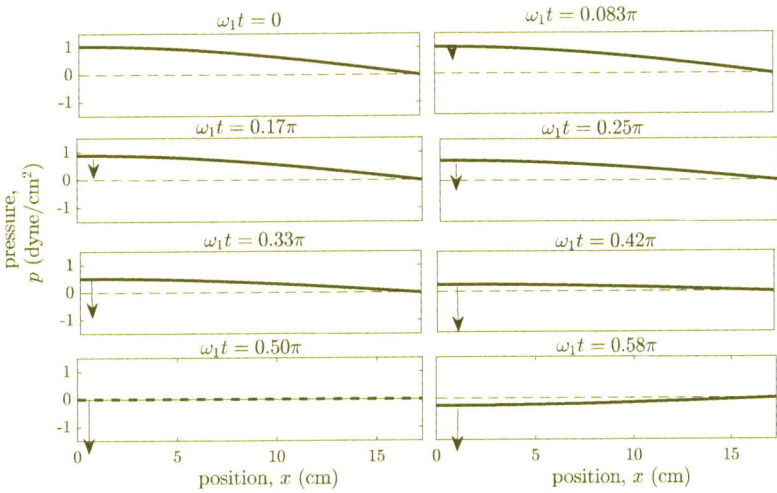

FIGURE 12. First pressure mode's movement, I

With Equation (53), Equation (51), which is the pressure distribution at time $t = 0$, can be written

$$p(x, t = 0) = P\left[1 - \left(\frac{x}{L}\right)^2\right]$$
$$= \frac{32P}{\pi^3}\left[\cos(k_1 x) - \frac{1}{27}\cos(k_2 x) + \frac{1}{125}\cos(k_3 x) - \ldots\right]. \quad (54)$$

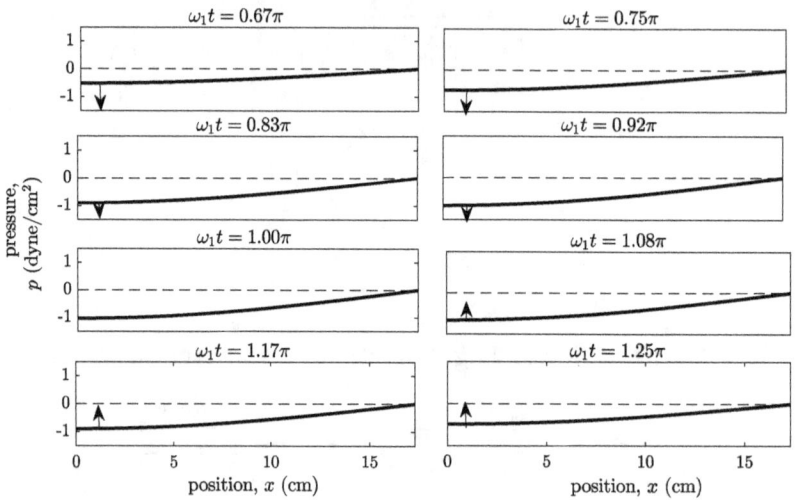

FIGURE 13. First pressure mode's movement, II

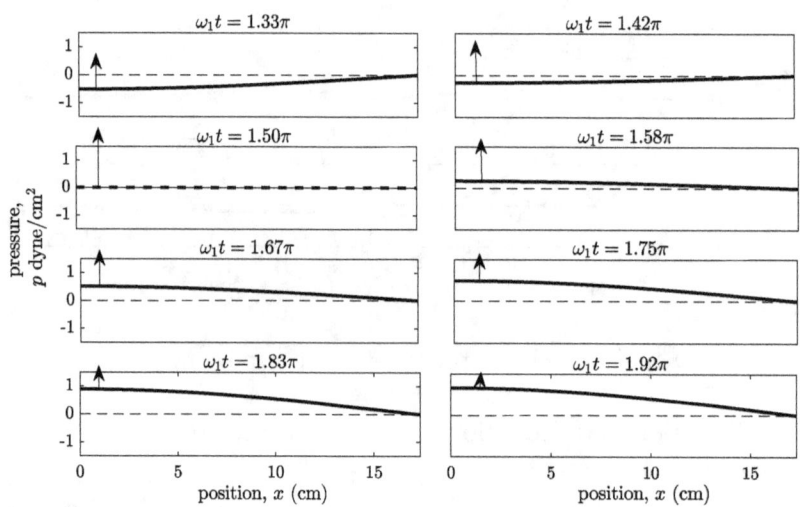

FIGURE 14. First pressure mode's movement, III

Acoustics of Speech Production

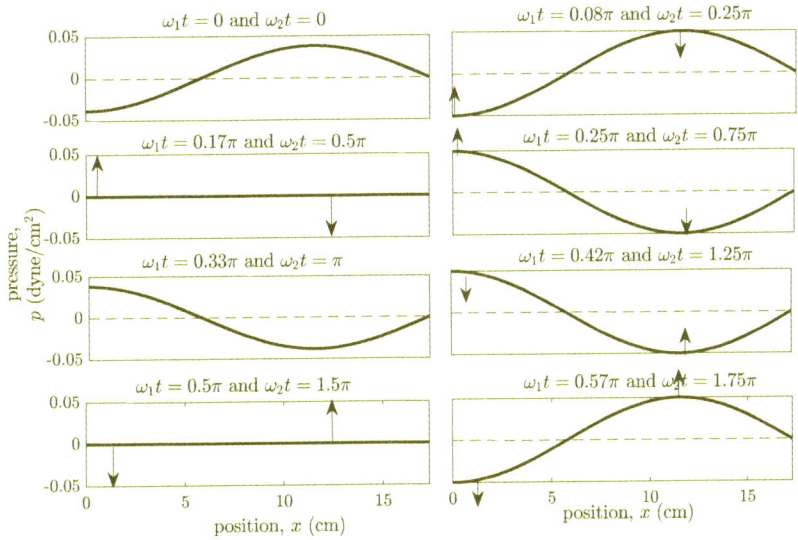

FIGURE 15. Second pressure mode's movement

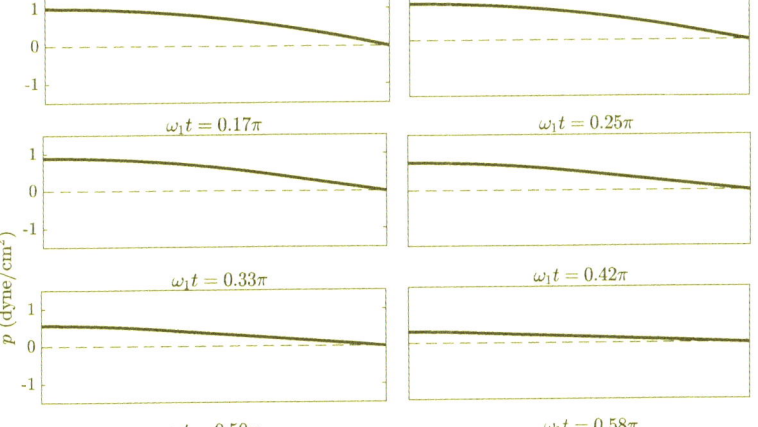

FIGURE 16. Combined pressure modes' movement, I

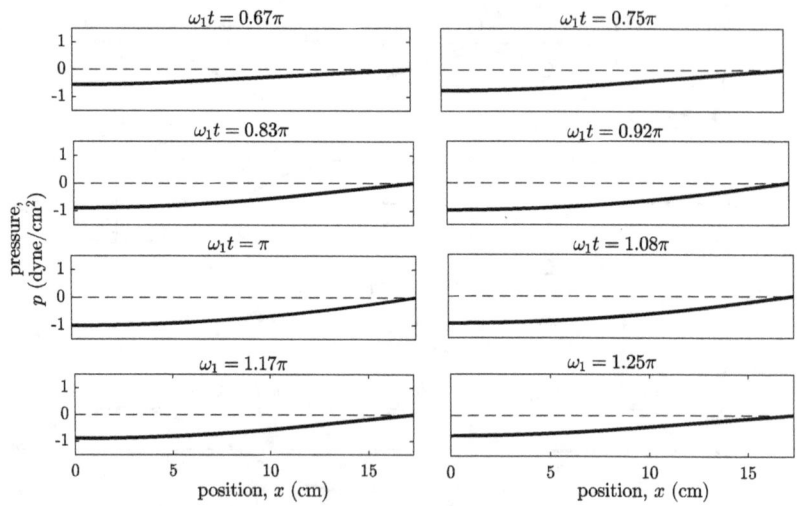

FIGURE 17. Combined pressure modes' movement, II

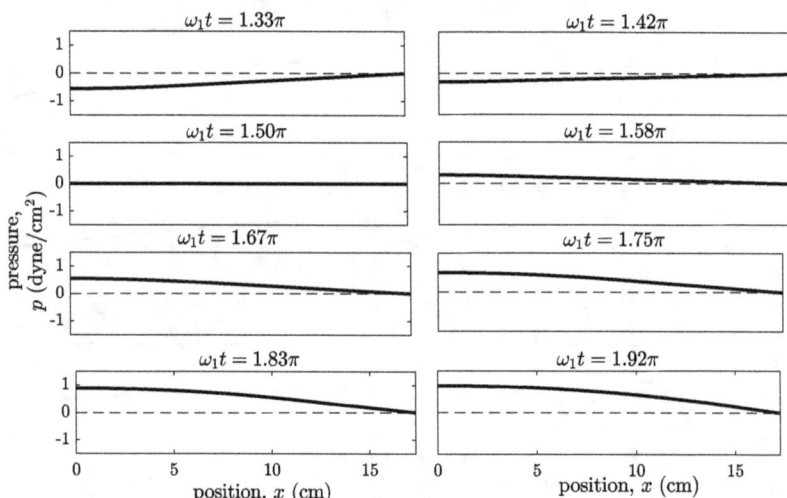

FIGURE 18. Combined pressure modes' movement, III

Therefore, Equation (48) can be written for any time $t > 0$

$$p(x,t) = \frac{32P}{\pi^3}\left[\cos(\omega_1 t)\cos(k_1 x) - \frac{1}{27}\cos(\omega_2 t)\cos(k_2 x) + \frac{1}{125}\cos(\omega_3 t)\cos(k_3 x) - \ldots\right]. \qquad (55)$$

We have solved the initial value problem using normal modes. Snapshots, or frames, of the first two normal modes in Equation (55) at various times are shown in the following series of figures, with $32P/\pi^3 = 1$ dyne/cm^2. Figures 12 through 14 show the pressure distribution of the first mode in the tube at 24 frames through one of its cycles. All the neighboring frames have equal durations between them. Thus, the phases of the first mode $\omega_1 t$ that are shown in these figures are given by $\omega_1 t = (n/24)2\pi$, for $n = 0, 1, 2, ..., 23$. The arrow in each frame denotes the magnitude and direction of the change in the pressure anti-node at $x = 0$ for the first mode. Figure 15 shows the changes in the second normal modes through one of its periods for 8 frames. The time phases of the second mode $\omega_2 t$ that are shown in this figure are $\omega_2 t = (n/8)2\pi$, for $n = 0, 2, ..., 7$. From Equations (15) and (16), $\omega_2 = 3\omega_1$. The pattern exhibited in Figure 15 for the second mode repeats itself three times while the first mode goes through one of its own cycles. Also, the differences between the vertical scales in Figures 12 through 14 and those in Figure 15, show the second mode to have substantially less amplitude than the first mode.

Finally, we compute the sum of the first ten normal modes, again with $32P/\pi^3 = 1$. The phasing for frames that was employed for the first normal mode in Figures 12 through 14 is used for this sum of normal modes. The result is shown in Figures 16 through 18. The series of frames looks very similar to those for the first normal mode in Figures 12 though 14. This is not too surprising, because of the rapid decrease in the P_m amplitudes, from the amplitude of P_1 shown in Figure 11. Also, the overall motion has a period, corresponding to that of the first mode.

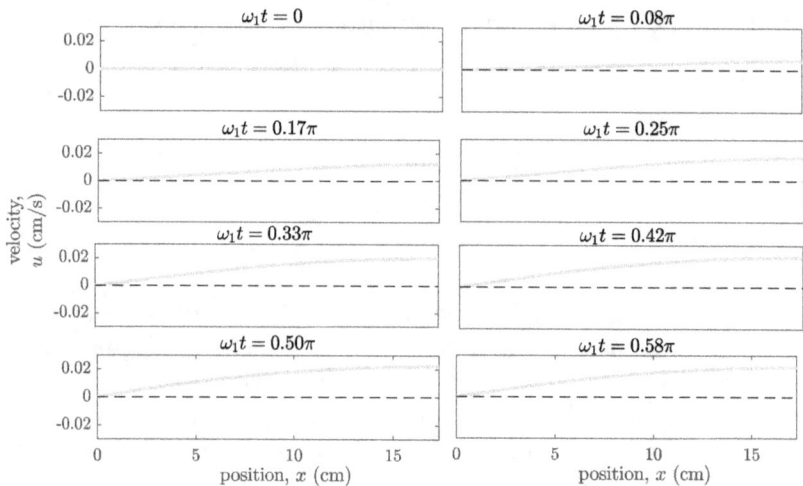

FIGURE 19. Combined particle velocity modes' movement, I

The initial value problem that we solved for pressure in Equation (55) has corresponding particle velocity and volume velocity solutions.

$$u(x,t) = -\frac{32P}{\rho_0 c_0 \pi^3}\left[\sin(\omega_1 t)\sin(k_1 x) - \frac{1}{27}\sin(\omega_2 t)\sin(k_2 x) + \frac{1}{125}\sin(\omega_3 t)\sin(k_3 x) - ...\right]$$

and (56)

$$Q(x,t) = -\frac{32P \cdot A}{\rho_0 c_0 \pi^3}\left[\sin(\omega_1 t)\sin(k_1 x) - \frac{1}{27}\sin(\omega_2 t)\sin(k_2 x) + \frac{1}{125}\sin(\omega_3 t)\sin(k_3 x) - ...\right]$$

Figures 19 through 21 show the standing wave pattern for particle velocity at the same times as those for pressure in Figures 16 though 18. It can be seen that the particle velocity standing wave lags behind the pressure standing wave in time by phase $\pi/2$. Otherwise, the

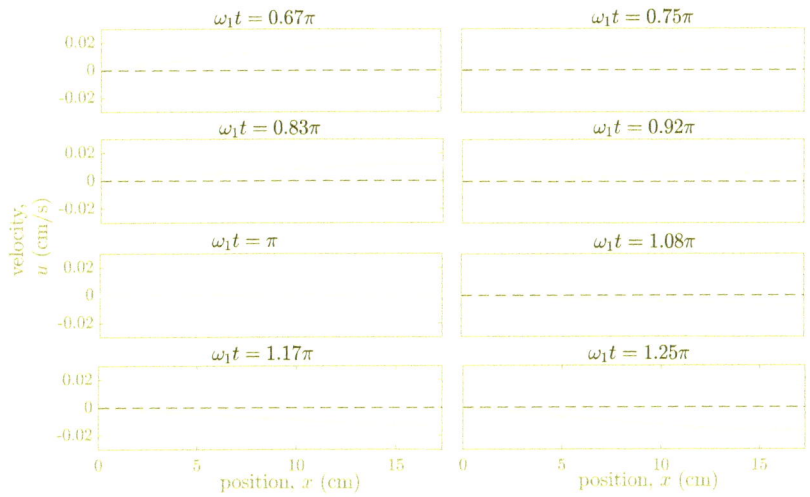

FIGURE 20. Combined particle velocity modes' movement, II

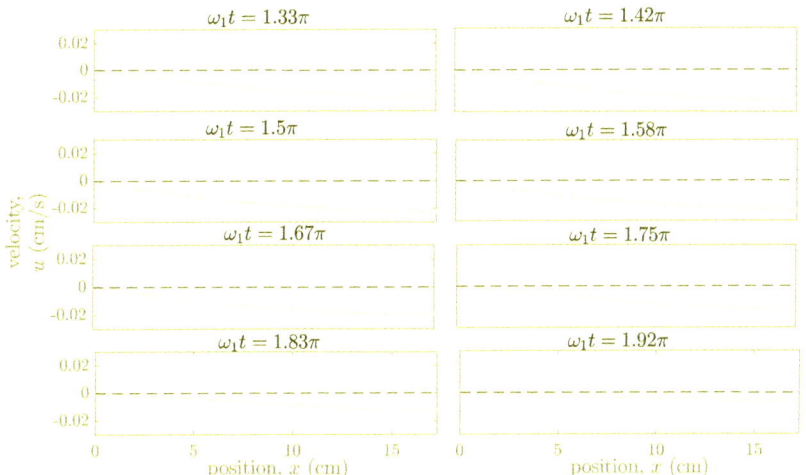

FIGURE 21. Combined particle velocity modes' movement, III

pressure and particle velocity standing waves are spatial mirror images of each other through the center of the tube.

The conservation equations in Equations (12) and (14) of Chapter 2 applied to the initial conditions for pressure, Equation (47), indicate that the initial particle velocity, $u(x, t = 0) = 0$, throughout the tube, that is for $0 < x < L$. This can be seen to be the case in the first frame of Figure 19. In these initial value problems, it is important not to attempt to over-specify initial conditions, because the equations of motion relate the physical quantities of pressure and particle velocity to one another.

There is very little practical use for initial value solutions without acoustic sources in speech production research. The main reason for this is that acoustic sources and sinks are almost always involved in speech production, because, usually, air flows from the lungs. However, one example of an initial value problem involving the vocal tract would be if the lips were closed and the air in the mouth were brought to subglottal pressure before the glottis is closed. If the lips were suddenly opened while the glottis remains closed, then we have an initial value problem. Initial value problems are much more common for stringed musical instruments that can be plucked.

Conclusion

Normal modes have been introduced, and their importance in the study of acoustics is great. The fact that acoustic disturbances in the finite-length tube without damping and without acoustic sources or sinks can be written as a sum of normal modes is very important. Intuitively, this means that we can view acoustic disturbances in a finite-length tube as an infinity of mass-spring-like elements, except that each element has its pressure and particle velocity disturbances distributed spatially through out the tube. This makes theoretically reasonable the observations of Chapters 3 and 4 that the acoustics of the piston-driven finite-length tube and the forced mass-spring system often behave in a similar way. [The reason that the mass-spring system with forcing frequency $\mathcal{F}_F = 1200$ Hz differs from the acoustics of the finite-length tube with piston velocity frequency $\mathcal{F}_{pst} = 1200$ Hz, is now seen to be the result of a normal mode frequency at $\mathcal{F}_2 = 1500$ Hz, as well as at $\mathcal{F}_1 = 500$ Hz, for the finite-length tube.] Simulations in Chapter 4 hinted at the importance of damping in a finite-length tube with a piston source in establishing steadiness. In that chapter, we saw more fully how this works in a mass-spring system. Now that we have established that the acoustics in a finite-length tube without acoustic sources or sinks and without damping behaves like many, or an ensemble of, mass-spring-like systems, we endeavor to carry the analysis forward in the next two chapters with the mass-spring system in mind.

We continue the study of acoustics of the finite-length tube with two important applications of normal modes in Chapter 6: *acoustic perturbation theory* and Green's functions. In Chapter 7, we pursue the normal mode representation to examine the similarities between the mass-spring system and acoustic propagation in a tube of finite length. In that chapter, damping and acoustic sources in the finite-length tube are examined with the help of what we studied for the mass-spring system in Chapter 4.

We have only considered the finite-length tube with constant cross-sectional area so far. Except for the specific normal mode frequencies and the particularly simple circular functions for spatial eigenfunctions, everything here still holds for the variable-area finite-length tube. This is addressed in latter half of the book.

References

Courant R. & Hilbert D. (1953). *Methods of Mathematical Physics, Volume I*. Interscience Publishers, Inc., New York. (pp 286-95, pp 358-362).

Chapter 6: Applications of Normal Modes: Acoustic Perturbation Theory and Green's Functions

Introduction

Acoustic perturbation theory and Green's functions are two topics that use normal modes. These topics are important in their own right, and they are used later in the book.

Acoustic perturbation theory permits us to predict the direction of change of normal mode frequencies when the cross-sectional areas of the finite-length tube is changed by a small amount in a region along the x-axis. While we have been focusing on a tube with constant cross-sectional area so far, the same considerations apply when a tube of variable cross-sectional area has a region of its cross-sectional areas changed by a small amount. Acoustic perturbation theory in variable cross-sectional area tubes is illustrated later in the book.

Green's functions are the analogues to impulse response functions that we introduced for the mass-spring system in Chapter 4. Green's functions are used for linear distributed systems, such as acoustic wave motion in the tube of finite length. They tell us how a distributed system responds to input that is localized in both time and space. As with the impulse response functions for mass-spring systems, the Green's function characterize the acoustic waves in a finite-length tube just as well as the equations of motion characterize acoustic behavior. Green's functions also provide a means for solving acoustics problems when acoustic sources and sinks are involved.

Acoustic perturbation theory

In this section we concern ourselves with changes in normal mode frequencies as the cross-sectional areas of the finite-length tube are altered by a small amount around certain positions x along the tube axis. The changes in cross-sectional area are supposed to be small. Expressed symbolically, if ΔA is the change in cross-sectional area A,

then
$$\frac{\Delta A}{A} \ll 1 \ . \tag{1}$$
We start with an informal argument that indicates the main result. After the informal argument is given, we proceed to explain the more rigorous theory that provides the same result.

Informal acoustic perturbation theory

We start with an intuitive argument, which involves the Helmholtz resonator introduced in Chapter 1. We can conceive of such a resonator as a tube made from two sub-tubes. A sub-tube to the left has a relatively large cross-sectional area and opens on the right into a sub-tube with relatively narrow cross-sectional area, which, in turn, opens on the right into the atmosphere. This is shown in Figure 11 of Chapter 1. [Recall that the solid mass in this figure has been replaced by air.]

We derived the spring constant κ that is provided by the air in the relatively large volume, and it is given in Equation (19) of Chapter 1, which is repeated here.
$$\kappa = \rho_0 c_0^2 \frac{A_c^2}{V_0} \ . \tag{2}$$
where A_c is the cross-sectional area of the constriction sub-tube on the right and V_0 is the volume of the larger sub-tube on the left that provides the springiness. The mass of the air in the constriction sub-tube m_c that is forced by this spring with spring constant κ is given by Equation (20) of Chapter 1, which is also repeated.
$$m_c = \rho_0 l_c A_c \ . \tag{3}$$
Finally, we rewrite the expression for natural frequency of the Helmholtz resonator given in Equation (21) of Chapter 1.
$$\mathcal{F}_H = \frac{1}{2\pi}\sqrt{\frac{\kappa}{m_c}} = \frac{1}{2\pi} c_0 \sqrt{\frac{A_c}{l_c V_0}} \ . \tag{4}$$

We examine what occurs when changes are made to the cross-sectional areas of either the left sub-tube with volume V_0, or the right constriction sub-tube with cross-sectional area A_c. If the cross-sectional area of the left sub-tube with volume V_0 is decreased, Equation (2) indicates that the spring constant κ increases. There is no change in the mass m_c according to Equation (3). Therefore, Equation (4) shows that the Helmholtz resonator natural frequency increases. The opposite occurs if we were to increase the cross-sectional area of the left sub-tube. If the cross-sectional area of the right sub-tube A_c is decreased, then Equations (2) and (3) indicate that both the spring constant κ and the mass m_c both decrease. These have opposite effects on the natural frequency $\mathcal{F}_H$, but the effect on the spring constant dominates the effect on the mass, as shown by Equation (4). Thus, the overall effect is for $\mathcal{F}_H$ to decrease. The opposite occurs if A_c is increased.

The large cross-sectional area sub-tube to the left with volume V_0 provides the springiness for the system, and the potential energy density of the entire system is mostly contained in this left sub-tube. The constriction sub-tube to the right with cross-sectional area A_c provides the mass for the system. Therefore, the great majority of kinetic energy density is contained in the right, constriction sub-tube. [The fact that the potential and kinetic energy densities are lumped as asserted is not proven until Chapter 12, but our description of the Helmholtz resonator in Chapter 1 makes this plausible, at least.]

It might be concluded that "squeezing" on a region of large potential energy density increases natural frequency, and squeezing on a region of high kinetic energy density decreases natural frequency. In fact, this turns out to be correct, except time-average energy densities are the quantities of concern. However, we have seen that acoustic systems are often characterized by having distributed potential energy and kinetic energy densities. The Helmholtz resonator is something of an anomaly with lumped potential and kinetic energy densities.

Let's consider the finite-length tube, which is a distributed system for its normal modes. In Chapter 5, we discovered that, in the time-average, the energy densities of the normal modes were independent of one another. Thus, we can examine the relative magnitudes of the time-average potential and kinetic energy densities for each normal mode separately. These are plotted for the first four normal modes in Figure 10 of Chapter 5. For a given normal mode, we might expect that if we decrease the cross-sectional area of the finite-length tube where the time-average potential energy density is greater than the time-average kinetic energy density for that mode, then that normal mode's frequency would increase. If we were to increase the cross-sectional area, then the frequency of that normal mode would decrease. The opposite would occur in regions where the time-average kinetic energy density is greater than the time-average potential energy density. That is, in regions where the time-average kinetic energy density dominates, a decrease in cross-sectional area would decrease the normal mode frequency, and an increase in cross-sectional area would increase the normal mode frequencies. All of this discussion is in reference to the the time-average energy densities.

For the remainder of the discussion on acoustic perturbation theory, we develop the more rigorous argument following Schroeder (1968). We need to cover two areas before concluding with the main result. These two areas are: *steady acoustic radiation pressure* and *Ehrenfest's theorem*.

Steady acoustic radiation pressure

We recall from Chapter 2 that the acoustic approximation essentially derives from considering small disturbances in an otherwise quiescent atmosphere. Mathematically, this means that we can ignore products of small quantities, nonlinear terms, in the conservation equations to derive equations for such things as pressure, $p = p(x,t)$ and particle velocity $u = u(x,t)$. However, we might question whether our solutions are precise enough.

One method that, sometimes, can make the solutions more precise, is the *method of successive approximation*. First equations are solved under the acoustic, or linear, approximation. Then, these solutions are used to approximate the previously neglected nonlinear terms. The resulting equations of motion can be solved again, resulting in a correction to the original linear approximation. The solutions derived under the acoustic approximation are called *first-order acoustic quantities*. The solutions in the next iteration of the method of successive approximation are called *second-order acoustic quantities*. The method of successive approximation can result in a better approximation over the original linear approximation in many circumstances.

Pressure can be calculated to second-order. In fact, the time-average of the second-order pressure is related to the time-average energy densities, $\langle e^{kin}(x)\rangle$ and $\langle e^{pot}(x)\rangle$, that are computed from the first-order quantities, as defined in Equation (38) of Chapter 5. It turns out that the time-average second-order pressure $\langle p(x)\rangle$ is

$$\langle p(x)\rangle = \langle e^{pot}(x)\rangle - \langle e^{kin}(x)\rangle . \tag{5}$$

This quantity is also known as the *steady acoustic radiation pressure*. The steady acoustic radiation pressure $\langle p \rangle$ is a second-order acoustic quantity, because it is written with terms that are squares of first-order acoustic quantities p and u in the kinetic and potential energy densities. The first-order pressure can have sinusoidal time dependence, $\sin(\omega t + \theta)$, in which case the time-average first-order pressure is zero. However, for the second-order approximation this is not the case for the time-average p in Equation (5), as seen for the time-averages of e^{kin} and e^{pot} in Equations (42), (43), and (46) of Chapter 5.

Chapter 5 tells us that it makes sense to define a steady acoustic radiation pressure for each normal mode m.

$$\langle p_m(x)\rangle \equiv \langle e_m^{pot}(x)\rangle - \langle e_m^{kin}(x)\rangle . \tag{6}$$

For the finite-length tube of constant cross-sectional area, Equation (44) of Chapter 5 gives

$$\begin{aligned}\langle p_m(x)\rangle &= \frac{P_m^2}{4\rho_0 c_0^2}\left(\cos^2(k_m x) - \sin^2(k_m x)\right)\\ &= \frac{P_m^2}{4\rho_0 c_0^2}(2\cos^2(k_m x) - 1)\\ &= \frac{P_m^2}{4\rho_0 c_0^2}\cos(2k_m x) ,\end{aligned} \tag{7}$$

where the final equality follows from Equation (2) of Chapter 5. The steady acoustic radiation pressure for the first four modes with $P_m^2/(4\rho_0 c_0^2) = 1$ dyne/cm^2 are shown in Figure 1 as functions of axial position along the tube.

As an aside, we can understand the relative magnitudes of first-order and second-order quantities with an example. Suppose that there is a high amplitude normal mode in the finite-length tube of, say, with a peak of 120 dB re 0.0002 dyne/cm^2. [Strictly speaking, there is no intensity, because energy is not traveling along the tube. However, we can still compute decibels based on the pressure amplitude.] This corresponds to a peak pressure of 200 dyne/cm^2. With the energy in

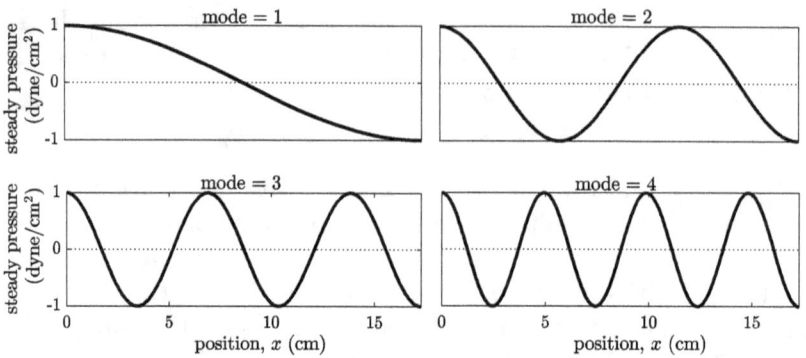

FIGURE 1. Steady acoustic radiation pressure

one of the normal modes, peak kinetic or potential energy density is about 0.013 dyne/cm^2, which is the maximum [over position] steady acoustic radiation pressure by Equation (7). Thus, the steady acoustic radiation pressure is more than four orders of magnitude smaller than the first-order acoustic pressure amplitude. Before we return to proving the main result in acoustic perturbation theory, we need to state Ehrenfest's theorem

Ehrenfest's theorem

Consider periodic motion with period $T = 1/\mathcal{F}$. The undamped mass-spring system provides such a motion, as is the undamped acoustic motion in the finite-length tube. In fact, we can consider the acoustics of the finite-length tube as a periodic system, with period equal to that of the first normal mode.

Suppose we were to add or take away energy, δE, to a periodic system in such a way that there is no heat flow into or out of the system. Changes are made slowly enough that the system passes through states of thermodynamic equilibrium, without generating heat flow. These changes are said to be part of an adiabatic process, because there is no change in entropy. Ehrenfest's theorem (Schroeder 1967) states that

$$\delta\left(\frac{E}{\mathcal{F}}\right) = 0 \ . \tag{8}$$

This says that the variation in the ratio of energy to frequency is stationary with respect to adiabatic changes. A quantity that is stationary when adiabatic changes are made is called an *adiabatic invariant*. If the system energy increases, by Ehrenfest's theorem, so must the frequency, $\mathcal{F}$. If the system energy decreases, so must $\mathcal{F}$ decrease.

The way that energy is added to, or taken away from, a system so as not to involve heat flow, is to do positive or negative *work* on the system. This is where the steady acoustic radiation pressure becomes a factor when applying Ehrenfest's theorem to the normal modes of the finite-length tube.

Acoustic perturbation theory, and examples for the finite-length tube of uniform cross-sectional area

Application of Ehrenfest's threorem to the finite-length tube involves posing hypothetical scenarios, or performing thought experiments. The walls of the finite-length tube are assumed to be made of a material so that its shape can be changed without force when there is no air motion in the finite-length tube. That is, no work needs to be done on the wall itself in order to change its shape when there is no acoustic disturbance present. Further, when there is no acoustic disturbance present, it is assumed that the rest pressure is uniform throughout space, so there is no pressure difference across the wall.

When there is acoustic wave motion in the tube, each normal mode m provides a steady acoustic radiation pressure, $\langle p_m(x) \rangle$, to the inside wall of the tube. We have established that the time-average kinetic and potential energy densities associated with each normal mode act independently, so that each normal mode can be considered to constitute is own system when steady acoustic radiation pressure is calculated. Also, independence means that Ehrenfest's theorem in Equation (8) can be applied separately to each normal mode, m. Symbolically,

$$\delta\left(\frac{E_m}{\mathcal{F}_m}\right) = 0 . \tag{9}$$

where E_m is the energy associated with the mth mode, and $\mathcal{F}_m$ is the mth mode's frequency.

It is easiest to present acoustic perturbation theory assuming a circular cross section for the tube. The same considerations apply to any shape for cross-sectional area. In the presence of the mth normal mode, an external agent is considered to slowly change the cross-sectional area A with radius $r = \sqrt{A/\pi}$, and width Δx near $x = x_0$. This is a disc of thickness Δx and circular cross-sectional area A. $2\pi r \Delta x$ is the surface area of the disc over which the cross-sectional area is changing. The steady force normal to the surface exerted by the steady acoustic radiation pressure of the mth mode on the external agent is

$$F_m = \langle p_m(x = x_0) \rangle 2\pi r \Delta x \ . \tag{10}$$

$F_m > 0$ when $\langle p_m(x_0) \rangle > 0$ and $F_m < 0$ when $\langle p_m(x_0) \rangle < 0$. Thus, F_m is positive if the radiation pressure tends to push outward on the walls of the tube, and it is negative if it tends to pull the walls inward.

For each mode m, the external agent must exert a force of infinitesimally greater magnitude than $|F_m|$, but of opposite sign in order to change the cross-sectional area of the disc around $x = x_0$ in a slow, adiabatic process. In changing the cross-sectional area the radius of the tube at $x = x_0$ changes from r to $r + \Delta r$. From the definition of work in classical physics, the work done by the external agent's force, moving the wall with the mth normal mode's steady acoustic radiation pressure present is

$$W_m = -F_m \Delta r = -\langle p_m(x = x_0) \rangle 2\pi r \Delta x \Delta r \ . \tag{11}$$

The minus sign indicates that shrinking the cross-sectional area, or decreasing radius, in the presence of a positive steady acoustic radiation pressure inside the tube requires positive work to be done by the external agent.

We now apply Equation (11). Because the amount of work done W_m is the same as the gain in the energy, E_m, of the mth normal mode, we have an increase in the frequency, $\mathcal{F}_m$, of the mth normal mode when $-\langle p_m(x = x_0) \rangle 2\pi r \Delta x \Delta r > 0$ and decrease when $-\langle p_n(x = x_0) \rangle 2\pi r \Delta x \Delta r < 0$. From Ehrenfest's theorem, both the case where $\langle p_m(x = x_0) \rangle > 0$ with a cross-sectional area decrease, or $\Delta r < 0$ near $x = x_0$, and the case where $\langle p_m(x = x_0) \rangle < 0$ with a cross-sectional area increase, or $\Delta r > 0$, near $x = x_0$ result in increased $\mathcal{F}_m$. Also, $\langle p_m(x = x_0) \rangle < 0$ and a cross-sectional area decrease,

Acoustics of Speech Production

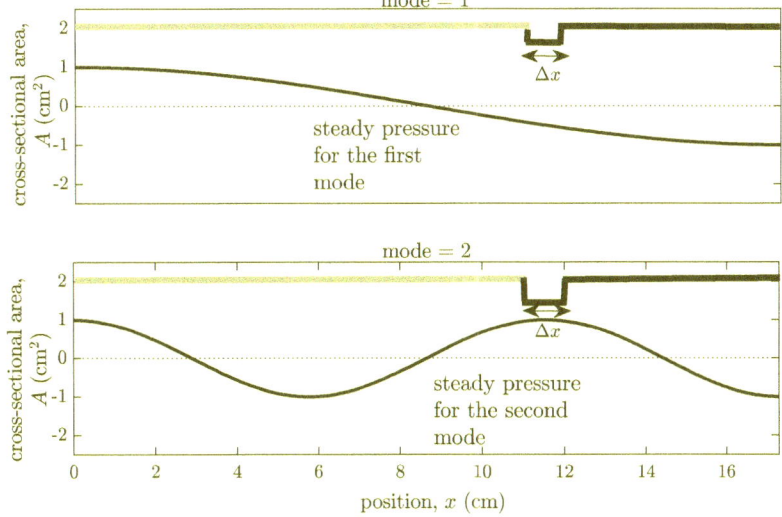

FIGURE 2. Example of small area change, I

$\Delta r < 0$, near $x = x_0$ and $\langle p_m(x = x_0) \rangle > 0$ and a cross-sectional area increase, or $\Delta r > 0$ near $x = x_0$ both result in decreased $\mathcal{F}_m$.

Figure 2 shows a straight finite-length tube of area $A = 2 \text{ cm}^2$, except for a small reduction of area at about $x = 11.5$ cm over a length of Δx. Thus, we are considering a disc-shaped region of thickness Δx being made slightly smaller in cross-sectional area. The top panel contains the trace of the steady acoustic radiation pressure associated with the first mode, and the bottom panel contains the same quantity for the second mode. These steady acoustic radiation pressure plots are represented in a non-dimensional way in Figure 2. The steady acoustic radiation pressure for the first mode is negative in the perturbed disc, or $\langle p_1(x = 11.5 \text{ cm}) \rangle < 0$, and it is positive for the second mode, or $\langle p_2(x = 11.5 \text{ cm}) \rangle > 0$. The prediction of acoustic perturbation theory is that $\mathcal{F}_1$ decreases from 500 Hz and $\mathcal{F}_2$ increases from 1500 Hz. [We should make the indentations to the area smooth functions of x-axis in order to avoid much deviation from our assumption of one-dimensional propagation. This is in contrast to what is shown in Figure 2.]

If we were to expand the tube at the location $x = 11.5$ cm, instead of contracting the tube, as shown in Figure 3, then we would obtain

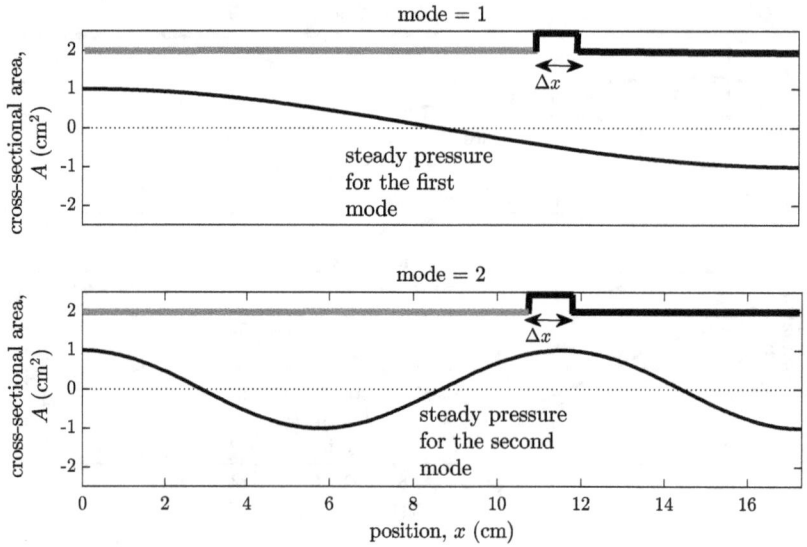

FIGURE 3. Example of small area change, II

the opposite effect on the normal mode frequencies. In this situation it is expected that $\mathcal{F}_1$ increases from 500 Hz and $\mathcal{F}_2$ decreases from 1500 Hz.

Figure 4 shows the situation when the perturbed area surrounds a *zero-crossing* for the steady acoustic radiation pressures for both the first and second normal modes. The average steady acoustic radiation pressure for both modes is such that the steady force they provide on the perturbed part of the surface is approximately zero. In this case, neither $\mathcal{F}_1$, nor $\mathcal{F}_2$ are expected to be sensitive to small changes to cross-sectional area.

We need to be clear that we are describing a perturbation theory, because energy densities change as changes to cross-sectional area are made. With large enough changes in cross-sectional area ΔA, what was once a situation with positive steady acoustic radiation pressure could reverse itself, for instance. The theory is not valid when large changes of area are made. Further, all that we have done here regarding acoustic perturbation theory applies to tubes that have variable cross-sectional area as the beginning shape to which small cross-sectional areas are applied. It turns out for tubes of variable cross-sectional area that there

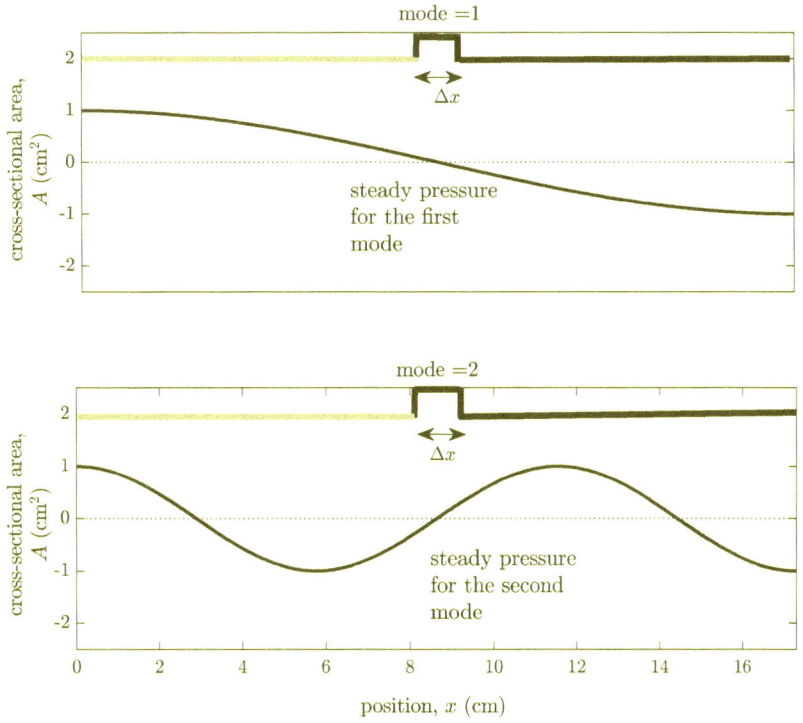

FIGURE 4. Example of small area change, III

are still discrete, independent normal modes, but the spatial eigenfunctions are not the circular functions. This is shown in Chapters 9 and 10.

Green's functions for the acoustics of the finite-length tube

We present an elementary development of Green's functions for acoustic propagation in the tube of finite length. Normal modes are essential in this development. Green's functions are analogous to impulse response functions for mass-spring systems, which were introduced in Chapter 4. They tell us about the acoustic wave motion in the finite-length tube in response to an external disturbance of unit magnitude that is localized in both time and space. However, Green's functions are a little more complicated than impulse response functions. The reason for this complication is that the acoustic response to an input depends on where and when that response is observed. Further, it also depends on the position of the input, as well as the time that the input is applied.

We derive Green's functions for two different kinds of inputs, and two different outputs. One type of input is an externally applied fluctuating volume velocity, and another possible input is externally applied fluctuating pressure. Also, we can observe the response in terms of either Q or p. We use a subscript to denote the externally applied input, or source type, and a superscript to denote the observed quantity. Thus, for an externally applied volume velocity and observed pressure, we would derive a Green's function denoted $\mathcal{G}_Q^p$. For an externally applied fluctuating pressure source and observed pressure, we derive Green's function $\mathcal{G}_p^p$.

Green's functions account for homogeneity in time, just as the impulse response function does. This means that we can write Green's functions as functions of the observer position x, the source position y, and the difference in observer time t and source time τ.

$$\mathcal{G}_Q^p = \mathcal{G}_Q^p(x,y,t-\tau) \text{ and } \mathcal{G}_p^p = \mathcal{G}_p^p(x,y,t-\tau) . \tag{12}$$

Green's functions for volume velocity sources and observed pressure

First we derive $\mathcal{G}_Q^p$ and therefore consider an externally applied, time-varying volume velocity. We let the externally applied volume flow be denoted $Q^{ext}(x,t)$, which is volume velocity that is over and above what would be expected from the conservation equations under the acoustic approximation for air in the tube. This is termed a *volume velocity source*. Again, we think of an external agent providing or taking away something from the system, and superscript ^{ext} denotes a quantity that is externally supplied. The moving piston is such a volume velocity source. We refer to acoustic source, meaning something that is externally applied and is a source of energy for acoustic wave motion. However, here we start to use the term acoustic source, or source, to cover both the situation where an input is a source of energy or a sink of energy. We examine this issue of energy input in Chapter 7. Incidentally, the volume velocity source can also be considered as mass source, $\rho_0 Q^{ext}$. The externally applied volume velocity Q^{ext} has units of volume velocity, which are cm^3/s, and $\rho_0 Q^{ext}$ has units of g/s.

Define the external volume velocity per unit length of the tube $\mathsf{q}^{ext}(x,t)$ as

$$\mathsf{q}^{ext}(x,t) = \frac{\Delta_x Q^{ext}(x,t)}{\Delta x} , \qquad (13)$$

The time rate-of-change of external volume velocity per unit length of tube $\dot{\mathsf{q}}^{ext}(x,t)$ is

$$\dot{\mathsf{q}}^{ext}(x,t) = \frac{\Delta_t \mathsf{q}^{ext}(x,t)}{\Delta t} . \qquad (14)$$

That is, the time rate-of-change of air mass per unit length of the tube is $\rho_0 \dot{\mathsf{q}}^{ext}(x,t)$.

In order to derive the Green's function, $\mathcal{G}_Q^p$, we need to examine the response of the acoustics of the finite-length tube to a brief and localized input of volume velocity. Suppose the $\dot{\mathsf{q}}^{ext}(x,t)$ is concentrated in time at τ and concentrated in position at y, for some y with $0 \le y \le L$. Symbolically,

$$\dot{\mathsf{q}}^{ext}(x,t) = \begin{cases} \dot{\mathsf{q}}_{y,\tau}^{ext} & \text{for } \tau - \Delta\tau/2 < t < \tau + \Delta\tau/2 \text{ and} \\ & y - \Delta y/2 < x < y + \Delta y/2 , \\ 0 & \text{otherwise} . \end{cases} \qquad (15)$$

where $\Delta\tau > 0$ and $\Delta y > 0$, and $\dot{q}_{y,\tau}^{ext}$ is a real number with dimensions cm^2/s^2. We also require that

$$\mathbb{Q}_{y,\tau}^{unit} \equiv \dot{\mathsf{q}}_{y,\tau}^{ext} \Delta y \Delta\tau = 1 . \qquad (16)$$

In addition to Equation (15), Equation (16) constrains the quantities $\dot{\mathsf{q}}_{y,\tau}^{ext}$, Δy, and $\Delta\tau$. For given $\Delta\tau > 0$ and $\Delta y > 0$, the $\dot{\mathsf{q}}_{y,\tau}^{ext}$ is chosen in order that $\mathbb{Q}_{y,\tau}^{unit}$ have the required unit magnitude. Instead of the two-dimensional spikes above the t axis in Figure 2 of Chapter 4, we can imagine three-dimensional spikes above the t-x plane. The *unit volume velocity* $\mathbb{Q}_{y,\tau}^{unit}$ plays a role analogous for the finite-length tube to the role played by the unit impulse $\mathbb{I}_\tau^{unit}$ for the mass-spring system in Chapter 4. The Green's function is found in the solution to this problem with brief unit magnitude input $\mathbb{Q}_{y,\tau}^{unit}$.

We solve this as an initial-value problem with the non-zero initial values specified at $t = \tau$. In other words, we solve the wave equation,

Equation (17) of Chapter 2, with initial conditions determined by the brief volume velocity specified in Equation (15), with constraints expressed in Equation (16). Thus, we find the pressure p^{uQ} that obeys

$$\frac{1}{c_0^2}\frac{\Delta_t^2 p^{uQ}(x,t)}{(\Delta t)^2} - \frac{\Delta_x^2 p^{uQ}(x,t)}{(\Delta x)^2} = 0$$

with initial conditions,

$$p^{uQ}(x, t=\tau) = 0, \text{ and}$$

$$\frac{\Delta_t p^{uQ}(x,t=\tau)}{\Delta t} = -\frac{\rho_0 c_0^2}{A}\dot{q}^{ext}(x,t)\Delta\tau = -\frac{\rho_0 c_0^2}{A}q^{ext}(x,t),$$

(17)

where $\dot{q}^{ext}(x,t)$ is given in Equation (14). [The second initial condition follows from modifying the mass conservation equation, Equation (14) of Chapter 2. This modification is in Equation (27) below.] The notation p^{uQ} is to express the fact that this is the solution for a brief volume velocity input that satisfies the constraint given in Equation (16). We can refer back to X_m^{ui} in Equations (10) through (14) of Chapter 4 for the analogous mass-spring solution for unit impulse.

We know from Equation (29) of Chapter 5, that the solution to Equation (17) is given as

$$p^{uQ}(x,t) = \sum_{m=1}^{\infty} C_m \cos(k_m x) \sin\left(\omega_m(t - (\tau + \Delta\tau/2))\right)$$
for $t > \tau + \Delta\tau/2$,

(18)

for some C_m. [The terms $-\omega_m \tau$ in the sine functions is just adding an initial phase ϕ_m.] From Equation (18) we obtain

$$\frac{\Delta_t p^{uQ}(x,t)}{\Delta t} = \sum_{m=1}^{\infty} \omega_m C_m \cos(k_m x) \cos\left(\omega_m(t - (\tau + \Delta\tau/2))\right)$$
for $t > \tau + \Delta\tau/2$.

(19)

The second initial condition in equation (17) can be written using Equation (19) as

$$\frac{\Delta_t p^{uQ}(x, t=\tau)}{\Delta t} \to \sum_{m=1}^{\infty} \omega_m C_m \cos(k_m x)$$

$$= -\frac{\rho_0 c_0^2}{A}\dot{q}^{ext}(x,t)\Delta\tau$$

(20)

$$= \begin{cases} -\frac{\rho_0 c_0^2}{A} \frac{\mathbb{Q}_{y,\tau}^{unit}}{\Delta y} & \text{for } y - \Delta y/2 < x < y + \Delta y/2 \,, \\ 0 & \text{otherwise} \,, \end{cases}$$

where $\mathbb{Q}_{y,\tau}^{unit} = 1$ is given in Equation (16).

If $\Delta \tau \to 0$ and $\Delta y \to 0$, then $\dot{\mathsf{q}}_{y,\tau}^{ext} \to \infty$ by Equation (16). This also means, that the infinite sum in Equation (20) $\to \infty$ when $x = y$, and it is zero for $x \neq y$, according to the third line in Equation (20). However, the infinite sum in Equation (20) approaches ∞ in the particular way given by the first line to right of the bracket in Equation (20). The consequence of these properties is

$$C_m = -\mathbb{Q}_{y,\tau}^{unit} \frac{2\rho_0 c_0^2}{AL\omega_m} \cos(k_m y) \,. \tag{21}$$

The descriptive reason for Equation (21), is that with this definition of C_m substituted into Equation (20), we obtain factors of the form $\cos(k_m x)\cos(k_m y)$, which can be written as $1/2\big[\cos\big(k_m(x+y)\big) + \cos\big(k_m(x-y)\big)\big]$ from Equation (2) of Chapter 5. When $x \neq y$, with $0 \leq x, y \leq L$, this causes the terms in Equation (19) to add to zero: there is complete destructive interference among the different cosine terms. But when $x = y$, the terms add to ∞, because $\cos\big(k_m(x-y)\big) = 1$ for $x = y$.

With Equation (21), Equation (18) becomes

$$p^{uQ}(x,t) = -\mathbb{Q}_{y,\tau}^{unit} \frac{2\rho_0 c_0^2}{AL} \sum_{m=1}^{\infty} \cos(k_m y)\cos(k_m x) \frac{\sin\big(\omega_m(t-\tau)\big)}{\omega_m} \tag{22}$$

for $t > \tau$.

The final step is completely analogous to our definition of the impulse response function in Equation (11) of Chapter 4. Write the solution

$$p^{uQ}(x,t) = \mathbb{Q}_{y,\tau}^{unit} \mathcal{G}_Q^p(x,y,t-\tau) \text{ for } t > \tau\,, \tag{23}$$

where

$$\mathcal{G}_Q^p(x,y,t-\tau) = \begin{cases} -\frac{2\rho_0 c_0^2}{AL}\sum_{m=1}^{\infty}\cos(k_m y)\cos(k_m x)\frac{\sin\big(\omega_m(t-\tau)\big)}{\omega_m} \\ \quad \text{for } t > \tau\,, \\ 0 \text{ for } t < \tau\,. \end{cases}$$

$$\tag{24}$$

This expression can be simplified with the *Heaviside step function* defined as

$$\mathcal{H}(t-\tau) = \begin{cases} 1 & \text{for } t > \tau \\ 0 & \text{for } t < \tau \end{cases}. \qquad (25)$$

With Equation (25), Equation (24) can be written

$$\mathcal{G}_Q^p(x,y,t-\tau) = -\frac{2\rho_0 c_0^2}{AL}\mathcal{H}(t-\tau)$$
$$\sum_{m=1}^{\infty} \cos(k_m y)\cos(k_m x)\frac{\sin(\omega_m(t-\tau))}{\omega_m}. \qquad (26)$$

A second method for deriving $\mathcal{G}_Q^p$

We now pursue an alternative method for deriving $\mathcal{G}_Q^p$ in Equation (26) that puts the brief external volume velocity input on the right-hand-side of the wave equation. In this case, both the pressure and its centered first difference in time are zero at time τ. That is, $p(x, t = \tau) = 0$ and $\Delta_t p(x, t = \tau)/\Delta t = 0$.

The first task is to modify the wave equation, Equation (17) of Chapter 2, to account for the external mass flow. We modify the mass conservation equation expressed in terms of pressure, Equation (14) of Chapter 2. Adding a term for externally supplied mass flow and dividing through by c_0^2 produces

$$\frac{1}{c_0^2}\frac{\Delta_t p(x,t)}{\Delta t} = -\rho_0 \frac{\Delta_x u(x,t)}{\Delta x} - \rho_0 \frac{\mathsf{q}^{ext}(x,t)}{A}, \qquad (27)$$

where A is the tube cross-sectional area. Taking the centered first difference in time of Equation (27) gives

$$\frac{1}{c_0^2}\frac{\Delta_t^2 p(x,t)}{(\Delta t)^2} = -\rho_0 \frac{\Delta_t \Delta_x u(x,t)}{(\Delta t)(\Delta x)} - \rho_0 \frac{\dot{\mathsf{q}}^{ext}(x,t)}{A}. \qquad (28)$$

Taking the first difference in space of the momentum conservation equation, Equation (15) of Chapter 2 provides

$$\rho_0 \frac{\Delta_x \Delta_t u(x,t)}{(\Delta x)(\Delta t)} = -\frac{\Delta_x^2 p(x,t)}{(\Delta x)^2}. \qquad (29)$$

Combining Equations (28) and (29), we obtain

$$\frac{1}{c_0^2}\frac{\Delta_t^2 p(x,t)}{(\Delta t)^2} - \frac{\Delta_x^2 p(x,t)}{(\Delta x)^2} = -\frac{\rho_0}{A}\dot{\mathsf{q}}^{ext}(x,t) \ . \tag{30}$$

This is the wave equation, as shown in Equation (16) of Chapter 2, except with a, possibly, non-zero term on the right-hand-side.

Again, the Green's function, $\mathcal{G}_Q^p$, is the solution to Equation (30) in the case that the source term is specified by Equations (15) and (16), and the initial conditions of zero pressure and zero first difference in time of pressure. In contrast to the initial value problem in Equation (17), we used to obtain $\mathcal{G}_Q^p$ above in Equation (26), we are now solving

$$\frac{1}{c_0^2}\frac{\Delta_t^2 \mathcal{G}_Q^p(x,y,t-\tau)}{(\Delta t)^2} - \frac{\Delta_x^2 \mathcal{G}_Q^p(x,y,t-\tau)}{(\Delta x)^2} = -\frac{\rho_0}{A}\dot{\mathsf{q}}^{ext}(x,t) \ ,$$

with initial conditions (31)

$$\mathcal{G}_Q^p(x,y,t-\tau=0) = 0 \text{ and } \frac{\Delta_t \mathcal{G}_Q^p(x,y,t-\tau=0)}{\Delta t} = 0 \ ,$$

where $\dot{\mathsf{q}}^{ext}(x,t)$ is given in Equations (15) and (16). [There is an important fact that we do not discuss in depth. This is the fact that the Green's function actually is the solution of something called the *adjoint problem*. The wave equation is a *self-adjoint equation*, which means that the wave equation and its adjoint are the same. This, along with the boundary conditions, makes for a *self-adjoint problem*. We can recommend two advanced books in mathematical physics to study Green's functions further, and where the concept of adjoint is addressed. These are Courant & Hilbert (1953) and Morse & Feshbach (1953). Not all problems in speech are self-adjoint (McGowan and Howe 2013).]

Because the volume velocity source is present only briefly, we might expect that acoustic motion occurs at all real number circular frequencies ω for the brief time the source is non-zero, as discussed at the end of Chapter 3. This is the same circumstance as for the second method discussed in deriving the impulse response function. However, here we also have spatial variation to consider with all frequencies present for that brief time. This means that all wavenumbers $k = \omega/c_0$ are present. Therefore, we consider a Green's function in the frequency domain $\tilde{\mathcal{G}}_Q^p(x,y,\omega)$.

We now find the form of $\tilde{\mathcal{G}}_Q^p(x,y,\omega)$. With all the wavenumbers k present during the brief pulse, it is convenient to use the left and right semi-eigenfunctions, as introduced in Chapter 5 to describe spatial variation of this Green's function. We cannot be sure to satisfy both the boundary condition at $x=0$ and $x=L$, along with accounting for a localized pulse with all the wavenumbers k. The left semi-eigenfunctions $\cos(kx)$ satisfy the boundary condition at $x=0$ for all wavenumbers k. The right semi-eigenfunctions $\sin\big(k(x-L)\big)$ satisfy the boundary condition at $x=L$ for all wavenumbers k. The brief source is concentrated at y, and we "join up" the left and right semi-eigenfunctions there through some conditions that we leave unspecified. [Again, the interested reader should refer to Courant & Hilbert (1953) or Morse & Feshbach (1953) for the "joining up" conditions.]

Before the left and right semi-eigenfunctions are joined at y, the Green's function, $\tilde{\mathcal{G}}_Q^p(x,y,\omega)$ can be written

$$\tilde{\mathcal{G}}_Q^p(x,y,\omega) = \begin{cases} B^l \cos(kx) & \text{for } x<y \\ B^r \sin\big(k(x-L)\big) & \text{for } x>y, \end{cases} \quad (32)$$

where $B^l = B^l(y,\omega)$ and $B^r = B^r(y,\omega)$. The "joining up" at $x=y$ conditions permit us to solve for B^l and B^r

$$\tilde{\mathcal{G}}_Q^p(x,y,\omega) = -\frac{\rho_0}{Ak\cos(kL)} \begin{cases} \sin\big(k(x-L)\big)\cdot\cos(kx) & \text{for } x<y \\ \cos(ky)\cdot\sin\big(k(x-L)\big) & \text{for } x>y. \end{cases} \quad (33)$$

From Equation (33), we obtain the time domain Green's function, $\mathcal{G}_Q^p(x,y,t-\tau)$.

$$\mathcal{G}_Q^p(x,y,t-\tau) = -\frac{2\rho_0 c_0^2}{AL}\mathcal{H}(t-\tau)\sum_{m=1}^{\infty}\frac{(-1)^m \sin\big(\omega_m(t-\tau)\big)}{\omega_m} \times \quad (34)$$

$$\begin{cases} \sin\big(k_m(y-L)\big)\cdot\cos(k_m x) & \text{for } x<y \\ \cos(k_m y)\cdot\sin\big(k_m(x-L)\big) & \text{for } x>y. \end{cases}$$

Using, $\sin(k_m(x-L)) = (-1)^m \cos(k_m x)$, we obtain for Equation (34)

$$\mathcal{G}_Q^p(x,y,t-\tau) = -\frac{2\rho_0 c_0^2}{AL}\mathcal{H}(t-\tau)$$
$$\sum_{m=1}^{\infty} \cos(k_m y)\cos(k_m x)\frac{\sin(\omega_m(t-\tau))}{\omega_m}. \qquad (35)$$

Again, while all frequencies are available from the short volume velocity pulse, the finite-length tube acoustics "grabs" only those at the normal mode frequencies. That is, the volume velocity pulse only activates at the lines in the discrete magnitude spectrum at $\mathcal{F} = \mathcal{F}_m$. Equations (26) and (35) are identical, of course. We have derived these expression using two different methods.

Equation (35) contains a weighted sum of normal modes, $\cos(k_m x)\sin(\omega_m(t-\tau))$. We emphasize that, by Equation (29) of Chapter 5, that Equation (35) is a solution to the wave equation without a source, as it needs to be for $t > \tau$, that is, when the brief pulse has ceased.

Solving with the source distributed over time and space

As in the case of the impulse response function, the Green's function $\mathcal{G}_Q^p$ can be used to solve problems where the right-hand-side of the first expression in Equation (31) is distributed in space and time. Instead of $\dot{\mathsf{q}}_{y,\tau}^{ext}$ denoting the amplitude of $\dot{\mathsf{q}}^{ext}$ concentrated at y and τ, we have $\dot{\mathsf{q}}^{ext} = \dot{\mathsf{q}}^{ext}(x,t)$ as a distributed function of space x and time t in the first expression in Equation (31). We also suppose that positions of the non-zero source elements are discretized as y_i in Δy increments and time is discretized as τ_j in $\Delta \tau$ increments. Recall the convolution sum for continuous forcing in Equation (26) of Chapter 4. The pressure at x and t, or $p(x,t)$, is a *double convolution sum* of the Green's function

$\mathcal{G}_Q^p(x, y, t - \tau)$ with input $\dot{\mathsf{q}}^{ext}(y, \tau)$.

$$p(x,t) \approx \sum_i \sum_j \mathcal{G}_Q(x, t; y_i, \tau_j) \dot{\mathsf{q}}^{ext}(y_i, \tau_j) \Delta y \Delta \tau$$

$$= -\frac{2\rho_0 c_0^2}{AL} \sum_{m=1}^{\infty} \cos(k_m x) \Bigg[\sum_i \sum_j \mathcal{H}(t - \tau_j) \frac{\sin(\omega_m(t - \tau_j))}{\omega_m} \times$$

$$\cos(k_m y_i) \dot{\mathsf{q}}^{ext}(y_i, \tau_j) \Delta y \Delta \tau \Bigg] . \tag{36}$$

Green's functions for pressure sources and observed volume velocity

Another input to consider is fluctuating external pressure $p^{ext}(x, t)$ that is over and above what would be expected from the acoustic approximation. Again, the superscript ^{ext} denotes that this is externally supplied, and p^{ext} has the dimensions of pressure, dyne/cm^2. Define the pressure per unit length of the tube, $\mathsf{p}^{ext}(x, t)$ as

$$\mathsf{p}^{ext}(x, t) = \frac{\Delta_x p^{ext}(x, t)}{\Delta x} . \tag{37}$$

The time-rate of change of pressure per unit length of tube is

$$\dot{\mathsf{p}}^{ext}(x, t) = \frac{\Delta_t \mathsf{p}^{ext}(x, t)}{\Delta t} . \tag{38}$$

We incorporate the external pressure fluctuations into the conservation equations of Chapter 2. Here we modify the momentum conservation equation, Equation (15) of Chapter 2, to account for the external pressure.

$$\frac{\Delta_t u(x, t)}{\Delta t} = -\frac{1}{\rho_0} \frac{\Delta_x p(x, t)}{\Delta x} - \frac{1}{\rho_0} \frac{\Delta_x p^{ext}(x, t)}{\Delta x}$$

$$= -\frac{\Delta_x p(x, t)}{\Delta x} - \frac{1}{\rho_0} \mathsf{p}^{ext}(x, t) . \tag{39}$$

Equation (39) implies that

$$\frac{1}{c_0^2} \frac{\Delta_t^2 u(x, t)}{(\Delta t)^2} = -\frac{1}{\rho_0 c_0^2} \frac{\Delta_t \Delta_x p(x, t)}{\Delta t \Delta x} - \frac{1}{\rho_0 c_0^2} \dot{\mathsf{p}}^{ext}(x, t) . \tag{40}$$

Equation (14) of Chapter 2, the mass conservation equation, implies that
$$-\frac{1}{\rho_0 c_0^2}\frac{\Delta_t \Delta_x p(x,t)}{\Delta t \Delta x} = \frac{\Delta_x^2 u(x,t)}{(\Delta x)^2} \ . \tag{41}$$

Combining Equations (40) and (41) results in a wave equation for $u(x,t)$, or volume velocity $Q(x,t)$, with a source term on the right-hand-side.

$$\frac{1}{c_0^2}\frac{\Delta_t^2 u(x,t)}{(\Delta t)^2} - \frac{\Delta_x^2 u(x,t)}{(\Delta x)^2} = -\frac{1}{\rho_0 c_0^2}\dot{\mathsf{p}}^{ext}(x,t) \ ,$$

or (42)

$$\frac{1}{c_0^2}\frac{\Delta_t^2 Q(x,t)}{(\Delta t)^2} - \frac{\Delta_x^2 Q(x,t)}{(\Delta x)^2} = -\frac{A}{\rho_0 c_0^2}\dot{\mathsf{p}}^{ext}(x,t) \ .$$

We suppose that an external pressure spike is applied to the finite-length tube in order to derive the appropriate Green's function for a fluctuating *pressure source* and observed Q, $\mathcal{G}_p^Q$. Suppose that the $\dot{\mathsf{p}}^{ext}(x,t)$ is concentrated at time τ and in position at y, with $0 \leq y \leq L$. Thus,

$$\dot{\mathsf{p}}^{ext}(x,t) = \begin{cases} \dot{\mathsf{p}}_{y,\tau}^{ext}, & \tau - \Delta\tau/2 < t < \tau + \Delta\tau/2 \text{ and} \\ & y - \Delta y/2 < x < y + \Delta y/2 \\ 0, & \text{otherwise} \ , \end{cases} \tag{43}$$

where $\Delta\tau > 0$ and $\Delta y > 0$ are very small, and $\dot{\mathsf{p}}_{y,\tau}^{ext}$ is a real number with dimensions dyne/(cm³ s). Let

$$\mathbb{P}_{y,\tau}^{unit} \equiv \dot{\mathsf{p}}_{y,\tau}^{ext}\Delta y \Delta\tau = 1 \ . \tag{44}$$

$\mathbb{P}_{y,\tau}^{unit}$ has unit magnitude in whatever units in which we are working. In the c-g-s system $\mathbb{P}_{y,\tau}^{unit}$ has the dimensions of pressure, dyne/cm².

We proceed through the same steps to derive $\mathcal{G}_p^Q$ as those indicated by Equations (30) through (32) in the derivation of $\mathcal{G}_Q^p$. The major difference between the derivation for $\mathcal{G}_Q^p$ and $\mathcal{G}_p^Q$ is that we are starting from Equation (40) for volume velocity $Q(x,t)$ in the derivation for $\mathcal{G}_p^Q$, where we started from Equation (28) for $p(x,t)$ in the derivation of $\mathcal{G}_Q^p$. Thus, we use left and right semi-eigenfunctions appropriate for $Q(x,t)$. These are $\sin(kx)$ for the left semi-eigenfunction because

$Q(x = 0, t) = 0$, and $\cos(k(x - L))$ for the right semi-eigenfunction because $\Delta_x Q(x = L, t)/\Delta x = 0$. The result of the derivation is

$$\mathcal{G}_p^Q(x, y, t - \tau) = -\frac{2A}{\rho_0 L} \mathcal{H}(t - \tau) \times$$

$$\sum_{m=1}^{\infty} \sin(k_m y) \sin(k_m x) \frac{\sin(\omega_m(t - \tau))}{\omega_m} . \qquad (45)$$

We have again expressed the Green's function $\mathcal{G}_p^Q$ as a sum of normal modes $\sin(k_m x) \sin(\omega_m(t - \tau))$, for volume velocity $Q(x, t)$, which are the same as those for the particle velocity $u(x, t)$.

We can write another double convolution for $Q(x, t)$ in the case that the source term $\dot{\mathsf{p}}^{ext}$ of Equation (40) is distributed over space and time.

$$Q(x, t) \approx \sum_i \sum_j \mathcal{G}_p^Q(x, t; y_i, \tau_j) \dot{\mathsf{p}}^{ext}(y_i, \tau_j) \Delta y \Delta \tau$$

$$= -\frac{2A}{\rho_0 L} \sum_{m=1}^{\infty} \sin(k_m x) \Bigg[\sum_i \sum_j \mathcal{H}(t - \tau_j) \frac{\sin(\omega_m(t - \tau_j))}{\omega_m} \times$$

$$\sin(k_m y_i) \dot{\mathsf{p}}^{ext}(y_i, \tau_j) \Delta y \Delta \tau \Bigg] . \qquad (46)$$

This is analogous to Equation (34) for a volume velocity source.

Finding $\mathcal{G}_p^p$ from $\mathcal{G}_p^Q$

Because pressure p and volume velocity Q are related to one another via the conservation equations, Equations (12) and (13) of Chapter 2, we can derive an expression for $p(x, t)$ from Equation (46) when a pressure source is present. Equation (12) of Chapter 2 and Equation (46) provide

$$\frac{\Delta_t p(x, t)}{\Delta t} = -\frac{\rho_0 c_0^2}{A} \frac{\Delta_x Q(x, t)}{\Delta x}$$

$$= -\frac{\rho_0 c_0^2}{A} \frac{\Delta_x}{\Delta x} \Bigg\{ -\frac{2A}{\rho_0 L} \sum_{m=1}^{\infty} \sin(k_m x) \Bigg[\sum_i \sum_j \mathcal{H}(t - \tau_j) \times$$

$$\frac{\sin(\omega_m(t - \tau_j))}{\omega_m} \sin(k_m y_i) \dot{\mathsf{p}}^{ext}(y_i, \tau_j) \Delta y \Delta \tau \Bigg] \Bigg\}$$

$$= \frac{2c_0^2}{L} \sum_{m=1}^{\infty} \frac{\Delta_x \sin(k_m x)}{\Delta x} \qquad (47)$$

$$\left[\sum_i \sum_j \mathcal{H}(t-\tau_j) \times \frac{\sin(\omega_m(t-\tau_j))}{\omega_m} \sin(k_m y_i) \dot{\mathsf{p}}^{ext}(y_i, \tau_j) \Delta y \Delta \tau \right]$$

$$= \frac{2c_0^2}{L} \sum_{m=1}^{\infty} k_m \cos(k_m x)$$

$$\left[\sum_i \sum_j \mathcal{H}(t-\tau_j) \times \frac{\sin(\omega_m(t-\tau_j))}{\omega_m} \sin(k_m y_i) \dot{\mathsf{p}}^{ext}(y_i, \tau_j) \Delta y \Delta \tau \right].$$

From Equation (47), it follows that

$$p(x,t) = -\frac{2c_0}{L} \sum_{m=1}^{\infty} \cos(k_m x) \left[\sum_i \sum_j \mathcal{H}(t-\tau_j) \frac{\cos(\omega_m(t-\tau_j))}{\omega_m} \times \sin(k_m y_i) \dot{\mathsf{p}}^{ext}(y_i, \tau_j) \Delta y \Delta \tau \right]. \qquad (48)$$

Because the source type in Equation (48) is a pressure source, Equation (48) differs from Equation (35), where a volume velocity source is involved. However, the spatial distribution functions, $\cos(k_m x)$ are the same, while the two functions of time, $\cos(\omega_m(t-\tau))$ and $\sin(\omega_m(t-\tau))$ are $\pi/2$ out of phase.

The physical nature of pressure sources and their connection to volume velocity sources

We have provided an example of a volume velocity source with the piston at the end of the finite-length tube. However, physical pressure sources have not been described. They exist in speech production whenever there is a fluctuating external pressure on a solid obstacle due to air flow. These fluctuating pressures are external in the sense that they are not related to particle velocity via the conservation equations under the acoustic approximation. For instance, the random fluctuations in pressure on the vocal folds and teeth due to *turbulent flow* provide external pressure sources.

Often, pressure sources confined to small regions can be thought of as volume velocity gradient sources (Lighthill 1978). [The idea of small is discussed more quantitatively later, but roughly, a small region needs to possess length dimensions that are than the shortest wavelength of sound under consideration.] In this circumstance, we can write an approximate mass conservation equation in the region of the source that looks much like the acoustic approximation, Equation (5) of Chapter 2,

$$\begin{aligned}\frac{\Delta_t p^{ext}(x,t)}{\Delta t} &= c_0^2 \frac{\Delta_t \rho^{ext}(x,t)}{\Delta t} \\ &= -\frac{\rho_0 c_0^2}{A} \frac{\Delta_x Q^{ext}(x,t)}{\Delta x}\end{aligned} \qquad (49)$$

Therefore, a pressure source, which must necessarily fluctuate, is equivalent to a volume velocity gradient source, or the movement of air from one place to another. There is no net mass addition or subtraction in this motion, just a fluctuation volume velocity gradient. It is much like putting two volume velocity sources that are π out of phase in time very close to one another. This kind of source is different from the volume velocity source, where there is actually fluctuating air mass flow into or out of the system, or, equivalently, changes in system volume. We return to these ideas in Chapter 13.

Conclusion

We have discussed two topics regarding the acoustic wave motion in a tube of finite length: acoustic perturbation theory and Green's functions. These otherwise unrelated topics use normal modes substantially. Both acoustic perturbation theory and Green's functions play roles in the book.

Acoustic perturbation theory is a useful tool for those who study speech production. However, when we consider tubes with variable cross-sections further in the second half of the book, we find that the current acoustic theory can make it difficult to apply the acoustic perturbation theory without some care. It turns out that steady acoustic radiation pressure $\langle p \rangle$ can be discontinuous in our models at the junctions of sub-tubes with different cross-sectional areas. This is a word of caution if we are to use acoustic perturbation theory with confidence.

Green's functions are of great use in the next chapter, Chapter 7. Recall that an impulse response function characterizes a mass-spring system equally well as its equation of motion. The same is true of the Green's function for the acoustics of a finite-length tube. We use this fact to modify the Green's function $\mathcal{G}_Q^p$ to account for damping, as we did for the impulse response function. In Chapter 4, we saw that damping is important for establishing steady conditions, or steadiness, rapidly in a mass-spring system. Further, simulations of acoustics in a tube of finite length seem to indicate a similar phenomenon when damping of the acoustic motion is considered.

Chapter 7 is a culmination of the first part of the book, where the similarities between the forced, damped mass-spring system and the damped acoustics of the finite-length tube with a piston source is made as complete as possible. Expressing $\mathcal{G}_Q^p$ as a sum of normal modes is essential to understanding the similarities. The similarities between the systems can be understood with greater depth by considering energy flow between the source and the system. Green's functions help us to understand this energy flow. We really have only scratched the surface of Green's functions, which have been ubiquitous tools in mathematical physics.

References

Courant, R. & Hilbert, D. (1953). *Methods of Mathematical Physics, Volume 1.* Interscience Publishers, Inc., New York. (pp 351 - 363).

Lighthill, M.J. (1978). *Waves in Fluids.* Cambridge University Press, Cambridge, England. (pp 23-30).

Morse, P.M. & Feshbach, H. (1953). *Methods of Theoretical Physics.* McGraw-Hill Book Company, New York. (pp 793 - 895).

Schroeder, M.R. (1967). "Determination of the geometry of the human vocal tract by acoustic measurements", *Journal of the Acoustical Society of America,* **41**, p 1002.

Chapter 7: Damped Acoustic Motion in a Finite-Length Tube with Sinusoidal Piston Motion

Introduction

Here we examine the acoustic wave motion in a tube of finite length, including damping effects and a piston source. The forced motion of a damped mass-spring system of Chapter 4 serves as a guide in these considerations. Numerical simulations in Chapter 4 showed that the forced, damped mass-spring system has much in common with the acoustics of the finite-length tube with damping and a piston source.

In the present chapter we start with the analytic properties of the solution to the sinusoidally forced mass-spring system with damping in Equation (44) of Chapter 4, after the transients have died away and the system is in its steady phase. [Unlike the friction damping, we can permit the damping constant to be a function of frequency, as discussed for in the text leading up to Equation (45) of Chapter 4.] The *relative phase* between the force and velocity of the mass is important for time-average energy exchange in the mass-spring system. It is seen that the relative phases of these quantities depends on the relation between the forcing frequency and the natural frequency of the mass-spring system, as well as the magnitude of the damping.

Normal modes, which were introduced in Chapter 5, are used to understand piston driven, damped acoustic motion in the finite-length tube. An approximate solution for the damped acoustics of the finite-length tube in its steady phase is provided using a Green's function written as a sum of modified normal modes, similar to one found in Chapter 6. The main result is that the finite-length acoustic tube with damping and piston motion behaves similarly to a group of an infinite number of mass-spring systems that are related to the normal modes. However, important differences between mass-spring systems and the acoustics of a finite-length tube are noted.

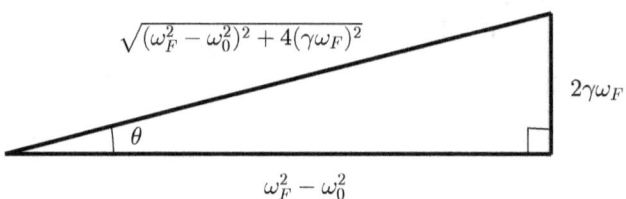

FIGURE 1. Definition of phase angle θ

Energy flow for the sinusoidally forced, damped mass-spring system

Our goal in this section is to compute the average power input by a sinusoidally varying external force to the mass-spring system with friction damping in the steady phase. From Equation (34) of Chapter 2, we need expressions for the external force and the velocity of the mass to compute power. We work with the external force we used in Chapter 4.

$$F(t) = F_0 \sin(\omega_F t) \text{ for } t > 0 . \tag{1}$$

This is Equation (27) of Chapter 4. The resulting velocity of the mass, with the transient negligible in the steady phase, is $v(t)$ given by Equation (44) of Chapter 4.

$$v(t) \approx \frac{-\omega_F F_0}{m\sqrt{\left(\omega_F^2 - \omega_0^2\right)^2 + 4\left(\gamma\omega_F\right)^2}} \cos(\omega_F t + \theta) \text{ for } t > 0 , \tag{2}$$

$$\text{where } \theta = \arctan\left[\frac{2\gamma\omega_F}{\omega_F^2 - \omega_0^2}\right] , \ 0 \leq \theta \leq \pi .$$

Recall that ω_0 is the natural circular frequency of the mass-spring system, and that ω_F is the circular frequency of the external force.

Let's examine the expression for θ in Equation (2). That expression means that

$$\tan(\theta) = \frac{\sin(\theta)}{\cos(\theta)} = \frac{2\gamma\omega_F}{\omega_F^2 - \omega_0^2} , \text{ where } 0 \leq \theta \leq \pi . \tag{3}$$

Equation (3) is true if

$$\sin(\theta) = \frac{2\gamma\omega_F}{\sqrt{\left(\omega_F^2 - \omega_0^2\right)^2 + 4\left(\gamma\omega_F\right)^2}},$$

$$\cos(\theta) = \frac{\omega_F^2 - \omega_0^2}{\sqrt{\left(\omega_F^2 - \omega_0^2\right)^2 + 4\left(\gamma\omega_F\right)^2}}, \quad \text{where } 0 \leq \theta \leq \pi. \quad (4)$$

These relations are shown with the right triangle in Figure 1 for $\omega_F > \omega_0$.

The phase θ is a function of ω_F, ω_0, and γ. Here we fix the parameters of the mass-spring system, ω_0 and γ, and consider θ to be a function of ω_F alone. Note that $\sin(\theta) \geq 0$ for $\gamma \geq 0$ and $\omega_F \geq 0$, and that $\cos(\theta)$ can be either negative or positive depending on the relative magnitudes of ω_F and ω_0. These observations are consistent with requiring $0 \leq \theta \leq \pi$.

It is convenient to define another phase α in terms of θ. [It should become clear why we have defined α once we express Equation (2) in terms of α instead of θ in Equation (7) below.]

$$\alpha = \theta - \frac{\pi}{2} \quad \text{or} \quad \theta = \alpha + \frac{\pi}{2}. \quad (5)$$

For $0 \leq \theta \leq \pi$, we must have $-\pi/2 \leq \alpha \leq \pi/2$. Further, with the identities in Equation(2) of Chapter 5, Equation (4) becomes, in terms of α

$$\sin(\alpha) = -\cos(\theta) = \frac{-(\omega_F^2 - \omega_0^2)}{\sqrt{\left(\omega_F^2 - \omega_0^2\right)^2 + 4\left(\gamma\omega_F\right)^2}},$$

$$\cos(\alpha) = \sin(\theta) = \frac{2\gamma\omega_F}{\sqrt{\left(\omega_F^2 - \omega_0^2\right)^2 + 4\left(\gamma\omega_F\right)^2}}. \quad (6)$$

Equation (2) can be written

$$v(t) \approx \frac{-\omega_F F_0}{m\sqrt{\left(\omega_F^2 - \omega_0^2\right)^2 + 4\left(\gamma\omega_F\right)^2}} \cos(\omega_F t + \alpha + \pi/2) \quad \text{for } t > 0,$$

$$= \frac{\omega_F F_0}{m\sqrt{\left(\omega_F^2 - \omega_0^2\right)^2 + 4\left(\gamma\omega_F\right)^2}} \sin(\omega_F t + \alpha) \quad \text{for } t > 0, \quad (7)$$

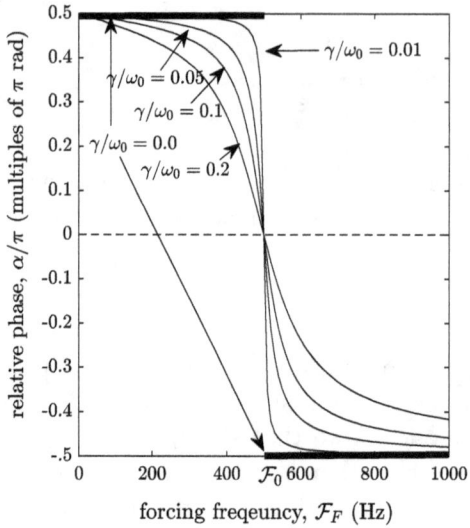

FIGURE 2. Relative phases α between F and v

$$\text{where} \quad \alpha = \arctan\left[\frac{2\gamma\omega_F}{-(\omega_F^2 - \omega_0^2)}\right], \quad -\frac{\pi}{2} \le \alpha \le \frac{\pi}{2}.$$

The second equality in Equation (7) follows from Equation (4) of Chapter 5. Comparing Equation (1) with the second equality in Equation (7), it is seen that α is the what we call the relative phase between the sinusoidal force and the velocity of the mass.

We plot relative phase α as a function of the forcing frequency $\mathcal{F}_F = \omega_F/(2\pi)$ for several values of γ/ω_0 using the first relation in Equation (6). These curves are shown in Figure 2. For forcing frequencies $\mathcal{F}_F$ very much less than the natural frequency $\mathcal{F}_0$, α is close to $\pi/2$. Similarly, for large forcing frequencies much greater than the natural frequency $\mathcal{F}_0$, α is close to $-\pi/2$. For $\mathcal{F}_F$ close to $\mathcal{F}_0$, α is close to 0, which means that the external force and the velocity of the mass are nearly in phase. The rate at which α makes transitions among these values as a function of $\mathcal{F}_F$ depends on the ratio γ/ω_0, the damping ration R. The larger the γ/ω_0, the more gradual the transition. For $\gamma/\omega_0 = 0$, there is a jump from $\alpha = \pi/2$ for $\mathcal{F}_F < \mathcal{F}_0$ through $\alpha = 0$ right at $\mathcal{F}_F = \mathcal{F}_0$ to $\alpha = -\pi/2$ for $\mathcal{F}_F > \mathcal{F}_0$.

We now relate these mathematical observations to energy flow in the steady phase, after transients have become negligible. The rate at which work is being done on, or energy is being input to, the mass-spring system by an external force is power $\mathcal{P}(t)$. We consider the time-average power over the infinite interval of $t > 0$ using the definition of power from Equation (34) of Chapter 2. From Equations (1) and (7)

$$\langle \mathcal{P} \rangle_{t > 0} = \langle F(t)v(t) \rangle_{t > 0} \tag{8}$$
$$= \frac{\omega_F F_0^2}{m\sqrt{(\omega_F^2 - \omega_0^2)^2 + 4(\gamma \omega_F)^2}} \langle \sin(\omega_F t + \alpha) \sin(\omega_F t) \rangle_{t > 0}.$$

The fact that we are averaging over positive times is indicated using the notation $\langle \cdot \rangle_{t > 0}$. The exclusion of negative times for time average has no effect on the identities involving time-averages of circular functions and their products in Equations (33) through (36) of Chapter 5. For instance $\langle \sin(\omega t) \cos(\omega t) \rangle_{t > 0} = \langle \sin(\omega t) \cos(\omega t) \rangle = 0$. These identities, along with the trigonometric identities of Equations (1) through (4) of Chapter 5 give

$$\langle \sin(\omega_F t + \alpha) \sin(\omega_F t) \rangle_{t > 0}$$
$$= \langle \big(\sin(\omega_F t) \cos(\alpha) + \cos(\omega_F t) \sin(\alpha) \big) \sin(\omega_F t) \rangle_{t > 0}$$
$$= \cos(\alpha) \langle \sin^2(\omega_F t) \rangle + \sin(\alpha) \langle \cos(\omega_F t) \sin(\omega_F t) \rangle_{t > 0} \tag{9}$$
$$= \cos(\alpha) \langle \tfrac{1}{2}[1 - \cos(2\omega_F t)] \rangle_{t > 0}$$
$$= \tfrac{1}{2} \cos(\alpha).$$

With Equation (9), Equation (8) can be rewritten

$$\langle \mathcal{P} \rangle_{t > 0} = \frac{\omega_F F_0^2}{2m\sqrt{(\omega_F^2 - \omega_0^2)^2 + 4(\gamma \omega_F)^2}} \cos(\alpha),$$
$$\text{where } -\frac{\pi}{2} \leq \alpha \leq \frac{\pi}{2}. \tag{10}$$

The relative phase between $F(t)$ and $v(t)$, which is α, plays an important role in the average power input to the mass-spring system, because a multiplicative factor of $\cos(\alpha)$ helps to determine its magnitude. With $-\pi/2 \leq \alpha \leq \pi/2$ and $\omega_F > 0$, the average power is

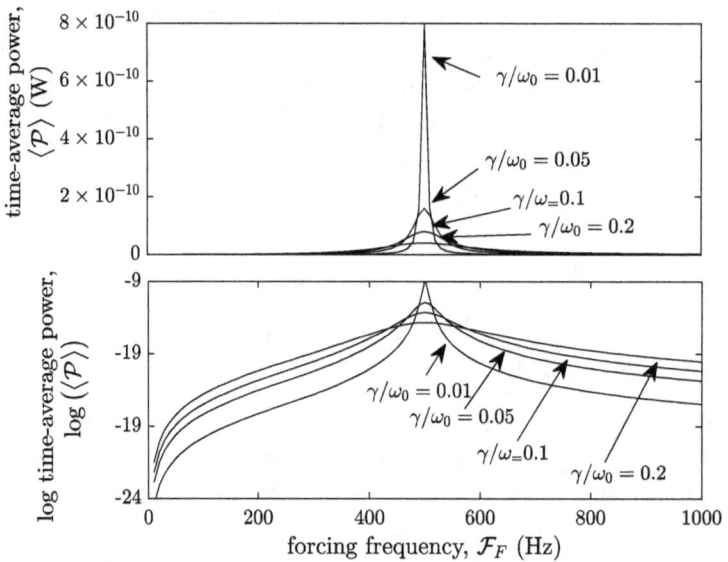

FIGURE 3. Power input by external forcing on a mass-spring system

always positive or zero. That is, energy, in the time-average, is never transferred from the mass-spring system to the external force. $\cos(\alpha)$ is maximum when $\alpha = 0$, which is when $\omega_F = \omega_0$ and steady state $v(t)$ is in phase with $F(t)$. $\omega_F = \omega_0$ also maximizes the factor to the left of $\cos(\alpha)$ in Equation (10).

In fact we, can rewrite Equation (10) by using the second expression in Equation (6).

$$\langle \mathcal{P} \rangle_{t\,>\,0} = \frac{\gamma \omega_F^2 F_0^2}{m\big((\omega_F^2 - \omega_0^2)^2 + 4(\gamma \omega_F)^2\big)} \,. \tag{11}$$

The time-average power input, $\langle \mathcal{P} \rangle_{t\,>\,0}$ ands its logarithm, for $F_0^2/(2m) = 1$ dyne^2/g, have been plotted as a function of ω_F in Figure 3 for the damping ratios shown in Figure 2, except for $\gamma/\omega_0 = 0$. The peaks of average power decrease in height, but get broader as the damping ratio γ/ω_0 increases. In general, the greater the damping, the greater the range of frequencies for which the external force $F(t)$ can input substantial energy into the mass-spring system, and the smaller the maximum energy input at $\omega_F^{res} = \omega_0$. Here, ω_F^{res} denotes the *resonance circular frequency* of the system. In the case of the friction damped mass-spring system, the *resonance frequency* equals the natural frequency of the system. That is, $\mathcal{F}_F^{res} = \mathcal{F}_0$. We need to be aware that there can be other definitions of resonance frequency in use. We have chosen the one that identifies the frequency for which there is maximum average power input to the mass-spring system.

If, in Equation (11), the case when $\gamma \to 0$ and $\omega_F \to \omega_0$ is considered we need to apply methods beyond the scope of this book to find power input. Of course, in nature there is always some damping so we never need to be practically concerned with the system with zero damping.

The finite-length tube with damping and a piston source

The Green's function

We are now ready to analyze the acoustics of a finite-length tube with damping and a piston source moving in a sinusoidal manner. Simulations of this situation in Chapter 4 revealed a close similarity to the response of the acoustics in a finite-length tube to piston movement as with the response of damped mass-spring system to external forcing. With the introduction of normal modes into the analysis of acoustic propagation in a finite-length tube, we have the knowledge to understand why there are similarities between the forced, damped mass-spring system and the damped acoustic waves in a tube of finite length with a moving piston. Because the details of the development of the mathematical description of acoustic motion in a finite-length tube with acoustic sources and damping requires substantial algebraic manipulation, we only state the results of analyses.

We include a small amount of damping from unspecified physical causes. Damping is presumed to be frequency dependent, so that for each normal mode frequency, or normal mode circular frequency ω_m, there is an associated exponential damping constant γ_m. By small we mean that the damping constants γ_m are small in relation to the normal mode circular frequency ω_m.

$$0 < \gamma_m \ll \omega_m \quad \text{or} \quad \frac{\gamma_m}{\omega_m} \ll 1 \ . \tag{12}$$

In other words, the damping ratios $R = \gamma_m/\omega_m$ are small.

Recall that the impulse response function for the mass-spring system was modified to account for unspecified damping in Equation (45) of Chapter 4, because impulse response functions can specify a mass-spring system just as well as equations of motion. The same principle applies in modifying the Green's function in Equation (26) of Chapter 6 to account for unspecified damping of acoustic propagation in a finite-length tube. The modified Green's function, for a volume velocity source at $y = 0$ is

$$\mathcal{G}_Q^p(x, y = 0, t - \tau) =$$

$$-\frac{2\rho_0 c_0^2}{AL}\mathcal{H}(t-\tau) \sum_{m=1}^{\infty} e^{-\gamma_m(t-\tau)} \times \qquad (13)$$

$$\left[\frac{\sin(\hat{\omega}_m(t-\tau))}{\hat{\omega}_m} \cdot p_m^{space}(x) - \frac{\cos(\hat{\omega}_m(t-\tau))}{\hat{\omega}_m} \cdot \bar{p}_m^{space}(x) \right].$$

The normal mode circular frequencies ω_m are perturbed to become *reduced normal mode circular frequencies* $\hat{\omega}_m = \sqrt{\omega_m^2 - \gamma_m^2}$. We explore this in Chapter 11. Also, each mode becomes an exponentially damped sinusoid in time, just as in Equation (36) for the impulse response function of Chapter 4 for the mass-spring system.

$p_m^{space}(x)$ and $\bar{p}_m^{space}(x)$ are functions that describe the spatial distribution of pressure. These functions are combined in Equation (13) so that they are $\pi/2$ out of phase in time. We can approximate these functions for small γ_m/ω_m.

$$p_m^{space}(x) \approx \cos(k_m x) \text{ for } \frac{\gamma_m}{\omega_m} \ll 1 ,$$

and $\qquad (14)$

$$\bar{p}_m^{space}(x) \approx \left(\frac{\gamma_m}{\omega_m}\right)\left[\cos(k_m x) + k_m(x-L)\sin(k_m x)\right] \text{ for } \frac{\gamma_m}{\omega_m} \ll 1 .$$

[$p^{space}(x)$ and $\bar{p}^{space}(x)$ satisfy the boundary condition at $x = L$, but not necessarily at $x = 0$, because they are right semi-eigenfunctions with the source at $y = 0$.] This means that we obtain the Green's function Equation (24) of Chapter 6 with source location $y = 0$ from the Green's function in Equation (13) when $\gamma_m \to 0$ for all n. Thus, the form of the Green's function is continuous going from the undamped to damped situation. On the other hand, there is a qualitative difference in behavior between the undamped and, even a slightly damped situation, as was seen for the mass-spring system. For one thing, the piston becomes a source of acoustic energy in the time-average

To be more precise in our approximation of the Green's function in Equations (13) and (14), we no longer have normal modes. Rather, the sum is in terms of something that we could call *modes* in this book. The terms in the sum are no longer independent in the two senses that were discussed in Chapter 5. The time averages of different modes making up the terms in Equation (13) are no longer zero. A related property of $\mathcal{G}_Q^p$ in Equation (13) is that its frequency domain representation, $\tilde{\mathcal{G}}_Q^p$, has a continuous spectrum, instead of a discrete spectrum when there is no damping. A property that follows is that a source, such as a piston oscillating at circular frequency ω_{pst} drives a broad band of frequencies composing the observed $p(x,t)$. The analogue to this phenomenon for the mass-spring system is the average power input as a function of forcing frequency $\mathcal{F}_F$ shown in Figure 3.

Sinusoidal piston movement

We assume that the piston on the left end of the tube is moving with volume velocity, $Q_{pst}(t)$. The task is to find the pressure $p(x,t)$ for $0 < x < L$ and $t > 0$ in the steady phase with a double convolution sum similar to Equation (36) of Chapter 6.

$$p(x,t) \approx \sum_i \sum_j \mathcal{G}_Q^p(x,t;y_i,\tau_j)\dot{\mathsf{q}}^{ext}(y_i,\tau_j)\Delta y \Delta \tau$$

$$= \sum_j \mathcal{G}_Q^p(x, y=0, t-\tau_j)\dot{Q}_{pst}(\tau_j)\Delta \tau$$

$$= -\frac{2\rho_0 c_0^2}{AL} \sum_{m=1}^{\infty} \left\{ \sum_j \mathcal{H}(t-\tau_j)\dot{Q}_{pst}(\tau_j)e^{-\gamma_m(t-\tau_j)} \right. \quad (15)$$

$$\left. \left[\frac{\sin\left(\hat{\omega}_m(t-\tau_j)\right)}{\hat{\omega}_m} p_m^{space}(x) - \frac{\cos\left(\hat{\omega}_m(t-\tau_j)\right)}{\hat{\omega}_m} \bar{p}_m^{space}(x) \right] \Delta \tau \right\},$$

where $\dot{Q}_{pst}(t) = \dfrac{\Delta_t Q_{pst}(t)}{\Delta t}$.

From Equation (15), we can derive an expression for pressure in the acoustic tube for the piston source $Q_{pst}(t) = \mathcal{Q}_{pst}\mathcal{H}(t)\sin(\omega_{pst}t)$. $\mathcal{Q}_{pst}$

is the amplitude of the volume velocity fluctuation at the piston. In the calculation of Equation (15) we take $\Delta \tau \to 0$ and $\Delta y \to 0$. The solution $p(x,t)$ for $0 < x < L$, can be written, without transients, as

$$p(x,t) \approx -2\frac{\rho_0 c_0^2 \mathcal{Q}_{pst}\mathcal{H}(t)}{AL} \sum_{m=1}^{\infty} \frac{\omega_{pst}}{\sqrt{(\omega_{pst}^2 - \omega_m^2)^2 + 4(\gamma_m \omega_{pst})^2}} \times \Bigg\{$$

$$\cos(\omega_{pst}t + \theta_m) p_m^{space}(x) - \frac{1}{\sqrt{(\omega_{pst}^2 - \omega_m^2)^2 + 4(\gamma_m \omega_{pst})^2}\sqrt{\omega_{pst}^2 - \gamma_m^2}} \times$$

(16)

$$\Big[\gamma_m(\omega_{pst}^2 + \omega_m^2)\cos(\omega_{pst}t) + \omega_{pst}(\omega_{pst}^2 - (\omega_m^2 - 2\gamma_m^2))\sin(\omega_{pst}t)\Big]$$

$$\bar{p}_m^{space}(x)\Bigg\} \;, \text{ where } \theta_m = \arctan\left(\frac{2\gamma_m \omega_{pst}}{\omega_{pst}^2 - \omega_m^2}\right) \;, \text{ with } 0 \leq \theta_m \leq \pi \;.$$

The phases θ_m are analogous to the θ of the mass-spring system in Equation 2.

We can now find the time-average power input for each mode m as a function of the piston frequency ω_{pst}. Call this $\langle \mathcal{P}_m \rangle_{t>0}$, and let $p_m(x,t)$ be the pressure in the mth mode according to Equation (16).

$$\langle \mathcal{P}_m \rangle_{t>0} = \langle Q_{pst} \cdot p_m(x=0,t) \rangle_{t>0}$$

(17)

$$\approx \frac{\rho_0 c_0^2 \mathcal{Q}_{pst}^2}{AL}\left(\frac{\gamma_m}{\omega_m}\right)\frac{\omega_{pst}^2}{\sqrt{\omega_m^2 - \gamma_m^2}} \times$$

$$\left\{\frac{\omega_m\sqrt{\omega_m^2 - \gamma_m^2} - (\omega_{pst}^2 - (\omega_m^2 - 2\gamma_m^2))}{(\omega_{pst}^2 - \omega_m^2)^2 + 4\gamma_m^2\omega_{pst}^2}\right\}.$$

The approximations in Equation (14) have been used in deriving Equation (17). The time-average involves only those terms in $p_m(x=0,t)$ with $\sin(\omega_{pst}t)$ time dependence from the properties of time-averages of circular functions given in Equations (35) and (36) of Chapter 5. There are components of the pressure not in phase with the piston volume velocity for the acoustics of the finite-length tube. These components have time dependence $\cos(\omega_{pst}t)$, do not contribute to time-average energy input to the system, and do not propagate as acoustic waves. These motions can be considered to be the behavior of a substance where acoustic wave motion is not possible (i.e. with $c_0 = \infty$). This is loosely termed the *sloshing behavior* of air.

From Equation (17) we also find the frequency of the piston for which there is maximum time-average power input for a mode m, ignoring the influence of the other modes. This is analogous to the calculation made in Equation (11) for the mass-spring system. Further, we would have the same result as we obtained for the mass-spring system that $\omega_{pst}^{res} = \omega_m$, except for the terms in the solution for pressure $p(x,t)$ in Equation (12) that involve $\bar{p}_m^{space}(x)$. The calculation of the circular frequency of maximum power input yields for the mth mode yields a resonance circular frequency and resonance frequency of

$$\omega_{pst}^{m\ res} \approx \omega_m \sqrt{1 - \frac{\gamma_m^2}{\omega_m^2}} \equiv \hat{\omega}_m \text{ and } \mathcal{F}_{pst}^{m\ res} \approx \hat{\mathcal{F}}_m . \qquad (18)$$

We use the approximation symbol in Equation (18), because we performed the calculation with the approximate $\bar{p}^{space}(x)$ in Equation (14). On the other hand, Equation (18) could well be the exact result. The effect of $\bar{p}^{space}(x)$ is to move the resonance frequency, as defined by maximum energy input, from the normal mode frequency $\mathcal{F}_m$ to the reduced normal mode frequency $\hat{\mathcal{F}}_m$.

However, so far we have neglected a mechanism that can cause a strong reduction in normal mode frequency. The pertinent physical phenomenon involves the boundary condition at the open end and the *radiation* of energy from this end. In essence, the Green's function $\mathcal{G}_Q^p$ does not fit a physically realistic boundary condition at the open end. In other words, $p(x = L, t) = 0$ is not the correct boundary condition. In later chapters, we see how to modify the Green's function in Equation (13) to produce more physically realistic results. This process is concluded in Chapter 11. The results in Chapter 11 leads us to change the *effective length* of the tube from L to $L' > L$ in the calculation of normal mode frequencies, as done in Chapter 5. Other than this, the results in Equations (13) through (18) provide some realistic details.

Simulations

In the following simulation we examine the time-average power input by the piston by summing the power inputs of the first five modes given in Equation (17). [At this point we drop the $\langle \cdot \rangle_{t\ >\ 0}$ notation for $\langle \cdot \rangle$.] The damping constants γ_m are set so that $\gamma_1 = 0.2\omega_1$, $\gamma_2 = 0.1\omega_2$, $\gamma_3 = 0.1\omega_3$, $\gamma_4 = 0.05\omega_4$, and $\gamma_5 = 0.1\omega_5$. The following is a list of the normal mode frequencies, $\mathcal{F}_m = \omega_m/(2\pi)$ and their associated reduced normal mode frequencies $\hat{\mathcal{F}}_m$.

$\mathcal{F}_m$ (Hz)	$\hat{\mathcal{F}}_m$ (Hz)
500	490
1500	1492
2500	2487
3500	3496
4500	4477

The cross sectional area of the tube is set to $A = 2$ cm^2.

To obtain plots that are familiar to those seen in speech production research, we take a multiple frequency piston that has frequencies at 120 Hz, as well as integral multiples of this fundamental frequency of 120 Hz, known as harmonics. The amplitude of the volume velocity at the fundamental frequency is taken to be $\mathcal{Q}_{pst} = 400$ cm^3/s. Thus, the piston velocity amplitude at the fundamental frequency is $U_{pst} = 200$ cm/s, because the cross-sectional area of the tube is $A = 2.0$ cm^2. In turn, this corresponds to a peak pressure of 8.3 cm H$_2$O. The subsequent harmonics have their volume velocities reduced at a rate of 1/4 for every doubling of frequency. This corresponds to a power declination of 12 dB per octave. We can use multiple piston frequencies because of the linearity of the system. Figure 4 shows the results in terms time-average power versus frequency.

Figure 5 shows the results in terms of *sound pressure level*, or *SPL*. This is the dB calculation according Equation (42) of Chapter 2, but with a root-mean square pressure, $p_{rms} = \sqrt{\langle p^2(t) \rangle}$, and not for peak pressure, $\max(|p|)$. For sinusoidal time dependence, $p_{rms} = \max(|p|)/\sqrt{2}$. Also, for a sinusoidally varying piston in a tube,

Acoustics of Speech Production 205

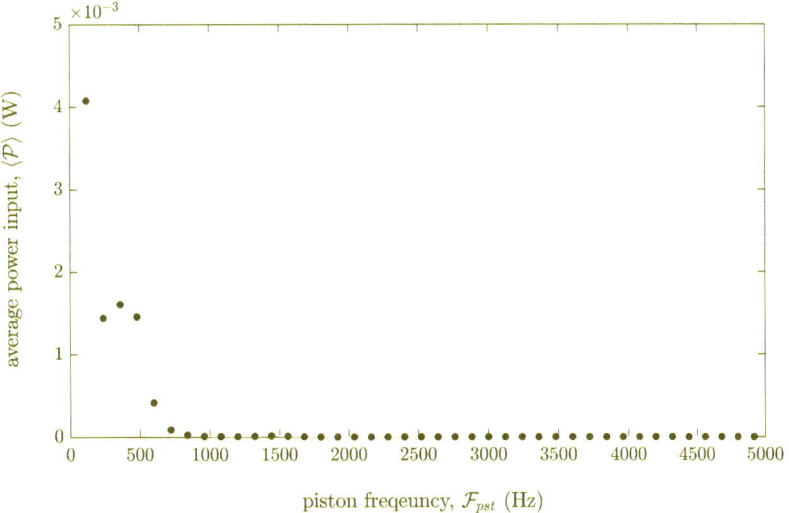

FIGURE 4. Time-average power input $<\mathcal{P}>$ versus piston frequency $\mathcal{F}_{pst}$ for harmonic piston frequencies

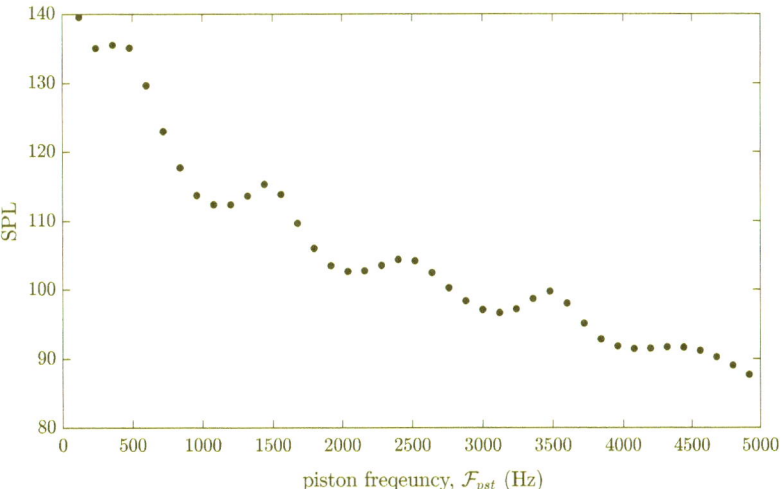

FIGURE 5. SPL versus piston frequency $\mathcal{F}_{pst}$ for harmonic piston frequencies.

$\langle \mathcal{P} \rangle A = p_{rms}^2/(\rho_0 c_0)$. From these considerations

$$\text{SPL} = 20 \log \left(\frac{p_{rms}}{p_{ref}} \right)$$
$$= 10 \log \left(\frac{p_{rms}^2}{p_{ref}^2} \right) \qquad (19)$$
$$= 10 \log \left(\frac{A \rho_0 c_0 \langle \mathcal{P} \rangle}{p_{ref}^2} \right).$$

We can also define something called *overall sound pressure level*, or *OASPL*. If the spectrum is discrete with frequencies $\mathcal{F}_k$ such as occurs for the harmonics of a voice source, then root-mean-square pressure is a function of frequency, or $p_{rsm} = p_{rms}(\mathcal{F}_k)$.

$$\text{OASPL} = 20 \log \left(\frac{\sum_k (p_{rms}(\mathcal{F}_k))}{p_{ref}} \right)$$
$$\approx 10 \log \left(\frac{\sum_k (p_{rms}(\mathcal{F}_k))^2}{p_{ref}^2} \right) \qquad (20)$$
$$= 10 \log \left(\frac{A \rho_0 c_0 \sum_k \langle \mathcal{P}(\mathcal{F}_k) \rangle}{p_{ref}^2} \right).$$

This is an approximation because the time-averages of products of quantities belonging to different modes are not necessarily zero, but they should be small.

The OASPL in Figure 5 is 141 dB. In Chapter 13 we discuss the reasons why this may not be indicative of the amplitude of sound just above the glottis and the rest of the supraglottal vocal tract in conversational speech.

Source location effects

It is clear that we obtain maximum response in pressure to the volume velocity of the piston when the piston moves sinusoidally at a frequency close to one of the normal mode frequencies. [The damping constants, and hence, the resonance frequencies, are not completely specified until later, but they will be somewhat close to each $\mathcal{F}_m$.]

We have seen this to be true for the volume velocity piston source at $y = 0$. The type of source, volume velocity source or pressure source, and the source location y does not have any effect the resonance frequencies. However, source location in conjunction with source type makes a difference on how strongly the acoustics of the finite-length tube responds.

In order to discuss source location effects we return to the undamped Green's functions derived in Chapter 6, $\mathcal{G}_Q^p$ and $\mathcal{G}_p^Q$. We write these again, along with $\mathcal{G}_Q^Q$ and $\mathcal{G}_p^p$ in the next two equations. Equations (35) and (45) of Chapter 6 are

$$\mathcal{G}_Q^p(x, y, t - \tau) = -\frac{2\rho_0 c_0^2}{AL} \mathcal{H}(t - \tau) \times$$
$$\sum_{m=1}^{\infty} \cos(k_m y) \cos(k_m x) \frac{\sin(\omega_m(t - \tau))}{\omega_m},$$

and (21)

$$\mathcal{G}_p^Q(x, y, t - \tau) = -\frac{2A}{\rho_0 L} \mathcal{H}(t - \tau) \times$$
$$\sum_{m=1}^{\infty} \sin(k_m y) \sin(k_m x) \frac{\sin(\omega_m(t - \tau))}{\omega_m}.$$

The conservation equations, Equations (12) and (13) of Chapter 2, can be applied to these expressions to obtain

$$\mathcal{G}_Q^Q(x, y, t - \tau) = \frac{2c_0}{L} \mathcal{H}(t - \tau) \times$$
$$\sum_{m=1}^{\infty} \cos(k_m y) \sin(k_m x) \frac{\cos(\omega_m(t - \tau))}{\omega_m},$$

and (22)

$$\mathcal{G}_p^p(x, y, t - \tau) = -\frac{2c_0}{L} \mathcal{H}(t - \tau) \times$$
$$\sum_{m=1}^{\infty} \sin(k_m y) \cos(k_m x) \frac{\cos(\omega_m(t - \tau))}{\omega_m}.$$

We see that both $\mathcal{G}_Q^p$ in Equation (21) and $\mathcal{G}_p^p$ in Equation (22) have their spatial variation in x as $\cos(k_m x)$. This is because we are observing $p(x, t)$, so that the spatial variation in x are expressed by the pressure eigenfunctions. Similarly for $\mathcal{G}_p^Q$ in Equation (21) and $\mathcal{G}_Q^Q$ in Equation (22), we find the spatial variations in the observed $Q(x,t)$ to be expressed by its eigenfunctions, $\sin(k_m x)$.

We also find commonality in the variation of source position y for similar sources. Equations (21) and (22) show that the spatial dependence in source location y for $\mathcal{G}_Q^p$ and $\mathcal{G}_Q^Q$ are the same, with the functions $\cos(k_m y)$. These are the eigenfunctions for pressure. The Green's functions $\mathcal{G}_p^Q$ and $\mathcal{G}_p^p$ have dependence $\sin(k_m y)$ in source location y. These are the eigenfunctions for volume velocity. Therefore, volume velocity sources are most effective where pressure in the finite-length tube has maximum amplitude. Similarly, pressure sources are most effective at places where volume velocity in the finite-length tube has maximum amplitude. This makes sense when we consider the fact that time-average power input is the time-average of pressure times volume velocity.

Let's examine the effect that volume velocity source location y has on the magnitude of the resulting acoustic motion in the tube in more detail. Recall that Equation (15) of Chapter 5 gives $k_m = (2m-1)\pi/(2L)$ for $m = 1, 2, 3, \ldots$. Suppose that we examine a single mode m. We note that $|\cos(k_m y)|$ is greatest when $k_m y = n\pi$ for some integer n, for n = 0, 1, ... (m - 1). Since $0 < y < L$, this means that $|\cos(k_m y)|$ is greatest when $y = 2nL/(2m-1)$, $n = 0, 1, \ldots (m-1)$. These are the locations along the tube where a volume velocity source would be most effective in exciting the mth normal mode. These are the locations for the anti-nodes for the pressure normal modes, which are the maxima of $|\cos(k_m y)|$. Even when the forcing frequency is not close to the mth normal mode frequency, a volume velocity source is in the right place to transfer energy into the mth mode at the pressure anti-nodes, because it is the time-average of the volume velocity of the source and the pressure of the acoustic wave that contributes to time-average power input. On the other hand, if a volume velocity source is located at a pressure node of the mth mode, that is where $\cos(k_m y) = 0$, then this mode will not be excited at any piston circular frequency ω_{pst}. These locations occur for $k_m y = (2n-1)\pi/2$ for $n = 1, 2, \ldots, m$. These are the positions are $y = (2n-1)L/(2m-1)$, $n = 1, 2, \ldots, m$.

Similar considerations can be applied to pressure sources. The functions of interest are $|\sin(k_m y)|$. Pressure sources are most effective where volume velocity sources are not, and they are the least effective where volume velocity sources are the most effective.

The most common volume velocity source in speech is the voice. As a first approximation, we model the voice source region as a piston. [In Chapter 13 we see that if the subglottal tract is included that we can characterize the voice source differently.] Thus, the fluctuating volume velocity at the glottis is optimally situated for exciting all the normal modes at $x = 0$, because all modes have a pressure anti-node at the piston. However, aspiration is not optimal in the glottal region, because it is a pressure source. In the real vocal tract, the glottal region does not provide an exact hard-wall boundary condition of a piston. So the volume velocity does not actually possess an exact node at the glottis. This is heightened by the fact that the vocal folds are abducted for aspiration. Still, the basic picture does suggest that the voice source is, potentially, highly effective in exciting all acoustic modes, while the aspiration pressure source is only weakly coupled to the acoustic motion.

About representations with modes

The non-acoustic nature of the solution with Green's functions

We have used the phrase: "satisfies the wave equation" in this book. The wave equation was derived from the conservation equations of Chapter 2. We need to keep in mind that the wave equation does not include the physics of acoustic sources. In Chapter 3 we worked out the acoustic evolution of waves that are created by an acoustic source, namely a moving piston. Acoustic waves are generated at the piston face with the time dependence of the piston. Their evolution can be tracked simply as waves: as objects that satisfied the wave equation. This procedure is correct, as long as one is careful to remember that the wave equation is valid everywhere in the tube other than at the piston itself. This is why we were so careful to separate the contribution of the boundary condition at the piston into the hard-wall condition [i.e. volume velocity $Q(x = 0, t) = 0$] and a source condition [i.e. volume velocity $Q(x = 0, t) = Q_{pst}(t)$]. We found a pattern when the piston moved for a brief time to create a pulse. We gave this procedure up when it came to the continuously moving piston, because the acoustic phenomena became too complicated for us to follow.

One reason for considering standing waves and normal modes for the finite-length tube is that they provide a humanly understandable picture of acoustic wave motion in a finite-length tube for the continuously moving piston source. However, using the Green's function that is composed of terms that represent modes can lead us a little astray in terms of understanding acoustics in a finite-length tube.

A double convolution sum with the volume velocity of the piston and the Green's function $\mathcal{G}_Q^p$ provides the solution for pressure in the tube with a volume velocity source. The steady phase of this solution is shown in Equation (16). It should be kept in mind that the individual terms of Equation (16) do not satisfy the wave equation. The source and the acoustics are tightly coupled in this representation, derived using the convolution sums. The standing waves oscillating at the piston circular frequency ω_{pst} becomes apparent with the double convolution sum, but the underlying acoustics in the finite-length tube is not apparent.

Implications for speech synthesis

There are two basic ways that one can build a speech synthesizer that transforms the shape of the vocal tract to speech. One way is to calculate formant frequencies and bandwidths from the vocal tract shape, and then to use digital filters with sources, such as a simulation of the voice source, to generate the output acoustics. This method is sometimes called *formant synthesis* (Klatt 1980), which is based on considerations of steady phase solution in the present chapter. This method does not account for transients in the signal.

The other way to build such a synthesizer is to do what we did initially in Chapter 3, and simply compute behavior in time. There are mathematical difficulties that one encounters when doing this, such as the frequency dependence of the damping constants. However, these difficulties are surmountable. Maeda (1982) provides such a synthesizer.

It can be argued that the latter approach to synthesis is to be preferred, because transients are naturally a part of the computation. In formant synthesis with digital filters, we are not necessarily taking account of the correct transient signal.

Connection with source-filter theory

The rudiments of the *source-filter theory* for the acoustics of speech production (Fant 1960; Flanagan 1965; Stevens 1998) are in the piston-in-a-finite-tube picture presented here, except, of course, the tube can be of various shapes. The voice source can be represented by the piston, where the movement of the piston is controlled by an external agent, which is unaffected by the acoustic waves in the tube. Not only is the piston a source of energy for acoustic wave motion, but it also is a hard-wall boundary where the particle velocity is zero when the piston is at rest.

We have insisted on expressing the boundary condition at the piston in terms of two components: one due to the hard-wall boundary, and the other due to the movement of the piston. The reason for this is that we wish to consider the tube to be an acoustic resonator without regard to a source of energy. The supraglottal vocal tract at the glottal end approximates a hard wall because the glottal opening is usually very small compared with the cross-sectional area of the vocal tract near that end. The source, during voicing, are the time-varying puffs of air through the glottis. In the source-filter theory, these puffs are unaffected by the acoustic motion in the supraglottal vocal tract. Thus, we have the piston-in-a-tube scenario in source-filter theory.

We have been considering the time domain aspects of the acoustics in a finite-length tube. Much of the source-filter theory of speech production has been derived in the frequency domain. In the frequency domain, mathematical entities called *transfer functions* are employed in the computation of acoustics. For instance, with a more physically realistic boundary condition at the open end of the finite-length tube, we often compute the volume velocity at this opening due to the volume velocity of the voice source or the pressure fluctuations of a fricative in the frequency domain. The volume velocity at the opening determines what a listener receives in terms of acoustics. This means that the Green's functions in the frequency domain, $\tilde{\mathcal{G}}_Q^Q(x,y,\omega)$ and $\tilde{\mathcal{G}}_p^Q(x,y,\omega)$ are of great use. In fact, we evaluate these for the observed Q at $x = L$. In the case of the voice source $y = 0$, and for a sibilant fricative, say, $y = y_0$, with $0 \leq y_0 \leq L$. Thus, the transfer function for the voice source is $\tilde{\mathcal{G}}_Q^Q(x = L, y = 0, \omega)$, and the transfer function for the sibilant fricative source is $\tilde{\mathcal{G}}_p^Q(x = L, y = y_0, \omega)$. What had been convolution sum in the time domain becomes multiplication in the frequency domain. This is known as the *convolution theorem*. Therefore

$$\tilde{Q}(x = L, \omega) = \tilde{\mathcal{G}}_Q^Q(x = L, y = 0, \omega) \cdot \tilde{Q}^{ext}(y = 0, \omega) ,$$

or (23)

$$\frac{\tilde{Q}(x = L, \omega)}{\tilde{Q}^{ext}(y = 0, \omega)} = \tilde{\mathcal{G}}_Q^Q(x = L, y = 0, \omega)$$

for the voice source. And

$$\tilde{Q}(x = L, \omega) = \tilde{\mathcal{G}}_p^Q(x = L, y = y_0, \omega) \cdot \tilde{p}^{ext}(y = y_0, \omega) ,$$

or (24)

$$\frac{\tilde{Q}(x = L, \omega)}{\tilde{p}^{ext}(y = y_0, \omega)} = \tilde{\mathcal{G}}_p^Q(x = L, y = y_0, \omega)$$

for a sibilant fricative, a pressure source, at $y = y_0$.

Conclusion

With damping comes time-average energy input to the system from the source. While perceptible acoustic waves do not require large amounts of energy, they do require some. Without explicitly identifying the damping mechanisms, we have discovered that solutions based on undamped normal modes can help to understand finite-length tube acoustics with damping. Analogies with the mass-spring system and single modes of the finite-length tube are strong, but there are three important differences between the tube acoustics and the mass-spring systems. Firstly, pressure and particle velocity fluctuations are distributed throughout the tube. Second, there is an infinity of normal mode frequencies. Third, there is a new spatial component to the pressure distributions for each mode m, $\bar{p}_m(x)$. One effect of this is to slightly alter the calculated resonance frequencies based on maximum power input as a function of piston frequency. It is shown later that there is a physical effect due to air motion that overwhelms the small alteration in resonance frequency just noted.

What has been discussed in terms of a finite-length tube with constant cross-sectional area generalizes to tubes with variable cross-sectional area. With variable cross-sectional area, normal mode frequencies are no longer 1000 Hz apart, and the spatial variation of the normal modes will no longer be circular functions, sine and cosine. However, normal modes exist in the case for variable-area tubes. It is just a matter of how one estimates the frequencies and spatial variation of pressure and volume velocity for such normal modes that is of interest.

Where do we proceed from here now that we have built a conceptual foundation in order to understand much of vocal tract acoustics? Tubes of variable cross-sectional area can be approximated with short sub-tubes, each of constant cross-sectional area. In order to perform meaningful calculations pertaining to normal modes in this situation, we assume sinusoidal time variation. In essence, we work in the frequency domain with a variable circular frequency ω. Thus, we "slide" into the frequency domain in such a way by examining the amplitudes A and phases ϕ in functions of the form $A\cos(\omega t + \phi)$, where we can consider both the A and ϕ to be functions of ω. That is, $A = A(\omega)$ and $\phi = \phi(\omega)$. This is the the way some physicists work in the frequency domain without having to consider Fourier transforms.

While much of our work is, or could be, done without complex variables, we introduce them in the next chapter. There are two reasons for doing this. Firstly, they are a convenient way to keep track of phase $\phi(\omega)$, as well as amplitude $A(\omega)$, when making transitions between many sub-tubes. Second, much of the relevant literature in acoustics uses complex variables.

References

Fant, G. (1960). *Acoustic Theory of Speech Production*. Mouton, The Hague.

Flanagan, J.L. (1965). *Speech Analysis, Synthesis, and Perception*. Springer-Verlag, New York.

Klatt, D.H. (1980). Software for a cascade/parallel formant synthesizer. *Journal of the Acoustical Society of America* **67**, p 971.

Maeda, S. (1982). A digital simulation method of the vocal-tract system. *Speech Communication* **1**, p 199.

Stevens, K.N. (1998). *Acoustic Phonetics*. MIT Press, Cambridge, Massachusetts.

Chapter 8: Introduction to Complex Variables for Acoustics

Introduction

In future chapters we discuss tubes whose cross-sectional area can change as a function of position along the tube axis. Many of the concepts developed in the previous chapters for the finite-length tube of constant cross-sectional area still apply for these variable-area tubes. However, to develop the concepts for variable-area, finite-length tubes, computations that iterate on short, constant-area sub-tubes are performed. In these computations, the relationships of physical quantities, such as pressure and volume velocity, among various sub-tubes need to be accounted for systematically. Complex variable representations are very useful for this task, as shown, for example, in Morse (1976). Computation can become unwieldy without complex variables.

The following is a short introduction to complex variables. We employ complex variables in a limited way, so that only the most elementary properties of this very powerful mathematical area are considered. The reader who wants more instruction in the elementary properties of complex variables can refer to a number of texts, such as the one by Polya & Latta (1974).

Complex numbers as two-space vectors

In essence, when complex variables are used, we add a dimension to the real number axis, and we call the second dimension the *imaginary numbers*. This two dimensional space is termed the *complex plane*. *Complex numbers* are points in the complex plane. [This is unfortunate terminology, but we are stuck with it: something imaginary residing in something that is complex.]

Complex numbers can be described in the complex plane using vectors. We introduce the symbol i to denote a vector in the imaginary direction, as shown in Figures 1 and 2. The real number 1 denotes a vector in the real direction. Both of these vectors have magnitude 1.

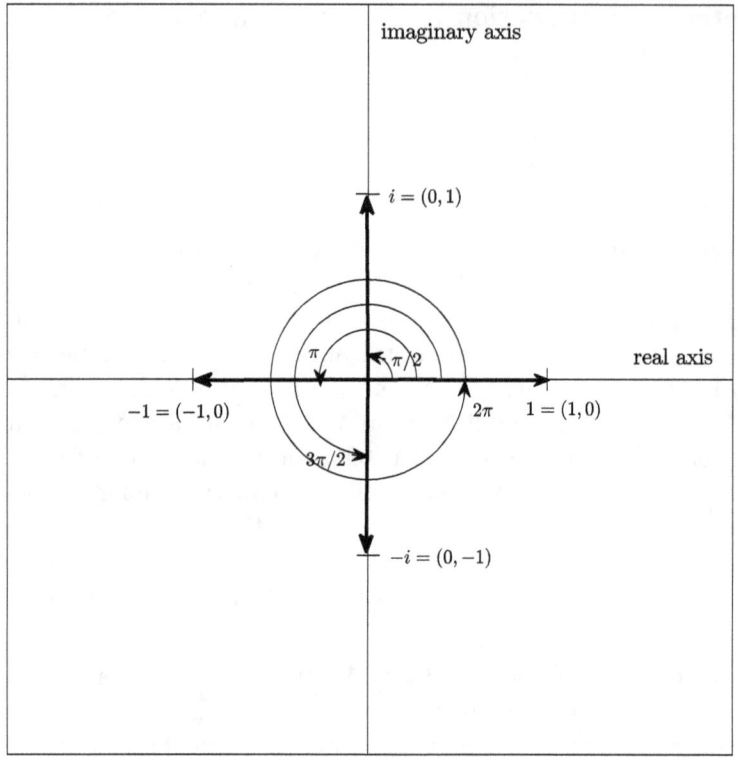

FIGURE 1. The complex plane, I

That is, $|1| = |i| = 1$. Because we are working in a two-dimensional space we can speak of angles. The angle between 1 and i is $\pi/2$ or $-3\pi/2$, the angle between 1 and -1 the angle is π or $-\pi$, and the angle between 1 and $-i$ the angle is $3\pi/2$ or $-\pi/2$. In general, any vector, or complex number, in the this two dimensional space can be written $a + ib$, where a and b are real numbers. The vector $a + ib$ has component a in the 1, or real, direction, and b is the component in the i, or imaginary, direction.

Figure 3 shows three examples of complex numbers represented as vectors in the complex plane. The real and imaginary components are indicated by the dashed lines. The complex number $1 + i$ has magnitude, $|1 + i| = \sqrt{1^2 + 1^2} = \sqrt{2}$. This is the length of the hypotenuse of a right triangle, whose two sides have length 1 along

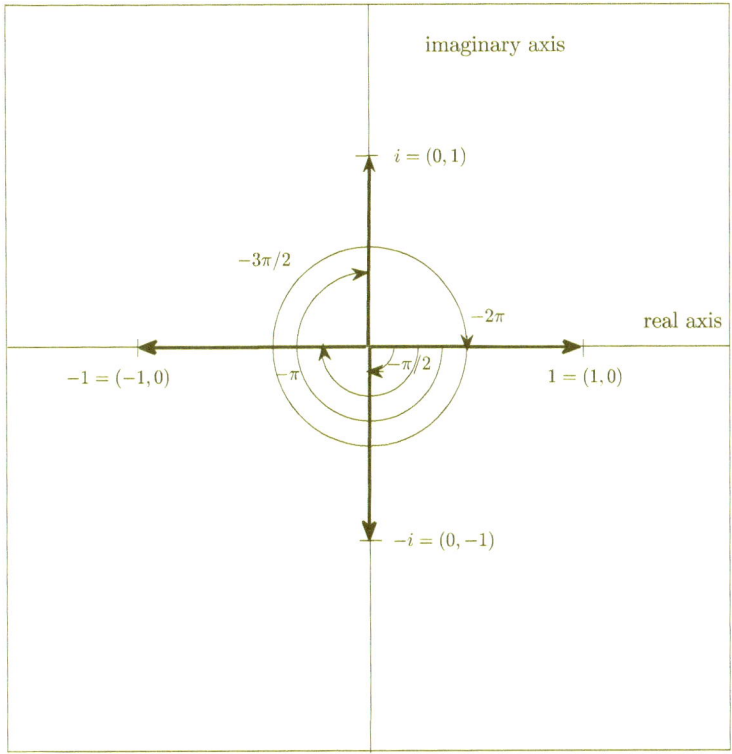

FIGURE 2. The complex plane, II

both the real and the imaginary axes. Let α be the angle between the hypotenuse and the adjacent side, which is along the real axis, as shown for the examples in Figure 3. The cosine of α is $\pm$ the ratio between the length of that adjacent side divided by the length of the hypotenuse. For $1 + i$ we have $\cos(\alpha) = \pm 1/\sqrt{2}$. When the vector is in the first or fourth quadrant, the real component is positive, and the plus sign is taken. [The quadrants are denoted by Roman numerals in Figure 3.] Therefore $\cos(\alpha) = 1/\sqrt{2}$, because $1 + i$ is in the first quadrant. $\sin(\alpha)$ is equal to $\pm$ the length of the opposite side over the length of the hypotenuse, where the opposite side is along the imaginary axis. So, for $1 + i$ we have $\sin(\alpha) = \pm 1/\sqrt{2}$. When the vector is in the first or second quadrant, the plus sign is taken, because in these quadrants the imaginary component is positive. For $1 + i$, we

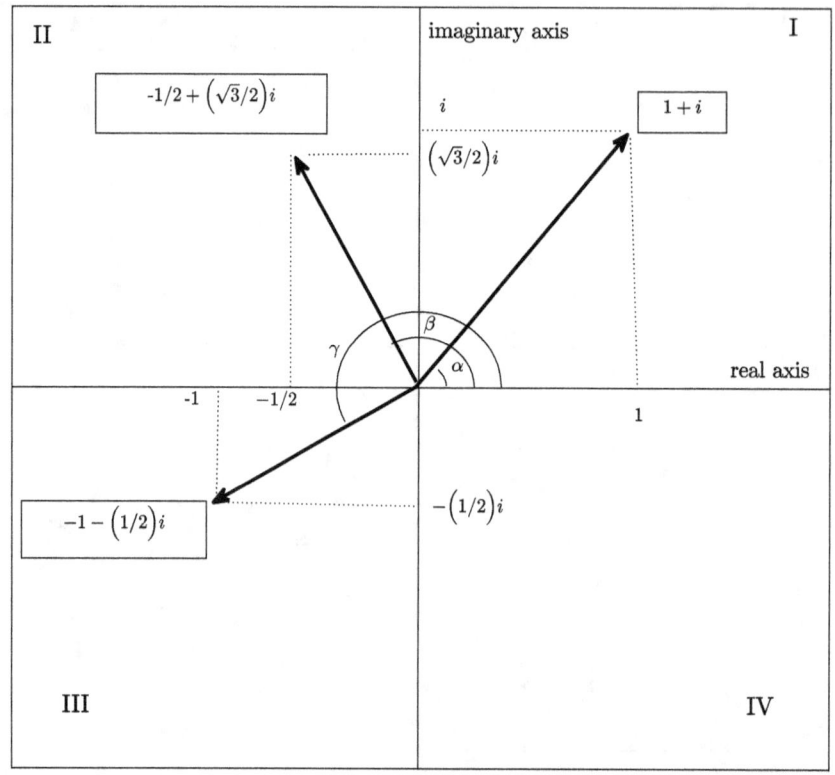

FIGURE 3. Complex numbers in the complex plane

have $\sin(\alpha) = 1/\sqrt{2}$. The unique α, with $0 \leq \alpha < 2\pi$, that is specified by these cosine and sine values is $\alpha = \pi/4$.

We briefly discuss the other examples in Figure 3. The complex number $-1/2 + (\sqrt{3}/2)i$ has magnitude, $\big|-1/2 + i(\sqrt{3}/2)i\big| = \sqrt{(-1/2)^2 + (\sqrt{3}/2)^2} = 1$. The angle that $-1/2 + i(\sqrt{3}/2)$ makes with the real axis, β, is such that $\cos(\beta) = -1/2$ and $\sin(\beta) = \sqrt{3}/2$. It turns out that $\beta = 2\pi/3$. The complex number $-1 - i/2$ has magnitude $|-1 - i/2| = \sqrt{(-1)^2 + (-1/2)^2} = \sqrt{5}/2$. The angle that this vector makes with the positive real axis, γ, is such that $\cos(\gamma) = -2/\sqrt{5}$ and $\sin(\gamma) = -1/\sqrt{5}$. This gives $\gamma \approx 1.14\pi$.

Let z be a complex variable. Then

$$z = Re(z) + iIm(z) \,, \tag{1}$$

where both $Re(z)$ and $Im(z)$ are real numbers, known as the *real part* and *imaginary part* of z, respectively. If we set $z = a + ib$, then $Re(z) = a$ and $Im(z) = b$. Let the magnitude of z be denoted $|z|$ and let $\arg(z)$ denote the angle that z makes with the positive real axis. It is true that

$$Re(z) = |z|\cos(\arg(z)) \text{ and } Im(z) = |z|\sin(\arg(z)) \,,$$
$$\text{so that } z = |z|\big(\cos(\arg(z)) + i\sin(\arg(z))\big) \,. \tag{2}$$

This can be verified by the reader for the examples in Figure 3. For $z = 1 + i$, $|z| = \sqrt{2}$ and $\arg(z) = \alpha = \pi/4$.

The angle for complex numbers has a number of names. The angle is the argument of the complex number, which is denoted arg. For physics, it represents the phase of the corresponding physical quantity. We have seen in the discussion of acoustic wave motion surrounding Equations (5) through (9) in Chapter 5 that phase has both time and spatial components.

The algebra of complex variables

The basics

Complex numbers and variables have mathematical power beyond what is afforded to them by their geometric interpretation, because a consistent algebra can be defined for these numbers. We stipulate that $i^2 = -1$ and permit the usual properties of the algebra of real numbers to continue to hold, such as the commutative property (i.e. $zw = wz$ for any two complex numbers z and w) and the associative property (i.e. $z(w + s) = zw + zs$ for any three complex numbers z, w and s). This enables a very powerful algebraic system.

Let's perform some simple operations symbolically, using complex variables z and w. What does the product of z and w look like in detail? We can write the product of z and w as another complex variable, s.

$$\begin{aligned}
s \equiv zw &= \big(Re(z) + iIm(z)\big)\big(Re(w) + iIm(w)\big) \\
&= Re(z)Re(w) + iRe(z)Im(w) + iIm(z)Re(w) + i^2 Im(z)Im(w) \\
&= Re(z)R(w) + i\big(Re(z)Im(w) + Im(z)Re(w)\big) - Im(z)Im(w) \quad (3) \\
&= \big(Re(z)Re(w) - Im(z)Im(w)\big) + i\big(Re(z)Im(w) + Im(z)Re(w)\big) \ .
\end{aligned}$$

Thus,

$$\begin{aligned}
s &= Re(s) + iIm(s), \text{ with} \\
Re(s) &= Re(z)Re(w) - Im(z)Im(w) \quad \text{and} \\
Im(s) &= Re(z)Im(w) + Im(z)Re(w) \ .
\end{aligned} \quad (4)$$

Complex exponential functions and Euler's theorem

Complex variables are objects that represent complex numbers. Complex functions are functions that take complex numbers and map them onto other complex numbers. Here, we introduce a specific complex function, the *complex exponential function*. The notation used for this function is e^z, where z is a complex variable and e is known as the natural number. ($e \approx 2.71$ is the natural number). We restrict our discussion to the case where $z = i\theta$, for real θ.

The usual rules for multiplication and division apply for complex exponential functions as they do for the real exponential functions.

$$e^{i\theta} e^{i\phi} = e^{i(\theta+\phi)} \ . \quad (5)$$

And

$$\frac{1}{e^{i\theta}} = e^{-i\theta} \ . \quad (6)$$

It follows from Equations (5) and (6) that

$$\frac{e^{i\phi}}{e^{i\theta}} = e^{i(\phi-\theta)} \ . \quad (7)$$

The reason that we are using the Greek letters usually associated with angle is that, for our purposes, they denote angles in the complex plane. This is discussed presently.

There is a theorem, Euler' theorem, that makes complex exponential functions particularly useful for representing complex numbers and variables. Euler's theorem (Polya & Latta 1974) is

$$e^{i\theta} = \cos(\theta) + i\sin(\theta) \ . \tag{8}$$

Note that $\left|e^{i\theta}\right| = \left|\cos(\theta) + i\sin(\theta)\right| = \sqrt{\cos^2(\theta) + \sin^2(\theta)} = 1$, for all real θ. Thus, $e^{i\theta}$ lies on the unit circle, which is the circle of radius one with the center at the origin. $e^{i\theta}$ is rotated angle θ from the positive real axis.

As we have seen in Equation (2), any complex variable z can be written as $z = |z|\bigl(\cos(\arg(z)) + i\sin(\arg(z))\bigr)$. With Euler's theorem we have another representation for z.

$$z = |z|e^{i\arg(z)} \ . \tag{9}$$

The algebra of complex variables is a little less cumbersome using complex exponential functions and Euler's theorem. For example, the product of w and z can be written as

$$\begin{aligned}
s = zw &= |z|e^{i\arg(z)} \, |w|e^{i\arg(w)} \\
&= |z||w|e^{i\arg(z)}e^{i\arg(w)} \\
&= |z||w|e^{i(\arg(z)+\arg(w))} \ .
\end{aligned} \tag{10}$$

Thus,

$$\begin{aligned}
s &= |s|e^{i\arg(s)} \\
&= |z||w|e^{i(\arg(z)+\arg(w))} \ .
\end{aligned} \tag{11}$$

Therefore,

$$\begin{aligned}
|s| &= |z||w| \ , \quad \text{and} \\
\arg(s) &= \arg(z) + \arg(w) \ .
\end{aligned} \tag{12}$$

It can be shown that Equations (3) and (11) provide the same result.

An interesting exercise is to equate the two representations of the product shown in Equations (4) and (12), with $z = e^{i\theta}$ and $w = e^{i\varphi}$ or $1/w = e^{-i\varphi}$. When this is done, and Euler's theorem, Equation (8), is invoked the trigonometric identities for the sine and cosine of the sum and difference of angles in Equation (1) of Chapter 5 can be proven. This is done in the Appendix to Chapter 8. This leads us to a way of calculating a *multiplicative inverse*. w is the multiplicative inverse of

z, if $wz = 1$. The multiplicative inverse is

$$z^{-1} = \frac{1}{z} = \left(\frac{1}{|z|}\right) e^{-i \arg(z)} . \tag{13}$$

Using the definition of magnitude and Euler' theorem in Equation (9), Equation (13) is rewritten

$$z^{-1} = \left(\frac{1}{\sqrt{Re(z)^2 + Im(z)^2}}\right) \left(\cos(\arg(z)) - i \sin(\arg(z))\right) . \tag{14}$$

Complex multiplication as an operator in the complex plane

Consider a positive real number a. If it is multiplied by i, then it becomes a vector of the same magnitude, but with a direction that is rotated $\pi/2$ counter-clockwise from the real axes to the positive imaginary direction. Multiply this by i, and we obtain the vector $-a$, which is ia rotated counter-clockwise by the angle $\pi/2$. Multiply $-a$ by i, and again we rotate counter-clockwise another $\pi/2$ to $-ia$. Finally, another multiplication by i brings us back to a. Thus, we can think of multiplication by i as an *operator* on complex variables, such that when a complex variables is multiplied by i, that result is a complex variable with the same magnitude as the original variable, but rotated counter-clockwise by $\pi/2$ from the original variable. Note that $i = e^{i\pi/2}$.

This is a general property of multiplication by $e^{i\theta}$, for real θ. If we take any complex variable, say w, and multiply it by $e^{i\theta}$, the result is a complex variable that is rotated counter-clockwise by the angle θ in the complex plane, and with the original magnitude $|w|$. This follows from Equation (12), where it is seen that multiplication of complex variables means that angles add.

Complex conjugate

For $z = Re(z) + iIm(z) = |z|e^{i \arg(z)}$, the *complex conjugate* of z, z^* is defined

$$z^* = Re(z) - iIm(z) = |z|e^{-i \arg(z)} . \tag{15}$$

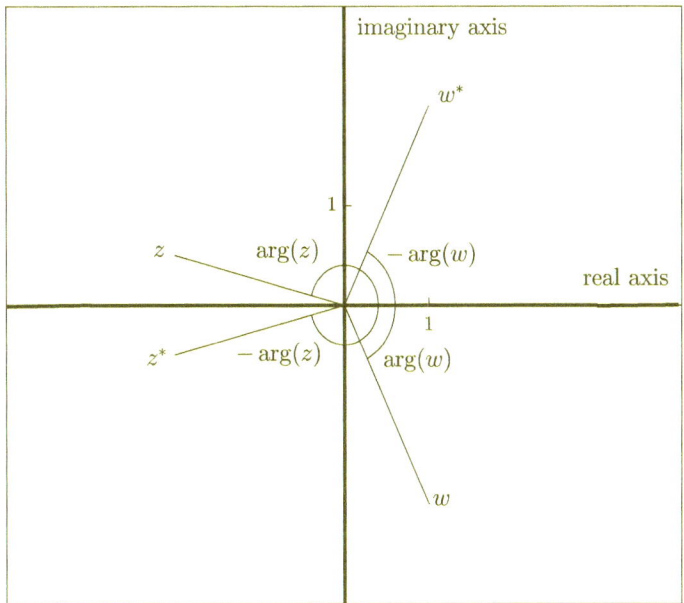

FIGURE 4. Complex conjugates

It follows from Equation (15) that, $|z^*| = |z|$ and $\arg(z^*) = -\arg(z)$. Geometrically, z^* is the image of z in the real axis. Figure 4 illustrates the geometric relation between complex conjugates for $z = -2 + i0.5$ and $w = 1 - i2$. In this case $z^* = -2 - i0.5$ and $w^* = 1 + i2$.

It also follows from the definition of complex conjugate that

$$zz^* = |z|^2 ,$$
and
$$\frac{1}{z} = \frac{z^*}{zz^*} = \frac{z^*}{|z|^2} .$$
(16)

Figure 5 shows the multiplicative inverses for the z and w, $1/z$ and $1/w$.

Physical quantities in complex notation

The acoustic variables that we have studied so far in terms of real variables, such as pressure and volume velocity, are now represented

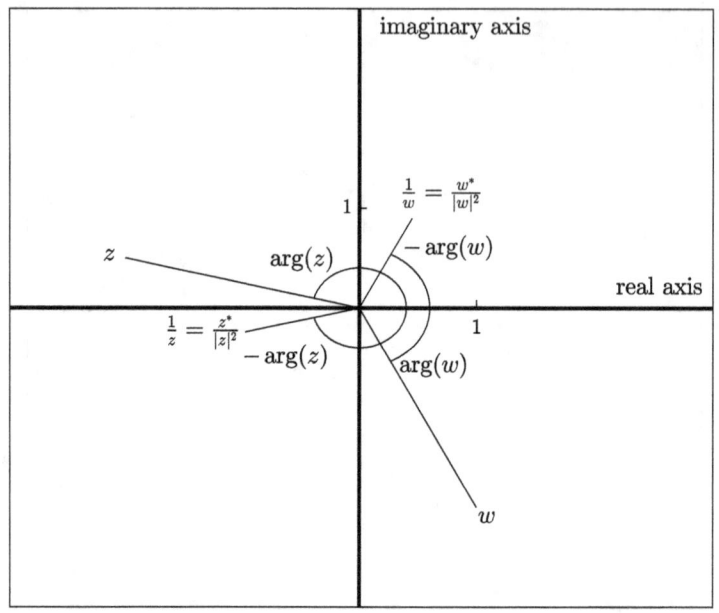

FIGURE 5. Multiplicative inverses

in terms of complex variables. Suppose $p(x,t) = B\cos(\omega t \pm kx + \phi)$ is the pressure of a traveling wave in terms of real quantities, B, ω, t, k, x, and ϕ, with $k = \omega/c_0$. This means that $p(x,t)$ is a real quantity. Note that $\cos(\omega t \pm kx + \phi) = Re(e^{-i(\omega t \pm kx + \phi)})$, so that $p(x,t) = Re(Be^{-i(\omega t \pm kx + \phi)})$, for real B. We use $e^{-i(\omega t + \phi)}$ for the complex representation of sinusoidal time dependence, and $e^{\pm ikx}$ for the complex representation for the spatial dependence.

Very often, authors simply write $p(x,t) = Be^{-i(\omega t \pm kx + \phi)}$, as, for example, is done in Morse (1976). While we write $p(x,t)$ as a complex function, it is understood that it is just the real part that represents the physical quantity. In general B may be complex, but if B is complex, the phase ϕ can be adjusted to ensure that B is real, and, further, that $B > 0$. [$B > 0$ implies that B is real.] We most often assume that ϕ has been adjusted for this to be the case.

One reason that the complex representation is employed is that it is easy to keep track of the arguments, or phases, like $-(\omega t - kx + \phi)$. For instance, suppose that a pressure wave, $p_I(x,t)$, with sinusoidal time dependence with circular frequency ω is an incident wave from the left on the opening of a tube at $x = L$.

$$p_I(x, t) = B e^{-i(\omega t - kx + \phi)}, \text{ with } B > 0. \tag{17}$$

where $\omega = c_0 k$. If $p_R(x,t)$ is the reflected wave that travels to the left then

$$p_R(x, t) = C e^{-i(\omega t + kx + \theta)}, \text{ with } C > 0, \tag{18}$$

for some real θ. For zero pressure at $x = L$

$$p_I(x = L, t) + p_R(x = L, t) = 0. \tag{19}$$

This implies that

$$B = C \text{ and } -kL + \phi = kL + \theta \text{ or } \theta = -2kL + \phi. \tag{20}$$

This calculation is made easy using the complex exponential function to represent pressure disturbances.

We note that we equated the complex representations of physical quantities in Equation (19). This is done all the time. All that is required physically is that the real parts of the complex equations hold. By stating equations in complex form, we require that the real parts be equal, as well as the imaginary parts. Thus, equality in complex representation requires equality in the real, physical representation.

Standing waves are often written in complex form as

$$p(x, t) = B f(x) e^{-i(\omega t + \phi)}. \tag{21}$$

where $B > 0$, ω, and ϕ are real, and $f(x)$ is a real-valued function of axial position along the tube. In the case of the finite-length tube with length L, $f(x) = \cos(k_m x)$ with $k_m = \big((2m-1)\pi/(2L)\big)$, as shown in Chapter 5.

Acoustic impedance, acoustic admittance, and power

In regions without damping and without acoustic sources, both pressure p, and volume velocity Q satisfy the wave equation. In the most general case of a combination of a right-going (or positive-going) wave and a left-going (or negative-going) wave, both with circular function time-dependence of circular frequency ω, pressure has a complex representation

$$p(x, t) = B_1 e^{-i(\omega t - kx + \phi_1)} + B_2 e^{-i(\omega t + kx + \phi_2)}, \tag{22}$$

with real constants $B_1 > 0, B_2 > 0, \phi_1$, and ϕ_2, and $\omega = c_0 k$. Algebra shows that Equation (22) can be written in the complex form

$$p(x,t) = B(x,\omega)e^{-i(\omega t + \phi(x,\omega))} , \qquad (23)$$

where $B(x,\omega) > 0$ and $\phi(x,\omega)$ are real-valued functions of x and ω. This is shown in the Appendix to Chapter 8. Similarly,

$$Q(x,t) = C(x,\omega)e^{-i(\omega t + \theta(x,\omega))} , \qquad (24)$$

for some real valued functions $C(x,\omega) > 0$ and $\theta(x,\omega)$.

On the other hand, the relationship between pressure and volume velocity is most often expressed in terms of the ratio of complex $p(x,t)$ to complex $Q(x,t)$, or its inverse. The former ratio is termed the *acoustic impedance*, Z, and its multiplicative inverse is termed the *acoustic admittance* Y.

$$\begin{aligned} Z &\equiv \frac{p(x,t)}{Q(x,t)} \\ &= \frac{B(x,\omega)e^{-i(\omega t + \phi(x,\omega))}}{C(x,\omega)e^{-i(\omega t + \theta(x,\omega))}} \\ &= \frac{B(x,\omega)}{C(x,\omega)} e^{-i(\phi(x,\omega) - \theta(x,\omega))} . \end{aligned} \qquad (25)$$

Note that, $Z = Z(x,\omega)$. The magnitude of Z is

$$\begin{aligned} |Z| &= \frac{B(x,\omega)}{C(x,\omega)} \\ &= \frac{|p|}{|Q|} . \end{aligned} \qquad (26)$$

The phase, or angle, of Z is the phase difference between p and Q.

$$\begin{aligned} \arg(Z) &= -\phi(x,\omega) + \theta(x,\omega) \\ &= \arg(p) - \arg(Q) . \end{aligned} \qquad (27)$$

The acoustic admittance Y is just the reciprocal of Z.

$$\begin{aligned}
Y &\equiv \frac{Q(x,t)}{p(x,t)} \\
&= \frac{C(x,\omega)e^{-i(\omega t+\theta(x,\omega))}}{B(x,\omega)e^{-i(\omega t+\phi(x,\omega))}} \\
&= \frac{C(x,\omega)}{B(x,\omega)}e^{-i(\theta(x,\omega)-\phi(x,\omega))} \ .
\end{aligned} \quad (28)$$

The magnitude of Y is

$$\begin{aligned}
|Y| &= \frac{C(x,\omega)}{B(x,\omega)} \\
&= \frac{|Q|}{|p|} \ .
\end{aligned} \quad (29)$$

The phase, or argument, of Y is the phase difference between Q and p.

$$\begin{aligned}
\arg(Y) &= -\theta(x,\omega) + \phi(x,\omega) \\
&= \arg(Q) - \arg(p) \ .
\end{aligned} \quad (30)$$

Both Z and Y provide information about the relative magnitudes, or amplitudes, and relative phase between pressure p and volume velocity Q as a function of axial position and frequency. They relate a thermodynamic property to a kinematic property. Unlike the physical quantities p and Q, both the acoustic impedance Z and acoustic admittance Y are understood to be inherently complex quantities: we do not take the real parts of these quantities as their physical definition. As a result, when we equate expressions that involve acoustic impedances and acoustic admittance, we are concerned that the equality holds for both the real parts and imaginary parts of the expressions.

We have seen in Equation (38) of Chapter 2, that the relationship between pressure and velocity, or volume velocity, is important for energy propagation in terms of power flow. The result of these two equations is that the rate at which acoustic energy travels to the right is power $\mathcal{P}(x,t)$ given by

$$\mathcal{P}(x,t) = p(x,t)Q(x,t) \ . \quad (31)$$

The real representations are assumed for $p(x,t)$ and $Q(x,t)$ in Equation (31). This is always the case when products of acoustic quantities are computed. If $\mathcal{P}(x,t)$ is positive then energy flows to the right, and if it is negative then energy flows to the left.

We now provide two examples of the use of complex variables in acoustics. Suppose pressure is a right-going wave with sinusoidal time dependence, $p(x,t) = Pe^{-i(\omega t - kx + \phi)}$, with $P > 0$ and $k = \omega/c_0$. We know, from Equation (21) of Chapter 2, that the relation between pressure p and particle velocity u for a positive-going, or right-going, wave is $p = \rho_0 c_0 u$. Recall that $\rho_0 c_0$ is the characteristic impedance of air. So the corresponding volume velocity is $Q = (A/\rho_0 c_0)p$ in a tube of cross-sectional area A.

$$Z = \frac{p}{Q} = \frac{\rho_0 c_0}{A}$$

$$Y = \frac{Q}{p} = \frac{A}{\rho_0 c_0} . \tag{32}$$

In this case the acoustic impedance and acoustic admittance are real and do not depend on position or frequency. The real representation of pressure is $p(x,t) = P\cos(\omega t - kx + \phi)$, so time-average power at position x can be computed, using the real representation of volume velocity $Q(x,t) = YP\cos(\omega t - kx + \phi)$, because Y is real.

$$\langle \mathcal{P} \rangle = \langle pQ \rangle$$
$$= \langle P^2 Y \cos^2(\omega t - kx + \phi) \rangle$$
$$= \frac{P^2}{2} \frac{A}{\rho_0 c_0} . \tag{33}$$

The time-average energy flow is independent of both spatial position and frequency for the traveling wave.

For a second example we consider individual normal modes, without damping, in the finite-length tube of length L. From Chapter 5, Equation (17), the real representation for the mth normal mode of pressure is

$$p_m(x,t) = P_m \cos(k_m x) \sin(\omega_m t + \phi_m) ,$$
$$\text{where, } P_m > 0, \ k_m = \frac{(2m-1)\pi}{2L} \text{ and } \omega_m = c_0 k_m . \tag{34}$$

The associated volume velocity from Chapter 5, Equation (24) is

$$Q_m(x,t) = -\frac{A}{\rho_0 c_0} P_m \sin(k_m x) \cos(\omega_m t + \phi_m) \ . \tag{35}$$

The time-average power in this case is

$$\begin{aligned}
\langle \mathcal{P}_m \rangle &= \langle p_m Q_m \rangle \\
&= \left\langle -\frac{A}{\rho_0 c_0} P_n^2 \cos(k_m x)\sin(k_m x)\sin(\omega_m t + \phi_n)\cos(\omega_m t + \phi_m) \right\rangle \\
&= -\frac{A}{\rho_0 c_0} P_n^2 \cos(k_m x)\sin(k_m x) \langle \sin(\omega_m t + \phi_m)\cos(\omega_m t + \phi_m) \rangle \\
&= 0 \ . \tag{36}
\end{aligned}$$

There is no energy flow anywhere along the finite-length tube, when there is no damping. The fact that power is zero can also be deduced by examining p_m and Q_m in complex representation.

$$p_m(x,t) = P_m \cos(k_m x) e^{-i(\omega_m t + \phi_m + \pi/2)}$$

and (37)

$$Q_m(x,t) = \frac{A}{\rho_0 c_0} P_m \sin(k_m x) e^{-i(\omega_m t + \phi_m + \pi)} \ .$$

So that $p_m(x,t)$ and $Q_m(x,t)$ are $\pi/2$ out of phase in time-average, which indicates zero time-average power.

The time domain and frequency domain

We now are in a position to explain the notation introduced at the end of Chapter 3 for a representation in the frequency domain. Recall that if $f(t)$ is a function of time t, then it has a frequency domain representation $\tilde{f}(\omega)$, which is known as its Fourier transform. [The $\tilde{f}(\omega)$ exists except for some pathological $f(t)$s.] We said that $\tilde{f}(\omega)$ contains two pieces of information, the amplitude, or magnitude, $|\tilde{f}(\omega)|$ and $\arg(\tilde{f}(\omega))$, or the phase of $\tilde{f}(\omega)$. In fact, we have

$$\tilde{f}(\omega) = |\tilde{f}(\omega)| e^{i \arg(\tilde{f}(\omega))} \ . \tag{38}$$

References

Morse, P.M. (1976). *Vibration and Sound*. Acoustical Society of America, Melville, New York.

Polya, G. & Latta, G. (1974). *Complex Variables*. John Wiley & Sons, Inc., New York.

Appendix to Chapter 8

In the following we use complex numbers and functions to derive the trigonometric identities in Equation (1) of Chapter 5. By Euler's theorem in Equation (8)

$$e^{i(\theta+\phi)} = \cos(\theta+\phi) + i\sin(\theta+\phi) \ . \tag{A8.1}$$

Also,
$$\begin{aligned} e^{i(\theta+\phi)} &= e^{i\theta} \cdot e^{i\phi} \\ &= \bigl[\cos(\theta) + i\sin(\theta)\bigr] \cdot \bigl[\cos(\phi) + i\sin(\phi)\bigr] \\ &= \bigl[\cos(\theta)\cos(\phi) - \sin(\theta)\sin(\phi)\bigr] + i\bigl[\sin(\theta)\cos(\phi) + \sin(\phi)\cos(\theta)\bigr] \ . \end{aligned} \tag{A8.2}$$

Equating the real and imaginary parts of Equations (A8.1) and (A8.2)

$$\begin{aligned} \cos(\theta+\phi) &= \cos(\theta)\cos(\phi) - \sin(\theta)\cos(\phi) \\ \sin(\theta+\phi) &= \sin(\theta)\cos(\phi) + \cos(\theta)\sin(\phi) \ . \end{aligned} \tag{A8.3}$$

These are the third and first relations of Equation (1) of Chapter 5. Again, Euler's theorem provides

$$e^{i(\theta-\phi)} = \cos(\theta-\phi) + i\sin(\theta-\phi) \ . \tag{A8.4}$$

Thus,
$$\begin{aligned} e^{i(\theta-\phi)} &= e^{i\theta} \cdot e^{-i\phi} \\ &= \bigl[\cos(\theta) + i\sin(\theta)\bigr] \cdot \bigl[\cos(\phi) - i\sin(\phi)\bigr] \\ &= \bigl[\cos(\theta)\cos(\phi) + \sin(\theta)\sin(\phi)\bigr] + i\bigl[\sin(\theta)\cos(\phi) - \cos(\theta)\sin(\phi)\bigr] \ . \end{aligned} \tag{A8.5}$$

Equating the real and imaginary parts of Equations (A8.4) and (A8.5) gives

$$\cos(\theta - \phi) = \cos(\theta)\cos(\phi) + \sin(\theta)\cos(\phi) \quad (A8.6)$$

$$\sin(\theta - \phi) = \sin(\theta)\cos(\phi) - \cos(\theta)\sin(\phi) \ .$$

These are the fourth and second relations of Equation (1) of Chapter 5.

In the following, we show that Equation (23) follows from Equation (22). Equation (24) is true for the same reasons. In complex representation, sinusoidal pressure variation that satisfies the wave equation is written

$$p(x,t) = B_1 e^{-i(\omega t - kx + \phi_1)} + B_2 e^{-i(\omega t + kx + \phi_2)} \ , \quad (A8.7)$$

with both $B_1, B_2 > 0$ and $k = \omega/c_0$. If the real part of Equation (A8.7) is taken, and recalling that $\cos(-\theta) = \cos(\theta)$ and with trigonometric identities in Equations (A8.3) and (A8.6), then

$$\begin{aligned} p(x,t) &= B_1 \cos(\omega t - kx + \phi_1) + B_2 \cos(\omega t + kx + \phi_2) \\ &= B_1\big(\cos(\omega t)\cos(-kx + \phi_1) - \sin(\omega t)\sin(-kx + \phi_1)\big) + \\ &\quad B_2\big(\cos(\omega t)\cos(kx + \phi_2) - \sin(\omega t)\sin(kx + \phi_2)\big) \\ &= \big(B_1\cos(-kx + \phi_1) + B_2\cos(kx + \phi_2)\big)\cos(\omega t) \\ &\quad - \big(B_1\sin(-kx + \phi_1) + B_2\sin(kx + \phi_2)\big)\sin(\omega t) \\ &= \mathcal{B}_1(x,\omega)\cos(\omega t) - \mathcal{B}_2(x,\omega)\sin(\omega t) \ . \end{aligned} \quad (A8.8)$$

where $\mathcal{B}_1 = B_1\cos(-kx + \phi_1) + B_2\cos(kx + \phi_2)$ and $\mathcal{B}_2 = B_1\sin(-kx + \phi_1) + B_2\sin(kx + \phi_2)$. Let $B = \sqrt{\mathcal{B}_1^2 + \mathcal{B}_2^2} > 0$, and let ϕ be such that $\cos(\phi) = \mathcal{B}_1/B$, and $\sin(\phi) = \mathcal{B}_2/B$. Then $B = B(x,\omega)$ and $\phi = \phi(x,\omega)$. Further $B > 0$. In this notation we obtain

$$\begin{aligned} p(x,t) &= B(x,\omega)\big(\cos(\phi(x,\omega))\cos(\omega t) - \sin(\phi(x,\omega))\sin(\omega t)\big) \\ &= B(x,\omega)\cos(\omega t + \phi(x,\omega)) \ . \end{aligned} \quad (A8.9)$$

The complex representation for p is

$$p(x,t) = B(x,\omega)e^{-i(\omega t + \phi(x,\omega))} \ . \quad (A8.10)$$

Chapter 9: Two Sub-Tubes of Unequal Cross-Sectional Area

Introduction

We return to undamped acoustic propagation, and consider the situation when a tube of finite length is composed of two shorter sub-tubes whose cross-sectional areas differ from one another. We stipulate that pressure and volume velocity be continuous across the junction between sub-tubes. It is not surprising that following acoustic wave motion in time for the two sub-tubes is even more complicated than for a single finite-length tube.

Before continuing onto the discussion of two sub-tubes, we finally explain under what circumstances that the assumption of one-dimensional, or plane wave, propagation is valid. We must have frequencies low enough so that the wavelengths are long in relation to the linear dimensions of the tube cross-section.

It turns out that many ideas that have been developed for the finite-length tube of constant cross-sectional area are applicable to the situation of two sub-tubes of different cross-sectional area. One of the central concepts that still holds is that of normal modes. Normal modes can be found using semi-eigenfunctions introduced in Chapter 5. However, we consider an alternative method for finding normal mode frequencies, using our knowledge of complex variables introduced in Chapter 8, along with the concept of *effective admittance*.

We examine energy densities and steady acoustic radiation pressure in the context of two sub-tubes. Discontinuity in kinetic energy density at the junction is seen to occur when continuity of volume velocity is enforced. The conceptual defect lies in the discontinuity of particle velocity at the junction. Further, we are unable to alleviate this defect without substantial mathematical effort that we deem to be unjustified in terms of practical gain in computation for acoustic perturbation theory. However, there is an adjustment to the condition of pressure continuity that can be made based on these observations.

Low-frequency acoustics

At the beginning of this book, we promised to explain the circumstances under which we need only consider plane wave propagation in the tube axis direction. It turns out that we have been working in what can be considered the *low-frequency approximation*. That is, the assumption of plane wave propagation requires that the frequency be sufficiently small, or low.

The determining criterion for when the frequency is sufficiently low is whether the corresponding wavelength is sufficiently large on the length scale of the distance between sides of the walls of the tube. Roughly speaking, if a substantial part of a wavelength can fit between the walls of a tube, then acoustic propagation can occur between the walls. If this happens, then acoustic propagation occurs in directions other than along the tube axis, and the plane wave propagation assumption is no longer valid. In this short discussion, we assume that the tube walls are hard in the sense that the piston is hard: there is no fluid velocity at the duct wall in the direction perpendicular to the wall. Further, the compliant walls of the vocal tract do not change these considerations substantially.

We have been using tubes with circular cross-section, but here we consider rectangular cross-section tubes. For a rectangular tube with dimensions d and h in the cross-sections, in order to maintain plane wave propagation $d < \lambda/2$ and $h < \lambda/2$, where λ is the wavelength of sound. In terms of frequency these criteria are: $\mathcal{F} < c_0/(2d)$ and $\mathcal{F} < c_0/(2h)$. So, if the maximum of d and h is 2 cm, then for just plane wave propagation, it is required that $\mathcal{F} < (34,100/4)$ Hz $\approx$ 8500 Hz. There are cases in speech where a dimension more on the order of 4 cm is appropriate. In that case it is required that $\mathcal{F} < 4300$ Hz for plane wave propagation. Tubes with other cross-sectional shapes have maximum frequencies for plane wave propagation as well, but we do not quote the results here.

Wave propagation considerations

Two sub-tubes, 1 and 2, with cross-sectional areas, $A^{(1)}$ and $A^{(2)}$, and lengths $l^{(1)}$ and $l^{(2)}$ are concatenated, as shown in Figure 1. These two sub-tubes now constitute the finite-length tube. Sub-tube 2 is to

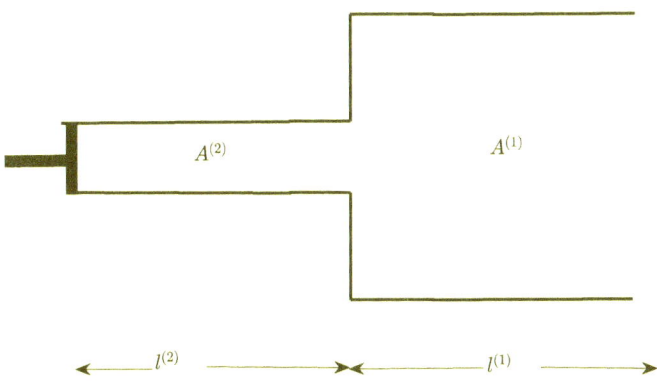

FIGURE 1. Two sub-tubes

the left of sub-tube 1, and the left end of sub-tube 2 has a piston that can act an acoustic source. Sub-tube 1 is open at its right end, and the concatenation of these tubes provides a variable area tube of length $L = l^{(1)} + l^{(2)}$. The coordinate system is such that $x = 0$ corresponds to the piston boundary and $x = L$ corresponds to the opening. The junction for the tubes is at $x = l^{(2)}$.

Suppose the piston has a volume velocity $Q_{pst}(t)$, where $Q_{pst}(t) = 0$ for $t < 0$. We follow the evolution of the acoustic wave through the tube. $Q_+^{(2)}(x = 0, t) = Q_{pst}(t)$, where $Q_+^{(2)}(x,t)$ is the right-going volume velocity in sub-tube 2, so that $Q_+^{(2)}(x,t) = Q_+^{(2)}(t - x/c_0)$. The pressure associated with this pulse is $p_+^{(2)}(t - x/c_0) = (\rho_0 c_0 / A^{(2)}) Q_+^{(2)}(t - x/c_0)$. This is the situation, at least for $0 < t < l^{(2)}/c_0$.

The acoustic wave with pressure $p_+^{(2)}(t - x/c_0)$ is incident on the discontinuity in cross-sectional area at $x = l^{(2)}$ for $t \geq l^{(2)}/c_0$. This is called the incident wave. A reflected wave, $p_-^{(2)}(t + x/c_0)$, and a *transmitted wave*, $p_+^{(1)}(t - x/c_0)$, are created for these times. These waves can be calculated in terms of $p_+^{(2)}$ when two continuity conditions are invoked. These conditions are continuity of pressure and continuity of volume velocity everywhere, but particularly across the junction at $x = l^{(2)}$, where there is a discontinuity of cross-sectional area. We emphasize that these conditions are only a first approximation to the actual conditions that are discussed later.

The volume velocities associated with the pressures are

$$Q_+^{(2)}(t - x/c_0) = \left(\frac{A^{(2)}}{\rho_0 c_0}\right) p_+^{(2)}(t - x/c_0) = Y^{(2)} p_+^{(2)}(t - x/c_0) ,$$

$$Q_-^{(2)}(t + x/c_0) = -\left(\frac{A^{(2)}}{\rho_0 c_0}\right) p_-^{(2)}(t + x/c_0) = -Y^{(2)} p_-^{(2)}(t + x/c_0) , \quad (1)$$

$$Q_+^{(1)}(t - x/c_0) = \left(\frac{A^{(1)}}{\rho_0 c_0}\right) p_+^{(1)}(t - x/c_0) = Y^{(1)} p_+^{(1)}(t - x/c_0) ,$$

where the $Y^{(n)} = A^{(n)}/(\rho_0 c_0)$, $n = 1, 2$ are known as the admittances of the sub-tubes, or simply, *tube admittances*.

The physical quantities satisfy the wave equation so that they are functions of either $t - x/c_0$ or $t + x/c_0$. However, when speaking of propagation in variable area tubes it is sometimes preferable to write the physical quantities as functions of both space and time, so that $p_+^{(2)} = p_+^{(2)}(x,t)$. This helps us to avoid some confusion in the discussion.

Now we are ready to find the transmitted wave, $p_+^{(1)}$, and reflected wave, $p_-^{(2)}$, using the continuity conditions. The continuity of pressure across the junction provides the equation

$$p_+^{(2)}(x = l^{(2)}, t) + p_-^{(2)}(x = l^{(2)}, t) = p_+^{(1)}(x = l^{(2)}, t) , \quad (2)$$

and continuity of volume velocity at $x = l^{(2)}$ gives

$$Q_+^{(2)}(x = l^{(2)}, t) + Q_-^{(2)}(x = l^{(2)}, t) = Q_+^{(1)}(x = l^{(2)}, t) ,$$

or \quad (3)

$$Y^{(2)} \left[p_+^{(2)}(x = l^{(2)}, t) - p_-^{(2)}(x = l^{(2)}, t) \right] = Y^{(1)} p_+^{(1)}(x = l^{(2)}, t) .$$

Solving for the reflected wave, $p_-^{(2)}(x = l^{(2)}, t)$, and the transmitted wave, $p_+^{(1)}(x = l^{(2)}, t)$, in terms of the incident wave, $p_+^{(2)}(x = l^{(2)}, t)$, we obtain

$$p_-^{(2)}(x = l^{(2)}, t) = \frac{Y^{(2)} - Y^{(1)}}{Y^{(2)} + Y^{(1)}} p_+^{(2)}(x = l^{(2)}, t) \quad (4)$$

$$\equiv \mathcal{R}^{(2,1)}\, p_+^{(2)}(x = l^{(2)}, t)\ .$$

Also,

$$p_+^{(1)}(x = l^{(2)}, t) = \frac{2Y^{(2)}}{Y^{(2)} + Y^{(1)}}\, p_+^{(2)}(l^{(2)}, t) \equiv \mathcal{T}^{(2,1)}\, p_+^{(2)}(x = l^{(2)}, t)\ , \quad (5)$$

where $\mathcal{R}^{(\cdot,\cdot)}$ and $\mathcal{T}^{(\cdot,\cdot)}$ are the *reflection coefficient* and *transmission coefficient*, respectively. The first superscript on $\mathcal{R}^{(\cdot,\cdot)}$ or $\mathcal{T}^{(\cdot,\cdot)}$ denotes the sub-tube from which the incident wave is coming and the second superscript the sub-tube in which the transmitted wave travels. Thus, the transmitted and reflected waves, $p_+^{(1)}(l^{(2)}, t)$ and $p_-^{(2)}(l^{(2)}, t)$, are functions of the tube admittances, $Y^{(2)}$ and $Y^{(1)}$, and the incident wave $p_+^{(2)}(l^{(2)}, t)$. This means that only the geometric properties of the tubes and the thermodynamic properties of the air determine the transmission and reflection properties between tubes. This seems very easy and simple. However, the considerations so far are only valid for $t < l^{(2)}/c_0 + \min\left(l^{(1)}/c_0, l^{(2)}/c_0\right)$, which is only up to a millisecond in a finite-length tube with $L = 17.1$ cm. The problem is that the transmitted and reflected waves themselves reach cross-sectional area discontinuities at $x = \left(l^{(1)} + l^{(2)}\right) = L$ and $x = 0$. The transmitted wave reaches $x = \left(l^{(1)} + l^{(2)}\right)$ at time, $t = \left(l^{(1)} + l^{(2)}\right)/c_0$ and the reflected wave reaches $x = 0$ at $t = 2l^{(2)}/c_0$.

When the transmitted wave from the junction of tubes 1 and 2 reaches the opening at $x = L$ at time $t = (l^{(2)} + l^{(1)})/c_0 = L/c_0$, reflected and transmitted waves are created at this cross-sectional area discontinuity. Using the same considerations used to derive Equation (4), we can write

$$p_-^{(1)}(x = l^{(2)} + l^{(1)}, t) = \frac{Y^{(1)} - Y^{(0)}}{Y^{(1)} + Y^{(0)}}\, p_+^{(1)}(x = l^{(2)} + l^{(1)}, t) \quad (6)$$

$$\equiv \mathcal{R}^{(1,0)}\, p_+^{(1)}(x = l^{(2)} + l^{(1)}, t)\ ,$$

where $Y^{(0)}$ is the acoustic admittance at the opening. So far $Y^{(0)} = Q^{(1)}(x = l^{(1)} + l^{(2)}, t)/p^{(1)}(x = l^{(1)} + l^{(2)}, t)$ is approximated $Y^{(0)} = \infty$, with no pressure fluctuation at the opening, as stipulated in Equation (2) of Chapter 3. [One can think of the "atmospheric tube", tube 0, as having area, $A^{(0)} = \infty$, giving $Y^{(0)} = A^{(0)}/(\rho_0 c_0) = \infty$.] The value for $Y^{(0)}$ is modified later to possess finite values, so for generality's sake, it will be written $Y^{(0)}$. In the special case that $Y^{(0)} = \infty$,

Equation (6) gives $p_-^{(1)}(x = l^{(2)} + l^{(1)}, t) = -p_+^{(1)}(x = l^{(2)} + l^{(1)}, t)$, while the volume velocity, $Q^{(1)}(x = l^{(2)} + l^{(1)}, t) = Y^{(1)}[p_+^{(1)}(x = l^{(2)} + l^{(1)}, t) - p_-^{(1)}(x = l^{(2)} + l^{(1)}, t)] = 2Y^{(1)}p_+^{(1)}(x = l^{(2)} + l^{(1)}, t)$. Unlike Equations (4) and (5), Equation (6) is valid at any time, because it is assumed that there is no wave coming from the atmosphere that is incident on the opening.

The acoustic admittance at $x = 0$ is denoted $Y^{(3)}$, and for the particular case of a hard-wall piston, $Y^{(3)} = 0$. [The presence of the piston means that $A^{(3)} = 0$.] However, the notation $Y^{(3)}$ is retained for more generality. Assuming that the piston has stopped moving by $t = 2l^{(2)}/c_0$

$$p_+^{(2)}(x = 0, t) = \frac{Y^{(2)} - Y^{(3)}}{Y^{(2)} + Y^{(3)}} p_-^{(2)}(x = 0, t) \equiv \mathcal{R}^{(2,3)} p_-^{(2)}(x = 0, t) , \quad (7)$$

for $t \geq 2l^{(2)}/c_0$.

Let's consider the time when either, or both, the reflected waves from $x = l^{(1)} + l^{(2)} = L$ and $x = 0$ return to the junction between sub-tubes 1 and 2, that is for $t > l^{(2)}/c_0 + 2\min(l^{(1)}/c_0, l^{(2)}/c_0)$. Continuity of pressure gives

$$p_+^{(2)}(x = l^{(2)}, t) + p_-^{(2)}(x = l^{(2)}, t) = \qquad (8)$$
$$p_+^{(1)}(x = l^{(2)}, t) + p_-^{(1)}(x = l^{(2)}, t) .$$

Continuity of volume velocity gives

$$Y^{(2)}\left(p_+^{(2)}(x = l^{(2)}, t) - p_-^{(2)}(x = l^{(2)}, t)\right) = \qquad (9)$$
$$Y^{(1)}\left(p_+^{(1)}(x = l^{(2)}, t) - p_-^{(1)}(x = l^{(2)}, t)\right) .$$

These equations are generalizations of Equations (2) and (3), and they are valid for all time.

If, instead of the piston ceasing to move, it continues to move indefinitely, the complexity of the situation is revealed. Let's consider a special case with $2l^{(1)} > l^{(2)} > l^{(1)}$.

$$p_-^{(2)}(x = l^{(2)}, t) = \mathcal{R}^{(2,1)} p_+^{(2)}(x = l^{(2)}, t) + \mathcal{T}^{(1,2)} p_-^{(1)}(x = l^{(2)}, t)$$
$$= \mathcal{R}^{(2,1)}\mathcal{R}^{(2,3)} p_-^{(2)}(x = 0, t - l^{(2)}/c_0) +$$
$$\mathcal{T}^{(1,2)}\mathcal{R}^{(1,0)} p_+^{(1)}(x = l^{(1)} + l^{(2)}, t - l^{(1)}/c_0) \qquad (10)$$

$$= \mathcal{R}^{(2,1)}\mathcal{R}^{(2,3)}\mathcal{R}^{(2,1)} p_+^{(2)}(x = l^{(2)}, t - 2l^{(2)}/c_0) +$$
$$\mathcal{T}^{(1,2)}\mathcal{R}^{(1,0)}\mathcal{T}^{(2,1)} p_+^{(2)}(x = l^{(2)}, t - 2l^{(1)}/c_0)$$
$$= \mathcal{R}^{(2,1)}\mathcal{R}^{(2,3)}\mathcal{R}^{(2,1)} \frac{Q_{pst}(t - 3l^{(2)}/c_0)}{Y^{(2)}} +$$
$$\mathcal{T}^{(1,2)}\mathcal{R}^{(1,0)}\mathcal{T}^{(2,1)} \frac{Q_{pst}(t - 2l^{(1)}/c_0 - l^{(2)}/c_0)}{Y^{(2)}},$$
$$\text{where } \mathcal{T}^{(m,n)} = \frac{2Y^{(m)}}{Y^{(m)} + Y^{(n)}} \text{ and } \mathcal{R}^{(m,n)} = \frac{Y^{(m)} - Y^{(n)}}{Y^{(m)} + Y^{(n)}}$$

for $l^{(2)}/c_0 + 4l^{(1)}/c_0 > t \geq l^{(2)}/c_0 + 2l^{(1)}/c_0$. The reason for the upper bound on the time interval is because of further reflections of $p_+^{(1)}$ at $x = L$ arriving at the junction $x = l^{(2)}$. Tube admittances are used to write transmission and reflection coefficients, $\mathcal{T}^{(2,1)}$, $\mathcal{T}^{(1,2)}$, $\mathcal{R}^{(1,0)}$, $\mathcal{R}^{(2,1)}$, $\mathcal{R}^{(1,2)}$, and $\mathcal{R}^{(2,3)}$. Thus, only the cross-sectional areas, sub-tube lengths, and the thermodynamic quantities ρ_0 and c_0 determine acoustic propagation for all time.

We could go on indefinitely in time in the two sub-tube case with a continuously moving piston by taking more and more reflected waves into account. As in Chapter 3, we soon run into the situation where it is humanly impossible to keep track of everything. However, this can be done algorithmically, either by symbolic manipulation or with a numerical procedure. Just as before, we look into a normal mode representation to keep the mathematics understandable.

Normal modes with two sub-tubes of unequal cross-sectional area

Finding normal modes with semi-eigenfunctions

The assumptions here are the same as in Chapter 5: 1) the wave motion has sinusoidal time dependence, $\sin(\omega t + \phi)$, and 2) there are no sources present. The geometry of Figure 1 still applies. We take a standing wave representation with time and spatial dependence described by circular functions

$$p(x,t) = g(x)\sin(\omega t + \phi), \qquad (11)$$

where $g(x)$ is described by circular functions in each sub-tube.

We rewrite the boundary conditions, which remain unchanged from the finite-length tube of constant cross-sectional area. These are given in Equation (13) of Chapter 5.

$$\frac{\Delta_x p}{\Delta x}(x = 0, t) = 0 ,$$

and (12)

$$p(x = L, t) = 0 .$$

With these boundary conditions, the same right and left semi-eigenfunctions that appeared in Chapter 5 can be used here. This means that Equation (11) becomes

$$p(x,t) = \begin{cases} p^{(2)}(x,t) & \text{for } 0 < x < l^{(2)} \\ p^{(1)}(x,t) & \text{for } l^{(2)} < x < l^{(1)} + l^{(2)} = L , \end{cases}$$

where (13)

$$p^{(2)}(x,t) = P^{(2)} \cos(kx) \sin(\omega t + \phi)$$
$$p^{(1)}(x,t) = P^{(1)} \sin(k(x - L)) \sin(\omega t + \phi)$$
$$= P^{(1)} \cos(kx + \theta^{(1)})) \sin(\omega t + \phi)$$

with $\theta^{(1)} = -(kL + \pi/2)$,

where $P^{(1)}$ and $P^{(2)}$ are pressure amplitudes with dimensions of pressure. The final equality In Equation (13) follows from Equation (4) of Chapter 5. $k = \omega/c_0$ in order that these expressions satisfy the wave equation, Equation (17) of Chapter 2. The corresponding expression for volume velocity is

$$Q(x,t) = \begin{cases} Q^{(2)}(x,t) & \text{for } 0 < x < l^{(2)} \\ Q^{(1)}(x,t) & \text{for } l^{(2)} < x < l^{(1)} + l^{(2)} = L , \end{cases}$$

where (14)

$$Q^{(2)}(x,t) = -P^{(2)} Y^{(2)} \sin(kx) \cos(\omega t + \phi)$$
$$Q^{(1)}(x,t) = -P^{(1)} Y^{(1)} \sin(kx + \theta^{(1)})) \cos(\omega t + \phi) .$$

We now enforce the continuity of pressure and volume velocity at $x = l^{(2)}$.

$$p^{(1)}(x = l^{(2)}, t) = p^{(2)}(x = l^{(2)}), t) , \text{ or}$$
$$- P^{(1)} \sin(kl^{(1)}) = P^{(2)} \cos(kl^{(2)}) ,$$

and (15)

$$Q^{(1)}(x = l^{(2)}, t) = Q^{(2)}(x = l^{(2)}), t) \text{ , or}$$
$$P^{(1)} A^{(1)} \cos\left(kl^{(1)}\right) = -P^{(2)} A^{(2)} \sin\left(kl^{(2)}\right) .$$

These relations are used to eliminate $P^{(1)}$ and $P^{(2)}$ to obtain an equation involving $k = \omega/c_0$ and the geometric parameters $A^{(1)}$, $A^{(2)}$, $l^{(1)}$, and $l^{(2)}$.

$$P^{(1)} = -P^{(2)} \frac{\cos(kl^{(2)})}{\sin(kl^{(1)})} ,$$

and (16)

$$\frac{A^{(2)}}{A^{(1)}} \tan\left(kl^{(1)}\right) \tan\left(kl^{(2)}\right) - 1 = 0 .$$

The first expression in Equation (16) for $P^{(1)}$ in terms of $P^{(2)}$, k, and the lengths $l^{(1)}$ and $l^{(2)}$. $P^{(2)}$ can be set to any real number with units of pressure. Given the geometry of the two sub-tubes, the second expression in Equation (16) can be solved for k_m, and then $\omega_m = k_m c_0$. [The second expression in Equation (16) can be used to solve for the various k_m using one of many numerical root-finding algorithms.] These are the eigenvalues of the two sub-tube configuration. The wavenumbers k_m are functions of the geometric parameters $A^{(1)}$, $A^{(2)}$, $l^{(1)}$, and $l^{(2)}$.

The eigenfunctions for pressure are

$$p_m(x, t) = \begin{cases} p_m^{(2)}(x, t) & \text{for } 0 < x < l^{(2)} \\ p_m^{(1)}(x, t) & \text{for } l^{(2)} < x < l^{(1)} + l^{(2)} = L \end{cases},$$

where (17)

$$p_m^{(2)}(x, t) = P_m^{(2)} \cos\left(k_m x\right) \sin(\omega_m t + \phi_m)$$
$$p_m^{(1)}(x, t) = -P_m^{(2)} \frac{\cos(k_m l^{(2)})}{\sin(k_m l^{(1)})} \cos\left(k_m x + \theta_m^{(1)}\right)) \sin(\omega_m t + \phi_m) .$$

where $\theta_m^{(1)} = -(k_m L + \pi/2)$. Similarly, the eigenfunctions for volume velocity are

$$Q_m(x, t) = \begin{cases} Q_m^{(2)}(x, t) & \text{for } 0 < x < l^{(2)} \\ Q_m^{(1)}(x, t) & \text{for } l^{(2)} < x < l^{(1)} + l^{(2)} = L \end{cases},$$

where (18)

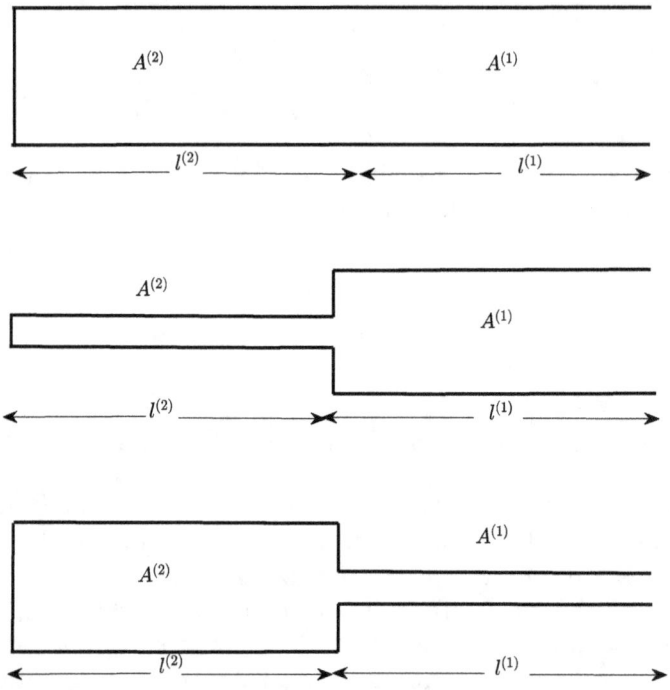

FIGURE 2. Three different two sub-tube configurations

$$Q_m^{(2)}(x,t) = -P_m^{(2)} Y^{(2)} \sin\left(k_m x\right) \cos(\omega_m t + \phi_m)$$

$$Q_m^{(1)}(x,t) = -P_m^{(2)} \frac{\cos(k_m l^{(2)})}{\sin(k_m l^{(1)})} Y^{(1)} \sin\left(k_m x + \theta_m^{(1)}\right) \cos(\omega_m t + \phi_m) \,.$$

Figure 2 shows three configurations for $l^{(1)} = l^{(2)} = L/2$ for $L = 17.1\ cm$. The top panel has $A^{(2)}/A^{(1)} = 1.0$, the middle panel has $A^{(2)}/A^{(1)} = 0.25$ and the bottom panel has $A^{(2)}/A^{(1)} = 4.0$. The left side of Equation (16) is plotted as a function of frequency $\mathcal{F} = \omega/(2\pi)$ in Figure 3. The top panel of Figure 3 corresponds to $A^{(2)}/A^{(1)} = 1$, the middle panel corresponds to $A^{(2)}/A^{(1)} = 0.25$, and the bottom panel corresponds to $A^{(2)}/A^{(1)} = 4$. We obtain the expected trends in terms of normal mode frequencies. For the top panel, $\mathcal{F}_1 = 500$ Hz, $\mathcal{F}_2 = 1500$ Hz, and $\mathcal{F}_3 = 2500$ Hz. For the middle panel, $\mathcal{F}_1 = 705$ Hz, $\mathcal{F}_2 = 1295$ Hz, and $\mathcal{F}_3 = 2705$ Hz. For the bottom panel, $\mathcal{F}_1 = 295$ Hz, $\mathcal{F}_2 = 1705$ Hz, and $\mathcal{F}_3 = 2295$ Hz.

Acoustics of Speech Production   243

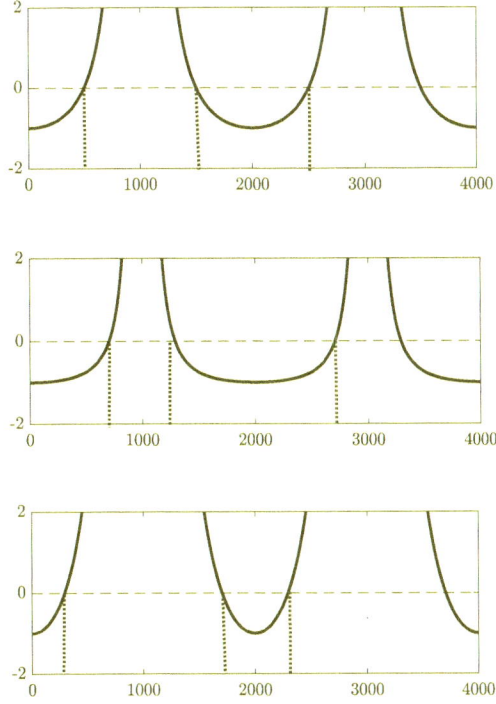

FIGURE 3. Normal mode frequencies for three different two sub-tube configurations

Figures 4 and 5 show the spatial dependence of the pressure amplitude and volume velocity amplitude for the first three modes in the case that $A^{(2)}/A^{(1)} = 0.25$. Figures 6 and 7 show the spatial distribution of the pressure amplitude and volume velocity amplitude for the first three modes in the case that $A^{(2)}/A^{(1)} = 4$. [We also have some phase information here, because we allow the amplitudes to be negative, as well as positive. We term these *signed amplitudes*.] These figures can be compared to Figures 4 and 9 of Chapter 5 for $A^{(2)}/A^{(1)} = 1$. Both the pressure signed amplitude and volume velocity signed amplitude are continuous throughout the tube for both area ratios, $A^{(2)}/A^{(1)} = 4$ and $A^{(2)}/A^{(1)} = 0.25$. As the mode number m increases there are more maxima and minima because the normal mode frequencies increase, so that wavelengths decrease. In this regard, recall the relations among frequency $\mathcal{F}$, circular frequency ω, wavenumber k,

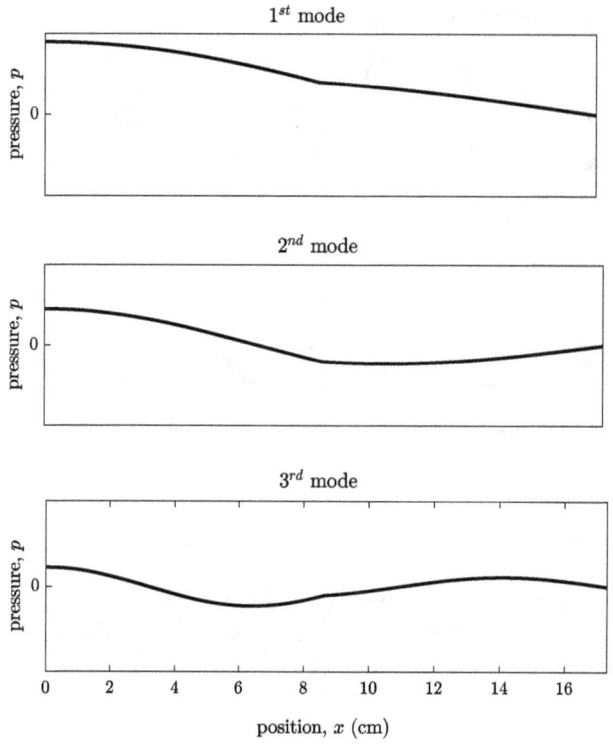

FIGURE 4. Pressure normal mode signed amplitudes for $A^{(2)}/A^{(1)} = 0.25$

and wavelength λ

$$\omega = 2\pi\mathcal{F}, \quad \lambda = \frac{c_0}{\mathcal{F}}, \quad k = \frac{\omega}{c_0} = \frac{2\pi\mathcal{F}}{c_0} = \frac{2\pi}{\lambda}. \tag{19}$$

Effective admittances

While the derivation above works well for two sub-tubes, it becomes unwieldy when there are more than a few sub-tubes. We introduce the concept of effective admittance that generalizes easily to many sub-tubes and allows us to calculate normal mode frequencies for many sub-tubes. The spirit and much of the substance of these considerations

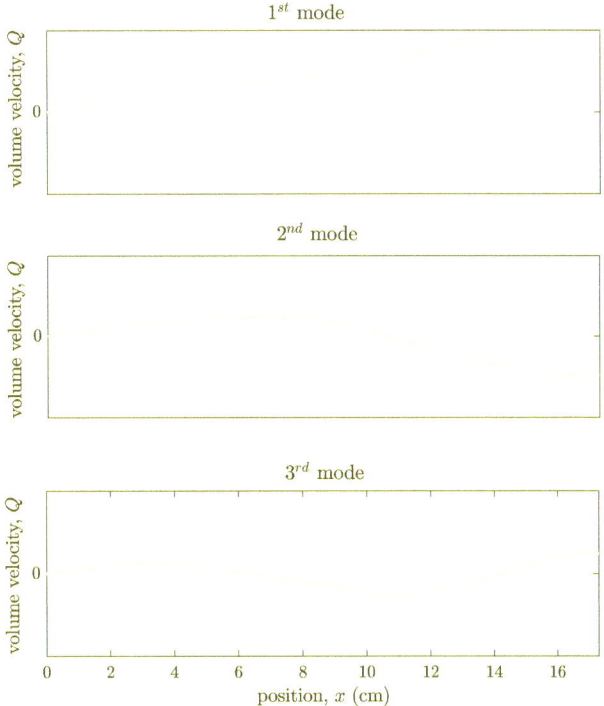

FIGURE 5. Volume velocity normal mode signed amplitudes for $A^{(2)}/A^{(1)} = 0.25$

relies on Chapter 2 of Lighthill's *Waves in Fluids* (Lighthill 1978). We use complex representations of physical quantities here.

We define objects called effective admittances when acoustic motion is sinusoidal with circular frequency ω. In tube n, with $n = 1$ or 2 we have

$$p^{(n)}(x,t) = p_+^{(n)}(x,t) + p_-^{(n)}(x,t) = P_+^{(n)} e^{-i\omega(t-x/c_0)} + P_-^{(n)} e^{-i\omega(t+x/c_0)},$$

and (20)

$$Q^{(n)}(x,t) = Y^{(n)} \left(p_+^{(n)}(x,t) - p_-^{(n)}(x,t) \right)$$
$$= Y^{(n)} \left(P_+^{(n)} e^{-i\omega(t-x/c_0)} - P_-^{(n)} e^{-i\omega(t+x/c_0)} \right),$$

where $P_+^{(n)}$ and $P_-^{(n)}$ are complex variables with units of pressure. It is understood that the real parts of the complex expressions in Equation (20) are the physical quantities of interest. With $k = \omega/c_0$, Equation

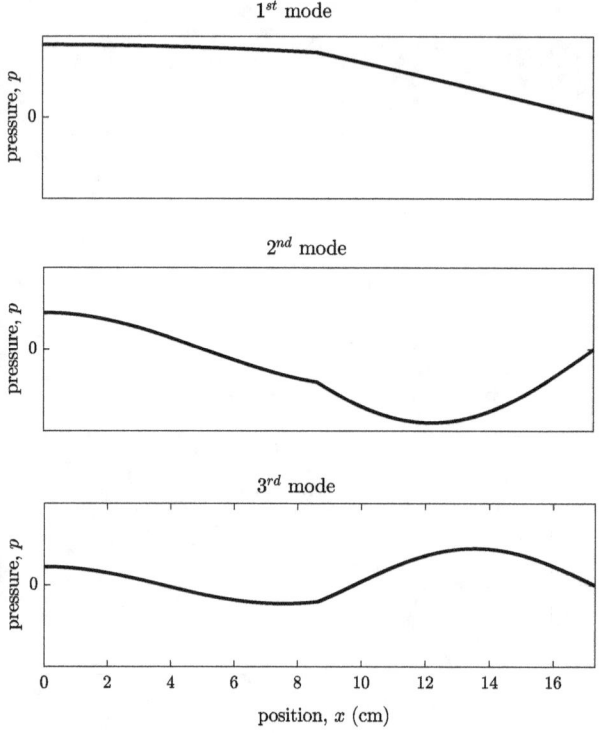

FIGURE 6. Pressure normal modes signed amplitudes $A^{(2)}/A^{(1)} = 4$

(20) is rewritten

$$p^{(n)}(x,t) = \left(P_+^{(n)} e^{ikx} + P_-^{(n)} e^{-ikx}\right) e^{-i\omega t},$$

and (21)

$$Q^{(n)}(x,t) = Y^{(n)} \left(P_+^{(n)} e^{ikx} - P_-^{(n)} e^{-ikx}\right) e^{-i\omega t}.$$

Define $p^{(n)}(x,t)\big|_+$ and $Q^{(n)}(x,t)\big|_+$ to be the pressure and volume velocity, respectively, looking right, taking account of all the reflections to the right of position x. This is the acoustic motion at position x that would result if the configuration is the given one, except that to the left of x the tube is a semi-infinite tube with cross-sectional area $A^{(n)}$, so that there would be no reflections from the left. Similarly,

Acoustics of Speech Production

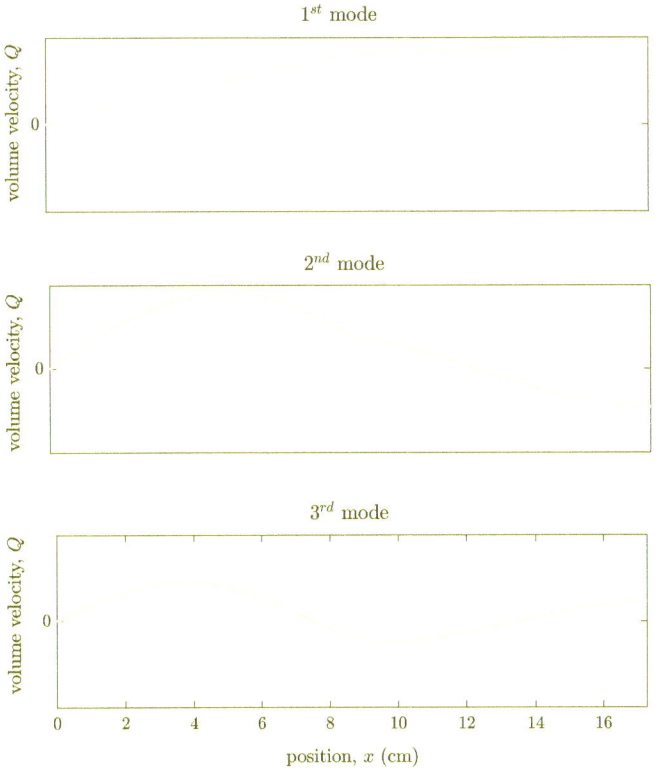

FIGURE 7. Volume velocity normal mode signed amplitudes for $A^{(2)}/A^{(1)} = 4$

define $p^{(n)}(x,t)|_-$ and $Q^{(n)}(x,t)|_-$ to be the pressure and volume velocity, respectively, looking left, taking account of all the reflections to the left of position x. This is the acoustic motion at x that would result if the configuration is the given one, except that to the right of x the tube is a semi-infinite tube with cross-sectional area $A^{(n)}$, so that there would be no reflections from the right. Effective admittances are defined in terms of $p^{(n)}(x,t)$ and $Q^{(n)}(x,t)$ looking right and looking left. Reference to Figue 8 can be made for the following discussion.

We proceed to show how effective admittances are computed. Because there are no waves approaching from the right of the opening, which is the right end of sub-tube 1, it follows that

$$\frac{Q^{(1)}(x=l^{(1)}+l^{(2)},t)}{p^{(1)}(x=l^{(1)}+l^{(2)},t)} = Y^{(0)} . \tag{22}$$

Substituting Equation (21) into Equation (22) gives

$$Y^{(0)} = Y^{(1)} \frac{P_+^{(1)} e^{ik(l^{(1)}+l^{(2)})} - P_-^{(1)} e^{-ik(l^{(1)}+l^{(2)})}}{P_+^{(1)} e^{ik(l^{(1)}+l^{(2)})} + P_-^{(1)} e^{-ik(l^{(1)}+l^{(2)})}} . \tag{23}$$

Equation (23) can be used to solve for $P_-^{(1)}$ in terms of $P_+^{(1)}$.

$$P_-^{(1)} = P_+^{(1)} e^{i2k(l^{(1)}+l^{(2)})} \frac{Y^{(1)} - Y^{(0)}}{Y^{(1)} + Y^{(0)}} . \tag{24}$$

We move to the left as far as the junction between sub-tubes 1 and 2. Define $p^{(1)}(x=l^{(2)},t)|_+$ and $Q^{(1)}(x=l^{(2)},t)|_+$ to be the pressure and volume velocity in sub-tube 1, respectively, looking right, from the junction between sub-tubes 1 and 2. These quantities do not account for any reflections to the left of the junction between sub-tubes 1 and 2. To realize this situation physically, sub-tube 2 would be extended indefinitely to the left as a semi-infinite tube. Effective admittance looking to the right from the junction is defined

$$\begin{aligned} Y_+^{(1)eff} &\equiv \frac{Q^{(1)}(x=l^{(2)},t)|_+}{p^{(1)}(x=l^{(2)},t)|_+} \\ &= Y^{(1)} \frac{P_+^{(1)} e^{ikl^{(2)}} - P_-^{(1)} e^{-ikl^{(2)}}}{P_+^{(1)} e^{ikl^{(2)}} + P_-^{(1)} e^{-ikl^{(2)}}} \\ &= Y^{(1)} \frac{Y^{(0)} - iY^{(1)} \tan(kl^{(1)})}{Y^{(1)} - iY^{(0)} \tan(kl^{(1)})} . \end{aligned} \tag{25}$$

The final equality follows from Equation (24). The following identities have also been used.

$$\begin{aligned} \cos(\phi) &= \frac{e^{i\phi} + e^{-i\phi}}{2} , \\ \sin(\phi) &= \frac{e^{i\phi} - e^{-i\phi}}{2i} , \\ \text{so that } \tan(\phi) &= \frac{\sin(\phi)}{\cos(\phi)} = -i \frac{e^{i\phi} - e^{-i\phi}}{e^{i\phi} + e^{-i\phi}} . \end{aligned} \tag{26}$$

The relations in Equation (26) can be derived from Euler's theorem, Equation (8) of Chapter 8. $Y_+^{(1)eff}$ in Equation (25) denotes effective admittance looking right from the junction of sub-tubes 1 and 2. It is the ratio of volume velocity to pressure at the junction between sub-tubes 1 and 2 taking account of reflections to the right of the junction, but not reflections to the left of the junction.

We move our attention to just left of the junction into sub-tube 2. We found the ratio of $P_+^{(1)}$ to $P_-^{(1)}$ in terms of tube admittances and tube lengths in Equation (24), and we do the same is done for the ratio of $P_+^{(2)}$ to $P_-^{(2)}$. Again we use the notation $Q^{(2)}(x,t)\big|_+$ and $p^{(2)}(x,t)\big|_+$ for volume velocity and pressure in sub-tube 2 when we are not considering reflections to the left of sub-tube 2; that is when we look right. By continuity of pressure and volume velocity across the junction between sub-tubes 1 and 2 and Equation (21)

$$Y_+^{(1)eff} = \frac{Q^{(1)}(x=l^{(2)},t)\big|_+}{p^{(1)}(x=l^{(2)},t)\big|_+}$$
$$= \frac{Q^{(2)}(x=l^{(2)},t)\big|_+}{p^{(2)}(x=l^{(2)},t)\big|_+} \qquad (27)$$
$$= Y^{(2)} \frac{P_+^{(2)} e^{-ikl^{(2)}} - P_-^{(2)} e^{ikl^{(2)}}}{P_+^{(2)} e^{-ikl^{(2)}} + P_-^{(2)} e^{ikl^{(2)}}} .$$

It follows that

$$P_-^{(2)} = P_+^{(2)} e^{-i2kl^{(2)}} \frac{Y^{(2)} - Y_+^{(1)eff}}{Y^{(2)} + Y_+^{(1)eff}} . \qquad (28)$$

We now move our attention further to the left. That is we move to the junction between sub-tubes 2 and 3, where sub-tube 3 is the piston, which is a tube of zero cross-sectional area. Again, we do not consider waves coming from the left of this junction, and define

$$Y_+^{(2)eff} \equiv \frac{Q^{(2)}(x=0,t)\big|_+}{p^{(2)}(x=0,t)\big|_+}$$
$$= Y^{(2)} \frac{P_+^{(2)} - P_-^{(2)}}{P_+^{(2)} + P_-^{(2)}} \qquad (29)$$

$$= Y^{(2)} \frac{Y_+^{(1)eff} - iY^{(2)} \tan(kl^{(2)})}{Y^{(2)} - iY_+^{(1)eff} \tan(kl^{(2)})} .$$

These equalities follow from Equation (28).

There is a nice pattern to these computations, if we set the most rightward effective admittance looking right to the acoustic admittance of the open end, $Y_+^{(0)eff} = Y^{(0)}$. With this definition substituted into Equation (25), and examination of Equations (25) and (29) indicates the following recursion

$$Y_+^{(n)eff} = Y^{(n)} \frac{Y_+^{(n-1)eff} - iY^{(n)} \tan(kl^{(n)})}{Y^{(n)} - iY_+^{(n-1)eff} \tan(kl^{(n)})} , \quad (30)$$

where $Y^{(n)}$ is the nth sub-tube's tube admittance, and $l^{(n)}$ is the length of the nth sub-tube for $n = 1, 2$. This provides an algorithm to compute all the effective admittances looking right.

Now we consider effective admittances looking left, while ignoring reflections to the right. We use the notation $Q^{(n)}(x,t)|_-$ to denote the volume velocity and $p^{(n)}(x,t)|_-$ to denote the pressure in sub-tube n ignoring reflections to the right of this tube. We can also start at the junction between 3 and 2 and proceed to the right, taking account of reflections to the left, and ignoring any reflections to the right of the junction. We start by setting the most leftward effective admittance looking left equal to the acoustic admittance of the piston.

$$Y_-^{(3)eff} = Y^{(3)} . \quad (31)$$

From Equation (21)

$$Y_-^{(3)eff} = Y^{(2)} \frac{P_+^{(2)} - P_-^{(2)}}{P_+^{(2)} + P_-^{(2)}} . \quad (32)$$

Solving for $P_-^{(2)}$ in terms of $P_+^{(2)}$

$$P_-^{(2)} = P_+^{(2)} \frac{Y^{(2)} + Y_-^{(3)eff}}{Y^{(2)} - Y_-^{(3)eff}} . \quad (33)$$

At the junction between sub-tubes 1 and 2

$$Y_-^{(2)eff} \equiv \frac{Q^{(2)}(x = l^{(2)}, t)|_-}{p^{(2)}(x = l^{(2)}, t)|_-}$$

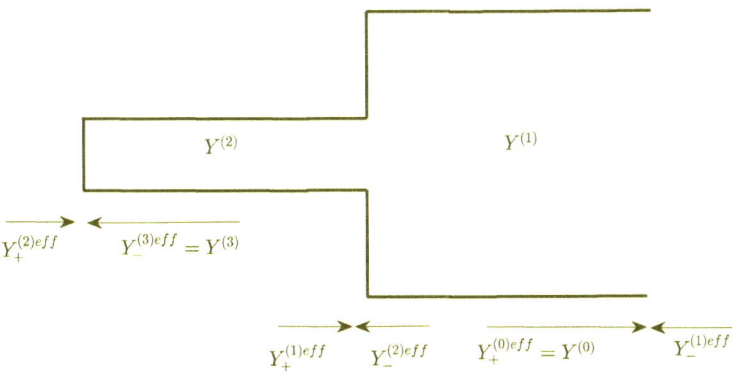

FIGURE 8. Effective admittances for two sub-tubes

$$= Y^{(2)} \frac{P_+^{(2)} e^{-ikl^{(2)}} - P_-^{(2)} e^{ikl^{(2)}}}{P_+^{(2)} e^{-ikl^{(2)}} + P_-^{(2)} e^{ikl^{(2)}}} \qquad (34)$$

$$= Y^{(2)} \frac{Y_-^{(3)eff} + iY^{(2)} \tan(kl^{(2)})}{Y^{(2)} + iY_-^{(3)eff} \tan(kl^{(2)})} \; .$$

$Y_-^{(2)eff}$ denotes the effective admittance looking left from the junction of sub-tubes 1 and 2. It accounts for all the reflections to the left of this junction.

This process can be continued so that the effective admittance looking left from the opening is

$$Y_-^{(1)eff} = Y^{(1)} \frac{Y_-^{(2)eff} + iY^{(1)} \tan(kl^{(1)})}{Y^{(1)} + iY_-^{(2)eff} \tan(kl^{(1)})} \; . \qquad (35)$$

The recursion suggested by Equations (34) and (35) is

$$Y_-^{(n)eff} = Y^{(n)} \frac{Y_-^{(n+1)eff} + iY^{(n)} \tan(kl^{(n)})}{Y^{(n)} + iY_-^{(n+1)eff} \tan(kl^{(n)})} \; . \qquad (36)$$

Figure 8 shows where and in what direction each of the effective admittances are defined. Note that $Y_+^{(n)}$ faces $Y_-^{(n+1)}$ for $n = 0, 1, 2$.

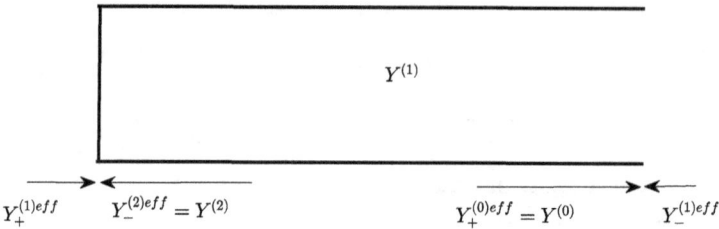

FIGURE 9. Normal mode frequencies for the tube of constant cross-sectional area

Conditions for normal modes using effective admittances

We have defined effective admittances, $Y_+^{(n)eff}$ and $Y_-^{(n)eff}$ at junctions between sub-tubes. We could have defined effective admittance looking right, $Y_+^{eff}(x)$, and looking left, $Y_-^{eff}(x)$ from any axial location within the tube, x. It is clear that we need to have continuity of volume velocity and pressure all along the tube, with the correct relative phases of these quantities. Thus, we must have

$$Y_+^{eff}(x) = Y_-^{eff}(x) , \qquad (37)$$

for all x , $0 < x < L$. In particular, this must be true at junctions.

$$Y_+^{(n)eff} = Y_-^{(n+1)eff} \text{ for } n = 0, 1, 2 , \qquad (38)$$

with $Y_+^{(0)eff} = Y^{(0)}$ and $Y_-^{(3)eff} = Y^{(3)}$.

Equation (38) defines the normal modes for tubes composed of constant cross-sectional area sub-tubes. In order for Equation (38) to hold for all junctions, it is sufficient that the equality hold at one junction in the tube. Each $Y_+^{(n)eff}$ takes account of the boundary condition the farthest to the right, expressed as an admittance $Y^{(0)}$. And each $Y_-^{(n)eff}$ takes account the boundary condition the farthest to the left, expressed as an admittance $Y^{(3)}$. Equation (38) ensures that both boundary conditions are satisfied. This is similar, in spirit, to matching left and right semi-eigenfunctions at a junction.

The criterion is illustrated for the case that we are familiar with: the finite-length tube with no changes in cross-sectional area. There is only one tube with length L, with tube admittance $Y^{(1)}$. The effective admittances are illustrated in Figure 9. The piston is now sub-tube 2 with zero cross-sectional area.

Consider $Y_+^{(1)eff}$ and $Y_-^{(1)eff}$, with the boundary conditions $Y_+^{(0)eff} = \infty$ and $Y_-^{(2)eff} = 0$.

$$Y_+^{(1)eff} = Y^{(1)} \frac{Y_+^{(0)eff} - iY^{(1)} \tan(kL)}{Y^{(1)} - iY_+^{(0)eff} \tan(kL)} \tag{39}$$

$$= -\frac{Y^{(1)}}{i \tan(kL)} = \frac{iY^{(1)} \cos(kL)}{\sin(kL)} = \frac{iY^{(1)} \cos(\omega L/c_0)}{\sin(\omega L/c_0)}.$$

$$Y_-^{(1)eff} = Y^{(1)} \frac{Y_-^{(2)eff} + iY^{(1)} \tan(kL)}{Y^{(1)} + iY_-^{(2)eff} \tan(kL)} \tag{40}$$

$$= iY^{(1)} \tan(kL) = \frac{iY^{(1)} \sin(kL)}{\cos(kL)} = \frac{iY^{(1)} \sin(\omega L/c_0)}{\cos(\omega L/c_0)}.$$

Either of the following criteria provides the normal mode frequencies $\mathcal{F}_m = \omega_m/(2\pi)$.

$$Y_+^{(1)eff} = Y_-^{(2)eff} = 0 \quad \text{and} \quad Y_-^{(1)eff} = Y_+^{(0)eff} = \infty . \tag{41}$$

These expressions mean that we meet both boundary conditions: the one at the opening and the one at the hard-wall, which happens to be a piston. By reference to Equations (38)-(40), the normal mode frequencies are

$$\omega_m \equiv 2\pi \mathcal{F}_m = \frac{(2m-1)\pi c_0}{2L}, \text{ for } m \text{ a positive integer,}$$

so that $\tag{42}$

$$\mathcal{F}_m = \frac{(2m-1)c_0}{4L}, \text{ for } m \text{ a positive integer.}$$

These are the normal mode frequencies that were derived in Chapter 5.

The following criterion can be applied to the two sub-tube case shown in Figure 8 at the junction between sub-tubes 1 and 2

$$Y_+^{(1)eff} = Y_-^{(2)eff} . \tag{43}$$

to obtain the following from Equations (30), (36), and (39)

$$Y^{(1)} \frac{Y^{(0)eff}_+ - iY^{(1)} \tan(kl^{(1)})}{Y^{(1)} - iY^{(0)eff}_+ \tan(kl^{(1)})} = Y^{(2)} \frac{Y^{(3)eff}_- + iY^{(2)} \tan(kl^{(2)})}{Y^{(2)} + iY^{(3)eff}_- \tan(kl^{(2)})} . \quad (44)$$

With the boundary conditions $Y^{(0)eff}_+ = \infty$ and $Y^{(3)eff}_- = 0$, this reduces to

$$iY^{(1)} \cot(kl^{(1)}) = iY^{(2)} \tan(kl^{(2)}) ,$$

or

$$\frac{Y^{(2)}}{Y^{(1)}} \tan(kl^{(1)}) \tan(kl^{(2)}) - 1 = 0 , \quad (45)$$

or

$$\frac{A^{(2)}}{A^{(1)}} \tan(kl^{(1)}) \tan\left(kl^{(2)}\right) - 1 = 0 .$$

This is the same as the first expression in Equation (16) derived above. This equation is used to find the permissible normal mode wavenumbers k_m and frequencies, $\mathcal{F}_m = k_m c_0/(2\pi), m = 1, 2, \ldots$.

There is enough information in Equations (22) through (34), and the solutions to Equation (45) solve for $P^{(1)}_-$, $P^{(1)}_+$, and $P^{(2)}_-$ in terms of $P^{(2)}_+$. This provides the complex representation of the normal modes in Equation (21) for the eigenvalues k_m and ω_m with complex constant $P^{(2)}_+$. Of course the real parts of these functions are the physical normal modes. This is analogous to solving the two real relations in Equation (16).

The next chapter is about using effective admittances with more than two sub-tubes. There we use the development of effective admittances to show how the spatial distributions of the normal mode pressure signed amplitudes can be computed as well.

Steady radiation pressure and acoustic perturbation theory

In this section we examine the acoustic perturbation theory that was introduced in Chapter 6. Recall the expressions for kinetic energy density e^{kin} and potential energy density e^{pot} in the acoustic approximation from Equations (28) and (30) of Chapter 2

$$e^{kin} = \frac{\rho_0}{2}u^2 = \frac{\rho_0}{2}\left(\frac{Q}{A}\right)^2$$

$$e^{pot} = \frac{1}{2\rho_0 c_0^2}p^2 .$$

(46)

where p is pressure, u is particle velocity, Q is volume velocity and A cross-sectional area. We also discussed how energy densities for each mode m can be considered separately in Chapter 6. From Equation (6) of Chapter 6, the steady acoustic radiation pressure for mode m is

$$\langle p_m \rangle = \langle e_m^{pot} \rangle - \langle e_m^{kin} \rangle = \frac{1}{2\rho_0 c_0^2}\langle p_m^2 \rangle - \frac{\rho_0}{2}\frac{\langle Q_m^2 \rangle}{A^2} . \quad (47)$$

where $\langle \cdot \rangle$ denotes time-average, as before.

We plot the time-average kinetic and potential energy densities and steady acoustic radiation pressure as a function of position in the tube for the first three normal modes for $A^{(2)}/A^{(1)} = 0.25$ and for $A^{(2)}/A^{(1)} = 4.0$. These are the configurations shown in the bottom two panels of Figure 2. Figures 10 and 11 show the quantities $\langle e_m^{kin} \rangle$, $\langle e_m^{pot} \rangle$, and $\langle p_m \rangle$ for the first three modes $m = 1, 2, 3$, in the case that $A^{(2)}/A^{(1)} = 0.25$. Figures 12 and 13 show the quantities $\langle e_m^{kin} \rangle$, $\langle e_m^{pot} \rangle$, and $\langle p_m \rangle$ for the first three modes $m = 1, 2, 3$, in the case that $A^{(2)}/A^{(1)} = 4$. The fact that $\langle \sin^2(\omega_m t) \rangle = \langle \cos^2(\omega_m t) \rangle = 1/2$ from Equation (36) of Chapter 5 is used here.

The $\langle e_m^{kin} \rangle$, and hence $\langle p_m \rangle$, are discontinuous at the junction between tubes 1 and 2. The particle velocities $u_m = Q_m/A$ are necessarily discontinuous across the junction of sub-tubes with different cross-sectional area A because the Q_m are continuous. This is a physically unrealistic part of the theory. Below, we go into detail on the meaning of the discontinuity of the u_m, as well as ways in which the continuity conditions might be amended to partially account for these discontinuities.

As might be expected, there are more maxima and minima in the time-average energy densities and steady acoustic radiation pressure amplitudes as the mode number m increases. Another observation is that the steady acoustic radiation pressure is both negative and positive

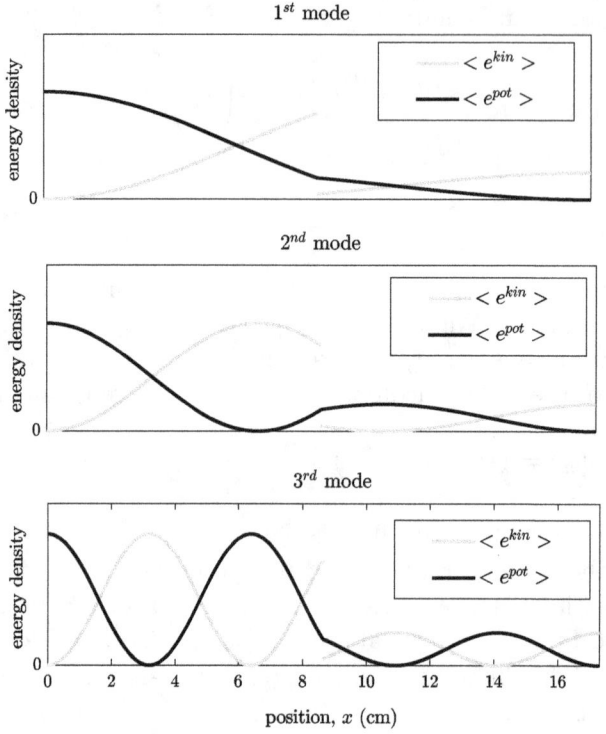

FIGURE 10. Time-average kinetic and potential energy densities for $A^{(2)}/A^{(1)} = 0.25$

in the half of the tube closest the opening, that is for $x > l^{(2)} = 8.55$ cm for the second mode, for both area ratios, $A^{(2)}/A^{(1)} = 0.25, 4.0$ as shown in Figures 11 and 13. In speech production, does tongue fronting always raise the second mode frequency, or second formant? What actually happens according to acoustic perturbation theory appears to depend on the base configuration of the tongue, and on which part of the tongue that is fronted.

Finally, we can note from the top panels in Figures 10 through 13 that there is more variation in time-average energy densities and steady acoustic radiation pressures of mode 1 in the back sub-tube 2 and front sub-tube 1 for $A^{(2)}/A^{(1)} = 0.25$ than for $A^{(2)}/A^{(1)} = 4.0$. In fact, the potential energy appears to be concentrated in the back sub-tube, and the kinetic energy concentrated in the front sub-tube

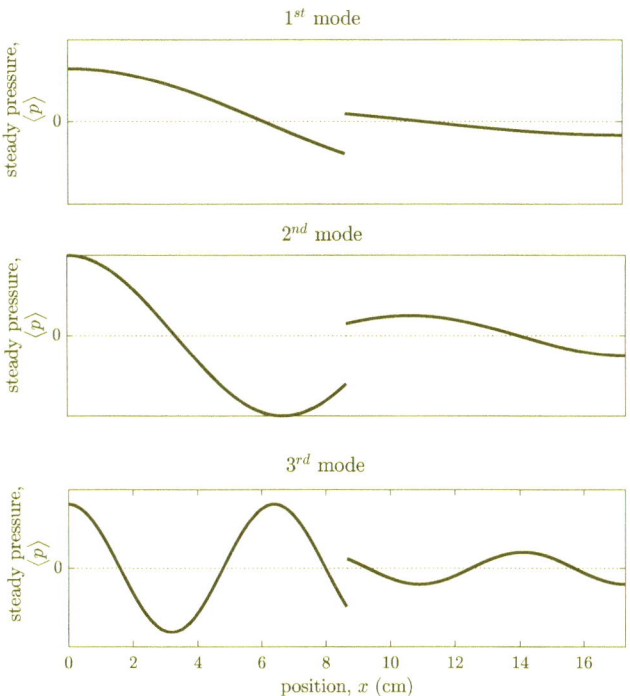

FIGURE 11. Steady acoustic radiation pressure for $A^{(2)}/A^{(1)} = 0.25$

for the $A^{(2)}/A^{(1)} = 4.0$ configuration. This is discussed in Chapter 12, where the Helmholtz resonators are examined in detail.

The continuity conditions and their amendment

Discontinuity in particle velocity at a junction between sub-tubes of different cross-sectional area results from imposing continuity of volume velocity across the junction. Also, consider the pressure and volume velocity plots at the junction between the sub-tubes 2 and 1 at $x = l^{(2)} = 8.55$ cm in Figures 4 through 7. While pressure and volume velocity are continuous at the junction, there are "kinks" in the curves so that the transition from one sub-tube to another is not smooth. This is because the waves of different amplitude from different sub-tubes possess the same wavelength, but they must meet at different phases

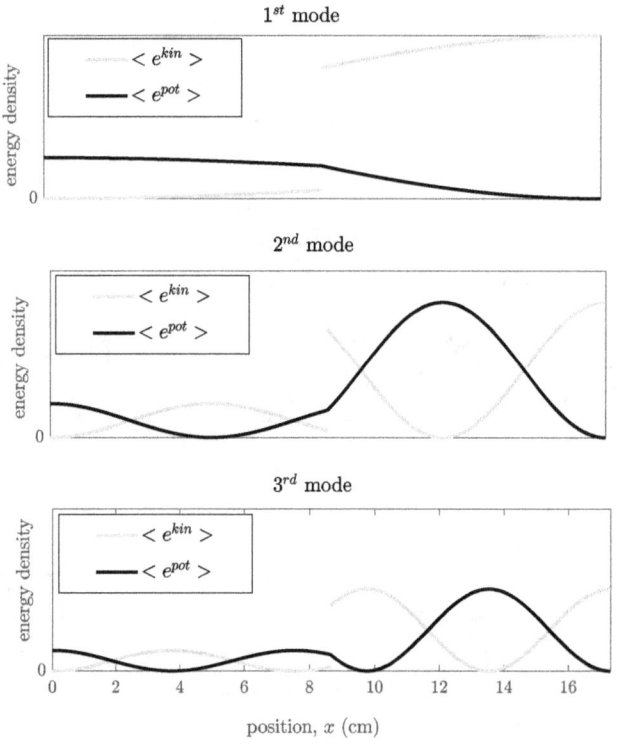

FIGURE 12. Time-average kinetic and potential energy densities for $A^{(2)}/A^{(1)} = 4$

along their wavelengths in order to represent a continuous function. The kinks, or non-smoothness, as well as the discontinuity in particle velocity result because continuity of pressure and volume velocity are approximate conditions that should be refined. The refinements require some difficult and sophisticated mathematics (Sen 1989). We deem most of these refinements not to be worth the effort at this point in our development. We pursue only the most basic refinement in terms of something called *lumped mass elements*.

Before proceeding to this refinement, we describe the physical reason that continuity of pressure and volume velocity at the junction is not quite correct. For the following discussion it is easiest to assume that the tubes have rectangular cross-sections of constant width, d, in the

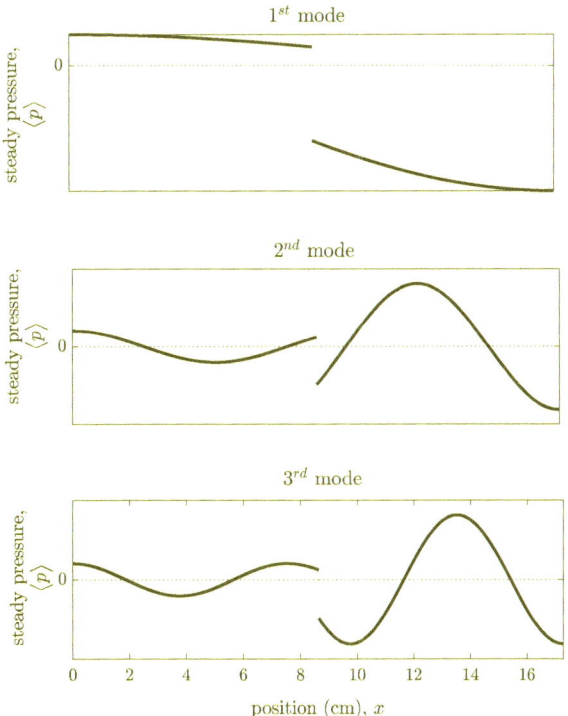

FIGURE 13. Steady acoustic radiation pressure for $A^{(2)}/A^{(1)} = 4$

dimension into the paper, so that all area changes are represented by width changes in the plane of the paper.

Figure 14a shows the *streamlines* of fluid particles as they oscillate near the junction under the assumption of continuous volume velocity across the junction and one-dimensional motion along the axis of the tube. The streamlines, in this case, are time independent and indicate paths along which air particles travel. Air also travels in regions between the streamlines without crossing them, and these regions are termed *stream tubes* (Batchelor 1967). [Actually the streamlines indicate two-dimensional surfaces, with the second dimension into the paper. These surfaces, along with the physical tube's surfaces, provide for the stream tubes.] Each stream tube can be thought of as a tube within the larger tube with solid walls. As drawn in Figure

14a, the distance between contiguous pairs of streamlines within a sub-tube is the same, which means that the cross-sectional areas of the stream-tubes are constants and equal to one another. Thus, the particle velocity amplitude in each stream tube within a sub-tube is the same. The stream tubes are narrower in sub-tube 2 than in sub-tube 1, so the particle velocity amplitude is larger in the former than in the latter. Further, as long as the length of the region under consideration in the x-direction is small compared to the wavelength of sound, the volume velocity throughout each stream tube is approximately constant at each instant in time. This is because there cannot be substantial variation of volume velocity Q at any time over a distance that is a fraction of a wavelength. Note that wavelength λ being large enough means that frequency $\mathcal{F} = c_0/\lambda$ is small enough.

Of course, we should be suspicious of the picture presented in Figure 14a, because discontinuities in streamline spacing at the junction do not occur in nature. Where have we gone wrong? When working in acoustics, we always need to remember that we are working in an approximation to fluid mechanics, so incorporating neglected aspects of fluid mechanics might be the answer. Here, we don't need to give up the acoustic approximation completely. All we need to do is to remember that air flow occurs in three dimensions, and not one dimension. As long as we are considering flow in straight tubes it is fine to work in the one spatial dimension along the axis of the tube. At junctions we need to consider three spatial dimensions.

Figure 14b shows the smooth transition of the streamlines between the two sub-tubes when the sub-tubes have rectangular cross-section. These streamlines still represent irrotational air motion at the junction. [The term irrotational motion was introduced at the beginning of Chapter 2. However, this idea will be clarified in Chapter 13.] In essence, the junction with the abrupt change in area requires a smooth transition in particle velocity from one sub-tube to another. The particle velocity changes continuously in space, albeit rapidly, in the neighborhood of the junction. It can be seen that the distances between streamlines in Figure 14b are not equal close to the junction, which means that the flow speed is not uniform in cross-sections of the tubes near the junctions. However, the streamlines are parallel to the x-axis away from the junction, just as in Figure 14a, indicating plane,

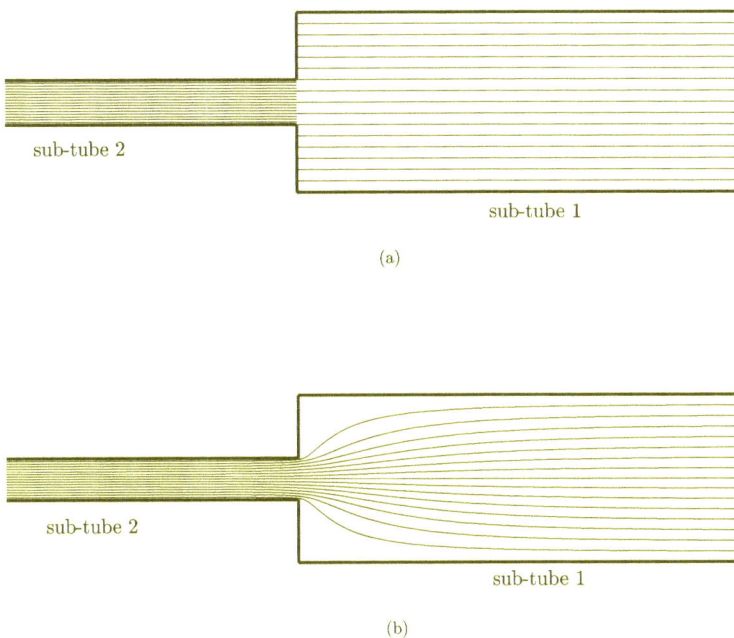

FIGURE 14. Flow streamlines near a junction between sub-tubes

or one-dimensional, wave motion away from the junction. The volume velocity does, indeed, remain approximately the same in each sub-tube.

The curving of the streamlines in Figure 14b means that they tend to be closer together in the larger sub-tube, sub-tube 1, near the junction than would otherwise be the case shown in Figure 14a. This means that the particle velocity is greater in this region of sub-tube 1 when the curving of the streamlines is taken into account than when it is not. Hence, there is more kinetic energy in the situation given by Figure 14b than the one given in Figure 14a.

This observation leads us to the first refinement to the conditions of continuity of volume velocity and continuity of pressure. When the wavelength of sound is much longer than the length scales of the region where the streamlines are curving, the increase in kinetic energy is modeled with a lumped mass element of air at the junction of sub-tubes 1 and 2, acting as a solid object with the density of air ρ_0 (Lighthill 1978; Pierce 1989). [We saw a lumped mass element when considering the Helmholtz resonator in Chapter 1.] The increase in kinetic energy in the curved streamline situation over the uncurved streamline situation is supposed to be incurred by this mass m of air. Further, we imagine that the mass m is contained in a volume of air V with cross-sectional area of the narrower tube, $A^{(2)}$ here, and length ℓ in the x-direction, contained in the region of sub-tube 1 that possesses the excess kinetic energy caused by the curving of the streamlines. The length ℓ has yet to be determined. The excess kinetic energy ΔKE in terms of the volume velocity at the junction Q is

$$\Delta KE \equiv \frac{m}{2}u^2 = \frac{\rho_0 V}{2}\left(\frac{Q}{A^{(2)}}\right)^2 = \frac{\rho_0 A^{(2)} \ell}{2}\left(\frac{Q}{A^{(2)}}\right)^2 = \frac{\rho_0 \ell}{A^{(2)}}\frac{Q^2}{2} . \quad (48)$$

Calculating the excess kinetic energy ΔKE due to the curving of the streamlines involves some fairly intricate mathematics. For the situation shown in Figure 14b, this calculation has been done by Morse & Ingard (1986) enabling us to draw Figure 14b. Their calculation also permits us to relate both ℓ and $\rho_0 \ell / A^{(2)}$ in Equation (48) to the ratio of areas of the tubes. These quantities are plotted in Figure 15 against $\log_{10}(A^{(1)}/A^{(2)})$. In these plots $d = 2$ cm and $A^{(1)} = 4$ cm^2.

How does this more detailed picture of what happens near the junction change the continuity conditions? Continuity of volume velocity is still a very good approximation, because the streamlines in Figure 14b straighten within a short distance from the junction to resemble those in Figure 14a. However, the increased kinetic energy of the volume of air with length ℓ and cross-sectional area $A^{(2)}$ requires that the air be accelerated through the junction. This requires a localized pressure gradient at the junction, which means that continuity of pressure through the junction is no longer valid. This can be seen mathematically from the momentum conservation equation, Equation (14) of Chapter 2.

$$\frac{\Delta_t u(x,t)}{\Delta t} = -\frac{1}{\rho_0}\frac{\Delta_x p(x,t)}{\Delta x},$$

or (49)

$$\frac{\Delta_t Q(x,t)}{\Delta t} = -\frac{A^{(2)}}{\rho_0}\frac{\Delta_x p(x,t)}{\Delta x},$$

where $A^{(2)}$ is cross-sectional area of the narrower of the sub-tubes.

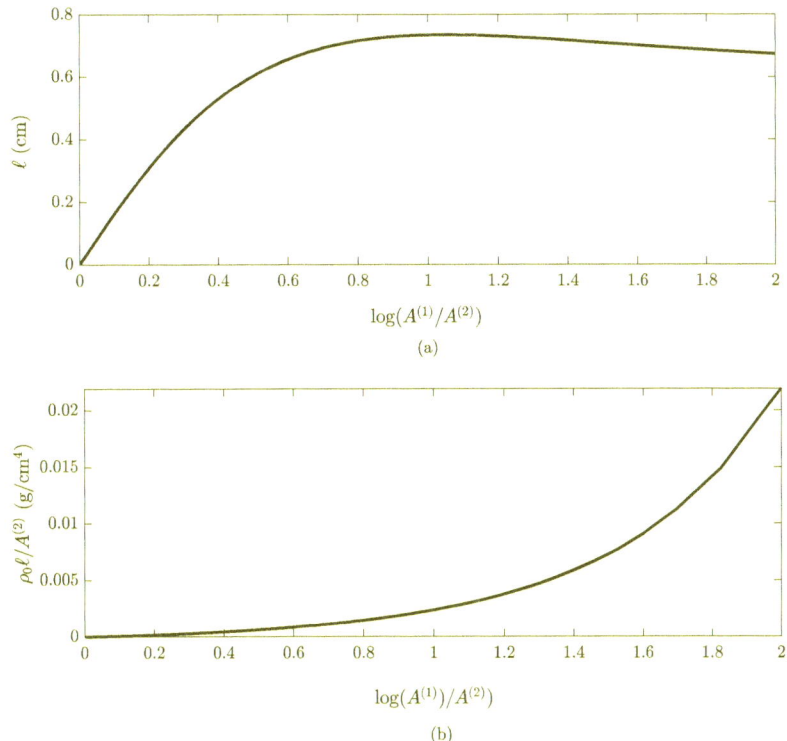

FIGURE 15. Lumped mass elements

Before continuing, we need the mathematical identity

$$\frac{\Delta_t e^{Kt}}{\Delta t} \approx K e^{Kt} . \qquad (50)$$

where K is a constant. The approximation improves as $\Delta t > 0$ gets smaller.

We apply Equation (49) across the lumped mass element of length $\Delta x = \ell$. For sinusoidal time dependence we write $Q(x,t) = \hat{Q}(x,\omega)e^{-i\omega t}$ and $p(x,t) = \hat{p}(x,\omega)e^{-i\omega t}$ in Equation (49), and then use Equation (50) to obtain

$$-i\omega \frac{\hat{Q}(x = x_J, \omega)}{A^{(2)}} \approx -\frac{1}{\rho_0} \frac{\Delta_x \hat{p}(x = x_J, \omega)}{\ell} ,$$

or (51)

$$iw\frac{\rho_0 \ell}{A^{(2)}}\hat{Q}(x=x_J,\omega) \approx \Delta_x \hat{p}(x=x_J,\omega) \,,$$

where $x = x_J$ is the position of the junction and $\Delta_x \hat{p}(x=x_J,\omega)$ is the change in pressure from right-to-left across the junction. With the factor i in Equation (51), $\Delta_x \hat{p}(x=x_J,\omega)$ is $\pi/2$ out of phase with the air volume velocity velocity at the junction. This means that no time-average work is done by $\Delta_x \hat{p}(x=x_J,\omega)$. This is sloshing behavior of air again. Note that the factor $\rho_0 \ell / A^{(2)}$ in Equation (51) is the same as the one appearing in the expression for excess kinetic energy in Equation (48).

In Equation (51) are considering the case where $A^{(2)} \leq A^{(1)}$. Suppose that $A^{(2)} > A^{(1)}$. Instead of Equation (49) we start with

$$\frac{\Delta_t Q(x,t)}{\Delta t} = -\frac{A^{(1)}}{\rho_0}\frac{\Delta_x p(x,t)}{\Delta x} \,. \tag{52}$$

With Equation (52)

$$iw\frac{\rho_0 \ell}{A^{(1)}}\hat{Q}(x=x_J,\omega) \approx \Delta_x \hat{p}(x=x_J,\omega) \,. \tag{53}$$

Let $\Delta \hat{p}_J \equiv \Delta_x \hat{p}(x=x_J,\omega)$. Both Equations (51) and (52) can be written

$$iw\frac{\rho_0 \ell}{A^{(nar)}}\hat{Q}(x=x_J,\omega) \approx \Delta \hat{p}_J \,, \tag{54}$$

where $A^{(nar)} = \min(A^{(1)}, A^{(2)})$ is the smaller cross-sectional are of the two sub-tubes that meet at the junction, $x = x_J$.

We insert a lumped mass element between sub-tubes 2 and 1 to account for excess kinetic energy and the required pressure gradient. This lumped mass element has an acoustic impedance and an acoustic admittance, respectively

$$Z_J \equiv \frac{\Delta \hat{p}_J(\omega)}{\hat{Q}_J(\omega)} \approx iw\frac{\rho_0 \ell}{A^{(nar)}} \quad \text{and} \quad Y_J \equiv \frac{\hat{Q}_J(\omega)}{\Delta \hat{p}_J(\omega)} \approx \left(iw\frac{\rho_0 \ell}{A^{(nar)}}\right)^{-1}, \tag{55}$$

where the subscript J denotes junction, and $\hat{Q}_J(\omega) = \hat{Q}(x=x_J,\omega)$. Defining $Y^{(nar)} = A^{(nar)}/\rho_0 c_0$ and $k = \omega/c_0$, Equation (55) can be rewritten

$$Z_J \equiv \frac{\Delta \hat{p}_J(\omega)}{\hat{Q}_J(\omega)} \approx i\frac{(k\ell)}{Y^{(nar)}} \quad \text{and} \quad Y_J \equiv \frac{\hat{Q}_J(\omega)}{\Delta \hat{p}_J(\omega)} \approx -i\frac{Y^{(nar)}}{(k\ell)} \,. \tag{56}$$

We have associated a length ℓ with this element, and this length is assumed to be very small on the scale of a wavelength. In fact, this is the nature of lumped elements in acoustics: they have small spatial extent in the direction of propagation compared to the wavelength of sound. This condition is written in terms of wavenumber k as

$$k\ell = 2\pi\frac{\ell}{\lambda} \ll 1 . \tag{57}$$

The lumped mass element is the first refinement that can be made to the continuity of pressure and volume velocity conditions at the junction. It is the only refinement that we consider in this book.

The work that we have done in inserting the lumped mass element does not fix the problem of discontinuity of physical quantities at the junction. In fact, this first amendment of the pressure and volume velocity continuity conditions makes pressure discontinuous across the junction. We have indicated that physical quantities are not really discontinuous across the junction. However, to further amend one-dimensional acoustic theory near junctions would require more work than is really necessary at this point. It would be necessary to go well beyond lumped elements into the details of the air flow shown in Figure 14b. On the other hand, we cannot apply acoustic perturbation theory in regions near large area changes because of discontinuities in $\langle p \rangle$. In the next chapter, we examine whether it helps to use a larger number of sub-tubes to avoid large discontinuities in cross-sectional area.

Conclusion

In this chapter we have explored the distribution of pressure signed amplitude and volume velocity signed amplitude in a tube with two sub-tubes using normal modes. As in the single tube system, these normal modes satisfy the wave equation, as well as the boundary conditions. The sub-tubes contain parts of sinusoidal waves with the same wavelength and frequency, but with different amplitudes, and, they meet at junctions at different spatial phases to permit continuity of pressure and volume velocity at the junctions between sub-tubes.

A problem arises when time-average kinetic energy density and steady acoustic radiation pressure are considered: they are discontinuous at the junction when continuity of volume velocity is assumed. It turns out that we have neglected the three-dimensional nature of the air flow at the junction. There is a continuous, but abrupt change in air particle velocity at the junction with curving of the flow streamlines. The curving of the streamlines means that we need to account for more air kinetic energy across the junction than with the original continuity conditions. This implies that there must be a local change in pressure near the junction to allow for this extra kinetic energy. We can amend the continuity of pressure across the junction with a lumped mass element to include a pressure change across the junction. In the latter part of Chapter 10, we explore what happens when the lumped mass elements and pressure gradients are incorporated into acoustics of a variable area tube. To account for a completely smooth transition of physical quantities across discontinuous junctions requires substantial theoretical effort.

References

Batchelor, G.K. (1967). *An Introduction to Fluid Dynamics.* Cambridge University Press: Cambridge, England (p 72).

Lighthill, M.J. (1978). *Waves in Fluids.* Cambridge University Press: Cambridge, England (Chapter 2).

Morse, P.M. & Ingard, K.U. (1986). *Theoretical Acoustics.* Princeton University Press: Princeton, NJ (p 488).

Pierce, A.D. (1989). *Acoustics: An Introduction to its Physical Principles and Applications.* Acoustical Society of America: Woodbury, NY (pp 324-30).

Sen, R. (1989). The asymptotics of evanescent fields in waveguides. *Journal of the Acoustical Society of America* **85**, p 1456.

Chapter 10: Multiple Sub-Tubes

Introduction

We examine what occurs when the number of sub-tubes is generalized from two to any finite number N with continuity of pressure and volume velocity along the tube axis. In terms of normal modes, we examine the spatial distributions of pressure and volume velocity signed amplitudes, time-average energy densities, and steady acoustic radiation pressure for multiple sub-tubes. We consider the abruptness of cross-sectional area changes, which are determined by the number and lengths of the sub-tubes, and the resulting discontinuities in steady acoustic radiation pressure calculations.

Toward the end of this chapter, we explain how to incorporate the lumped mass elements introduced in Chapter 9. Their effect on tube normal modes and their frequencies are considered. The idea of lumped mass elements has been used in other areas of acoustics, such as muffler design. We start examining lumped mass elements by introducing such an element at the opening of the tube to the atmosphere. This element, which is completely analogous with those at internal junctions, changes the acoustic admittance at the opening from infinity to a finite, imaginary value. We go on to introduce lumped mass elements at the junctions between all sub-tubes, and we consider their effects on the normal modes.

Multiple sub-tubes with pressure and volume velocity continuity conditions

Normal modes with semi-eigenfunctions

We perform the same steps to derive normal modes for N sub-tubes that we described for the two sub-tube case using semi-eigenfunctions. Initially, lumped mass elements are not included. The conditions of continuity of pressure and volume velocity at junctions provide the means to construct the normal mode pressure and volume velocity distributions. As pictured in Figure 1a, we assume that the tube of

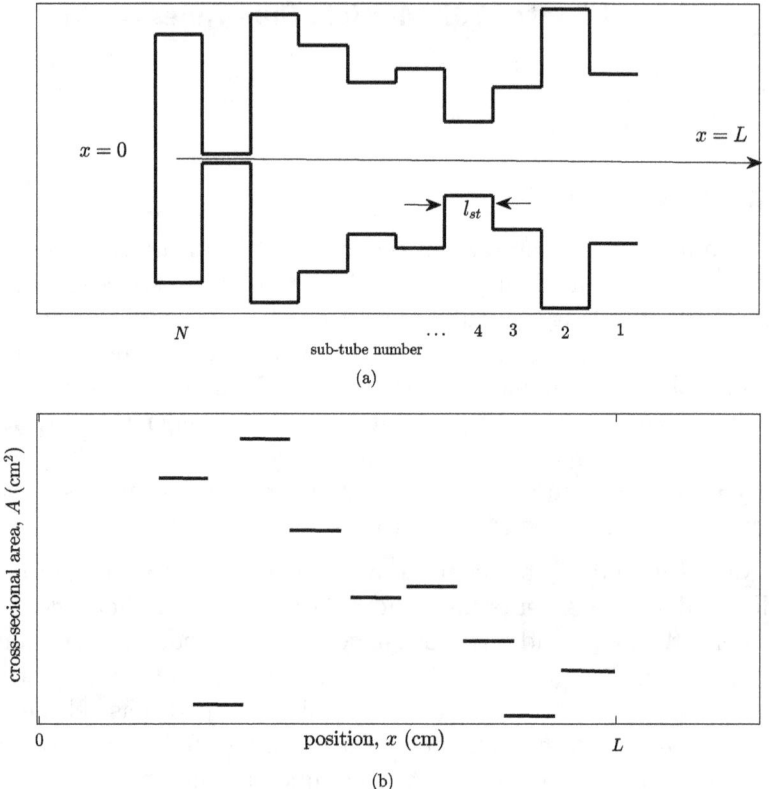

FIGURE 1. (a) A finite-length tube composed of sub-tubes of equal length l_{st}, and (b) the corresponding area function

length L has been divided into N sub-tubes of equal length l_{st}. This is not the most general case, but it keeps the results manageable. The results could easily be generalized to sub-tubes of unequal length. Instead of starting with the left semi-eigenfunction in sub-tube 2, we start with the same semi-eigenfunction in sub-tube N. We use the amplitudes $P^{(n)}$ and phases $\theta^{(n)}$, $1 \leq n \leq N$, to ensure continuity of volume velocity and pressure between sub-tubes. An equation for the wavenumber k is used to find the eigenvalues that enforce the boundary condition at the opening. We plot the cross-sectional areas A of the tube shown in Figure 1a as a function of position x in Figure 1b. This is a plot of the *area function*.

Instead of Equation (13) of Chapter 9, we have

$$p(x,t) = \begin{cases} p^{(N)}(x,t) & \text{for } 0 < x < l_{st} \\ p^{(N-1)}(x,t) & \text{for } l_{st} < x < 2l_{st} \\ \ldots, \end{cases}$$

where (1)

$$p^{(N)}(x,t) = P^{(N)} \cos(kx) \sin(\omega t + \phi) \text{ for } 0 < x < l$$
$$p^{(N-1)}(x,t) = P^{(N-1)} \cos(kx + \theta^{(N-1)}) \sin(\omega t + \phi)$$
$$\ldots,$$

where $P^{(n)}$ for $n = 1, 2, \ldots, N$ have units of pressure. In place of Equation (18) of Chapter 9, we have

$$Q(x,t) = \begin{cases} Q^{(N)}(x,t) & \text{for } 0 < x < l_{st} \\ Q^{(N-1)}(x,t) & \text{for } l_{st} < x < 2l_{st} \\ \ldots \end{cases}$$

where, (2)

$$Q^{(N)}(x,t) = -Y^{(N)} P^{(N)} \sin(kx) \cos(\omega t + \phi)$$
$$Q^{(N-1)}(x,t) = -Y^{(N-1)} P^{(N-1)} \sin(kx + \theta^{(N-1)}) \cos(\omega t + \phi)$$
$$\ldots.$$

We can proceed as we did for the two sub-tube situation and use continuity of pressure and volume velocity at the junction between sub-tubes N and $N-1$ to find $P^{(N-1)}$ in terms of $P^{(N)}$, wavenumber k, and lengths l_{st}. $\theta^{(N-1)}$ is also found with the continuity conditions at the junction. This is seen below for the general situation between any two sub-tubes. This procedure can be continued until the opening is reached.

In general, between sub-tubes $N - n$ and $N - (n+1)$ for $n = 0, 1, 2, \ldots N - 2$. we suppose that $P^{(N-n)}$ and $\theta^{(N-n)}$ are known in terms of $P^{(N)}$, and we find $P^{(N-(n+1))}$ and $\theta^{(N-(n+1))}$ in terms of $P^{(N)}$ by invoking continuity of pressure and volume at the junction. The pertinent expressions for pressure are

$$p(x,t) = p^{(N-n)}(x,t)$$

$$= P^{(N-n)} \cos(kx + \theta^{(N-n)}) \sin(\omega t + \phi)$$
$$\text{for } nl_{st} < x < (n+1)l_{st},$$

and (3)

$$p(x,t) = p^{(N-(n+1))}(x,t)$$
$$= P^{(N-(n+1))} \cos(kx + \theta^{(N-(n+1))}) \sin(\omega t + \phi)$$
$$\text{for } (n+1)l_{st} < x < (n+2)l_{st}.$$

The pertinent expressions for volume velocity are

$$Q(x,t) = Q^{(N-n)}(x,t)$$
$$= -Y^{(N-n)} P^{(N-n)} \sin(kx + \theta^{(N-n)}) \cos(\omega t + \phi)$$
$$\text{for } nl_{st} < x < (n+1)l_{st},$$

and (4)

$$Q(x,t) = Q^{(N-(n+1))}(x,t)$$
$$= -Y^{(N-(n+1))} P^{(N-(n+1))} \sin(kx + \theta^{(N-(n+1))}) \cos(\omega t + \phi)$$
$$\text{for } (n+1)l_{st} < x < (n+2)l_{st}.$$

The result of invoking the continuity conditions at the junction between $N-n$ and $N-(n+1)$ is

$$P^{(N-(n+1))} = P^{(N-n)} \frac{\cos\left(k(n+1)l_{st} + \theta^{(N-n)}\right)}{\cos\left(k(n+1)l_{st} + \theta^{(N-(n+1))}\right)}, \quad (5)$$

$$\text{where } \theta^{(N-(n+1))} = \arctan\left[\frac{Y^{(N-n)}}{Y^{(N-(n+1))}} \tan\left(k(n+1)l_{st} + \theta^{(N-n)}\right)\right]$$
$$- k(n+1)l_{st}.$$

$P^{(1)}$ and $\theta^{(1)}$ are determined in terms of $P^{(N)}$, wavenumber k, and the length l_{st} so that

$$p(x,t) = p^{(1)}(x,t) = P^{(1)} \cos(kx + \theta^{(1)}) \sin(\omega t + \phi), \quad (6)$$
$$\text{for } (N-1)l_{st} < x < Nl_{st} = L.$$

The wavenumbers of the normal modes k_m are determined using the boundary condition at the opening. The k_m are such that

$$0 = p(x=L,t) = p^{(1)}(x=L,t) = \quad (7)$$
$$P^{(1)} \cos(k_m L + \theta^{(1)}) \sin(\omega_m t + \phi_m),$$

where $\omega_m = c_0 k_m$ and $\phi_m = \phi(\omega_m)$. From Equation (5) it follows that the k_m are functions of l_{st} and the tube admittances $Y^{(n)}$ for $0 \leq n \leq N$ only. Therefore, the k_m are functions of the cross-sectional areas, the sub-tube lengths l_{st}, and thermodynamic quantities ρ_0 and c_0 of the undisturbed air. In the next section we explore how to derive the same results using effective admittances.

The mth eigenfunction, or normal mode, $p_m(x,t)$ can be written in various ways.
$$p_m(x,t) = p_m^{space}(x) p_m^{time}(t) = p_m^{space}(x) \sin(\omega_m t + \phi_m) . \qquad (8)$$
We have just shown that,
$$p_m^{space}(x) = \begin{cases} P_m^{(N)} \cos(k_m x) & \text{for } 0 < x < l_{st} \\ P_m^{(N-1)} \cos(k_m x + \theta_m^{(N-1)}) & \text{for } l_{st} < x < 2 l_{st} \\ \quad \cdot \\ \quad \cdot \\ \quad \cdot \\ P_m^{(n)} \cos(k_m x + \theta_m^{(n)}) & \text{for } (N-n) l_{st} < x \\ & \qquad < (N-n+1) l_{st} \\ \quad \cdot \\ \quad \cdot \\ \quad \cdot \\ P_m^{(1)} \cos(k_m x + \theta_m^{(1)}) & \text{for } (N-1) l_{st} < x < N l_{st} , \end{cases} \qquad (9)$$
with $k_m = \omega_m / c_0$.

The functions $p_m^{space}(x)$ and $p_j^{space}(x)$ are orthogonal, or normal, functions when $m \neq j$, just as for the finite-length tube of constant cross-section. We repeat the mathematical statement of normality, or orthogonality, of functions that was stated in Equation (31) of Chapter 5 for $m \neq j$.
$$\sum_{i=0}^{K} p_m^{space}(x_i) p_j^{space}(x_i) \approx 0 . \qquad (10)$$
for any fine division of the x-axis from 0 to L, $0 = x_0 < x_1 < x_2 < \ldots < x_{K-2} < x_{K-1} < x_K = L$, for a large integer K. The functions $p_m(x,t)$ and $p_j(x,t)$ are also uncorrelated in time for $m \neq j$, as for the finite-length tube of constant cross-sectional area. That is, $\langle p_m(x,t) p_j(x,t) \rangle = 0$ for $m \neq j$.

Normal modes using effective admittances

Equation (30) of Chapter (9) can be generalized to the case where $1 \leq n \leq N$ to give effective admittances looking to the right.

$$Y_+^{(n)eff} = Y^{(n)} \frac{Y_+^{(n-1)eff} - iY^{(n)} \tan(kl_{st})}{Y^{(n)} - iY_+^{(n-1)eff} \tan(kl_{st})} , \qquad (11)$$

with $Y^{(0)eff} = Y^{(0)}$. Effective admittances looking to the left are generalizations for $1 \leq n \leq N$ of Equation (36) of Chapter 9.

$$Y_-^{(n)eff} = Y^{(n)} \frac{Y_-^{(n+1)eff} + iY^{(n)} \tan(kl_{st})}{Y^{(n)} + iY_-^{(n+1)eff} \tan(kl_{st})} , \qquad (12)$$

with $Y_-^{(N+1)eff} = Y^{(N+1)}$.

At any given point along the axis of the tube, in order for a wave to "fit" into a given tube system there must be continuity of volume velocity and pressure everywhere. Also, these quantities need to be in the correct phase relation everywhere. [Please note that this is not the case when acoustic sources or sinks are present. Discontinuities occur at acoustic sources and sinks.] In particular, we are assuming that this is the case at junctions. This gives us Equation (38) of Chapter 9

$$Y_+^{(n)eff} = Y_-^{(n+1)eff} . \qquad (13)$$

for $0 \leq n \leq N$. In particular, if there are N tubes, then

$$Y_+^{(N)eff} = Y_-^{(N+1)eff} = Y^{(N+1)} . \qquad (14)$$

For a hard-wall piston at the end opposite to the opening, Equation (14) reduces to $Y_+^{(N)eff} = 0$.

For a fixed tube geometry, the effective admittances can considered to be a function of wavenumber k or of frequency $\mathcal{F} = \omega/(2\pi) = kc_0/(2\pi)$. Solving Equation (14) produces a series of $k_m = \omega_m/c_0 = 2\pi\mathcal{F}_m/c_0$, with $m = 1, 2, 3, \ldots$. The $\mathcal{F}_m$ are eigenvalues, or normal mode frequencies.

Now we present an alternative to the the method of the previous section for calculating pressure and volume velocity distributions throughout the N sub-tubes that correspond to the normal mode frequencies and the corresponding wavenumbers. We saw in the previous section that the spatial distributions of pressure signed amplitude and volume velocity signed amplitude in a tube composed of multiple sub-tubes are sinusoids in each sub-tube. Each of these sinusoids have different amplitudes and connect at the junction with different spatial phases, but they all have the same wavenumber k_m.

Using the complex form, the pressure disturbance for the mth normal mode in the nth tube can be written

$$p_m^{(n)}(x,t) = \hat{p}_m^{(n)}(x)e^{-i\omega_m t} \qquad (15)$$
$$= \left(P_{m+}^{(n)}e^{ik_m(x-L)} + P_{m-}^{(n)}e^{-ik_m(x-L)}\right)e^{-i\omega_m t},$$

where $P_{m+}^{(n)}$ and $P_{m-}^{(n)}$ are complex, so that $P_{m+}^{(n)}$ and $P_{m-}^{(n)}$ contain amplitude and phase information for the right-going traveling wave and the left-going traveling wave of the mth mode in the nth tube, respectively. It is the real part of the pressure expressed in Equation (15) that gives the physical pressure of interest. $x - L$ is written in the exponentials instead of x as a matter of convenience. It can be shown from Equation (12) or (13) of Chapter 2 that the volume velocity of the mth mode in the nth tube is

$$Q_m^{(n)}(x,t) = \hat{Q}_m^{(n)}(x)e^{-i\omega_m t}$$
$$= \frac{A^{(n)}}{\rho_0 c_0}\left(P_{m+}^{(n)}e^{ik_m(x-L)} - P_{m-}^{(n)}e^{-ik_m(x-L)}\right)e^{-i\omega_m t} \qquad (16)$$
$$= Y^{(n)}\left(P_{m+}^{(n)}e^{ik_m(x-L)} - P_{m-}^{(n)}e^{-ik_m(x-L)}\right)e^{-i\omega_m t}.$$

It remains to find $P_{m+}^{(n)}$ and $P_{m-}^{(n)}$. By the definition of acoustic admittance

$$\frac{Q_m^{(1)}(x=L,t)}{p_m^{(1)}(x=L,t)} = Y^{(1)}\frac{P_{m+}^{(1)} - P_{m-}^{(1)}}{P_{m+}^{(1)} + P_{m-}^{(1)}} = Y^{(0)}. \qquad (17)$$

Solving for $P_{m-}^{(1)}$ in terms of $P_{m+}^{(1)}$

$$P_{m-}^{(1)} = \frac{Y^{(1)} - Y^{(0)}}{Y^{(1)} + Y^{(0)}}P_{m+}^{(1)}. \qquad (18)$$

Because we do not have an acoustic source, nor an initial condition, the complex variable $P_{m+}^{(1)}$ remains undetermined.

We proceed to find the other complex variables, $P_{m+}^{(n)}$ and $P_{m-}^{(n)}$, in terms of $P_{m+}^{(1)}$. At the junction between sub-tube 2 and sub-tube 1 at $x = L - l_{st}$, we have continuity of pressure and volume velocity for each mode, m

$$p_m^{(2)}(x = L - l_{st}, t) = p_m^{(1)}(x = L - l_{st}, t)$$
$$\left(P_{m+}^{(2)} e^{ik_m l_{st}} + P_{m-}^{(2)} e^{-ik_m l_{st}}\right) = \left(P_{m+}^{(1)} e^{ik_m l_{st}} + P_{m-}^{(1)} e^{-ik_m l_{st}}\right),$$
and (19)
$$Q_m^{(2)}(x = L - l_{st}, t) = Q_m^{(1)}(x = L - l_{st}, t)$$
$$Y^{(2)}\left(P_{m+}^{(2)} e^{ik_m l_{st}} - P_{m-}^{(2)} e^{-ik_m l_{st}}\right) = Y^{(1)}\left(P_{m+}^{(1)} e^{ik_m l_{st}} - P_{m-}^{(1)} e^{-ik_m l_{st}}\right).$$

From Equation (19) it follows that

$$P_{m+}^{(2)} = \frac{Y^{(2)} + Y^{(1)}}{2Y^{(2)}} P_{m+}^{(1)} + \frac{Y^{(2)} - Y^{(1)}}{2Y^{(2)}} P_{m-}^{(1)} e^{-i2k_m l_{st}},$$
and (20)
$$P_{m-}^{(2)} = P_{m-}^{(1)} + \left(P_{m+}^{(1)} - P_{m+}^{(2)}\right) e^{i2k_m l_{st}}.$$

With Equations (18) and (20), we now have $P_{m+}^{(2)}$ and $P_{m-}^{(2)}$ in terms of $P_{m+}^{(1)}$.

This process can be continued using the formulas

$$P_{m+}^{(n+1)} = \frac{Y^{(n+1)} + Y^{(n)}}{2Y^{(n+1)}} P_{m+}^{(n)} + \frac{Y^{(n+1)} - Y^{(n)}}{2Y^{(n+1)}} P_{m-}^{(n)} e^{-i2nk_m l_{st}},$$
and (21)
$$P_{m-}^{(n+1)} = P_{m-}^{(n)} + \left(P_{m+}^{(n)} - P_{m+}^{(n+1)}\right) e^{i2nk_m l_{st}}.$$

Ultimately, all the $P_{m+}^{(n)}$ and $P_{m-}^{(n)}$ are known in terms of $P_{m+}^{(1)}$, tube admittances $Y^{(n)}$ and integer multiples of the spatial phase factor $k_m l_{st}$. The pressure of the mth mode in the nth sub-tube is given by the real part of Equation (15) and the corresponding volume velocity is given by the real part of Equation (16).

We have solved for all of the $P_{m+}^{(n)}$ and $P_{m-}^{(n)}$ for $1 \leq n \leq N$ in terms of in terms of $P_{m+}^{(1)}$. We could have solved in terms of $P_{m+}^{(N)}$ instead. This is closer to what was done for the first method in this chapter. In that method we solved for real amplitudes $P_m^{(n)}$ and real phases $\theta_m^{(n)}$ in terms of real $P_m^{(N)}$ and real $\theta_m^{(N)} = 0$. Both amplitude and phase information are contained in the complex $P_{m+}^{(n)}$ and $P_{m-}^{(n)}$ in the present section.

Examples of computing normal modes from area functions

We present two examples of tube shapes and the properties of their normal modes. We pay particular attention to the number of sub-tubes that may be needed to approximate the shape of the tube in order to characterize its normal modes properly. Tube shape is expressed by cross-sectional area as a function of tube position x in an area function. The computations were performed using the procedures outline in the section above on effective admittances beginning wth Equations (11) and (12). The eigenvalues k_m can be found at either the piston or mouth boundary by using a root-finding algorithm. While the shapes of functions, and not their physical magnitudes, are of concern here. $P_+^{(1)} = 1$ was used in the follwoing examples.

Example I

The first example to be considered has area function

$$A(x) = A_0 + A_{amp} \cos\left(\frac{2\pi x}{L}\right). \tag{22}$$

This area function is shown in Figure 2 for $A_0 = 2.0$ cm^2 and $A_{amp} = 1.0$ cm^2, along with the 100 sub-tube approximation in the top panel, 30 sub-tubes in the middle panel, and 10 sub-tubes in the bottom panel.

We use Equation (11) with $Y_+^{(0)eff} = Y^{(0)} = \infty$ to find $Y_+^{(N)eff}$ for $N = 100, 30$, and 10 sub-tubes with a root-solving algorithm to find the first three normal mode frequencies, $\mathcal{F}_1$, $\mathcal{F}_2$, and $\mathcal{F}_3$, according to the criterion given in Equation (14) for $N = 100, 30$, and 10. For 100 sub-tubes $\mathcal{F}_1 = 388$ Hz, $\mathcal{F}_2 = 1675$ Hz, and $\mathcal{F}_3 = 2632$ Hz. For 30 sub-tubes $\mathcal{F}_1 = 388$ Hz, $\mathcal{F}_2 = 1673$ Hz, and $\mathcal{F}_3 = 2629$ Hz. For 10 sub-tubes $\mathcal{F}_1 = 389$ Hz, $\mathcal{F}_2 = 1661$ Hz, and $\mathcal{F}_3 = 2603$ Hz. [Recall that $Y_-^{(N+1)eff} = Y^{(N+1)} = 0$ because of the piston.] This shows that the higher normal mode frequency estimation depend more on the details

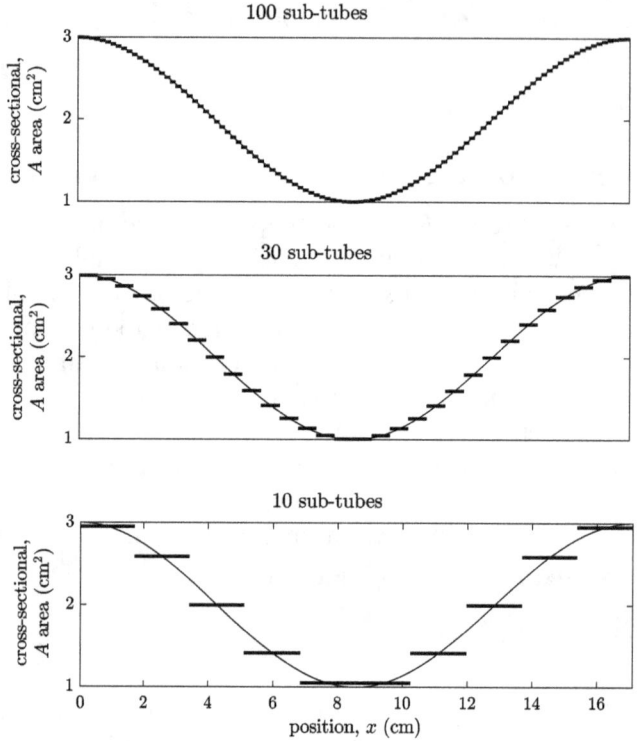

FIGURE 2. A sinusoidal area function

of the sub-tube approximations, at least in terms of absolute frequency. However, there is not much variation in any of these normal mode frequencies in percentage terms among the sub-tube approximations.

There is an imortant caveat regarding the use of very short sub-tubes. This is that we cannot allow sub-tubes to become so short that *evanescent waves* created at one juntion interact with neightboring junctions. Enanescent waves are waves created at frequencies for acoustic modes in a duct that do not porpgate according the wave equation, but decay exponentially with distance (Pierce 1989).

However, caution must be exercised in regard to very short sub-tubes. Evanescent waves are created at each junction between sub-tubes. These waves decay exponentially with axial distance from the junction where they are created. These waves are not accounted for at neighboring junctions in the current development, and this means that we are assuming that junctions between sub-tubes are far enough apart that evanescent waves are damped out between junctions to a good approximation.

Five quantities: pressure signed amplitude, volume velocity signed amplitude, time-average potential and kinetic energy densities, and steady acoustic radiation pressure have been plotted at each of the first three normal modes and for each of the number of sub-tubes: $N = 100, 30$, or 10 sub-tubes.[1] Equation (6) of Chapter 6 is used to compute the steady acoustic radiation pressure for each mode m. Quantities of interest are plotted in Figures 3 through 14. Each of the Figures shows one or two quantities for a single normal mode for 100 sub-tubes in the top panel, 30 sub-tubes in the middle panel, and 10 sub-tubes in the bottom panel. Figures 3 through 6 are for the first normal mode, Figures 7 through 10 are for the second normal mode, and Figures 11 through 14 are for the third normal mode.

In Figures 3, 4, 7, 8, 11, and 12 we see that even the 10 sub-tube approximation to the pressure and volume velocity signed amplitude distributions compare well to the 100 sub-tube approximation.[2] There are a few more apparent kinks in the rougher approximation, but the overall shapes of the volume velocity and pressure signed amplitudes are preserved from the 100 sub-tube approximation in the 10 sub-tube approximation. The same cannot be said for the time-average kinetic energy densities (Figures 5, 9, and 13), and the steady acoustic radiation pressure distributions (Figures 6, 10, and 14). [The intermittent breaks in the plots are due to the way the plots are rendered. There is no difference between a break and a vertical line: they both represent discontinuities in either time-average kinetic energy density or steady acoustic radiation pressure.] The discontinuities become larger going from 100 sub-tubes to 30 sub-tubes, and on to 10 sub-tubes, because the differences in the cross-sectional areas of neighboring sub-tubes become greater when fewer sub-tubes are used to approximate the area function.

---

[1] There is some phase information in the amplitudes, because they are both positive and negative. They are termed signed amplitudes.
[2] We must be aware that the 100-tube approximation could be inappropriate because the sub-tube lengths are too short and would permit evanescent waves to travel between junctions of sub-tubes.

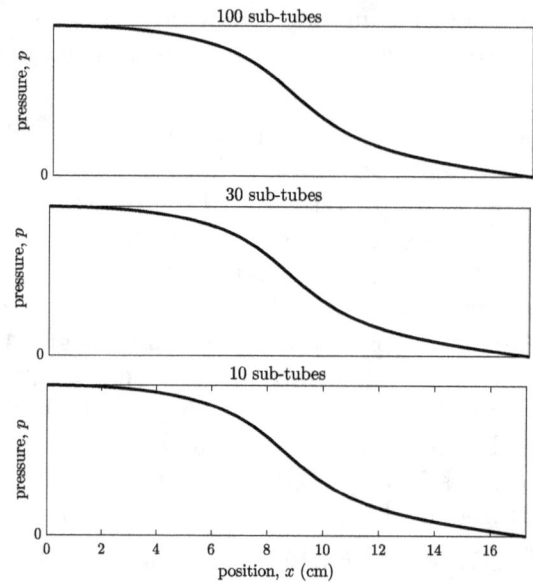

FIGURE 3. First mode pressure signed amplitude

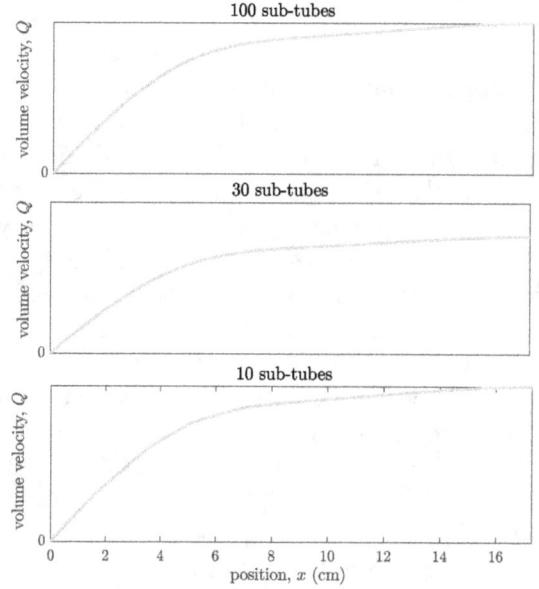

FIGURE 4. First mode volume velocity signed amplitude

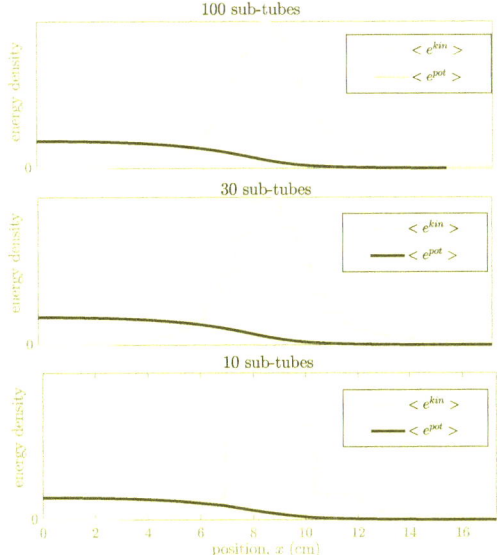

FIGURE 5. First mode time-average energy densities

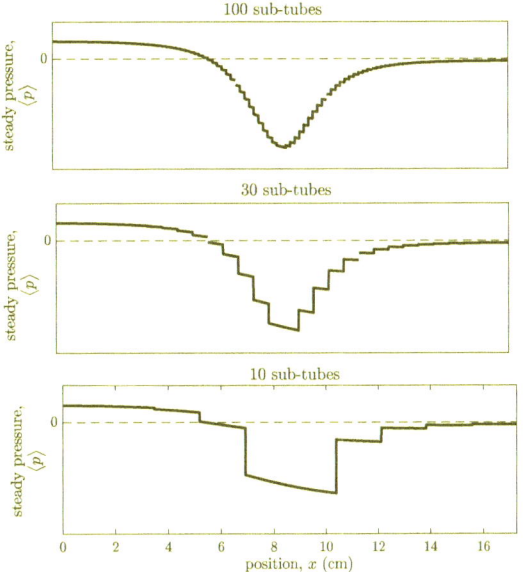

FIGURE 6. First mode steady pressure

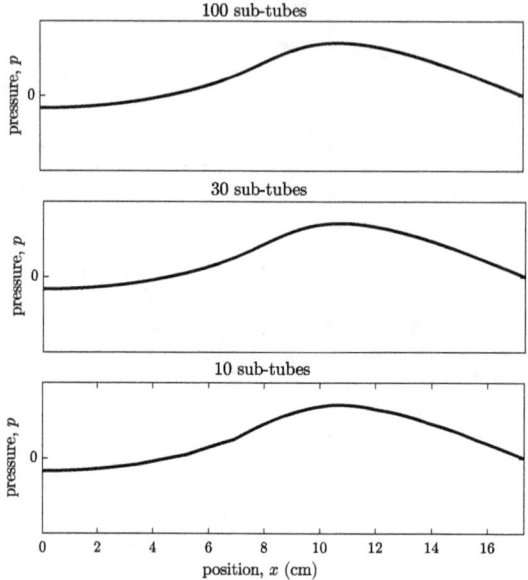

FIGURE 7. Second mode pressure signed amplitude

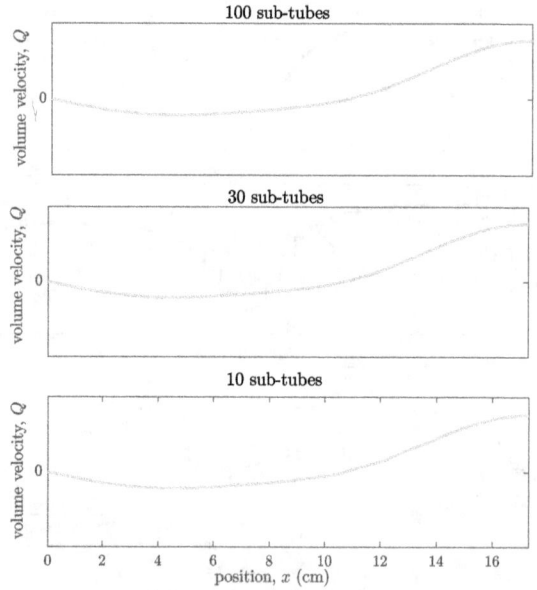

FIGURE 8. Second mode volume velocity signed amplitude

Acoustics of Speech Production

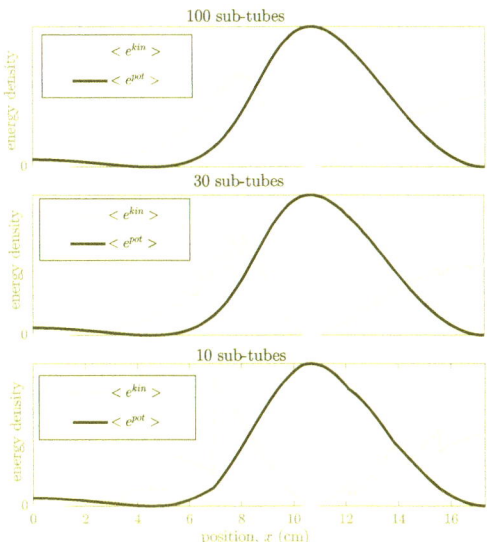

FIGURE 9. Second mode time-average energy densities

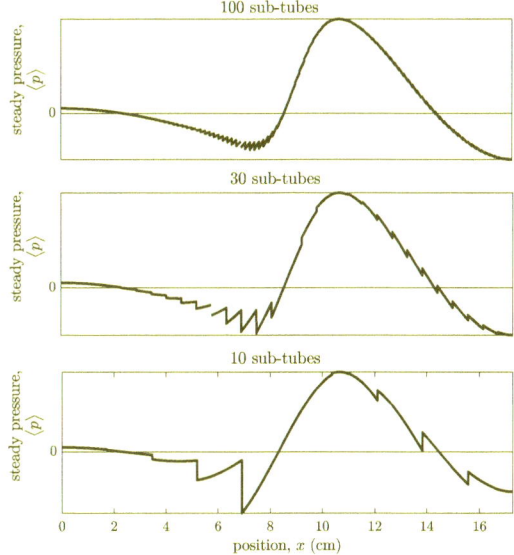

FIGURE 10. Second mode steady pressure

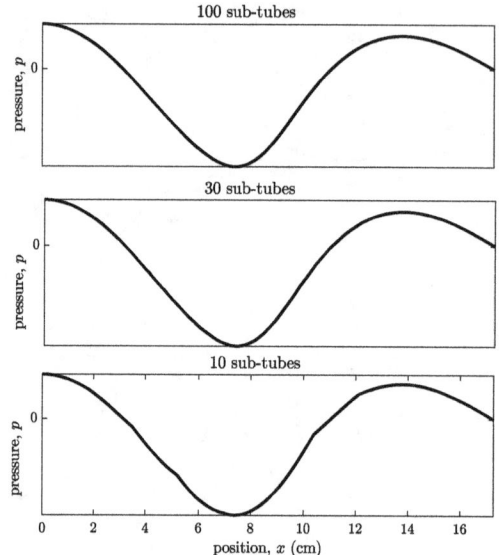

FIGURE 11. Third mode pressure signed amplitude

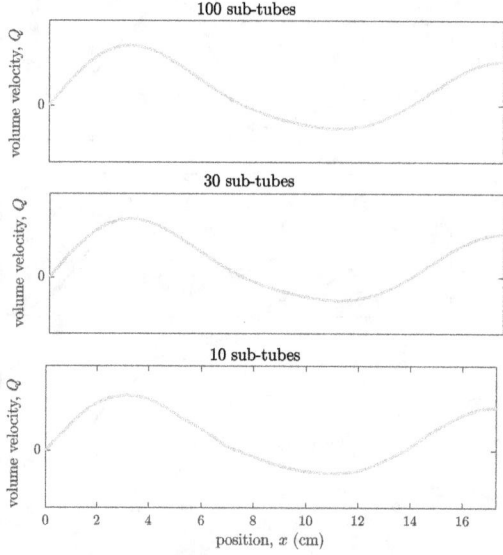

FIGURE 12. Third mode volume velocity signed amplitude

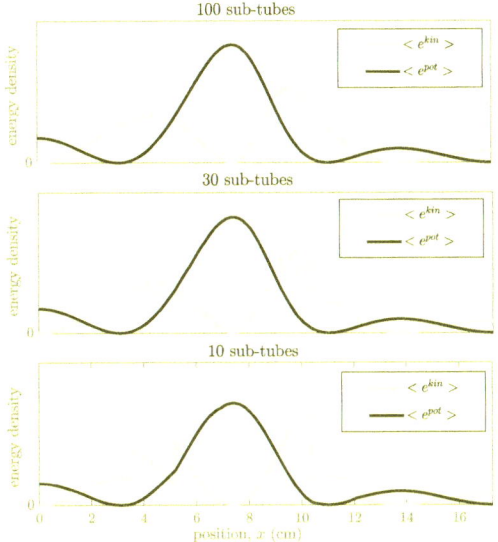

FIGURE 13. Third mode time-average energy densities

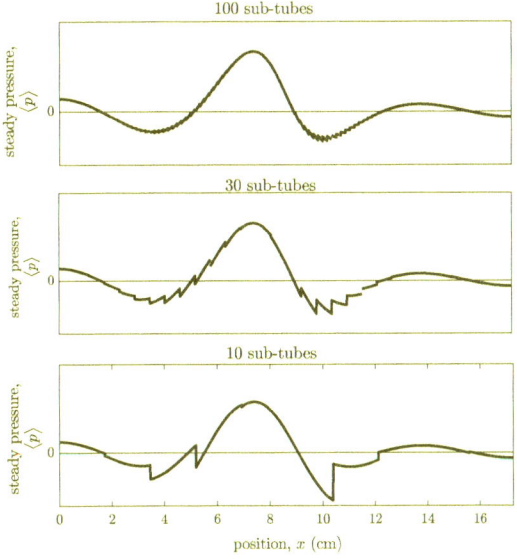

FIGURE 14. Third mode steady pressure

The discontinuities in steady acoustic radiation pressure $\langle p \rangle$ are particularly critical if one is attempting to use the perturbation theory introduced in Chapter 6 to predict what would happen if a small change were made to an area function. Acoustic perturbation theory predicts that normal mode frequency increases if an area reduction is made where the steady acoustic radiation pressure is positive, and that the opposite occurs if the steady acoustic radiation pressure is negative. This makes accurate knowledge of where the steady acoustic radiation pressure is close to zero important. Any location in the tube where the steady acoustic radiation pressure actually crosses from positive to negative, or negative to positive, is called a zero-crossing. For example, consider the steady acoustic radiation pressure for the first normal mode in Figure 6. Note that for 100 sub-tubes there is a zero-crossing at a tube position a little greater than $x = 6$ cm. In middle panel of Figure 6 with 30 sub-tubes, this zero-crossing gets blurred over the length of the sub-tube, which is about 0.58 cm. Similarly, in the bottom panel of Figure 6 for 10 sub-tubes, the zero-crossing is blurred over the length of a sub-tube, which is 1.73 cm. For the second normal mode steady acoustic radiation pressure in Figure 10, there appears to be less of a problem. However, for the 10 sub-tube case in the bottom panel of Figure 10, we should consider the sub-tube that includes $x = 14$ cm. With the magnitude of the discontinuities at both ends of this sub-tube, there is ambiguity in the location of the zero-crossing across the length of the sub-tube. In this case, the zero-crossing for the 10 sub-tube approximation is close to the zero-crossing in the 100 sub-tube approximation, but we would not know this if all we had was enough information to make a 10 sub-tube approximation. For the third normal mode in Figure 14, there is a zero-crossing in steady acoustic radiation pressure at about $x = 5$ cm, according to the top panel of Figure 14. In the 10 sub-tube case in the bottom panel of Figure 14, there are three zero-crossings that blurs the actual zero-crossing. The zero-crossing at $x \approx 12.5$ cm is similarly ambiguous for the 10 sub-tube approximation.

Sometimes speech production researchers refer to an idea of *cavity affiliation* in speech production research. Certain regions of the vocal tract are somehow supposed to contain most of the energy of a mode or formant frequency. We do not regard the concept of cavity affiliation to be useful. If we consider the time-average energy densities in the

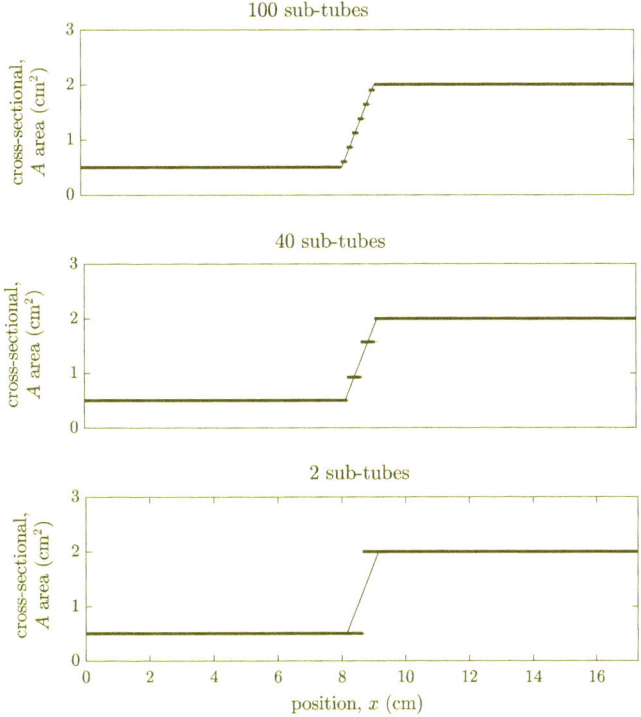

FIGURE 15. Two sub-tubes connected by a short sub-tube

top panels in Figures 5, 9, and 13, we cannot conclude that any of the normal modes is a "back cavity" or "front cavity" resonance. In terms of the sum of time average kinetic and potential energy densities, they all involve, at least, either the front or rear large cross-sectional area regions and the constriction region.

Example II

In a second example, we reexamine the two sub-tube example from Chapter 9, as though it was an approximation of an area function with three sections. The back section is a narrow tube and the front section is a wide tube, but between them is a third section that connects the two with a rapidly changing area, as shown in Figure 15. This connecting section is 1 cm long and goes from a cross-sectional area of

0.5 cm^2 to 2 cm^2 in that length. This time we approximate the shape of the area function using 100 sub-tubes in the top panel, 40 sub-tubes in the middle panel, and 2 sub-tubes in the bottom panel. With the 100 sub-tube approximation there are 7 sub-tubes in the middle connecting region, for the 40 sub-tube approximation there are 3 sub-tubes in the middle connecting region, and for 2 sub-tubes there are no sub-tubes in the middle connecting region. Thus, this latter approximation is the two-tube model. For 100 sub-tubes $\mathcal{F}_1 = 710$ Hz, $\mathcal{F}_2 = 1307$ Hz, and $\mathcal{F}_3 = 2717$ Hz. For 40 sub-tubes $\mathcal{F}_1 = 710$ Hz, $\mathcal{F}_2 = 1307$ Hz, and $\mathcal{F}_3 = 2716$ Hz. For 2 sub-tubes $\mathcal{F}_1 = 705$ Hz, $\mathcal{F}_2 = 1295$ Hz, and $\mathcal{F}_3 = 2705$ Hz.

The steady acoustic radiation pressure at the first three normal modes are shown for the three sub-tube approximations in Figures 16 through 18. Even in the cases of the 100 sub-tube approximations in the top panels of Figures 16 through 18 there is some ambiguity regarding zero-crossings in the vicinity of the connecting region, with x slightly greater than 8 cm. For the first and second normal modes in Figures 16 and 17, the uncertainty in the zero-crossing for 100 sub-tubes is about the length of a two sub-tubes, or about 0.36 cm. The same can be said for the first and second normal modes for the 40 sub-tube approximation in the middle panels of Figures 16 and 17, so that the region of uncertainty is about 0.86 cm long. For the 2 sub-tube approximation it is difficult to estimate the region of uncertainty in the zero-crossing for $x > 8$ cm for the second normal mode in Figure 17. This figure appears to show that the 2 sub-tube approximation in the bottom panel has a zero-crossing about 0.5 cm to the right of that shown for the 100 tube approximation in the top panel. The third normal mode in Figure 18 is even more problematic, because there are indications that there are two zero-crossings in the transition region. Even the 100 sub-tube approximation is not very useful in resolving the locations of the zero-crossings.

Accounting for lumped mass elements

We now account for the lumped mass elements at the junctions of sub-tubes that were introduced in Chapter 9. These lumped mass elements account for the excess kinetic energy of the bending of fluid particle velocity streamlines when an area change is encountered. We

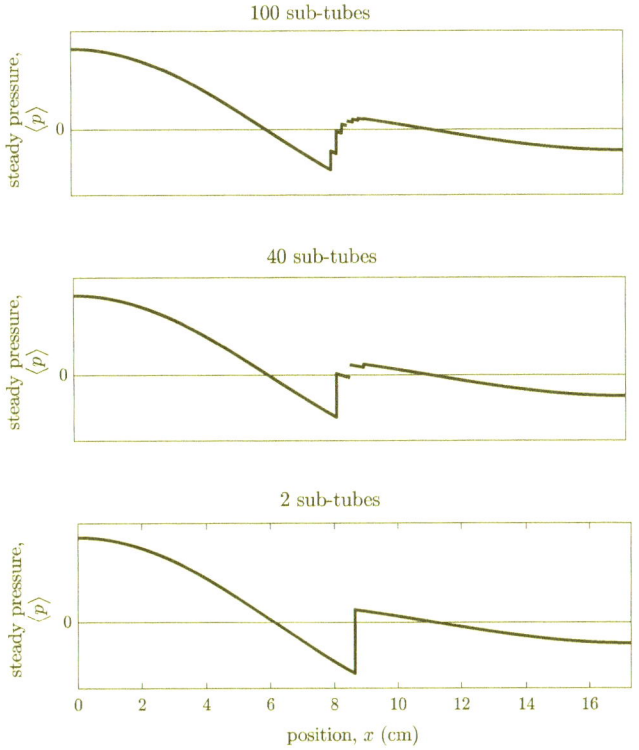

FIGURE 16. First mode steady acoustic radiation pressure distributions

saw that this results in a pressure change $\Delta\hat{p}_J(\omega)$ over a length ℓ, which is much shorter than the shortest wavelength of sound under consideration. The volume velocity through the junction $\hat{Q}_J(\omega)$ remains approximately constant through the junction. We repeat the expression for the resulting acoustic impedance and acoustic admittance at the junction given in Equation (56) of Chapter 9.

$$Z_J \equiv \frac{\Delta\hat{p}_J(\omega)}{\hat{Q}_J(\omega)} \approx i\frac{(k\ell)}{Y^{(nar)}} \quad \text{and} \quad Y_J \equiv \frac{\hat{Q}_J(\omega)}{\Delta\hat{p}_J(\omega)} \approx -i\frac{Y^{(nar)}}{(k\ell)} \ . \quad (23)$$

$Y^{(nar)} = A^{(nar)}/\rho_0 c_0$ is the tube admittance of the narrower of the two tubes. Equation (23) tells us that $\Delta\hat{p}_J(\omega)$ is $\pi/2$ out of phase with $\hat{Q}_J(\omega)$, because Z_J and Y_J are purely imaginary quantities. Thus, there is no time-average work, or time-average power exchange between

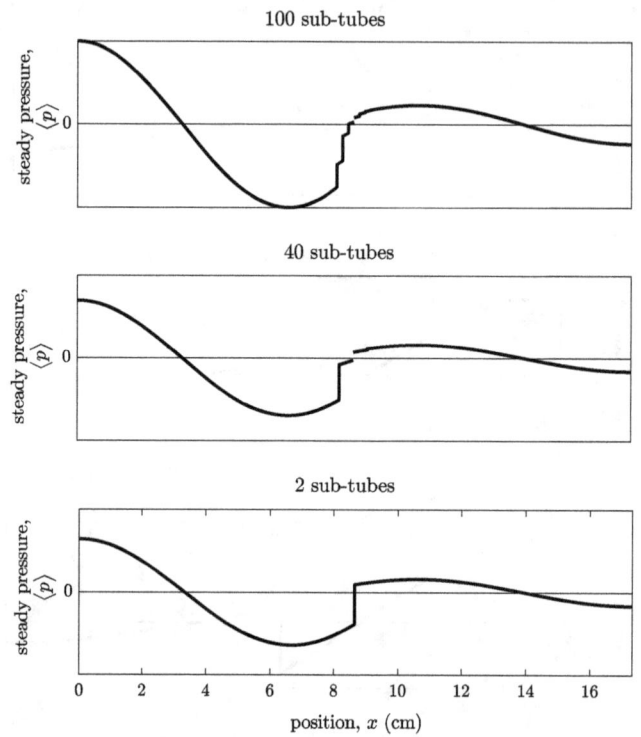

FIGURE 17. Second mode steady acoustic radiation pressure distributions

$\Delta \hat{p}_J(\omega)$ and the acoustic motion represented by $\hat{Q}_J(\omega)$. On the other hand, because we account for more mass and kinetic energy with lumped mass elements, we can expect that the normal mode frequencies decrease. This is analogous to increasing mass in a mass-spring system, which would result in its natural frequency decreasing, as shown in Equation (15) of Chapter 1.

Length scales, again

At the beginning of Chapter 9, we described the conditions for one-dimensional acoustic motion in a tube. In order to assume one-dimensional motion, or plane wave motion, the smallest wavelength

Acoustics of Speech Production

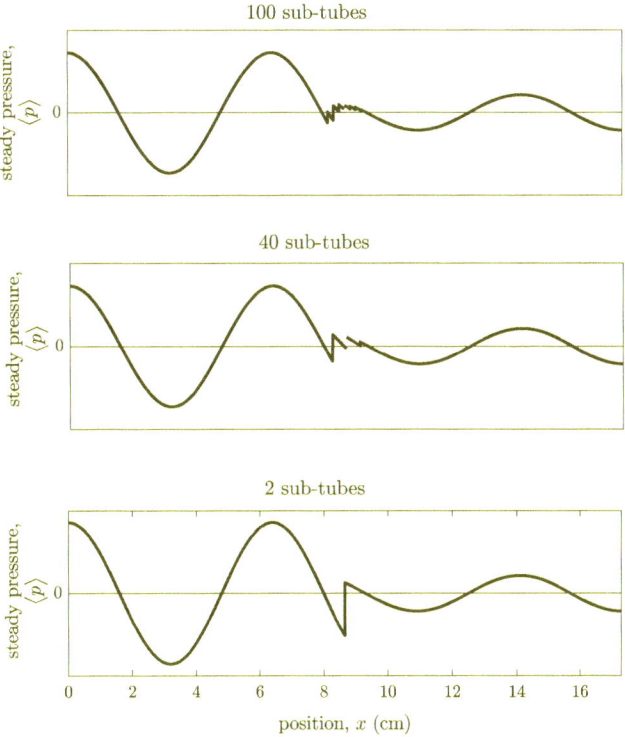

FIGURE 18. Third mode steady acoustic radiation pressure distributions

under consideration must be longer than the about twice the cross-dimensions of the tube. Because wavelength λ and frequency $\mathcal{F}$ are inversely related this puts an upper limit on the frequencies under consideration.

Restriction on frequency also apply to lumped elements. Lumped elements are objects that are much shorter, in the direction of acoustic wave propagation, compared to wavelengths of sound under consideration. Lumped elements are not just somewhat shorter than the shortest wavelength under consideration, they are at least on order-of-magnitude smaller. This is expressed by the relation given in Equation (54) of Chapter 9. The symbol $\ll$ in this Equation means "much less than", and has a somewhat qualitative definition, although a factor of at least ten seems to be meant in most cases.

An object of dimension ℓ, say, is much smaller compared to wavelength λ only if the frequency $\mathcal{F}$ is sufficiently low, because wavelength and frequency are reciprocally related through the speed of sound, c_0: $\lambda = c_0/\mathcal{F}$. So in order that $\ell \ll \lambda$, we must have $\mathcal{F} \ll c_0/\ell$.

The lumped mass element for the two-tube example in Chapter 9, Figure 14 has maximum extent along the x-axis of about $0.5\ cm$. In order for this to be considered lumped, we require that $\mathcal{F} \ll (34,100/0.5)\ \text{Hz} \approx 70,000\ \text{Hz}$. Certainly, frequencies up to $20,000$ Hz would meet the criterion to make the element a lumped element. Note that the use of lumped elements is consistent with the low-frequency approximation for plane wave, or one-dimensional, motion.

Because the length scales of the lumped element in the direction of propagation are much smaller than the wavelength, the volume velocity is approximately uniform throughout the lumped element at each instant of time, so that the element moves as a block. Further, we ignore the spatial extent of lumped elements in the propagation direction when performing calculations.

Effect of a lumped mass element at the open end of the tube

Until now, we have assumed that the pressure fluctuations at the opening are zero. However, it seems that this cannot be the case. After all, there is a large change in cross-sectional area at this end: the cross-sectional area goes from finite to infinite at the open end, which means that a lumped mass element and fluctuation pressure needs to included at the open end of the tube at $x = L$. However, unlike the junctions between tubes of finite cross-sectional area, it is shown in Chapter 11 that there must be a component of the pressure fluctuation in phase with the volume velocity fluctuation at the opening. Thus, the acoustic impedance and acoustic admittance at the opening, known as the *radiation impedance*, Z_{rad}, and *radiation admittance*, Y_{rad}, respectively, have real and imaginary parts: $Z_{rad} = Re(Z_{rad}) + iIm(Z_{rad})$ and $Y_{rad} = Re(Y_{rad}) + iIm(Y_{rad})$. We await Chapter 11 to examine the real parts of the radiation impedance and admittance. In this chapter, we discuss only the lumped mass element at the opening, which means that we only discuss the imaginary parts of the radiation impedance and admittance.

Because the mathematics necessary to find the imaginary parts of the radiation impedance and admittance is advanced, we simply quote the result for the case that the opening has flanged circular cross-section with radius $a = \sqrt{A^{(1)}/\pi}$. [Recall from Figure 1 of Chapter 3, that we are assuming that the opening is a hole in a infinite plane, which is a flanged opening. This is supposed to represent the head.] We quote the result for small ratio of radius-to-wavelength, $a/\lambda = \mathcal{F}a/c_0 \ll 1$. Recall that $k = 2\pi \mathcal{F}/c_0 = \omega/c_0$ is the wavenumber, and that $Y^{(1)} = A^{(1)}/(\rho_0 c_0)$, where $A^{(1)}$ is the cross-sectional area of the sub-tube adjacent to the opening.

$$Im(Z_{rad}) \approx -\frac{8}{3\pi}\frac{(ka)}{Y^{(1)}},$$
and (24)
$$Im(Y_{rad}) \approx \frac{3\pi}{8}\frac{Y^{(1)}}{(ka)}.$$

where the approximation improves as ka gets small (Rayleigh 1945; Morse 1976). Equation (24) was first derived by Rayleigh in the late nineteenth century and appears in his *Theory of Sound*.

Recall that the lumped mass element is associated with a length ℓ in the direction of propagation, while cross-sectional area of the element in the case of the opening is $A^{(1)}$. For the opening, we denote this length ℓ_{rad}. The pressure difference $\Delta \hat{p}_J(\omega)$ is such that the pressure is fluctuation is zero at a distance ℓ_{rad} to the right of the opening. If we compare Equation (58) of Chapter 9 and Equation (24), we obtain ℓ_{rad}

$$\ell_{rad} = \frac{8}{3\pi}a. \qquad (25)$$

Again, this formula appears in Rayleigh's *Theory of Sound* (Rayleigh 1945).

Apparenty, there are two methods that we can use to calculate normal mode frequencies. They both involve using Equation (11) to find $Y_+^{(N)eff}$, and finding the frequencies for which Equation (14), $Y_+^{(N)eff} = Y^{(N+1)}$, is satisfied. For the piston boundary, $Y^{(N+1)} = 0$. The first method is to start with $Y_+^{(0)eff} = Im(Y_{rad})$, with $Im(Y_{rad})$ given in Equation (24). The second method is to lengthen the tube from L to an effective length $L' = L + \ell_{rad}$, and then apply $Y_+^{(0)eff} = \infty$.

We examine the effect that radiation admittance has on normal mode frequencies using both methods. Without accounting for the lumped mass element at the end of the straight tube the first three normal mode frequencies are 500 Hz, 1500 Hz, and 2500 Hz. When we do account for the lumped mass element and use the first method, we obtain 481 Hz, 1444 Hz, and 2408 Hz for the first three normal mode frequencies. Using the second method, the first three normal mode frequencies are 481 Hz, 1443 Hz, and 2406 Hz.

We also consider the 30 sub-tube approximation to the sinusoidal area function shown in the middle panel of Figure 2. Without accounting for the excess kinetic energy at the opening, the first three normal mode frequencies are 388 Hz, 1673 Hz, and 2629 Hz. Using the first method to account for the excess kinetic energy at the opening, the first three normal mode frequencies are 383 Hz, 1511 Hz, and 2558 Hz. With the second method the first three normal mode frequencies of 383 Hz, 1508 Hz, and 2554 Hz

Finally, the configuration with a narrow rear sub-tube, a broad front sub-tube, and a short region of rapid transition is considered with the 100 sub-tube approximation shown in the top panel of Figure 14. The first three normal mode frequencies calculated without radiation admittance at the opening had been 710 Hz, 1307 Hz, and 2716 Hz. The first method that accounts for a lumped mass element gives these frequencies as 682 Hz, 1262 Hz, and 2601 Hz, while the second method gives 682 Hz, 1262 Hz, and 2597 Hz.

The first and second methods do not agree exactly, but they provide results that are close to one another. Both methods depend on approximations that are are good approximations when the freqeuncies are low. In all cases, normal mode frequencies decrease when accounting for excess knetic energy.

Effect of a lumped mass element at sub-tube junctions

Here we include the pressure jumps at junctions that are caused by lumped mass elements in the calculations of effective admittances and the acoustic wave motion within the tube. This will change the normal mode frequency calculations, as well as the pressure and volume velocity signed amplitude distributions within the tube. The complex pressure amplitude change from the narrower sub-tube to the wider sub-tube at the junction from the $(n+1)^{th}$ sub-tube to the nth sub-tube is denoted $\Delta\hat{p}_J^{(n)}(\omega)$ for a given complex volume velocity amplitude, $\hat{Q}_J^{(n)}(\omega)$, through the junction. Because $\Delta\hat{p}_J^{(n)}(\omega)$ is complex, it accounts for both magnitude and phase jumps in pressure across the junction due to the lumped mass element. The time dependence for both the volume velocity and pressure is sinusoidal, given by $e^{-i\omega t}$.

The junction between the $(n+1)^{th}$ and n^{th} sub-tubes is denoted $x^{(n)}$. $\hat{p}^{(n+1)}(x = x^{(n)}, \omega)$ denotes the complex pressure in the $(n+1)^{th}$ sub-tube at the junction and $\hat{p}^{(n)}(x = x^{(n)}, \omega)$ denotes the complex pressure in the nth sub-tube at the junction. We replace the continuity of pressure across the junction with

$$\hat{p}^{(n)}(x = x^{(n)}, \omega) = \hat{p}^{(n+1)}(x = x^{(n)}, \omega) + \Delta\hat{p}_J^{(n)}(\omega)$$

$$\Rightarrow \frac{\hat{p}^{(n)}(x = x^{(n)}, \omega)}{\hat{Q}_J^{(n)}(x = x^{(n)}, \omega)} = \frac{\hat{p}_J^{(n+1)}(x = x^{(n)}, \omega)}{\hat{Q}^{(n+1)}(x = x^{(n)}, \omega)} + \frac{\Delta\hat{p}_J^{(n)}(\omega)}{\hat{Q}^{(n)}(\omega)},$$ (26)

where $\Delta\hat{p}_J^{(n)}(\omega)$ is the pressure jump across the junction, and $\hat{Q}^{(n+1)}(x = x^{(n)}, \omega) = \hat{Q}^{(n)}(x = x^{(n)}, \omega) = \hat{Q}_J^{(n)}(\omega)$ by continuity of volume velocity.

Equation (56) of Chapter 9 can be generalized to

$$Z_J^{(n)} = Z_J^{(n)}(\omega) \equiv \frac{\Delta\hat{p}_J^{(n)}(\omega)}{\hat{Q}_J^{(n)}(\omega)} = i\frac{(k\ell^{(n)})}{Y^{(nar)}},$$

and (27)

$$Y_J^{(n)} = Y_J^{(n)}(\omega) = \frac{\hat{Q}_J^{(n)}(\omega)}{\Delta\hat{p}_J^{(n)}(\omega)} = -i\frac{Y^{(nar)}}{(k\ell^{(n)})},$$

where $Y^{(nar)} = \min(Y^{(n)}, Y^{(n+1)})$, and $\ell^{(n)}$ is the length of the lumped mass element at the junction between sub-tube $n+1$ and sub-tube n. The relationship between $\Delta\hat{p}_J^{(n)}(\omega)$ and $\hat{Q}_J^{(n)}(\omega)$ across the junction is written as

$$\Delta\hat{p}_J^{(n)}(\omega) = Z_J^{(n)}\hat{Q}_J^{(n)}(\omega) = \frac{\hat{Q}_J^{(n)}(\omega)}{Y_J^{(n)}}.$$ (28)

where $Z_J^{(n)}$ is the acoustic impedance of the lumped mass element at the junction, and $Y_J^{(n)}$ is its acoustic admittance.

Let $[Y]_-^{(n+1)eff}$ be the effective admittance in the nth sub-tube looking left at the junction with the $(n+1)^{th}$ sub-tube, but not accounting for the acoustic admittance due to the lumped mass element at the junction $Y_J^{(n)}$. Similarly, let $[Y]_+^{(n)eff}$ be the effective admittance in the $(n+1)^{th}$ sub-tube looking right at the junction with the nth tube, where $Y_J^{(n)}$ is not taken into account.

If the unbracketed effective admittances are understood to take account of the acoustic admittance of the lumped mass element at the junction, then by the definition of the bracketed admittances

$$\frac{1}{Y_+^{(n)eff}} = \frac{1}{[Y]_+^{(n)eff}} + \frac{1}{Y_J^{(n)}}. \tag{29}$$

Starting with $Y_+^{(0)eff} = Y^{(0)}$ Equation (29) is used in conjunction with

$$[Y]_+^{(n)eff} = Y^{(n)} \frac{Y_+^{(n-1)eff} - iY^{(n)}\tan(kl_{st})}{Y^{(n)} - iY_+^{(n-1)eff}\tan(kl_{st})} \tag{30}$$

to obtain a recursion that replaces Equation (11). The same argument applies looking left

$$\frac{1}{Y_-^{(n+1)eff}} = \frac{1}{[Y]_-^{(n+1)eff}} - \frac{1}{Y_J^{(n)}}. \tag{31}$$

Starting with $Y_-^{(N+1)eff} = Y^{(N+1)}$, Equation (30) can be used in conjunction with

$$[Y]_-^{(n)eff} = Y^{(n)} \frac{Y_-^{(n+1)eff} + iY^{(n)}\tan(kl_{st})}{Y^{(n)} + iY_-^{(n+1)eff}\tan(kl_{st})}. \tag{32}$$

to replace the recursion in Equation (12).

Previously, we used continuity of effective admittances at junctions to derive expressions from which normal mode frequencies can be calculated, as in Equation (13). Two expressions can replace Equation (13) when there is a lumped mass element at the junction

$$Y_+^{(n)eff} = [Y]_-^{(n+1)eff} \quad \text{or} \quad [Y]_+^{(n)eff} = Y_-^{(n+1)eff}. \tag{33}$$

At the left end of the tube with the piston boundary, $0 = Y^{(N+1)} = Y_-^{(N+1)eff}$.

We now use complex pressure amplitudes, $P_{m+}^{(n)}$ and $P_{m-}^{(n)}$ to calculate normal modes for pressure and volume velocity when lumped mass elements are present. For any n, the following equations hold

$$\hat{p}_m^{(n+1)}(x = x^{(n)}) = \hat{p}_m^{(n)}(x = x^{(n)}) + \Delta\hat{p}_{Jm}^{(n)}, \text{ or}$$

$$P_{m+}^{(n+1)} e^{ik_m x^{(n)}} + P_{m-}^{(n+1)} e^{-ik_m x^{(n)}} =$$

$$P_{m+}^{(n)} e^{ik_m x^{(n)}} + P_{m-}^{(n)} e^{-ik_m x^{(n)}} + \Delta\hat{p}_{Jm}^{(n)},$$

and (34)

$$\hat{Q}_{Jm}^{(n)} \equiv \hat{Q}_m^{(n+1)}(x = x^{(n)}) = \hat{Q}_m^{(n)}(x = x^{(n)}), \text{ or}$$

$$Y^{(n+1)}\left(P_{m+}^{(n+1)} e^{ik_m x^{(n)}} - P_{m-}^{(n+1)} e^{-ik_m x^{(n)}}\right) =$$

$$Y^{(n)}\left(P_{m+}^{(n)} e^{ik_m x^{(n)}} - P_{m-}^{(n)} e^{-ik_m x^{(n)}}\right),$$

where $\hat{p}_m^{(n)}(x = x^{(n)}) = \hat{p}^{(n+1)}(x = x^{(n)}, \omega = \omega_m)$,
$\hat{Q}_m^{(n)}(x = x^{(n)}) = \hat{Q}^{(n)}(x = x^{(n)}, \omega = \omega_m)$, and
$\Delta\hat{p}_{Jm}^{(n)} = \Delta\hat{p}_J^{(n)}(\omega = \omega_m)$.

The latter quantity denotes the pressure change in the mth mode at the junction between sub-tubes $n+1$ and n from left to right at the junction. From Equations (28) and (34)

$$\Delta\hat{p}_{Jm}^{(n)} = \frac{\hat{Q}_{Jm}^{(n)}}{Y_{Jm}^{(n)}} = Y^{(n)} \frac{\left(P_{m+}^{(n)} e^{ik_m x^{(n)}} - P_{m-}^{(n)} e^{-ik_m x^{(n)}}\right)}{Y_{Jm}^{(n)}}, \quad (35)$$

where $Y_{Jm}^{(n)} = Y_j^{(n)}(\omega_m)$. Thus, Equation (34) can be written

$$\left(P_{m+}^{(n+1)} e^{ik_m x^{(n)}} + P_{m-}^{(n+1)} e^{-ik_m x^{(n)}}\right) =$$

$$\left(P_{m+}^{(n)} e^{ik_m x^{(n)}} + P_{m-}^{(n)} e^{-ik_m x^{(n)}}\right) +$$

$$\frac{Y^{(n)}}{Y_{Jm}^{(n)}}\left(P_{m+}^{(n)} e^{ik_m x^{(n)}} - P_{m-}^{(n)} e^{-ik_m x^{(n)}}\right),$$

and (36)

$$Y^{(n+1)}\left(P_{m+}^{(n+1)} e^{ik_m x^{(n)}} - P_{m-}^{(n+1)} e^{-ik_m x^{(n)}}\right) =$$

$$Y^{(n)}\left(P_{m+}^{(n)} e^{ik_m x^{(n)}} - P_{m-}^{(n)} e^{-ik_m x^{(n)}}\right).$$

Solving for $P_{m+}^{(n+1)}$ and $P_{m-}^{(n+1)}$

$$P_{m+}^{(n+1)} = \frac{1}{2}\left[1 + Y^{(n)}\left(\frac{1}{Y^{(n+1)}} + \frac{1}{Y_{Jm}^{(n)}}\right)\right]P_{m+}^{(n)} +$$

$$\frac{1}{2}\left[1 - Y^{(n)}\left(\frac{1}{Y^{(n+1)}} + \frac{1}{Y_{Jm}^{(n)}}\right)\right]P_{m-}^{(n)}e^{-i2k_m x^{(n)}}, \quad (37)$$

where

$$P_{m-}^{(n+1)} = \frac{Y^{(n)}}{Y^{(n+1)}}P_{m-}^{(n)} + \left(P_{m+}^{(n+1)} - \frac{Y^{(n)}}{Y^{(n+1)}}P_{m+}^{(n)}\right)e^{i2k_m x^{(n)}}.$$

The computation begins by relating $P_{m-}^{(1)}$ to $P_{m+}^{(1)}$. Because there are no acoustic sources or initial conditions the complex amplitude $P_{m+}^{(1)}$ is not determined.

Example I again, with and without lumped mass elements

We examine the effect of the lumped mass elements on the acoustics of tubes that consist of a finite number of sub-tubes. As before, these sub-tubes can often be considered to approximate a continuous area function. We consider the sinusoidal area function shown in Figure 2 and expressed in Equation (22). Comparisons in normal mode signed amplitudes are made to a base condition without the lumped mass elements at internal junctions. Both conditions, with and without internal lumped mass elements, assume that the radiation admittance is purely imaginary and given by Equation (23). The results for the 10 sub-tube approximation are shown in Figures 19 through 22 with $P_{m+}^{(1)} = 1$.

Figure 19 shows the pressure signed amplitude distributions of the first three normal mode frequencies in the 10 sub-tube approximation for both conditions. The solid line is the pressure signed amplitude distribution with lumped mass elements at junctions, and the broken line is the pressure signed amplitude distribution without lumped mass elements at junctions. Figure 19, top panel shows the first normal mode, the middle panel of Figure 19 shows the second normal mode, and the bottom panel of Figure 19 shows the third normal mode. It is often difficult to make out a difference between the solid and dashed

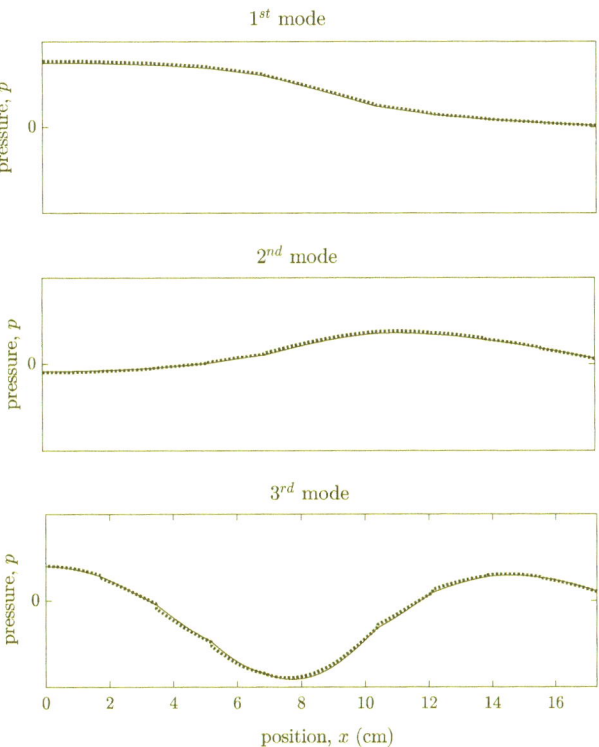

FIGURE 19. Pressure signed amplitude distributions for the first three normal modes with lumped mass elements at junctions (solid) and without (broken) in the 10 subtube approximation of the sinusoidal area function

lines, particularly for the first two normal modes. There are discernible jumps in the pressure signed amplitude distribution in the third normal mode in the bottom panel with the lumped mass elements. On the other hand, there are noticeable differences between the conditions when these plots are viewed more closely. Figure 20 shows the pressure signed amplitude distributions for the second normal mode with (solid) and without (dashed) lumped mass elements at junctions. Pressure signed amplitude jumps are noticeable at least at $x = 7\ cm$ and at $x = 14\ cm$ in the condition with lumped mass elements.

Figure 21 shows the volume velocity signed amplitude distribution for the first three normal mode frequencies with and without lumped mass

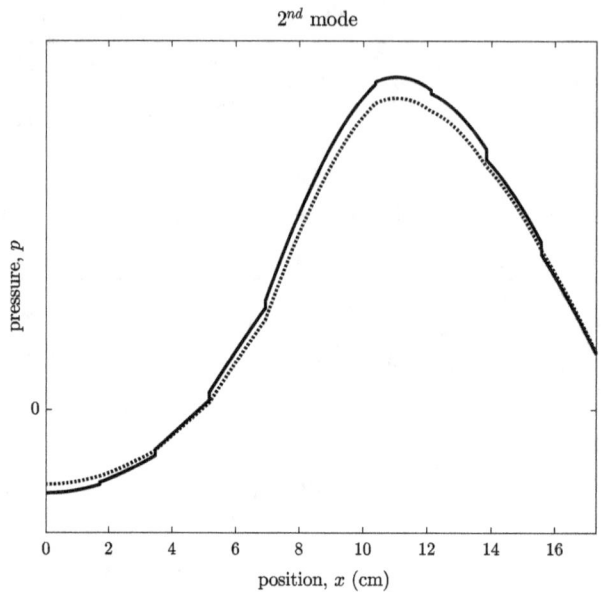

FIGURE 20. Pressure signed amplitude distributions for the second normal mode with lumped mass elements at junctions (solid) and without (broken) in the 10 sub-tube approximation of the sinusoidal area function

elements at junctions. Again, it is difficult to make out any differences between the with and without lumped mass element conditions. Figure 22 shows both conditions in more detail for the volume velocity signed amplitude distribution of the second normal mode. There are no jumps in the volume velocity amplitude, as is expected because continuity of volume velocity is still in effect. However, kinks in the volume velocity signed amplitude at junctions are visible. There may be a very slight increase in the absolute magnitude of the difference of maximum to minimum volume velocity for the second normal mode frequency in the condition with lumped mass elements compared to the condition without these elements.

Figure 23 shows the first three normal mode frequencies plotted against the number of approximating sub-tubes when the lumped mass elements at junctions are accounted for (solid line) and the case when they are not (broken line). The top panel of Figure 23 is for the first

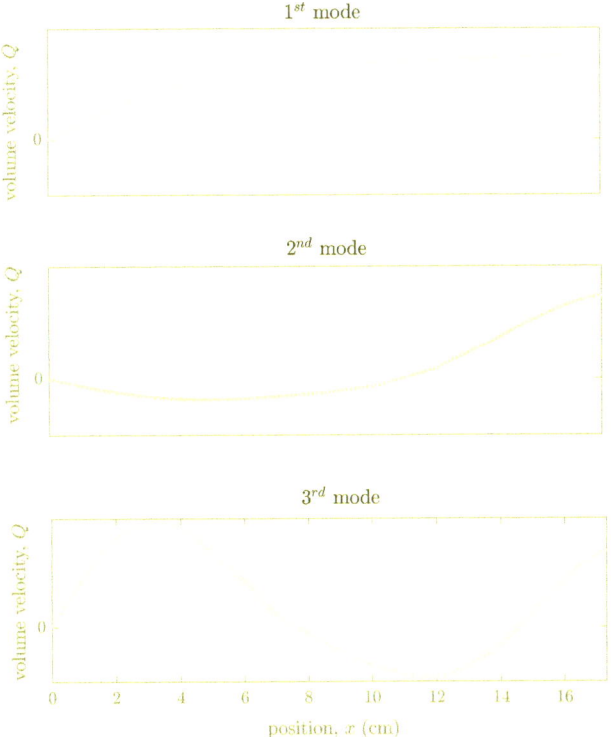

FIGURE 21. Volume velocity signed amplitude distributions for the first three normal modes with lumped mass elements at junctions (solid) and without (broken) in the 10 sub-tube approximation of the sinusoidal area function

normal mode, the middle panel Figure 23 is for the second normal mode, and Figure 23 is for the third normal mode. The differences between the normal mode frequencies with and without internal lumped mass elements increase both with the number of approximating sub-tubes and with normal mode number. The latter effect is not surprising because the magnitudes of the acoustic impedances of lumped mass elements increase with frequency, as seen in Equation (27). This leads to larger pressure amplitude jumps for nearly the same volume velocity flows.

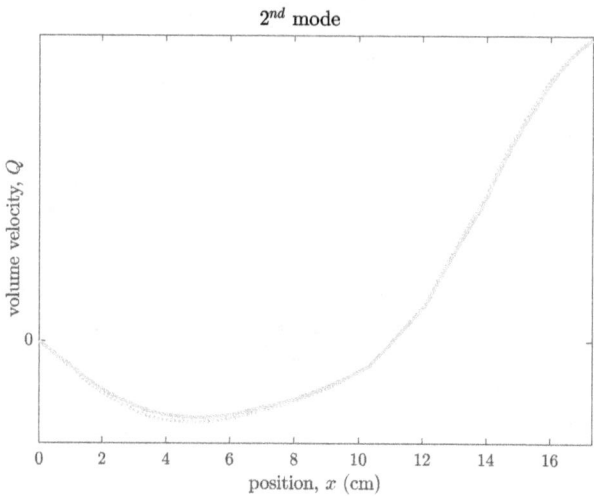

FIGURE 22. Volume velocity signed amplitude distributions for the second normal mode with lumped mass elements at junctions (solid) and without (broken) in the 10 sub-tube approximation of the sinusoidal area function

The reason that normal mode frequencies decrease when more lumped mass elements are added when more sub-tubes are added can be seen from Figure 14b of Chapter 9. While an increase in the number of sub-tubes should lead to a steady decrease in the ratios of the areas of neighboring wider and narrower tubes, this figure shows that $\rho_0 \ell / A^{(nar)}$ decreases as the logarithm of the ratio of the larger-to-smaller cross-sectional area. Thus, it is entirely plausible that the addition of more sub-tubes adds more lumped mass elements faster than the effect of each element diminishes. In fact, while the normal mode frequencies reach an asymptote in the case when lumped mass elements are not included, the normal mode frequencies continue to decrease with an increasing number of sub-tubes. This is an indication that the straight sub-tube approximations may not be a good approximation in which to study the behavior of the normal mode frequencies when many sub-tubes are used with lumped mass elements. We have not mentioned that there is an additional restriction on the use of lumped mass elements of length ℓ. This restriction is that the lumped mass element length must be much less than the length of the

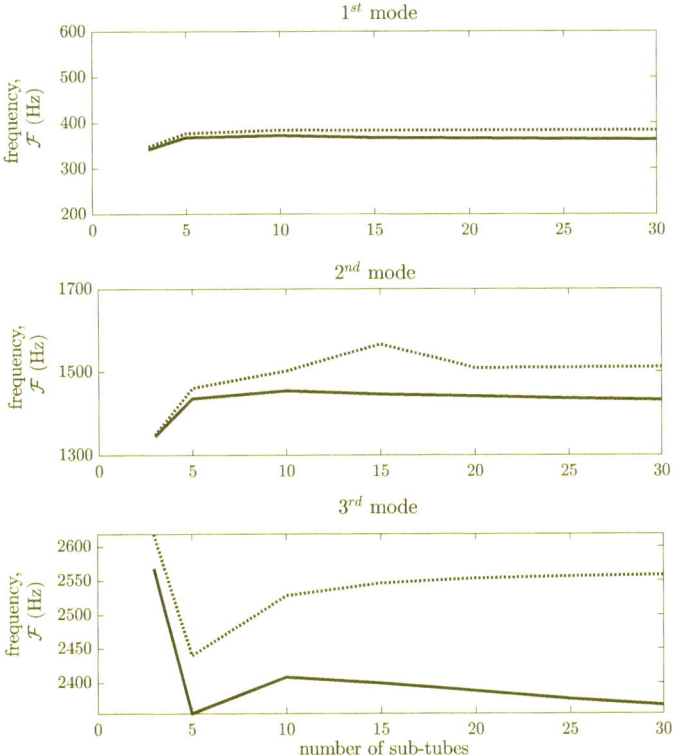

FIGURE 23. Normal mode frequencies as a function of the number of sub-tubes with lumped mass elements at junctions (solid) and without (broken)

sub-tubes.
$$\ell \ll l_{st} \,. \tag{38}$$
Here we list our estimated upper bounds on ℓ and lengths l_{st} for different numbers of sub-tubes for sinusoidal area function used in the example. For 10 sub-tubes $\max(\ell) < 0.16$ cm and $l_{st} = 1.71$ cm. For 30 sub-tubes $\max(\ell) < 0.046$ cm and $l_{st} = 0.57$ cm. For 100 sub-tubes $\max(\ell) < 0.008$ cm and $l_{st} = 0.171$ cm. We seem to have the ℓ is at least an order of magnitude smaller than l_{st} in all of these cases. Here we are assuming rectangular cross-sections with $d = 1$ cm in the dimension into the paper, and using the equation that generated Figure 15a of Chapter 9.

In general, the addition of lumped mass elements models local increases in kinetic energy near junctions. When the ratio of kinetic to potential energies increase in acoustic system, we can expect the normal mode frequencies to decrease.

Discussion of lumped mass elements and multiple sub-tube approximations

To our knowledge, this is the first time that a systematic study of lumped mass elements at junctions in a one-dimensional acoustic system with straight sub-tubes has been conducted. We have computed the changes to the normal mode frequencies of a tube composed of straight sub-tubes when lumped mass elements are included. These changes can be substantial, particularly for the second and third normal mode frequencies. On the other hand, we believe that we have overestimated the effect lumped mass elements on these normal mode frequencies for the underlying sinusoidal area function for which the sub-tubes are an approximation. Our reading of Rayleigh's *Theory of Sound* (Rayleigh 1945) leads us to believe that approximating a smooth area function by sub-tubes that jump in cross-sectional area over-estimates the total excess kinetic energy. Rayleigh provides an upper bound on a quantity similar to lumped mass element admittance, which is particularly useful when the area function does not change rapidly with position along the tube axis.

We suggest that using a continuous area function instead of concatenated straight sub-tubes may be a better way to calculate the normal modes. This would eliminate the problem of discontinuity of particle velocity at the junctions. One possibility is to use concatenated sections of cones, for which analytic properties of acoustic wave propagation are known, including the quantity of excess kinetic energy involved. One problem that would need to be overcome with this departure from plane wave propagation is the fact that different parts of an acoustic wave front arrive at the junctions at different times. Thus, a change from straight sub-tubes would involve some mathematical research.

Conclusion

It seems that multiple sub-tube approximations to area functions are useful in discovering the properties of the normal modes of a tube. In general, the more sub-tubes the better, particularly in relation to steady acoustic radiation pressure and the acoustic perturbation theory. On the other hand, in the instances that are examined here, properties such as normal mode frequencies and the pressure and volume velocity signed amplitude distributions in the vocal tract do not suffer much degradation when the number of sub-tubes is reduced. Thus, the appropriate number of sub-tubes depends on the research question. For normal mode frequency estimation, signed amplitude distribution information, or even formant synthesis, a crude approximation to the area function may suffice, although more examples should be examined. On the other hand, for acoustic perturbation theory analyses in certain regions of the vocal tract, it may be necessary to obtain a very good approximation to the area function. On the other hand, the limitations that are placed on sub-tube lengths because of evanescent waves must be considered.

We introduced the imaginary parts of the radiation impedance and radiation admittance as an extension of the idea of lumped mass elements at junctions between sub-tubes of unequal cross-sectional area. We still need to consider the real parts of the radiation impedance and radiation admittance, because the real parts of the radiation impedance and admittance play the essential role of permitting energy to radiate from the lip opening.

References

Morse, P.M. (1976). *Vibration and Sound*. Acoustical Society of America: Melville, NY. (p 333).

Pierce, A.D. (1989). *Acoustics: An Introduction to Its Physical Principles and Applications*. Acoustical Society of America: Melville, NY. (pp 316-318).

Rayleigh, J.W.S. (1945). *Theory of Sound: Volume II*. Dover Publications, New York. (Articles 302-313).

Chapter 11: Damping and Green's Function Modifications

Introduction

In Chapter 7 we invoked damping of acoustic waves in the straight tube in order to derive meaningful results when there are sinusoidally varying acoustic sources. Damping is important in such circumstances because the transients die away exponentially fast, leaving an infinite number of modes with distributed kinetic and potential energies that are driven by the source in a steady manner. These modes have the spatial signed amplitude distributions very nearly the same as those of the undamped normal modes, but their time dependence is that of the source. The source provides energy to each mode, so that in order for the system to become steady, the energy in each mode must flow elsewhere. The mechanisms that provide this energy flow are the same ones that damp the transient normal modes. In this chapter important damping mechanisms are identified and discussed.

The five mechanisms that we discuss are radiation, *jetting, wall vibration damping, acoustic momentum boundary layer damping*, and *acoustic thermal boundary layer damping*. While acoustic energy from speech always ends up as heat [and, sometimes, for a while, as human or machine memories], this does not happen immediately for that portion of the energy lost from the vocal tract by radiation. After all, in order for us to be heard, some energy must leave the vocal tract in a way that allows for acoustic propagation to the listener in the atmosphere. However, as far as the vocal tract is concerned, it loses energy by radiation, so radiation contributes to the damping, which results in loss of energy from the vocal tract. We calculate the amount of energy that is lost from the finite-length tube by radiation in the next section.

Identifying damping mechanisms permits us to calculate how quickly the transient terms become negligible, as well as to calculate the details of the responses as a function of piston circular frequency ω_{pst} near the normal mode circular frequencies [e.g. Figures 4 and 5 of Chapter 7]. Conversely, we can estimate these damping constants by examining the bandwidths of the peaks in magnitude spectra (e.g. Fant 1960). In fact, empirical determination of damping constants γ_m might be preferred to attempting to calculate them using the ideas in this chapter. The mathematical models discussed in this chapter are idealized, making accurate estimates of damping for a human vocal tract difficult. On the other hand, these models do indicate trends of the effectiveness of each mechanism as a function of frequency. Therefore, they are illuminating. Unfortunately, we do not even provide a mathematical model for jetting. Part of the reason for this is that it is a phenomenon that needs concepts developed in Chapter 13. The other reason is that this damping by jetting depends on the precise details of the situation.

At the end of the current chapter we return to the Green's function solution for pressure with a volume velocity piston source in Chapter 7. The things that have been discovered since Chapter 7 point a way to construct a more realistic Green's function.

Radiation damping

In order that sound be heard outside the finite-length tube, it is necessary that energy be radiated from its open end. We noted in Chapter 3 that requiring the pressure disturbance to be zero at the opening of the finite-length tube can only be an approximation. In Chapter 10 we noted that the change from finite to infinite cross-sectional area at the end of the tube requires a lumped mass element. This accounts for the acceleration of fluid particles required by the change in cross-sectional area. The resulting imaginary part for radiation admittance $Im(Y_{rad})$ is stated in Equation (24) of Chapter 10, and repeated here.

$$Im(Y_{rad}) \approx \frac{3\pi}{8} \frac{Y^{(1)}}{(ka)} , \qquad (1)$$

where $Y^{(1)}$ is the tube admittance of the sub-tube that opens to the atmosphere, a is the radius of the circular opening, and $k = \omega/c_0$ is the wavenumber. This is for a flanged opening, so that a wall of infinite extent is presumed to surround the opening. We are restricting ourselves to sinusoidal motion, with circular frequency ω and the above result is only valid for small ka, i.e. for $ka \ll 1$. Because $k = 2\pi/\lambda$, this means that wavelength must be large with respect to the radius of the opening, or $a \ll \lambda$. Therefore, we are restricted to long wavelengths or

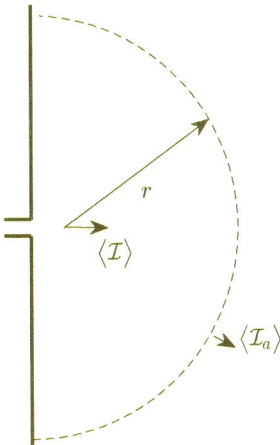

FIGURE 1. Geometry for calculation of power output to the atmosphere

low frequencies, and we say that we are working in the low-frequency limit, or under the low-frequency approximation.

Equation (1) says that there must be a sinusoidal pressure variation at the opening, if there is a sinusoidal volume velocity. However, Equation (1) provides a pressure variation that is $\pi/2$ out-of-phase with the volume velocity at the opening, because this admittance is purely imaginary. It is a *reactive admittance*. If there was only the imaginary part to the radiation admittance, then there would not be any acoustic energy transmitted from the opening into the atmosphere in the time-average. A non-zero real part, or a *resistive admittance*, of the radiation admittance would mean that there is oscillating pressure in phase with volume velocity at the opening, and acoustic energy would leave, or enter, the tube. We know that acoustic energy leaves the tube in the time-average.

We use an indirect method to find the real part of the radiation admittance. In particular, we consider the fact that a time-varying volume velocity at the mouth acts as an acoustic source for the atmosphere. The tube opening is approximated as a volume velocity source concentrated at a point at the lips in the low-frequency approximation. Also, we consider regions in the atmosphere that are many wavelengths away from the tube opening. This is known as the *far-field approximation*. Both the low-frequency and far-field approximations can be made when there are multiple frequencies and wavelengths present. In that case, one considers the "worst" cases to find whether the criterion for either approximation has been satisfied. For the low-frequency approximation one considers the shortest wavelength, or highest frequency, in relation to the radius of the tube opening. For the far-field approximation one considers the longest wavelength, or lowest frequency, in relation to the distance from the tube opening to the observer or listener. As a reminder, we are working with a flanged opening, so the source is radiating into an infinite half-space, as shown in Figure 1.

The volume velocity at the tube opening is denoted $Q(x = L, t)$, so that $Q(x = L, t) = u(x = L, t)A^{(1)}$. Recall the relation between acceleration $a(x, t) \approx a_E(x, t)$, and particle velocity $u(x, t)$ given in Equation (8) of Chapter 2. We use another notation, $\dot{u}(x, t)$

$$a(x, t) \approx \dot{u}(x, t) \approx \frac{\Delta_t u(x, t)}{\Delta t} . \qquad (2)$$

In an analogous way volume acceleration $\dot{Q}(x, t)$ is

$$\dot{Q}(x, t) \approx \frac{\Delta_t Q(x, t)}{\Delta t} . \qquad (3)$$

Let r denote the distance of the listener from the tube opening and $p_a(r, t)$ denote the pressure disturbance at a distance r from the tube opening. The subscript "a" denotes that the quantity is in the atmosphere outside the tube. It can be shown, under the approximations that we have made, that

$$p_a(r, t) = K_s \rho_0 \frac{\dot{Q}(x = L, t - r/c_0)}{4\pi r} , \qquad (4)$$

where $K_s = 2$ for the flanged opening. [In more generality, $1 \leq K_s \leq 2$ is a factor that accounts for scattering of acoustic waves by the head. In the low-frequency approximation, as frequency increases from 0 Hz to 2000 Hz, K_s increases from 1 to 1.7 (Fant 1960). At very low frequencies, the head is small compared to the wavelength of radiated acoustics and has little effect, and $K_s \approx 1$. At higher frequencies, but still low frequency, the head acts more like a plane baffle, and $K_s \approx 2$.]

There are some things to observe about Equation (4). The pressure disturbance amplitude diminishes as $1/r$ as it travels from the tube opening. This is what is known as *spherical spreading*. In order that acoustic energy be conserved as the area of the wave front increases, it is necessary that the amplitude of the disturbance decrease as $1/r$.

Also, the pressure disturbance that the listener hears at time, t, is related to the time-rate-of-change of the volume velocity at the tube opening at the retarded time $t - r/c_0$. r/c_0 is the time it takes for an acoustic disturbance to travel from the opening to the listener. The term r/c_0 is the acoustic delay time.

Some speech production researchers say that it is the time-rate-of-change of the volume velocity $\dot{Q}(x = 0, t)$ at the glottis that is the acoustic source during voiced sounds. From the physical point of view, this is not a correct statement. The source of acoustics for the voice source in the tube is $Q(x = 0, t)$, and not $\dot{Q}(x = 0, t)$, because we are assuming one-dimensional propagation. However, when the sound from the lip opening is radiating into the three-dimensional atmosphere, time-rate-of-change of the volume velocity at the lips, $\dot{Q}(x = L, t - r/c_0)$ is the source. This has the practical effect of transmitting the time-rate-of-change of the volume velocity at the glottis $\dot{Q}(x = 0, t - r/c_0 - L/c_0)$ to the listener. This is the reason that speech researchers are concerned with the time rate-of-change of the volume velocity at the glottis.

Let's go back and calculate the acoustic power in the far-field of the spherically spreading wave with a plane baffle surrounding the tube opening. In the far field, the spherical wave behaves locally like a plane wave, so that

$$u_a(r, t) \approx \frac{p_a(c, t)}{\rho_0 c_0} = \frac{K_s}{c_0} \frac{\dot{Q}(x = L, t - r/c_0)}{4\pi r}, \quad (5)$$

where $u_a(r, t)$ is the air particle velocity in the radial direction. Using the definition of intensity given in Equation (38) of Chapter 2, the time-average intensity at any point on the half-spherical surface at a distance r from the tube opening is directed radially outward and has magnitude

$$\langle \mathcal{I}_a(r, t) \rangle = \langle p_a(r, t) u_a(r, t) \rangle$$
$$= \frac{\langle p_a^2(r, t) \rangle}{\rho_0 c_0} \quad (6)$$

$$= \frac{K_s^2 \rho_0}{c_0 16\pi^2 r^2} \langle \dot{Q}^2(x=L, t-r/c_0) \rangle .$$

If the time-average intensity is summed over the area of the surface of the half-sphere of radius r, we obtain the time-average power $\langle \mathcal{P}_a \rangle$, which is the time-average energy flow into the half-spherical surface.

$$\langle \mathcal{P}_a(r,t) \rangle = -\langle \mathcal{I}_a(r,t) \rangle (2\pi r^2) = -\frac{\rho_0}{c_0 2\pi} \langle \dot{Q}^2(x=L, t-r/c_0) \rangle , \quad (7)$$

because $K_s = 2$. The minus sign indicates that energy is leaving the half-sphere that encloses the end of the tube. For sinusoidal motion at circular frequency ω, with complex amplitude $\hat{Q}(x=L)$, we have $Q(x=L, t-r/c_0) = Re\big[\hat{Q}(x=L)e^{-i\omega(t-r/c_0)}\big]$, and

$$\langle \dot{Q}^2(x=L, t-r/c_0) \rangle = \frac{\omega^2}{2}|\hat{Q}(x=L)|^2 . \quad (8)$$

With Equation (8), Equation (7) can be written

$$\langle \mathcal{P}_a(r,t) \rangle = -\frac{\omega^2 \rho_0}{c_0 4\pi}|\hat{Q}(x=L)|^2 . \quad (9)$$

Because there is time-average energy leaving the region bounded by the baffle and the half-spherical surface through that half-spherical surface, there must be time-average energy entering that region from the tube opening. The time-average intensity at the opening is

$$\langle \mathcal{I}(x=L, t) \rangle = \langle p(x=L,t)u(x=L,t) \rangle$$
$$= \langle Re[\hat{p}(x=L)e^{-i\omega t}] Re[\hat{u}(x=L)e^{-i\omega t}] \rangle \quad (10)$$
$$= \frac{1}{A^{(1)}} \langle Re[\hat{p}(x=L)e^{-i\omega t}] Re[\hat{Q}(x=L)e^{-i\omega t}] \rangle .$$

Pressure at the tube opening is related to the volume velocity by

$$\hat{p}(x=L)e^{-i\omega t} = Z_{rad}\hat{Q}(x=L)e^{-i\omega t}$$
$$= \big(Re[Z_{rad}] + iIm[Z_{rad}]\big)\hat{Q}(x=L)e^{-i\omega t} \quad (11)$$
$$= \big(Re[Z_{rad}]Re[\hat{Q}(x=L)e^{-i\omega t}] - Im[Z_{rad}]Im[\hat{Q}(x=L)e^{-i\omega t}]\big) +$$
$$i\big(Re[Z_{rad}]Im[\hat{Q}(x=L)e^{-i\omega t}] + Im[Z_{rad}]Re[\hat{Q}(x=L)e^{-i\omega t}]\big) .$$

Combining Equations (10) and (11) results in

$$\langle \mathcal{I}(x=L,t) \rangle = \frac{1}{A^{(1)}} \langle Re[\hat{p}(x=L)e^{-i\omega t}] Re[\hat{Q}(x=L)e^{-i\omega t}] \rangle \quad (12)$$

$$= \frac{Re[Z_{rad}]}{2A^{(1)}}|\hat{Q}(x=L)|^2 .$$

The average power entering the region between the baffle and the half-sphere from the tube opening is

$$\langle \mathcal{P}(x=L,t)\rangle = \langle \mathcal{I}(x=L,t)\rangle A^{(1)} = \frac{Re[Z_{rad}]}{2}|\hat{Q}(x=L)|^2 . \quad (13)$$

By energy conservation, using Equations (9) and (13) gives

$$\langle \mathcal{P}_a(r,t-r/c_0)\rangle + \langle \mathcal{P}(x=L,t)\rangle = 0 ,$$

or

$$-\frac{\omega^2 K_s^2 \rho_0}{c_0 16\pi}|\hat{Q}(x=L)|^2 + \frac{Re[Z_{rad}]}{2}|\hat{Q}(x=L)|^2 = 0 , \quad (14)$$

or

$$Re[Z_{rad}] = \frac{\omega^2 \rho_0}{2\pi c_0} = \frac{\omega^2}{c_0^2} \frac{A^{(1)}}{2\pi} \frac{\rho_0 c_0}{A^{(1)}} = \frac{(ka)^2}{2} \frac{\rho_0 c_0}{A^{(1)}} = \frac{(ka)^2}{2} \frac{1}{Y^{(1)}} ,$$

where we have set $K_s = 2$, and $A^{(1)} = \pi a^2$ for the circular opening of radius a.

Combining this result with the quoted result in Equation (1) with Equation (14) we obtain radiation impedance.

$$Z_{rad} = Re[Z_{rad}] + iIm[Z_{rad}] \approx \frac{1}{2}\frac{(ka)^2}{Y^{(1)}} - i\frac{8}{3\pi}\frac{(ka)}{Y^{(1)}} . \quad (15)$$

The radiation admittance is

$$\begin{aligned} Y_{rad} &= \frac{1}{Z_{rad}} \\ &= \frac{Z_{rad}^*}{|Z_{rad}|^2} \\ &= \frac{Y^{(1)}}{(ka)^2} \cdot \left[\frac{\frac{1}{2} + i\frac{8}{3\pi(ka)}}{\frac{1}{4} + \left(\frac{8}{3\pi(ka)}\right)^2}\right] . \end{aligned} \quad (16)$$

The results provided here for radiation impedance and radiation admittance for a flanged opening are for the low-frequency approximation, $ka \ll 1$. There is a closed-form expression for the radiation impedance that is valid at all frequencies, as given by Rayleigh (1945), and re-derived by many later authors (e.g. Morse 1976; Pierce 1989). The closed-form expression involves transcendental functions, so it is beyond the scope of this book. The real part of the radiation impedance or admittance accounts for the energy of acoustic wave motion that leaves the tube by radiation, while the imaginary part accounts for extra oscillatory lumped mass element at the tube opening that lowers the mode frequencies.

Jetting

Acoustic wave motion can lose energy to rotational air motion (Ingard & Ising 1967; Bechert 1980; Howe 1980; Davies, McGowan, & Shadle 1993). This is a damping mechanism that is usually not considered in speech acoustics, but is well-known in other areas of duct acoustics, such as muffler design. We cannot go into depth here, because the conceptual material needed for a good discussion is in Chapter 13.

These losses are largest when there is a mean flow, which is almost always the case in speech with a net flow of air from the lungs to the atmosphere. The oscillating air velocity of acoustic motion that occurs at abrupt cross-sectional area expansions or at sharp solid edges is then transformed into rotational air motion that is convected downstream by the the mean flow. We would expect this to be strongest during the production of voiced, sibilant fricatives in the region of the tongue tip and teeth. There is also the possibility of jetting at the teeth and lips for any acoustic wave during speech. There is a closed-form solution for this loss and a related added mass for the special case of a circular aperture in a thin plate in Howe (1998).

In Chapter 13 we discuss jetting, where rotational air motion is created at the vocal folds. This is not an acoustic damping mechanism at the vocal folds because it is a part of the mechanics of the voice source.

Wall vibration damping

Many of the solid surfaces in the vocal tract are compliant and, thus, respond to pressure fluctuations. We model this in the simplest way possible, which is with *locally reacting walls* for the tube inner surfaces whose mechanical properties are uniform throughout the tube. This means that each point on the surface responds only to the air pressure at that point, so that the surface does not possess wave motion independent of the acoustic wave motion.

The sub-tubes that compose the whole tube are supposed here to be cylindrical. We work out the results for a single sub-tube of cross-sectional area, $A_0 = \pi\zeta_0^2$, where ζ_0 is the radius of the cross-section when there is no acoustic disturbance. With an acoustic disturbance, the cross-sectional area, $A = A(t)$ is time-dependent, so that

$$A(t) = \pi\zeta^2(t) \text{ , with } \left|\frac{A(t) - A_0}{A_0}\right| \ll 1 \text{ and } \left|\frac{\zeta(t) - \zeta_0}{\zeta_0}\right| \ll 1 \text{ .} \quad (17)$$

That is, the amplitude of wall vibration is small.

Under the small amplitude approximation the wall vibration is assumed to be linear. Thus, the equation of motion for small surface area element of the locally reacting wall is essentially the same as that of the forced, friction damped mass-spring system, Equation (40) of Chapter 4. However, coefficients, such as mass and spring constant, are expressed as per surface area, because the force is provided by the pressure fluctuations of the acoustic field $p(x,t)$. We suppress the spatial variable x for now, and assume that $p(t)$ is the pressure at some specific small area element on the surface of the wall. Let $\Delta\zeta(t) = \zeta(t) - \zeta_0$. Recall that $p(t)$ is the perturbation pressure around p_0.

$$\bar{m}\frac{\Delta_t^2(\Delta\zeta(t))}{(\Delta t)^2} + \bar{\mu}\frac{\Delta_t(\Delta\zeta(t))}{\Delta t} + \bar{\kappa}(\Delta\zeta(t)) \approx p(t) \text{ ,}$$

or (18)

$$\frac{\Delta_t^2(\Delta\zeta(t))}{(\Delta t)^2} + 2\gamma_w\frac{\Delta_t(\Delta\zeta(t))}{\Delta t} + \omega_w^2\Delta\zeta(t) \approx \frac{p(t)}{\bar{m}} \text{ ,}$$

where the ¯ means that per-unit-area values of the parameters are taken, and $\gamma_w = \bar{\mu}/2\bar{m}$, and $\omega_w = \sqrt{\bar{\kappa}/\bar{m}}$. Equation (18) tells us that $\zeta = \zeta(p(x,t))$, after the transients are substantially reduced. In other words, steady radius variation depends on position x and time t only through the value of pressure p at x and t. Thus, $A = A(p(x,t))$ as well.

The data for the values of the mechanical parameters $\bar{m}$, $\bar{\mu}$ and $\bar{\kappa}$ are sparse. The most consistently cited source for values is the article by Ishizaka, French, & Flanagan (1975). They determined the mechanical parameters for the cheek (relaxed and tensed), the neck, and the forearm of a human subject. Their results provide wall natural frequencies, $\mathcal{F}_w = \omega_w/(2\pi)$ from 32 Hz for the relaxed cheek to 93 Hz for the forearm. Typical values are $\bar{m} = 2$ g/cm^2, $\bar{\mu} = 1000$ g/cm^2 s, and $\bar{\kappa} = 100,000$ dyne/cm^3. These particular values result in $\omega_w = 224$ rad/s or $\mathcal{F}_w = 36$ Hz, and $\gamma_w = 250$ rad/s. The wall motion is a heavily damped motion because γ_w/ω_w is close to one.

With this model for wall motion we proceed to re-derive the wave equation, Equation (17) of Chapter 2, except we now permit small changes in cross-sectional area (Lighthill 1978). With cross-sectional area time variable, we replace Equation (3) of Chapter 2 with

$$\text{time rate-of-change of mass} = \frac{\Delta_t[A(x,t)\rho(x,t)]}{\Delta t}\Delta x \ . \qquad (19)$$

The net inflow of mass is still given by Equation (4) of Chapter 2. Thus, mass conservation with time-variable cross-sectional area is given by

$$\frac{\Delta_t[A(x,t)\rho(x,t)]}{\Delta t} \approx -\rho_0 \frac{\Delta_x Q(x,t)}{\Delta x} \ . \qquad (20)$$

We endeavor to replace $A(x,t)\rho(x,t)$ in Equation (20) with $p(x,t)$. This can be done because we have $A(x,t) = A(p(x,t))$, and in the acoustic approximation, $\rho(x,t) = \rho(p(x,t))$. First, define c_w

$$\frac{1}{c_w^2} \equiv \frac{\rho_0}{A_0}\frac{\Delta_p A}{\Delta p} \ . \qquad (21)$$

c_w has the dimensions of velocity, and c_w^2 is inversely proportional to the change of cross-sectional area to changes in air pressure. Thus, c_w^2 is a measure of wall stiffness.

In the linear approximation

$$\begin{aligned}\frac{\Delta_p[A\rho]}{\Delta p} &\approx A_0 \frac{\Delta_p \rho}{\Delta p} + \rho_0 \frac{\Delta_p A}{\Delta p} \\ &= A_0 \frac{1}{c_0^2} + \rho_0 \frac{\Delta_p A}{\Delta p} \\ &= A_0\left(\frac{1}{c_0^2} + \frac{1}{c_w^2}\right),\end{aligned} \qquad (22)$$

where Equation (9) of Chapter 1 and Equation (21) have been invoked. Equation (22) tells us that changes in $A\rho$ compared to the rest value,

$A_0\rho_0$, depends on the distensibility provided by air $(\rho_0 c_0^2)^{-1}$ and on the distensibility provided by the wall $(\rho_0 c_w^2)^{-1}$. With Equation (22), the left-hand-side of the mass conservation equation, Equation (20), can be written

$$\frac{\Delta_t A(x,t)\rho(x,t)}{\Delta t} = \frac{\Delta_p[A\rho]}{\Delta p}\frac{\Delta_t p}{\Delta t}$$

$$= A_0\left(\frac{1}{c_0^2} + \frac{1}{c_w^2}\right)\frac{\Delta_t p}{\Delta t} \qquad (23)$$

$$= A_0 \frac{1}{c^2}\frac{\Delta_t p}{\Delta t} \ , \ \text{where} \ \frac{1}{c^2} = \frac{1}{c_0^2} + \frac{1}{c_w^2} \ .$$

With Equation (23), mass conservation, Equation (20), can be written

$$\frac{1}{c^2}\frac{\Delta_t p}{\Delta t} = -\frac{\rho_0}{A_0}\frac{\Delta_t Q(x,t)}{\Delta x} \ . \qquad (24)$$

Equation (13) of Chapter 2 is still the applicable momentum conservation equation.

$$\frac{\rho_0}{A}\frac{\Delta_t Q(x,t)}{\Delta t} = -\frac{\Delta_x p(x,t)}{\Delta x} \ . \qquad (25)$$

In the linear approximation, A can be replaced by A_0 in Equation (25).

$$\frac{\rho_0}{A_0}\frac{\Delta_t Q(x,t)}{\Delta t} = -\frac{\Delta_x p(x,t)}{\Delta x} \ . \qquad (26)$$

To derive a new wave equation, we apply $\Delta_x/\Delta x$ to Equation (26) and $\Delta_t/\Delta t$ to Equation (24) to obtain

$$\frac{1}{c^2}\frac{\Delta_t^2 p}{(\Delta t)^2} - \frac{\Delta_x^2 p}{(\Delta x)^2} = 0 \ , \ \text{with} \ \frac{1}{c^2} = \frac{1}{c_0^2} + \frac{1}{c_w^2} \ . \qquad (27)$$

This is the wave equation that was derived in Equation (17) of Chapter 2, except that c replaces c_0. c is known as the *phase speed* of the acoustic wave. Originally, we had $c = c_0$, the adiabatic speed of sound, which is a thermodynamic property of the air that is constant as long as other thermodynamic properties such as the rest pressure and rest temperature remain constant.

Before examining c, however, we examine c_w when there is sinusoidal time dependence, with $\zeta(x,t) = \hat{\zeta}(x)e^{-i\omega t} = \hat{\zeta}e^{-i\omega t}$. In analogy with the definition of c_w given in Equation (21), we write

$$\hat{c}_w^2 \equiv \frac{A_0}{\rho_0} \left(\frac{\Delta_{\hat{p}} \hat{A}}{\Delta \hat{p}} \right)^{-1}, \quad \text{where } A = \hat{A} e^{-i\omega t}. \tag{28}$$

The definition of $\hat{A}$ is consistent in the linear approximation, because

$$\frac{\Delta_{\hat{p}} \hat{A}}{\Delta \hat{p}} = \pi \frac{\Delta_{\hat{p}} \hat{\zeta}^2}{\Delta \hat{p}} \approx 2\pi \zeta_0 \frac{\Delta_{\hat{p}} \hat{\zeta}}{\Delta \hat{p}} = s_0 \frac{\Delta_{\hat{p}} \hat{\zeta}}{\Delta \hat{p}}, \tag{29}$$

where s_0 = circumference of the circle that defines the rest cross-sectional area A_0. [Note that by assuming $e^{-i\omega t}$ time dependence we have eliminated the transient part of the solution.] We rewrite Equation (18) using the identity in Equation (50) of Chapter 9, $\Delta_t e^{Kt}/\Delta t \approx K e^{Kt}$.

$$-\omega^2 (\Delta_{\hat{p}} \hat{\zeta}) - i 2 \gamma_w \omega (\Delta_{\hat{p}} \hat{\zeta}) + \omega_w^2 (\Delta_{\hat{p}} \hat{\zeta}) = \frac{\Delta \hat{p}}{\bar{m}},$$

or

$$\left[(\omega^2 - \omega_w^2) + i 2 \gamma_w \omega \right] (\Delta_{\hat{p}} \hat{\zeta}) = -\frac{\Delta \hat{p}}{\bar{m}},$$

or $\hspace{10cm} (30)$

$$\frac{\Delta_{\hat{p}} \hat{\zeta}}{\Delta \hat{p}} = -\frac{1}{\bar{m}\left[(\omega^2 - \omega_w^2) + i 2 \gamma_w \omega \right]},$$

or

$$\left(\frac{\Delta_{\hat{p}} \hat{\zeta}}{\Delta \hat{p}} \right)^{-1} = \left| \frac{\Delta_{\hat{p}} \hat{\zeta}}{\Delta \hat{p}} \right|^{-1} e^{-i\phi_w} = -\bar{m}\left[(\omega_w^2 - \omega^2) - i 2 \gamma_w \omega \right], \text{ where}$$

$$\left| \frac{\Delta_{\hat{p}} \hat{\zeta}}{\Delta \hat{p}} \right|^{-1} = \bar{m}\sqrt{(\omega^2 - \omega_w^2)^2 + 4(\gamma_w \omega)^2}$$

and $\phi_w = \arctan\left(\frac{2 \gamma_w \omega}{(\omega_w^2 - \omega^2)} \right).$

Using Equations (29) and (30), Equation (28) can be written

$$\hat{c}_w^2 = \frac{A_0}{\rho_0} \left(\frac{\Delta_{\hat{p}} \hat{A}}{\Delta \hat{p}} \right)^{-1}$$

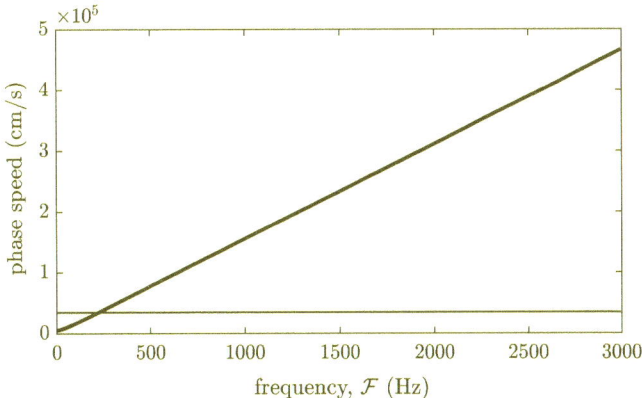

FIGURE 2. Magnitude of $\hat{c}_w$ (thick dark line) as functions of frequency $\mathcal{F} = \omega/(2\pi)$. The adiabatic speed of sound c_0 is shown in the horizontal (thin dark) line

$$= \frac{A_0}{s_0 \rho_0} \left| \frac{\Delta_{\hat{p}} \hat{\zeta}}{\Delta \hat{p}} \right|^{-1} e^{-i\phi_w} \qquad (31)$$

$$= \frac{A_0 \bar{m}}{s_0 \rho_0} \sqrt{(\omega^2 - \omega_w^2)^2 + 4(\gamma_w \omega)^2}\, e^{-i\phi_w}\ .$$

For a 2 cm² circular cross section $(A_0 \bar{m}/s_0 \rho_0) \approx 800$ cm². Let $\mathcal{E} = (A_0 \bar{m}/s_0 \rho_0)$, so that Equation (31) can be rewritten

$$\hat{c}_w^2 = \mathcal{E} \sqrt{(\omega^2 - \omega_w^2)^2 + 4(\gamma_w \omega)^2}\, e^{-i\phi_w}\ , \qquad (32)$$

where $\phi_w = \arctan\left(\dfrac{2\gamma_w \omega}{(\omega_w^2 - \omega^2)} \right)$.

Thus,
$$\hat{c}_w = \pm \sqrt{\mathcal{E}} \left[(\omega^2 - \omega_w^2)^2 + 4(\gamma_w \omega)^2 \right]^{\frac{1}{4}} e^{-i\frac{\phi_w}{2}}\ . \qquad (33)$$

$|\hat{c}_w|$ is plotted as a function of frequency in Figures 2 and 3. [The positive sign in Equation (33) has been taken in these figures.] In both figures, the thick dark line is the magnitude of $\hat{c}_w$, and the thin dark horizontal line shows the adiabatic speed of sound $c_0 = 34{,}100$ cm/s. In Figure 2, it can be seen that the magnitude of $\hat{c}_w$ grows with frequency and surpasses c_0 substantially for frequencies greater than 1500 Hz. The details of the behavior of $\hat{c}_w$ for frequencies below

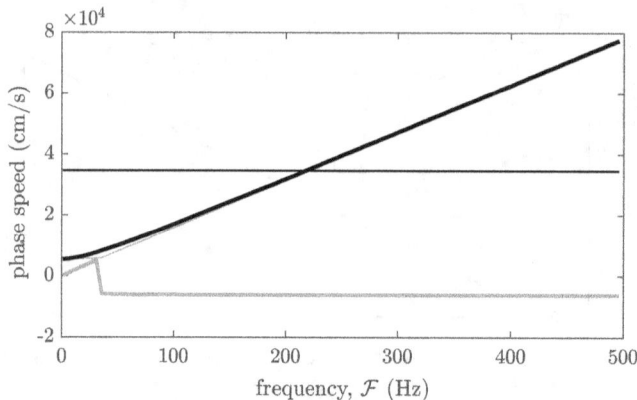

FIGURE 3. With wall vibration, magnitude of $\hat{c}_w$ (thick dark), real part of $\hat{c}_w$ (thin dark), and imaginary $\hat{c}_w$ (thick gray) as functions of frequency $\mathcal{F} = \omega/(2\pi)$. The adiabatic speed of sound c_0 is shown in the horizontal (thin dark) line

500 Hz are shown in Figure 3. The real part of $\hat{c}_w$ is represented by the thin dark line that blends with the magnitude of $\hat{c}_w$, except near $\mathcal{F} = 50$ Hz. The imaginary part of $\hat{c}_w$, is shown by the thick gray line. This quantity is positive from zero frequency until the wall vibration natural frequency is attained at $\mathcal{F}_w = 36$ Hz, where the imaginary part of $\hat{c}_w$ becomes negative.

For frequencies below about 400 Hz a substantial amount of energy is propagated through wall vibration, because the magnitude of $\hat{c}_w$ is less than, or comparable to, c_0. This is indicated by the relation among distensibilities $1/(\rho_0 \hat{c}^2) = 1/(\rho_0 c_0^2) + 1/(\rho_0 \hat{c}_w^2)$. With parallel paths for propagation, the majority of energy uses the least stiff path. But, as frequency increases, the wall gets stiffer and more of the energy is carried in the acoustic wave of the air contained within the tube.

It is clear that $\hat{c}$ depends on frequency, because $\hat{c}_w$ depends on frequency. When the phase speed of wave motion depends on frequency, the wave motion is called *dispersive wave motion*. By the definition of $\hat{c}$ in Equation (23)

$$\hat{c} = \pm \frac{c_0 \hat{c}_w}{\sqrt{c_0^2 + \hat{c}_w^2}} = \pm c_0 \frac{1}{\sqrt{1 + (c_0/\hat{c}_w)^2}} . \tag{34}$$

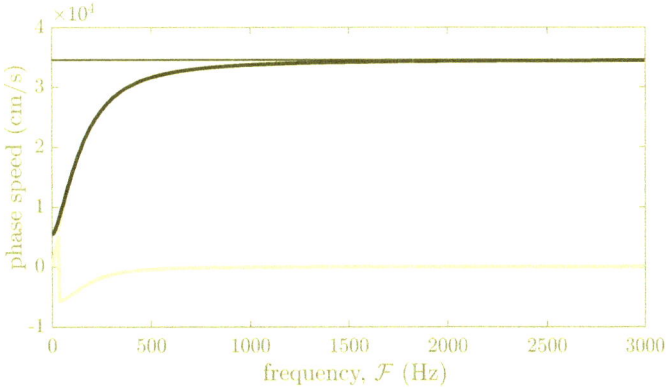

FIGURE 4. With wall vibration, magnitude of $\hat{c}$ (thick dark), real part of $\hat{c}$ (thin gray), and imaginary part of $\hat{c}$ (thick gray) as functions of frequency $\mathcal{F} = \omega/(2\pi)$. The adiabatic speed of sound c_0 is shown in the horizontal (thin dark) line

Figure 4 shows the phase speed $\hat{c}$ corresponding to the + sign in Equation (34) as a function of frequency in terms of its magnitude (thick solid line), real part (thin gray line), and imaginary part (thick gray line). The thin dark horizontal line shows the adiabatic speed of sound c_0. Wall vibration has a substantial effect on the phase speed below 1500 Hz, for the wall mechanical parameters used here. The reason that the effects of wall vibration spread from its resonance frequency of 36 Hz up to 1500 Hz is that the damping of the wall is large. The real part of $\hat{c}$ differs from the magnitude of $\hat{c}$ in the lowest frequencies, where the imaginary part makes a substantial contribution to the phase speed. At 500 Hz, there is still at least a 10% reduction in phase speed magnitude from the c_0 value. This has a substantial effect on *mode frequencies*, as will be seen as we consider wavenumbers after we examine damping due to *viscous friction* and heat conduction. [Recall that normal modes become modes in the presence of damping.]

Acoustic momentum boundary layer and acoustic thermal boundary layer damping

We add two other damping mechanisms that provide dissipative damping for the acoustic wave motion in the tube. When an acoustic wave interacts with a solid surface, there are two phenomena that cause dissipation of energy that are not included in the acoustic approximation that need to be taken into account. These phenomena are: 1) viscous friction that results in the production of rotational air motion near the solid surface, and 2) heat conduction between the air in acoustic motion and the solid surface, and the resulting production of entropy. In most instances for the vocal tract these phenomena occur in regions that are thin compared to the linear dimensions of the cross-section. The region close to the solid surface where rotational air motion is important is called the *acoustic momentum boundary layer* or the *Stokes layer*. The region where heat conduction produces excess entropy is known as the *acoustic thermal boundary layer*.

The acoustic approximation is not valid in either the acoustic momentum or thermal boundary layers. We treat them separately from the regions where the acoustic approximation is valid, and endeavor to find their effects on the acoustic motion outside the boundary layers. For both of these phenomena, we consider a plane wave traveling in a direction that is parallel to a solid surface. This is a good approximation, even when the surface is the interior of a tube, when the acoustic momentum and thermal boundary layers are thin in comparison to the linear dimensions of the cross-section. The circular frequency of the plane wave is ω.

With an acoustic wave $Pe^{i\omega(x/\hat{c}-t)}$ traveling along the surface of the solid plane, air particles move back and forth as the wave passes, as shown in Chapter 2. [The phase speed $\hat{c}$ can include the effects of wall vibration.] However, particles of air cannot move at the surface of the solid plane because of viscous friction, and this is known as the *no-slip condition*. This means that there must be a zone of transition, the acoustic momentum boundary layer, where the amplitude of the air particle velocity increases from zero at the planar surface to very close to the value it would have if the plane had not been present. [With the solid plane present, theory shows that the particle velocity never actually attains the value it would have without the plane, but that it rapidly approaches this value as the distance from the plane increases.]

This change in fluid particle velocity necessitates the formation of rotational air motion, or *vorticity*. While this vorticity oscillates at circular frequency ω, it is not governed by the wave equation, and the vast majority of the vorticity remains in the acoustic momentum boundary layer. If the acoustic wave were to cease, this rotational motion would quickly dissipate into heat.

The parameter for air friction is *kinematic viscosity*, ν. The thickness of the acoustic momentum boundary layer, δ_ν, is taken to be the distance from the plane where the velocity amplitude attains 86.5% of its value outside the boundary layer. It can be shown that (Howe 2007)

$$\delta_\nu = 2\sqrt{\frac{\nu}{\omega}}. \tag{35}$$

For air in normal atmospheric conditions, $\nu \approx 0.14$ cm^2/s (Batchelor 1967). At $\mathcal{F} = \omega/(2\pi) = 100$ Hz, $\delta_\nu = 0.03$ cm.

As the traveling wave passes there is compression and expansion of air at the surface of the solid plane, thus creating temperature fluctuations in the air in contact with the solid surface. In the acoustic approximation, these temperature fluctuations are adiabatic. However, it is not possible for these temperature fluctuations to be a part of an adiabatic process near the solid surface, because the solid is assumed to have a large heat capacity, and, thus, remain at approximately constant temperature. The temperature of the air adjacent to the plane will tend toward this temperature due to heat conduction, which means that its temperature variation near the surface is not adiabatic. Entropy is produced in this acoustic thermal boundary layer, so the acoustic approximation does not apply in this layer.

The parameters that gives the strength of heat conduction in air is *thermal diffusivity*, χ. The thickness of the acoustic thermal boundary layer, δ_χ, is taken to be

$$\delta_\chi \approx 2(\gamma - 1)\sqrt{\frac{\chi}{\omega}}, \tag{36}$$

where γ is the ratio of specific heats for air. For air at 15° C, $\chi \approx 0.20$ cm^2/s, and the ratio of specific heats for air is about 1.4 (Batchelor 1967). At $\mathcal{F} = \omega/(2\pi) = 100$ Hz, $\delta_\chi = 0.014$ cm.

Both the acoustic momentum and thermal boundary layers get thinner as frequency increases. Both the acoustic momentum and thermal boundary layers are quite thin, even at the relatively low frequency of $\mathcal{F} = 100$ Hz.

All of these considerations are for plane wave propagation over a plane, but, as stated above, we can imagine the considerations to apply to the inner walls of a tube, as long as the acoustic momentum and thermal boundary layers are thin compared to the linear dimensions of the cross-section. [Suppose that the plane has a finite width s_0, and imagine that it is rolled into a tube with cross-section A_0, so that the length of the perimeter of the cross-section is s_0. Here, we consider the case where the cross-sectional shapes are circular, but these considerations can be generalized to other shapes as well, with $r_0 = 2A_0/s_0 \gg \delta_\nu, \delta_\chi$. For a circular cross-section r_0 is the radius.]

As for wall vibration, some of the effects of the acoustic momentum and thermal boundary layers are instantiated in the phase speed $\hat{c}$. We quote the results from Howe (2007) without derivation. In order to include the effects of the acoustic momentum and thermal boundary layer damping into the phase speed, we replace the phase speed $\hat{c}$, which contains wall vibration effects with

$$\pm\hat{c} \to \pm\hat{c}\left[1 - \left(\frac{s_0}{2A_0\sqrt{\omega}}\right)\left(\sqrt{\nu} + \frac{\beta c_0^2}{c_p}\sqrt{\chi}\right)e^{i\frac{\pi}{4}}\right], \qquad (37)$$

where β is the coefficient of expansion at constant pressure and c_p is specific heat at constant pressure. This replacement of phase speeds is valid for $\delta_\nu, \delta_\chi \ll 2A_0/s_0$. For an ideal gas $\beta c_0^2/c_p = \gamma - 1$, where γ is the ratio for specific heats.

Figure 5 shows the $\hat{c}$ just for acoustic momentum and thermal boundary layer damping. That is, we take the $\hat{c}$ to the right of the arrow in Equation (37) to be c_0 for Figure 5. The effect of the acoustic momentum and thermal boundary layers on phase speed appears to be much less than the effect of wall vibration. In fact, Figure 6 shows that the acoustic viscous and thermal boundary layer effects are restricted to low frequencies. The reader is cautioned that the boundary layer theory breaks down at extremely low frequencies and for very narrow tubes. Figure 7 showing the effects of both wall vibration, the acoustic momentum boundary layer, and the thermal boundary layer differ very little from the effect of wall vibration alone in Figure 4.

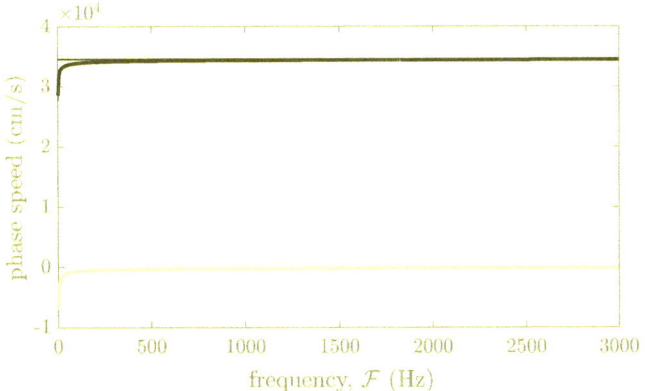

FIGURE 5. Acoustic momentum and thermal boundary layer effects on: magnitude of $\hat{c}$ (thick dark), real part of $\hat{c}$ (thin gray), and imaginary part of $\hat{c}$ (thick gray) functions of frequency $\mathcal{F} = \omega/(2\pi)$ up to 3000 Hz. The adiabatic speed of sound c_0 is shown in the horizontal (thin dark) line.

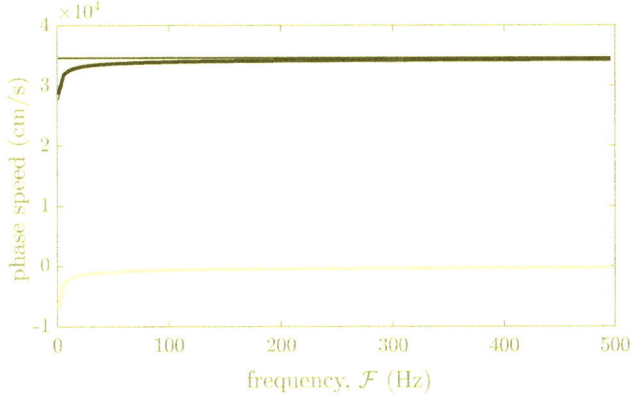

FIGURE 6. Acoustic momentum and thermal boundary layer effects on: magnitude of $\hat{c}$ (thick dark), real part of $\hat{c}$ (thin gray), and imaginary part of $\hat{c}$ (thick gray) functions of frequency $\mathcal{F} = \omega/(2\pi)$ up to 500 Hz. The adiabatic speed of sound c_0 is shown in the horizontal (thin dark) line.

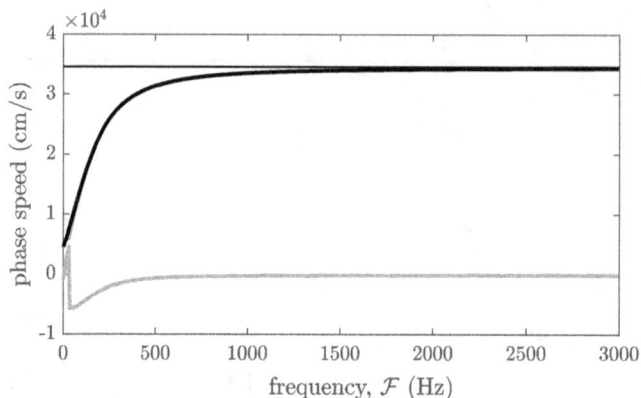

FIGURE 7. With wall vibration, and acoustic momentum and thermal boundary layer effects, magnitude of $\hat{c}$ (thick dark), real part of $\hat{c}$ (thin gray), and imaginary part of $\hat{c}$ (thick gray) functions of frequency $\mathcal{F} = \omega/(2\pi)$. The adiabatic speed of sound c_0 is shown in the horizontal (thin dark) line.

Lighthill shows, in the presence of the boundary layers and for $\delta_\nu, \delta_\chi \ll 2A_0/s_0$, that tube admittances Y should be replaced by

$$Y \to Y\left[1 - \left(\frac{s_0}{2A_0\sqrt{\omega}}\right)\left(\sqrt{\nu} + \frac{\beta c_0^2}{c_p}\sqrt{\chi}\right)e^{i\frac{\pi}{4}}\right]. \qquad (38)$$

in the absence of wall vibration. We need to investigate whether a similar modification can be made in the presence of wall vibration. This would change our calculations of mode frequencies in the presence of the damping mechanism, when we calculate effective admittances, as we did in Chapters 9 and 10. Alternatively, we investigate how this could be done by modifying the Green's function $\mathcal{G}_Q^p$ of Chapter 7 below. Before doing this, we examine the relation among damping mechanisms and their effect on wavenumbers and frequencies.

Damping mechanisms that affect phase speed

Wall vibration, viscous friction, and heat conduction all affect the phase speed of acoustic waves. Let's examine how the phase speed $\hat{c}$

affects wave propagation. For a wave traveling in the x-direction

$$p(x,t) = Pe^{i\omega(x/\hat{c}-t)} = Pe^{i(\hat{k}x-\omega t)} , \qquad (39)$$

where wavenumber $\hat{k} = \omega/\hat{c}$, but now $\hat{c} = \hat{c}(\omega)$, and $\hat{c}$ is complex. Thus,

$$\hat{k} = Re(\hat{k}) + iIm(\hat{k}) \equiv \frac{\omega}{\hat{c}} = \frac{\omega \hat{c}^*}{\hat{c}\hat{c}^*} = \frac{\omega}{|\hat{c}|^2}\hat{c}^* . \qquad (40)$$

So that

$$Re(\hat{k}) = \frac{\omega}{|\hat{c}|^2}Re(\hat{c}) \text{ and } Im(\hat{k}) = -\frac{\omega}{|\hat{c}|^2}Im(\hat{c}) . \qquad (41)$$

If we take the $+$ sign in Equation (37), then Figure 7 shows that $Re(\hat{c}) > 0$ and $Im(\hat{c}) < 0$ for frequencies greater than about 36 Hz. From Equation (41), this means that $Re(\hat{k}) > 0$ and $Im(\hat{k}) > 0$ for frequencies greater than 36 Hz. The traveling wave in Equation (39) is written

$$p(x,t) = Pe^{-Im(\hat{k})x}e^{i(Re(\hat{k})x-\omega t)} . \qquad (42)$$

which describes a damped wave traveling in the rightward direction for $x > 0$. If we choose the $-$ sign in Equation (37), then we get a wave traveling to the left that is damped for $x < 0$. In either case, $|Re(\hat{c})| < c_0$ means that the wavelength of waves are longer in the presence of wall vibration, viscous friction, and heat conduction.

From the definition of wavenumber $\hat{k}$ in Equation (40)

$$\omega^2 = |\hat{k}|^2|\hat{c}|^2 = \left(Re(\hat{k})^2 + Im(\hat{k})^2\right)|\hat{c}|^2 . \qquad (43)$$

Define

$$\hat{\omega} \equiv Re(\hat{k})|\hat{c}| \text{ and } \gamma' \equiv Im(\hat{k})|\hat{c}| . \qquad (44)$$

Equation (43) can be written

$$\omega^2 = \hat{\omega}^2 + \gamma'^2 \text{ or } \hat{\omega} = \sqrt{\omega^2 - \gamma'^2} . \qquad (45)$$

$\hat{\omega}$ is the *reduced circular frequency* and γ' is the damping constant. [The prime is used to denote that it includes only wall vibration, viscous friction, and thermal conduction damping.]

When we use the methods of Chapter 10 to find the eigenvalues for wavenumber without damping we are finding normal mode wavenumbers k_m and the corresponding $\omega_m = c_0 k_m$. With the damping mechanisms that affect phase sped, eigenvalues k_m are modified to $\hat{k}_m$, so that mode wavenumber $Re(\hat{k}_m)$ replaces k_m, and

$\hat{\omega}_m$ replaces ω_m, where
$$\hat{\omega}_m = \sqrt{\omega_m^2 - \gamma_m'^2} \ . \tag{46}$$
These modified circular frequencies are reduced normal mode circular frequencies.

Damping mechanisms that work locally

We have encountered two damping mechanisms that work locally, that is, at particular places in the vocal tract. The two that we have considered are radiation at the opening and jetting.[3]

radiation damping

We have radiation from the open end of the finite-length tube as a damping mechanism, as we derived in the first part of this chapter. We also know from Equation (23) of Chapter 10, that the open end provides a lumped mass element. This is equivalent to lengthening the tube, and Equation (24) of Chapter 10 gives an approximation to the length that should be added to the length of the tube L. When there is a flanged circular opening of radius a this extra length is
$$\ell_{rad} = \frac{8}{3\pi} a \ . \tag{47}$$
Let normal mode wavenumber for a tube of length L that accounts for the additional length in Equation (47) is
$$k_m^{rad} = \frac{k_m}{1 + \ell_{rad}/L} = \frac{k_m}{1 + \frac{8a}{3\pi L}} \ , \tag{48}$$
where k_m is the normal mode wavenumber in a tube of length L without accounting for the extra length ℓ_{rad} Equation (48) would mean that normal mode circular frequencies are perturbed from ω_m to approximately
$$\omega_m^{rad} = \frac{\omega_m}{1 + \ell_{rad}/L} = \frac{\omega_m}{1 + \frac{8a}{3\pi L}} \ . \tag{49}$$

Consider the damping caused by radiation and the change in wavenumber due to radiation. Neither phenomenon is the result of

---

[3]These mechanisms may, in fact, be considered to be localized sinks of acoustic energy.

changes in phase speed inside the tube. Rather, radiation damping, corresponding to the real part of the radiation impedance, and lumped mass elements at the opening, corresponding to the imaginary part of the radiation impedance, are boundary phenomena. As a result they are localized in a plane perpendicular to the direction of propagation.

Strictly speaking, we should add a traveling wave propagating to the right to the standing waves to account for the radiation of energy from the flanged opening. However, in steady conditions, we can approximate this behavior as a continuous damping of the standing wave, with damping constants γ_m^{rad}. An estimate can be provided for these damping constants for small γ_m^{rad}/ω_m. Howe (2007) follows a method devised by Rayleigh to show that the fraction of incident power lost at the opening by a wave traveling right with wavenumber k is approximately $(ka)^2/2$, for $ka \ll 1$, where we have modified the calculation to account for the flanged opening. With this result, we have the damping constants due to radiation

$$\gamma_m^{rad} = \omega_m \left(\frac{k_m a}{\sqrt{2}} \right). \tag{50}$$

jetting

We have not quantified the damping due to jetting, as this depends on the specific geometry of the constriction and sharp edge. On the other hand, if the lumped mass elements between sub-tubes are accounted for when the calculations of eigenvalues and eigenfunctions are done, then this should account for making mode frequencies less than normal mode frequencies due to jetting (Luong, Howe, & McGowan 2005).

Generalities and comparisons of mode frequency reductions that accompany damping

There is a property that all the the identified damping mechanisms possess: with the damping comes a reduction of normal mode frequency values for mode frequency values. The reason for this is shared by all damping mechanisms is that the acoustics of a finite-length tube constitutes a stable, causal system. These damping mechanism can be represented either as impedances, like radiation, or by complex phase speeds, for other damping mechanisms. For stable causal systems, the real and imaginary parts of the impedances and phase speeds must be *Hilbert transform* pairs. These pairs of real and imaginary parts are said to be related by the *Kramer-Kronig relations* (Howe 1998), or by the *Plemelj relations* (Levine 1978). All of the damping mechanisms here satisfy these relations, and, thus, lower mode frequencies couple with damping.

In service to the final section on modifying the Green's function, we estimate the relative sizes of normal mode frequency changes due to each damping mechanism. We exclude jetting from this, but note that part of the modification of mode frequencies due to jetting are accounted for with the lumped mass elements according to the methods of Chapter 10. In all of the following we assume that $\gamma'_m/\omega_m \ll 1$ and $\gamma_m^{rad}/\omega_m \ll 1$. Under this assumption, Equation (46) for wall vibration, viscous friction, and heat conduction can be approximated

$$\hat{\omega}_m = \sqrt{\omega_m^2 - \gamma_m'^2} \approx \omega_m \left(1 - \frac{\gamma_m'^2}{2\omega_m^2}\right). \tag{51}$$

In the case of radiation damping, if we further have $a/L \ll 1$, then from Equations (49) and (50) we obtain

$$\omega_m^{rad} = \frac{\omega_m}{1 + \frac{8a}{3\pi L}} \approx \omega_m \left(1 - \frac{8a}{3\pi L}\right)$$

$$= \omega_m \left(1 - \left(\frac{8\sqrt{2}a}{3\pi k_m L}\right) \left(\frac{\gamma_m^{rad}}{\omega_m}\right)\right) \tag{52}$$

Comparing Equations (51) and (52) we see that the damping mechanisms that affect phase speed also change the normal mode frequencies as the square of the ratio (γ'_m/ω_m), $(\gamma'_m/\omega_m)^2$, whereas the radiation damping is associated with normal mode frequencies changes that vary simply as the first power of $(\gamma_m^{rad}/\omega_m)$.

Modifying the Green's functions

We amend the Green's function evaluated at $y = 0$ for the voice source in Equation (13) of Chapter 7 employing the knowledge that has been accumulated over the past few chapters. Strictly speaking these steps are valid only for the tube of constant cross-sectional area. Similar considerations should apply when Green's functions for variable cross-sectional areas without damping are derived.

These are the steps that should be followed to use the Green's function $\mathcal{G}_Q^p$ in Equation (13) of Chapter 7. Other Green's function that include damping mechanisms also follow these steps.

1) Normal mode wavenumbers and frequencies need to be modified to take account of the added mass at the tube opening. Equation (48) can be used to account for the lumped mass element at the open end. That is, replace ω_m with ω_m^{rad} and k_m with [We have been using the specific flanged opening in this book. Others are possible.]

2) If the damping that occurs via propagation, including wall vibration and boundary layer damping, are to be included, modify the normal mode circular frequencies ω_m^{rad} and wavenumbers k_m^{rad} obtained in step 1), according to Equation(46). ω_m^{rad} replaces ω_m in Equation (46) and wavenumbers $Re(\hat{k}_m) = \hat{k}_m$, which has replaced k_m^{rad} in step 1). [With empirically determined γ_m, we cannot perform this operation, unless we are willing to estimate the part of γ_m due to radiation and subtract these off to obtain γ'_m.]

3) Use the total damping constant, $\gamma_m = \gamma_m^{rad} + \gamma'_m$, where they appear in Equation (13) of Chapter 7.

4) The functions $p_m^{space}(x)$ will be given in step 1). We have not examined how to calculate the functions $(\gamma_m/\omega_m)\bar{p}^{space}(x)$. This would involve following Lighthill's suggestion of modifying the tube admittances as in Equation (38), but including wall vibration.

Conclusion

We have reviewed ways that acoustic motion in the tube is damped: loss of energy via radiation into the atmosphere, as well as damping due to jetting, wall vibration, and acoustic momentum and thermal boundary layers. The damping caused by wall vibration and acoustic boundary layers are most important for frequencies less than 1000 Hz. On the other hand, the damping caused by radiation increases with frequency, because the real part of radiation impedance increases with frequency, as seen in Equation (15).

Interestingly, each of these damping mechanism is accompanied by phenomena that either tends to lengthen the effective length of the tube in the case of radiation, or that decreases the real part of the phase speed below the adiabatic speed of sound, in the case of the other mechanisms. So accompanying the loss of energy in acoustic wave motion, there are increases in the real part of the wavenumber, and, hence, longer wavelengths for a given frequency. Mode frequencies are less than normal mode frequencies as a result.

References

Batchelor, G.K. (1967). *An Introduction to Fluid Dynamics.* Cambridge University Press, Cambridge, England. (p 594).

Bechert, D.W. (1980). Sound absorption caused by vorticity shedding, demonstrated with a jet flow. *Journal of Sound and Vibration*, **70**. p 389.

Davies, P.O.A.L., McGowan, R.S., & Shadle, C.H. (1993). Practical flow duct acoustics applied to the vocal tract. In (I.R. Titze, ed.) *Vocal Fold Physiology: Frontiers in Basic Science.* Singular Publishing Group, Inc., San Diego, California. (pp 93-142).

Fant, G. (1960). *The Acoustic Theory of Speech Production.* Mouton, The Hague. (p 35).

Howe, M.S. (1980). The dissipation of sound at an edge. *Journal of Sound and Vibration*, **70**. p 407.

Howe, M.S. (1998). *Acoustics of Fluid-Structure Interactions.* Cambridge University Press, Cambridge, England. (pp 360-65, 376-79).

Howe, M.S. (2007). *Hydrodynamics and Sound.* Cambridge University Press, Cambridge, England. (pp 421-3, pp 446-50).

Ingard, K.U. & Ising, H. (1967). Acoustic nonlinearity of an orifice. *The Journal of the Acoustical Society of America*, **42**. p 6.

Ishizaka, K., French, J.C., & Flanagan, J.L. (1975). Direct determination of vocal tract wall impedance. *IEEE Transactions on Acoustics, Speech, and Signal Processing*, **ASSP-23**. p 370.

Levine, H. (1978). *Unidirectional Wave Motions*. North-Holland Publishing Company, Amsterdam. (pp 308-13).

Lighthill, J. (1978). *Waves in Fluids*. Cambridge University Press, Cambridge, England. (Chapter 2).

Luong, T., Howe, M.S., & McGowan, R.S. (2005). On the Rayleigh conductivity of a bias-flow aperture. *Journal of Fluids and Structures*, **21**. p 769.

Morse, P.M. (1976). *Vibration and Sound*. Acoustical Society of America: Melville, NY. (p 333).

Pierce, A.D. (1989). *Acoustics: An Introduction to Its Physical Principles and Applications*. Acoustical Society of America, Woodbury, NY. (pp 220-25).

Rayleigh, J.W.S. (1945). *Theory of Sound: Volume II*. Dover Publications, New York. (Articles 302, 312).

Chapter 12: Helmholtz Resonators and Side Branches

Introduction

In this chapter we discuss two topics on acoustic wave propagation that are relevant to speech production. The first topic is that of the Helmholtz resonator, and the second is that of acoustic *side branches*.

The acoustic situation that is most similar to the mass-spring system is the Helmholtz resonator. We have already discussed this kind of resonator in Chapter 1, although some important details are missing in that discussion. In Chapter 9 we encountered a two-tube system where, for the first normal mode, the kinetic energy is concentrated in the front narrow tube, and the potential energy is concentrated in the rear wide tube shown in Figure 12 of Chapter 9. For sufficiently low frequencies, this can be approximated by a combination of a lumped mass element and a *lumped spring element*, which are the components of a Helmholtz resonator. The example of [ɹ] production completes the discussion of Helmholtz resonators.

The second part of the chapter explores the consequences of including side branches in the vocal tract. Side branches are present during the production of nasal and laterals. The piriform sinuses near the larynx also provide side branches.

Helmholtz resonators

Lumped mass elements

We have seen lumped mass elements at the junctions between sub-tubes and the junction between a sub-tube and the atmosphere. Let's consider a sub-tube with cross-sectional area A_c and length l_c. We suppose that this sub-tube is connected with sub-tubes with much larger cross-sectional areas on both of its ends. A schematic area function for a lumped mass element is shown in Figure 1. Of course, the atmosphere is always a wider sub-tube. We need to be careful if the wide neighboring sub-tubes are themselves connected

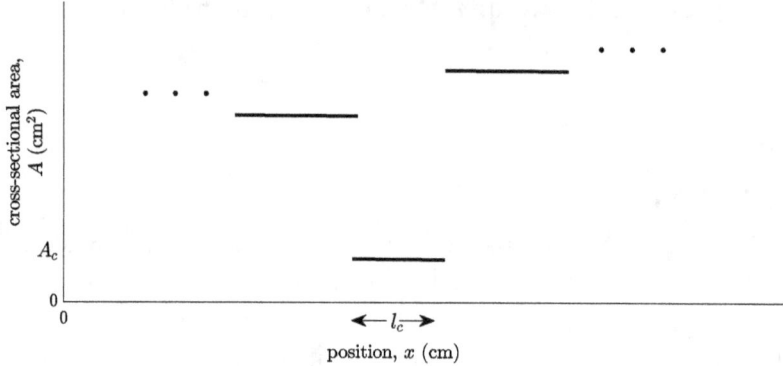

FIGURE 1. Schematic area function of lumped mass element sub-tube

to even wider sub-tubes. In this case we may need to include the neighboring sub-tube or sub-tubes as part of the lumped mass element, and the length of the constriction tube is l_c = total length of all the contiguous narrow sub-tubes, while we take A_c to be an average of the cross-sectional areas of all the narrow sub-tubes. The other requirement for a lumped mass element is that the length of the narrow sub-tube be much shorter than the shortest wavelength of sound under consideration. This is expressed as

$$kl_c \ll 1 \quad \text{or} \quad l_c \ll \frac{\lambda}{2\pi}, \quad \text{because} \quad k = \frac{2\pi}{\lambda}, \tag{1}$$

where $\lambda = Re(\hat{c})/\mathcal{F} = (2\pi Re(\hat{c})/\omega$ is the shortest wavelength of interest. The condition expressed in Equation (1) is most likely to be met at frequencies of the lowest order modes in speech.

Because the ratios of cross-sectional areas of the neighboring wide sub-tubes to the cross-sectional area of the constriction tube, the particle velocities in the wide tubes are negligible compared to those of the narrow tube by continuity of volume velocity. In this case, the particle velocity is substantial only in the constriction tube. Therefore, kinetic energy density is concentrated in this tube. Further, because the length of the constriction tube is short compared to the acoustic wavelength, the air in the tube can be considered to move in-phase, like a single lumped mass.

The lumped mass element in the constriction tube has a velocity $u = \hat{U}e^{-i\omega t}$ and an acceleration $a = -i\omega \hat{U}e^{-i\omega t}$. The volume velocity is $Q = \hat{Q}e^{-i\omega t}$, with $\hat{Q} = \hat{U}A_c$, where A_c is the cross-sectional area of the constriction tube. In order for the air in the constriction tube to oscillate at circular frequency ω, an oscillatory pressure difference must be applied across the ends of the tube, $\Delta p = \Delta \hat{P}e^{-i\omega t}$. The mass of air in the constriction tube is $m_c = \rho_0 A_c l_c$. The net force applied to the mass of air in the constriction tube is $\Delta p A_c = \hat{F}e^{-i\omega t} = \Delta \hat{P} A_c e^{-i\omega t}$. Newton's second law expressed in Equation (5) of Chapter 1 states that force equals mass times acceleration

$$\hat{F}e^{-i\omega t} = m_c(-i\omega)\hat{U}e^{-i\omega t} ,$$

or

$$\Delta \hat{P} A_c e^{-i\omega t} = (\rho_0 l_c A_c)(-i\omega)\hat{U}e^{-i\omega t} , \qquad (2)$$

or

$$\Delta \hat{P} A_c e^{-i\omega t} = -(\rho_0 l_c)(i\omega)\hat{Q}e^{-i\omega t} .$$

Solving Equation (2) for $\Delta \hat{P}$ gives

$$\Delta \hat{P} = -i\omega \frac{\rho_0 l_c}{A_c} \hat{Q} . \qquad (3)$$

Considerations of Chapter 9 imply that the length l_c needs to be supplemented by lengths ℓ^- and ℓ^+ at each of its ends. Including these lumped mass element effects produces

$$\Delta \hat{P} = -i\omega \frac{\rho_0 l'_c}{A_c} \hat{Q} , \qquad (4)$$

where $l'_c = l_c + \ell^- + \ell^+$.

Lumped spring elements

Another type of lumped element is a lumped spring element. Groups of contiguous sub-tubes that form a lumped spring element are associated with regions whose acoustic energy content can be approximated to be exclusively potential energy. Suppose that a group of contiguous sub-tubes has average cross-sectional area A_V and length l_V, that their neighboring sub-tubes are have much smaller cross-sectional areas compared to A_V. We require that l_V satisfies the

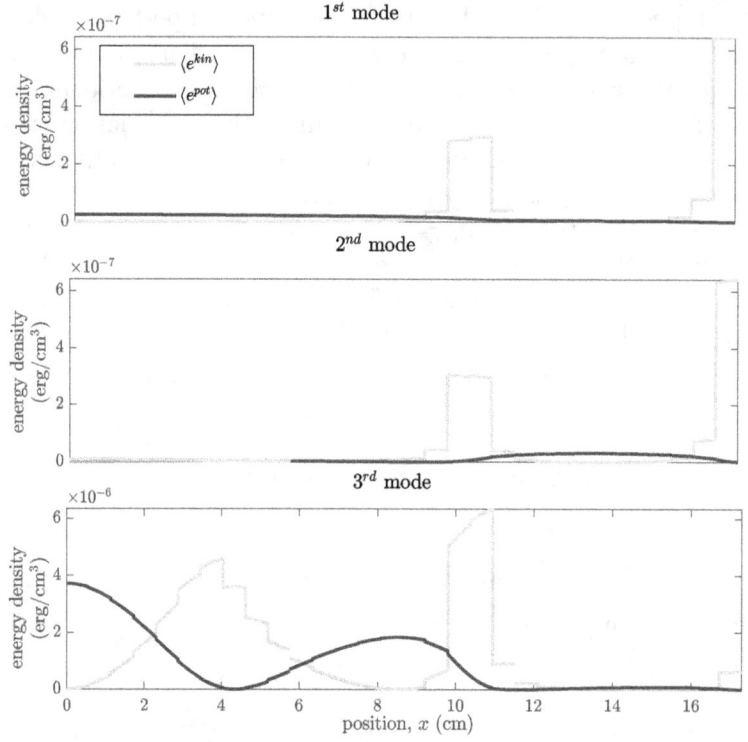

FIGURE 2. Schematic area function of lumped spring element sub-tube

same relation as l_c does in Equation (1). We call this kind of element made of wide sub-tubes an expansion tube, and it acts as a lumped spring element.

Consider sinusoidal motion at circular frequency ω. Because we are considering only one-dimensional motion in the x-direction and the expansion tube of average cross-sectional area A_V is a lumped element, the oscillating pressure is approximately in-phase throughout in this expansion and can be written as $p = \hat{P}e^{-i\omega t}$. That is, the pressure is the approximately uniform in the expansion tube at any time. An oscillating pressure means that there is an oscillating density, and according to Equation (9) of Chapter 1, $\rho = \left(\hat{P}/c_0^2\right)e^{-i\omega t}$. There is a change in volume velocity across the length the the sub-tube necessitated by mass conservation expressed in Equation (5) of Chapter

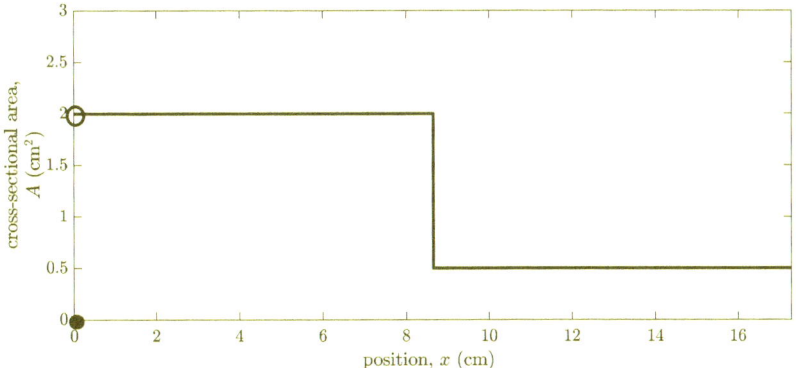

FIGURE 3. A two sub-tube area function

2. Let the change in volume velocity across the sub-tube be denoted $\Delta Q = \Delta \hat{Q} e^{-i\omega t}$. Mass conservation gives

$$\frac{\Delta_t \rho}{\Delta t} = -\frac{\rho_0}{A_V} \frac{\Delta_x Q}{\Delta x} ,$$

or

$$\frac{-i\omega}{c_0^2} \hat{P} e^{-i\omega t} = -\frac{\rho_0}{A_V} \frac{\Delta \hat{Q}}{l_V} e^{-i\omega t} , \qquad (5)$$

or

$$\Delta \hat{Q} = \frac{i\omega V_0}{\rho_0 c_0^2} \hat{P} ,$$

where $V_0 = A_V l_V$ is the volume of the wide sub-tube.

The single Helmholtz resonator

Figure 3 shows the area function for a two-tube configuration. The closed circle at $x = 0$ indicates that the area is zero there, and so there is hard wall at $x = 0$. It is seen that the right sub-tube in Figure 3 could be a lumped mass element and the left sub-tube in Figure 3 could be a lumped spring element at sufficiently low frequencies. In particular, this seems to be the case for the first normal mode frequency, with the right sub-tube associated with kinetic energy density and the left sub-tube with potential energy density, as shown in Figure 12 of Chapter 9.

This configuration was termed a Helmholtz resonator in Chapter 1, and its first normal mode frequency can be estimated assuming that the left sub-tube corresponds to a spring and the right sub-tube to a mass in a mass-spring system.

The natural frequency for this Helmholtz resonator $\mathcal{F}_H$ was derived in Equation (21) of Chapter 1. We use $V_0 = l^{(2)} A^{(2)}$, $A_c = A^{(1)}$, and $l'_c = l^{(1)'}$ for

$$\mathcal{F}_H = \frac{c_0}{2\pi} \sqrt{\frac{A_c}{l'_c V_0}} = \frac{c_0}{2\pi} \sqrt{\left(\frac{A^{(1)}}{l^{(1)'}}\right)\left(\frac{1}{A^{(2)} l^{(2)}}\right)}. \qquad (6)$$

where the $A^{(1)}$ is the cross-sectional area of the front sub-tube, $A^{(2)}$ is the cross-sectional area of the left sub-tube, and $l^{(2)}$ is the left sub-tube length. $l^{(1)'}$ is the length of the right tube with the lengths of lumped mass elements appended at both ends, just as after Equation (4) of the present chapter. If, however, we wish a crude approximation, we use $l^{(1)'} \approx l^{(1)}$. Importantly, the Helmholtz resonator here approximates the first normal mode, and $\mathcal{F}_H \approx \mathcal{F}_1$.

We can simulate an approximate Helmholtz resonator with two sub-tubes of the acoustic tube. Let $A^{(2)} = 2.0$ cm^2, $A^{(1)} = 0.5$ cm^2, and $l^{(2)} = l^{(1)} = L/2 = 8.55$ cm in Figure 3. The Helmholtz resonator natural frequency from Equation (6) is, $\mathcal{F}_H = 318$ Hz, ignoring the added lengths due to the added lumped mass elements between sub-tubes. Normal mode frequencies can also be calculated using the methods of Chapter 10, with the added lumped masses elements at the ends of the sub-tube on the right. The first normal mode frequency from this calculation is $\mathcal{F}_1 = 288$ Hz, which can be compared to $\mathcal{F}_H = 318$ Hz. The former value is more accurate because 1) it does not assume lumped mass and spring elements, in the right and left sub-tubes, respectively, 2) and it accounts for the lumped mass elements at the junction and at the opening. However, Equation (6) makes it clear how to change the natural frequency $\mathcal{F}_1 \approx \mathcal{F}_H$ of a Helmholtz resonator.

Figure 4 shows the kinetic energy densities and potential energy densities as a function of position along the tube axis at each of the normal mode frequencies. The top plot corresponds to $\mathcal{F}_1$, the

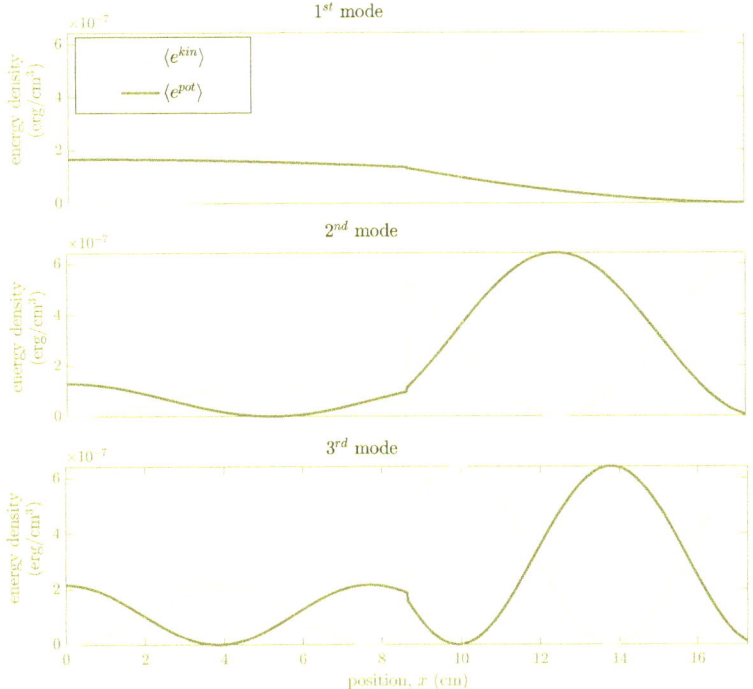

FIGURE 4. Energy densities for the area function in Figure 3

middle plot to $\mathcal{F}_2$, and bottom plot to $\mathcal{F}_3$. [There is no source, so the numerical values of energy density have no meaning. The shapes of the curves in the plots are what are important.] For $\mathcal{F}_1$ potential energy dominates in the rear expansion tube, and kinetic energy dominates in the front constriction tube. Further, there appears to be little change in the levels of these energy densities in each of the sub-tubes where they dominate. This means that the first normal mode can be approximated with a Helmholtz resonator. The plots for the energy densities corresponding $\mathcal{F}_2$ and $\mathcal{F}_3$ show patterns of distributed potential and kinetic energy densities throughout the two sub-tubes.

Figure 5 exhibits a configuration with a narrow left sub-tube, an expansion in the middle sub-tube, and a narrow right sub-tube. The normal mode frequencies calculated according to the methods of Chapter 10 are $\mathcal{F}_1 = 379$ Hz, $\mathcal{F}_2 = 1474$ Hz, and $\mathcal{F}_3 = 2508$ Hz. Figure 6 shows the kinetic (black lines) and potential energy (gray lines)

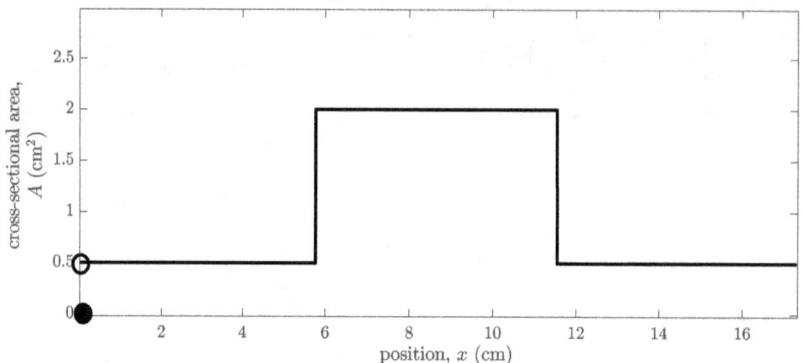

FIGURE 5. Two narrow tubes neighboring a wide tube

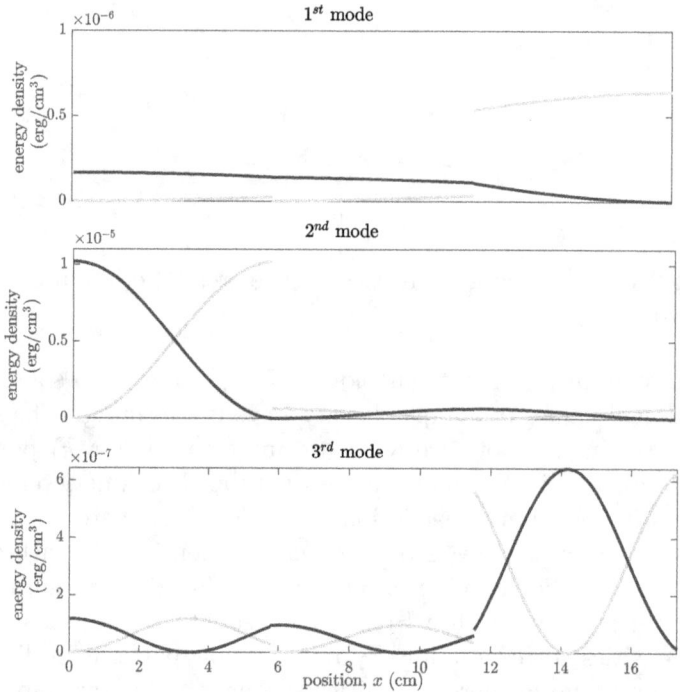

FIGURE 6. Energy densities for the area function shown in Figure 5

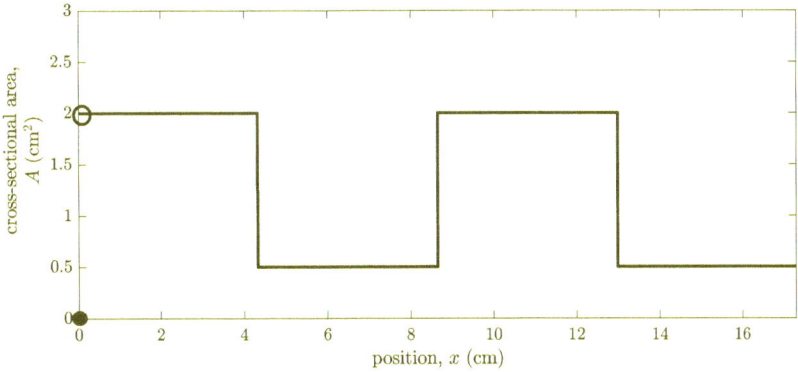

FIGURE 7. Area function for a double Helmholtz resonator

densities for the first three normal mode frequencies, top to bottom. According to the top plot of Figure 6, $\mathcal{F}_1 = 379$ Hz corresponds to an approximate Helmholtz resonator, with the lumped spring element corresponding to the middle, expanded sub-tube and the rear left sub-tube. The right sub-tube provides the lumped mass element for the Helmholtz resonator. Again, the plots of the kinetic and potential energy densities corresponding to $\mathcal{F}_2$ and $\mathcal{F}_3$ indicate that these normal mode frequencies are not those of a lumped system. Perhaps it is surprising that the narrow left sub-tube is a part of the volume of the lumped spring element. This has to do with the hard-wall boundary condition at the left end of that sub-tube. In order for this sub-tube to act as a lumped mass element it would need to open into wide sub-tubes at both ends. Thus, if the hard-wall boundary condition were to change, we would need to reconsider the nature of the left sub-tube. In the next section we complicate matters by coupling two Helmholtz resonators.

The double Helmholtz resonator

Figure 7 shows an area function that contains, from left to right, an expansion, a narrowing, another expansion, and another narrowing at the right. It turns out that this configuration provides two normal modes that can be approximated by coupled Helmholtz resonators.

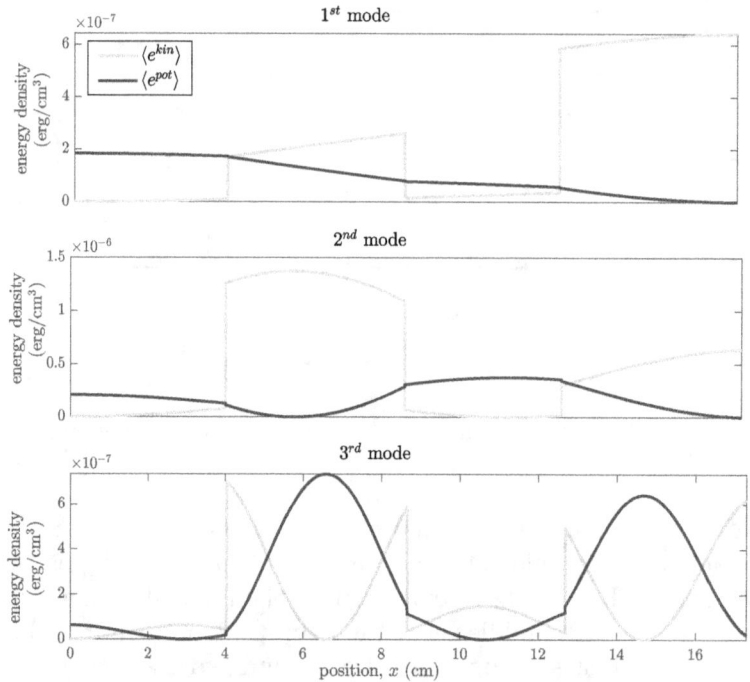

FIGURE 8. Energy densities for the area function in Figure 7

The normal mode frequencies are $\mathcal{F}_1 = 332$ Hz, $\mathcal{F}_2 = 910$ Hz, and $\mathcal{F}_3 = 2936$ Hz, as calculated by the methods of Chapter 10.

The plots of energy densities are shown in Figure 8. Both the plots for the mode 1 and mode 2 show that the potential energy densities dominates the kinetic energy densities in the wide sub-tubes, and the opposite dominance occurs in the narrow sub-tubes. The effect is not large for the left-most constriction and right-most expansion for mode 1. However, the patterns of dominance mean that the first two normal modes can be approximated by two lumped spring elements and two lumped mass elements that are coupled.

Let's explore this using facts that we know about the uncoupled Helmholtz resonators. The combination of the left expansion and the left narrowing constitutes one Helmholtz resonator, and the combination of the right expansion and the right narrowing, constitutes another Helmholtz resonator. We have made these resonators identical in this example, and if we use Equation (14), neglecting added lumped masses at the ends of constriction tubes, to estimate the natural frequency of just one of these resonators, we obtain, $\mathcal{F}_H = 637$ Hz for each resonator.

Because the two Helmholtz resonators are coupled, they produce two natural frequencies. The lowest normal mode, corresponding to $\mathcal{F}_1 = 332$ Hz, is the result of the two lumped mass elements in the two constriction tubes moving closely together in time: approximately in-phase. That is, when the mass in the left constriction is moving rightward, the mass in the right constriction is also moving rightward, most of the time. With the lumped mass elements moving in phase there is not much spring compression and expansion supplied by the right expansion. The next normal mode, corresponding to $\mathcal{F}_2 = 910$ Hz, is the result of the two lumped mass elements moving nearly completely out-of-phase. When the mass in the left constriction tube is moving rightward, the mass in the right constriction tube is moving leftward, for the most part. This, in turn, means that there is maximal change in volume in the right expansion tube, so that it is providing more spring compression and expansion than when the masses in the constriction tubes are moving in-phase. With the added compression and expansion and no more mass, the system oscillates at a higher frequency.

We use the theory of coupled oscillators to calculated the natural frequencies of coupled Helmholtz resonators, given our single Helmholtz resonator natural frequency of $\mathcal{F}_H = 637$ Hz. The two natural frequencies for the coupled Helmholtz resonators are: $\mathcal{F}_{H1} \approx 0.62 \mathcal{F}_H = 395$ Hz, and $\mathcal{F}_{H2} \approx 1.62 \mathcal{F}_H = 1032$ Hz. There are two reasons for the discrepancy between these two natural frequencies and the frequencies for the first two normal modes, with $\mathcal{F}_1 = 332$ Hz and $\mathcal{F}_2 = 910$ Hz. Firstly, lumped elements provide an approximation to the acoustic situation, and second we have not used added lumped mass elements at the ends of the constriction tubes in the coupled Helmholtz resonator calculation.

[ɹ] production

We end the topic on Helmholtz resonators with application to the production of English [ɹ]. Figure 9 shows a stylized area function that could be one that is used to produce an [ɹ] with a lip constriction and

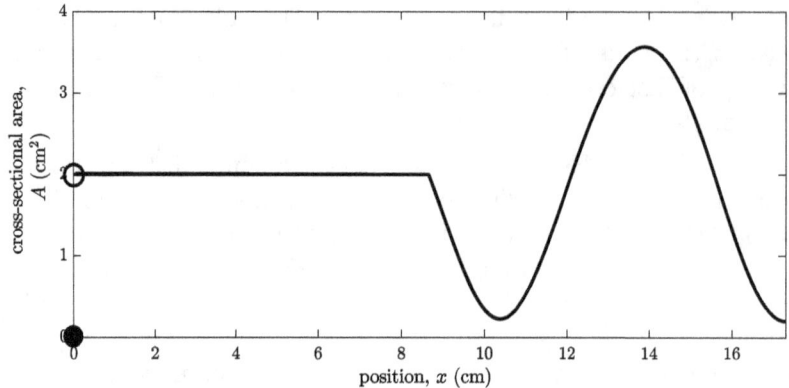

FIGURE 9. An area function for [ɹ] production

a palatal constriction. The normal mode frequencies corresponding to this area function are computed to be $\mathcal{F}_1 = 316$ Hz, $\mathcal{F}_2 = 910$ Hz, and $\mathcal{F}_3 = 1975$ Hz. All of these are substantially lower than their values for a finite-length tube of constant cross-sectional area, which are 500 Hz, 1500 Hz, and 2500 Hz, respectively, without accounting for the lumped mass element at the opening. The energy densities corresponding to these three frequencies are shown in Figure 10. For $\mathcal{F}_1$ and $\mathcal{F}_2$ the regions near the constrictions behave as lumped mass elements. Although it is difficult to see in the plots, the expansions are dominated by potential energy density. For $\mathcal{F}_1$ the potential energy density is greatest in the rear expansion, and for $\mathcal{F}_2$ the potential energy density is greatest in the front expansion. These are the patterns seen for the double Helmholtz resonator in Figure 8. The description of the way that lumped mass elements move with with respect one another for the first two normal modes is the same as that given for the configuration in Figure 7. Thus, Figure 9 can be approximated as a double Helmholtz resonator for the first two modes.

The [ɹ] in prevocalic position in a stressed syllable can be particularly strong, with a very low $\mathcal{F}_3$. Consider the energy density plot in the bottom panel of Figure 10, which corresponds to $\mathcal{F}_3$. According to the acoustic perturbation theory in Chapter 6, if we expand the tube from about $x = 7$ cm to about $x = 10$ cm along the tube axis, and narrow the tube from about $x = 3$ cm to about $x = 7$ cm, $\mathcal{F}_3$ should decrease. In terms of the vocal tract, we are describing an expansion

Acoustics of Speech Production

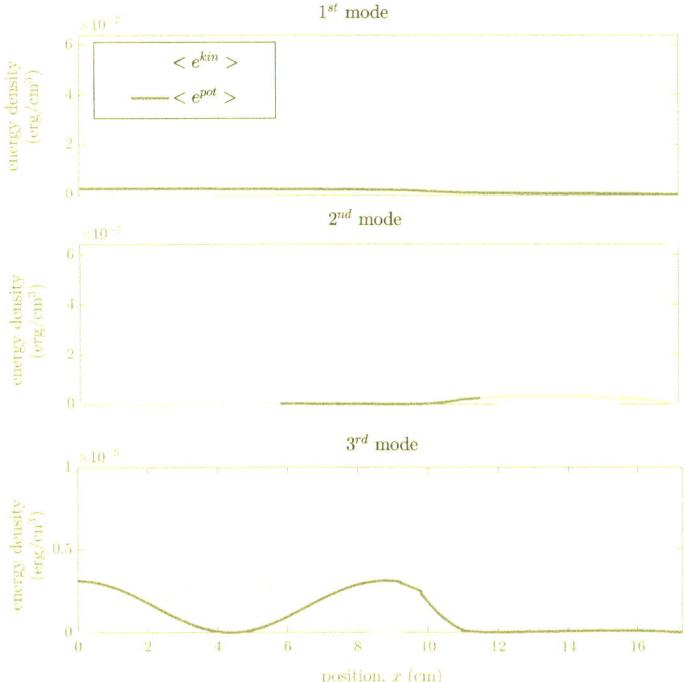

FIGURE 10. Energy densities the area function in Figure 9

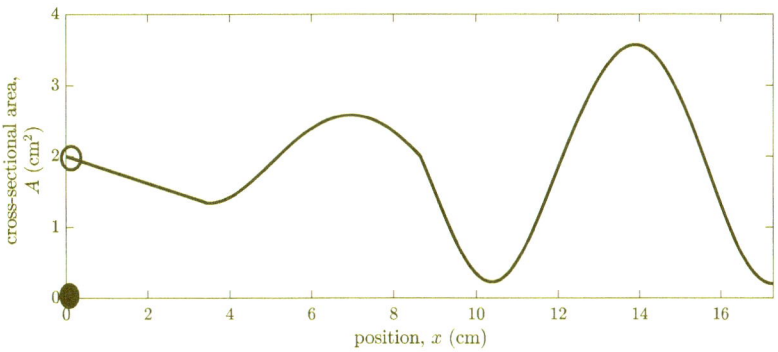

FIGURE 11. An area function for strong [ɹ] production

in the uvular and upper pharyngeal region, and a narrowing just below this region as shown in Figure 11. There should not be be much effect

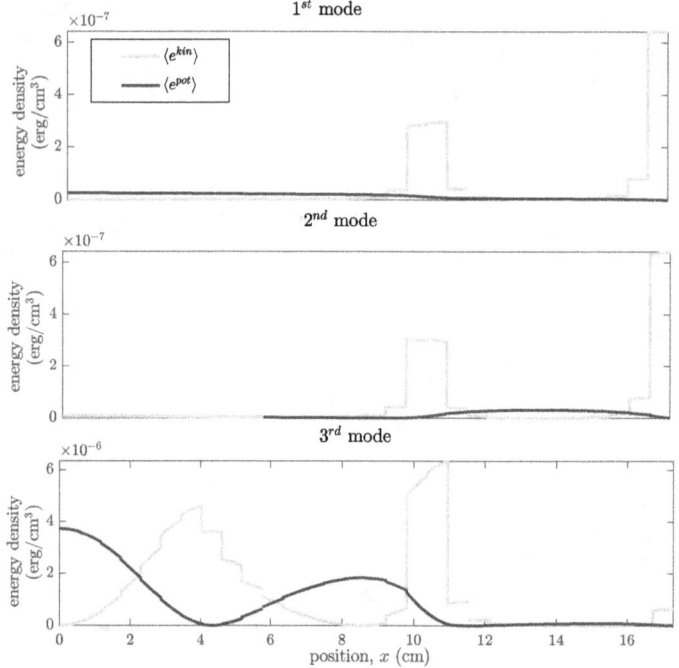

FIGURE 12. Energy densities for the area function in Figure 11

of these changes on $\mathcal{F}_1$ and $\mathcal{F}_2$, because their energy densities are small and approximately constant in the region with the new expansion and new narrowing.

The normal mode frequencies corresponding to the area function shown in Figure 11 are $\mathcal{F}_1 = 319$ Hz, $\mathcal{F}_2 = 913$ Hz, and $\mathcal{F}_3 = 1749$ Hz. This is a 1% increase in $\mathcal{F}_1$, a 0.3% increase in $\mathcal{F}_2$, and a 11% decrease in $\mathcal{F}_3$ from the original area function in Figure 9. Thus, we obtain a substantial decrease in $\mathcal{F}_3$ for negligible changes in $\mathcal{F}_1$ and $\mathcal{F}_2$ in the transition from the area function shown in Figure 9 and to the area function shown in Figure 11. Figure 12 shows the energy densities corresponding to the new normal mode frequencies resulting from the area function in Figure 11. The basic patterns for the energy densities have not changed from those of Figure 10 for the first two normal mode frequencies. The same can be said for $\mathcal{F}_3$, except that it appears that the kinetic and potential energy densities are beginning to separate in

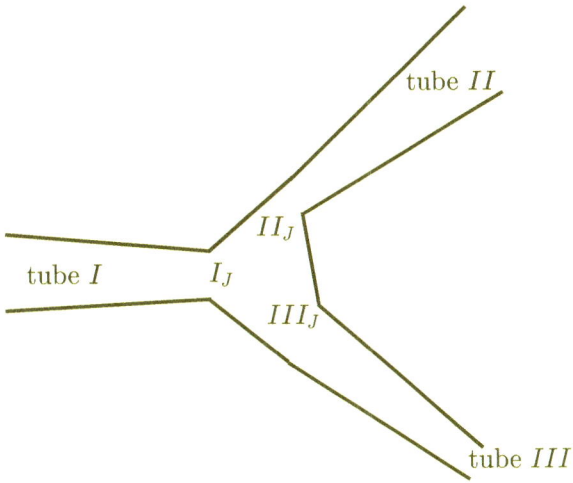

FIGURE 13. Region of intersection for 3 tubes

the rear of the vocal tract. If we were to constrict further in the rear, then we could obtain an approximate triple Helmholtz resonator.

Summary of Helmholtz resonators

Approximate Helmholtz resonators occur within the theory that has been developed over the past several chapters. In other words, the lumped system natural frequencies are derived as special cases from the theory that provides normal modes with distributed potential and kinetic energies. Thus, they are a part of phenomena that are predicted from the acoustic approximation in a finite-length tube. While approximate Helmholtz resonators can be derived from the general theory, they possess a particular property: the kinetic and potential energy densities are separated and show little phase change across their regions of concentration.

Side branches

We consider the situation of a junction where three tubes come together, as shown in Figure 13. We assume that the junction region is small enough that pressure and volume velocity are continuous through the junction. [We do not consider the problem of additional kinetic energy in this section.] The three tubes and junction form an acoustic system, and we continue to assume one-dimensional motion.

For tube I, we associate positive volume velocity with flow into the junction, and for tubes II and III we associate positive volume velocity with flow out of the junction. The reason that it is necessary to be clear about directionality is that it determines how we write the continuity conditions for volume velocity. We let the subscript J attached to the number of the tube denote the end of the tube at the junction.

We apply continuity of pressure and continuity of volume velocity at the junction. Continuity of pressure gives

$$p^{(I_J)} = p^{(II_J)} = p^{(III_J)} \ . \tag{7}$$

For continuity of volume velocity

$$Q^{(I_J)} = Q^{(II_J)} + Q^{(III_J)} \ . \tag{8}$$

Equation (7) says that all tube ends experience the same pressure. Equation (8) says that the volume velocity entering the junction equals the volume velocity leaving the junction. The combination of Equations (7) and (8) gives

$$\frac{Q^{(I_J)}}{p^{(I_J)}} = \frac{Q^{(II_J)}}{p^{(II_J)}} + \frac{Q^{(III_J)}}{p^{(III_J)}} \ ,$$

or $\hspace{10em}$ (9)

$$Y^{(I_J)eff} = Y^{(II_J)eff} + Y^{(III_J)eff} \ ,$$

where $Y^{(I_J)eff} = Q^{(I_J)}/p^{(I_J)}$ and so on.

We can calculate the three effective admittances in Equation (9) using the methods of Chapter 10. $Y^{(I_J)eff}$ is the effective admittance looking into tube I from the junction. Here, Equation (12) of Chapter 10 applies. $Y^{(II_J)eff}$ and $Y^{(III_J)eff}$ are the effective admittances looking into their respective tubes from the junction. In these cases, Equation (11) of Chapter 10 applies. With these interpretations of $Y^{(I_J)eff}$, $Y^{(II_J)eff}$, and $Y^{(III_J)eff}$, Equation (9) is seen as a continuity condition that determines the modes of acoustic propagation in the branching system.

We consider the special cases of $Y^{(II_J)eff} = \infty$ or $Y^{(III_J)eff} = \infty$, exclusively. In these cases it is best to work with impedances: $Z = 1/Y$.

Equation (9) can be written

$$\frac{1}{Z^{(I_J)eff}} = \frac{1}{Z^{(II_J)eff}} + \frac{1}{Z^{(III_J)eff}},$$

or (10)

$$Z^{(I_J)eff} = \frac{Z^{(II_J)eff} Z^{(III_J)eff}}{Z^{(II_J)eff} + Z^{(III_J)eff}}.$$

If $Y^{(II_J)eff} = \infty$, then $Z^{(II_J)eff} = 0$, and if $Y^{(III_Jeff)} = \infty$, then $Z^{(III_J)eff} = 0$. In either case Equation (10) gives $Z^{(I_J)eff} = 0$, which means that $Y^{(I_J)eff} = \infty$.

$$Y^{(I_J)eff} = \infty \text{ when either } Y^{(II_J)eff} = \infty \text{ or } Y^{(III_J)eff} = \infty. \quad (11)$$

Thus, the junction end of tube I is continuous with whichever of the junction ends of tubes II or III has an infinite effective admittance or zero effective impedance. In this case all of the energy of acoustic wave motion is transmitted between the two tubes that meet with infinite effective admittance. The other tube with finite effective admittance or non-zero effective impedance has no energy for acoustic wave motion. We do not consider the case when $Y^{(II_J)eff}$ and $Y^{(III_J)eff}$ are simultaneously infinite.

Nasals

For a nasal sound, the junction at the velum divides the vocal tract into two parts. One part is to the left of the junction containing the pharynx and the one to the right of the junction contains the mouth, and these are called tubes I and II, respectively, in Figure 14. Tube III is the nasal tract. We need to assign left and right ends to all of these tubes. That is easy for the case of the nasal: the right end of tube I is at the junction and the right end of tube II is at its opening to the atmosphere at the lips. The right end of tube III also opens to the atmosphere at the nares. Thus, the junction is formed with the right end of tube I and the left ends of tube II and tube III.

We consider the case when tube I has constant cross-sectional area $A^{(I)}$ and tube II has constant cross-sectional area $A^{(II)}$. Figure 14 shows the situation when $A^{(I)} = A^{(II)}$. For nasals, $Y^{(III_J)eff}$ is the effective admittance at the velum junction looking toward the nares, which is the tube III opening, and $Y^{(II_J)eff}$ is the effective admittance

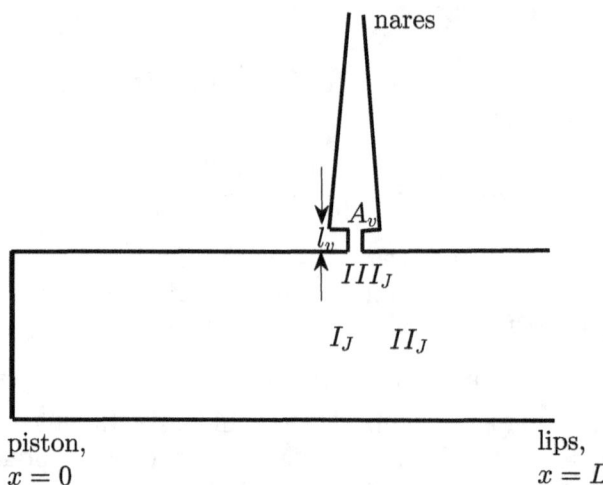

FIGURE 14. Configuration for nasals

at the velum junction looking toward the lips, which is the tube II opening. For illustrative purposes, we examine the special case when tube II has length l_{II} and constant cross-sectional area, A_{II}. From Equation (11) of Chapter 10 we obtain

$$Y^{(II_J)eff} = Y^{(II)} \frac{Y_{rad} - iY^{(II)} \tan(kl_{II})}{Y^{(II)} - iY_{rad} \tan(kl_{II})} , \qquad (12)$$

where $Y^{(II)} = A^{(II)}/\rho_0 c_0$ is the tube admittance for tube II, and Y_{rad} is the radiation admittance at the lips. We return to the simplified case where $Y_{rad} = \infty$. Equation (12) becomes

$$Y^{(II_J)eff} = iY^{(II)} \cot(kl_{II}) . \qquad (13)$$

Let $Y^{(nasal)eff}$ be the effective admittance looking toward the nares, at the velum, but not including the velar tube. $Y^{(nasal)eff}$ can be calculated using published data of the nasal area function using the methods of Chapter 10 (e.g Dang, Honda, & Suzuki 1994). For the nasal tube III, the variable part is the degree of velar opening, which is given by a cross-sectional area $A^{(v)}$ in Figure 14. Let $Y^{(v)} = A^{(v)}/\rho_0 c_0$. We assume that the length of the velar tube, l_v, is small on the scale of the wavelength of sound, so that $kl_v \ll 1$. Again, invoking Equation

(11) of Chapter 10 it follows that

$$Y^{(III_J)eff} = Y^{(v)}\frac{Y^{(nasal)eff} - iY^{(v)}\tan(kl_v)}{Y^{(v)} - iY^{(nasal)eff}\tan(kl_v)} \qquad (14)$$
$$\approx Y^{(v)}\frac{Y^{(nasal)eff} - iY^{(v)}(kl_v)}{Y^{(v)} - iY^{(nasal)eff}(kl_v)},$$

because $\tan(kl_v) \approx kl_v$ for $kl_v \ll 1$.

$Y^{(I_J)eff}$ is the same as the effective admittance looking from the junction toward the piston. Assume that this tube has cross-sectional area A_I and length l_I. From Equation (12) of Chapter 10, and because $Y^{(piston)eff} = Y^{(piston)} = 0$, it is the case that

$$Y^{(I_J)eff} = Y^{(I)}\frac{Y^{(piston)eff} + iY^{(I)}\tan(kl_I)}{Y^{(I)} + iY^{(piston)eff}\tan(kl_I)} \qquad (15)$$
$$= iY^{(I)}\tan(kl_I).$$

Consider the case when the velum is closed, or $Y^{(v)} = A^{(v)} = 0$. From Equation (14), when $A_v = 0$, $Y^{(III_J)} = 0$, and Equation (9) reduces to

$$Y^{(I_J)eff} = Y^{(II_J)eff}. \qquad (16)$$

With Equations (13) and (15), Equation (16) can be rewritten

$$Y^{(I)}\tan(kl_I) = Y^{(II)}\cot(kl_{II}),$$

or

$$\frac{A^{(I)}}{\rho_0 c_0}\tan(kl_I) = \frac{A^{(II)}}{\rho_0 c_0}\cot(kl_{II}), \qquad (17)$$

or

$$\frac{A^{(I)}}{A^{(II)}}\tan(kl_I)\tan(kl_{II}) - 1 = 0.$$

This is the same as Equation (45) of Chapter 9 for the two sub-tube situation. [Note that the tube closest to the piston is I in the present notation and 2 in the notation of Chapter 9.]

For the remainder of the section on nasals, we compute the frequencies for which the continuity condition, Equation (9), is true. There are three separate situations. The first one is when $Y^{(I_J)eff} \neq \infty$, which means that $Y^{(II_J)eff} \neq \infty$ and $Y^{(III_J)eff} \neq \infty$, so that acoustic energy is in both tube II and tube III, as well as in tube I when there

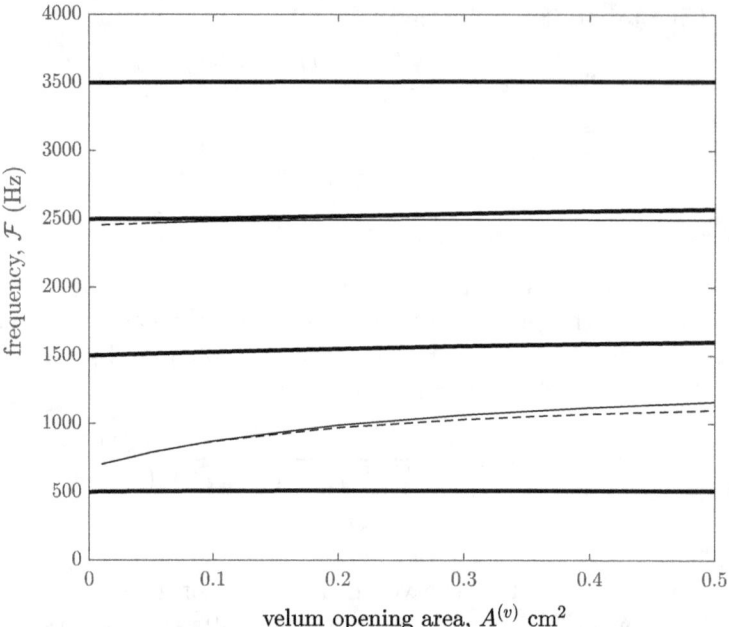

FIGURE 15. Mode frequencies (thick and thin solid lines) and single-path mode frequencies where there is no radiation from the mouth (dashed lines) when the lips are open

is a piston source. The frequencies found by invoking Equation (9) are simply termed mode frequencies here. The second and third situations occurs when $Y^{(I_J)eff} = \infty$, so that either $Y^{(II_J)eff} = \infty$ and there is no acoustic energy in tube III for a piston sources, or $Y^{(III_J)eff} = \infty$ and there is no acoustic energy in tube II for a piston source. The frequencies that result are termed *single-path mode frequencies* here.

In the following calculations, tubes I and II are presumed to both possess a cross-sectional area of 4 cm^2. Tube I has length $l_I = 11.0$ cm and tube II has length $l_{II} = 6.1$ cm. We use the data from Dang, Honda, & Suzuki (1994) for the nasal tract to calculate $Y^{(nasal)eff}$. The first four mode frequencies below or at 4000 Hz for the case without nasal coupling, i.e. with $A^{(v)} = 0.0$ cm^2, are the expected 500 Hz, 1500 Hz, 2500 Hz, and 3500 Hz. The mode frequencies that are continuous with these frequencies as $A^{(v)}$ is increased from zero

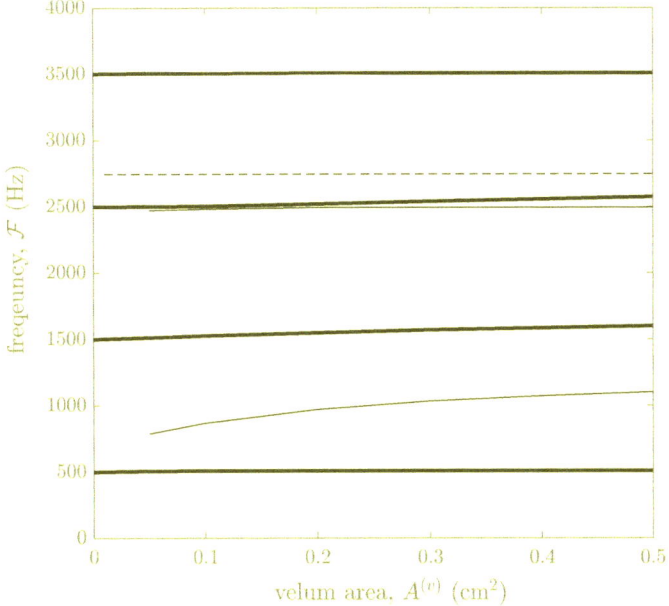

FIGURE 16. Mode frequencies (thick and thin solid lines) and single-path mode frequencies where there is no radiation from the nares (dashed lines) when the lips are open

are shown in the bold, solid lines in Figures 15 and 16. The mode frequencies that are not continuous with these mode frequencies when $A^{(v)} = 0$ are drawn as thin solid lines in Figures 15 and 16.

Figures 15 also shows the single-path mode frequencies when all acoustic energy flows into the nasal tact, tube *III*. There is no energy radiated from the lips for these frequencies. These single-path mode frequencies are shown as dashed lines, and they are close to mode frequencies in drawn with the thin lines that arise only when $A^{(v)} > 0$. Figure 16, in addition to the mode frequencies, shows the single-path mode frequencies when all energy flows into the mouth, tube *II* in a dashed line. There is no energy radiated from the nares for these frequencies. Because Figure 16 accounts only for variation in $A^{(v)}$, the frequency of this single-path mode frequency does not change in this figure.

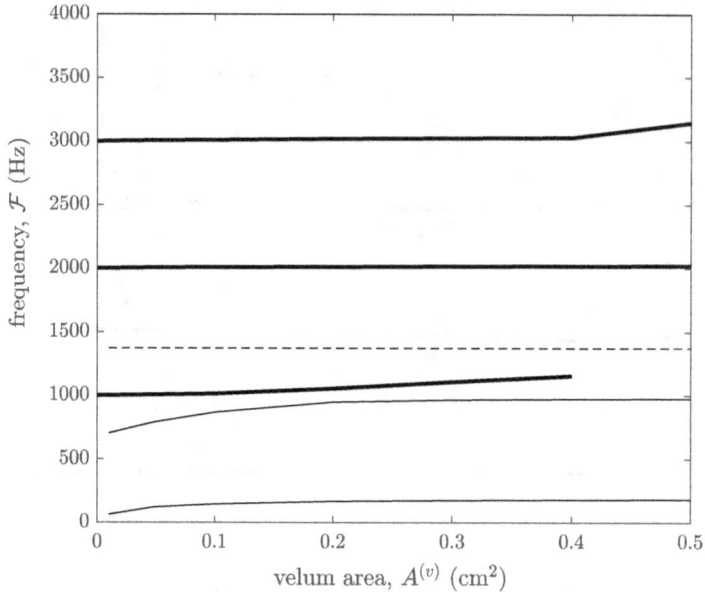

FIGURE 17. Mode frequencies (thick and thin solid lines) and frequencies where $Y^{(II_J)}$ is infinite (dashed lines) when the lips are closed.

Now we consider the case when the lips are closed, so that the right end of tube II has boundary condition of zero particle velocity or zero volume velocity. First we consider the vocal tract without nasal coupling, or $A^{(v)} = 0$ cm^2, but with the lips closed (closed tube II on the right). The four mode frequencies at or below 4000 Hz are now 1000 Hz, 2000 Hz, 3000 Hz, and 4000 Hz. That is, the vocal tract of constant cross-sectional area is a *half-wave resonator* when the lips are closed. The mode frequencies that are continuous with these frequencies as $A^{(v)}$ is increased from zero are shown in the bold, solid lines in Figure 17.

There are two mode frequencies that arise when $A^{(v)} > 0$ that are shown in thin solid lines in Figure 17. There is energy in both the nasal tact and the mouth, but there is radiation only from the nares, because the lips are closed. Both of these mode frequencies are below the first mode frequency that is continuous with 1000 Hz.

Also, $Y^{(II_J)eff} = \infty$ at 1374 Hz. In the case that $Y^{(II_J)eff} = \infty$ there is no radiation of sound to the atmosphere, because there is no energy in the nasal tube and the lips are closed. This frequency is not a function of $A^{(v)}$, and this single-tube mode frequency is drawn as a dashed line in Figure 17.

A closed channel in the tongue anterior

It is possible to form a closed channel with the margins of the front portion of the tongue blade in contact with the palate as shown schematically in Figure 18(a). This is one possible configuration for a lateral consonant. This configuration is abstracted further to the tube system shown in Figure 18(b). Tube *I* corresponds to the vocal tract up to the posterior of the tongue-palate contact. Tube *II* corresponds to the vocal tract from this posterior point of contact and lateral to the tongue contact to the lips. Thus, tube *II* is split in two by the margins of the tongue in contact with the palate. Tube *III* is the tube formed in the middle of the tongue anterior between between the sides of the tongue-palate contact. With the closed end tube *III* Equation (11) of Chapter 10 gives

$$Y^{(III_J)eff} = -iY^{(III)} \tan(kl_{III}) , \qquad (18)$$

where $Y^{(III)} = A^{(III)}/\rho_0 c_0$, and $A^{(III)}$ is the cross-sectional area of tube *III*.

As an example, we take the cross-sectional area of the vocal tract to be $A^{(I)} = A^{(II)} = 4$ cm^2, the length of tube *III* to be $l_{III} = 4$ cm (Stevens 1998), and the area $A^{(III)} = 1$ cm^2. Five mode frequencies appear below 4000 Hz: 499 Hz, 1445 Hz, 1995 Hz, 2631 Hz, and 3549 Hz. This is in contrast to the four mode frequencies at 500 Hz, 1500 Hz, 2500 Hz, and 3500 Hz for the neutral vocal tract without a side branch. The frequency below 4000 Hz for which $Y^{(III_J)eff} = \infty$ is 2163 Hz, a single-path mode frequency. There is no radiation from the mouth at this frequency, and this frequency is close to the mode frequency at 1995 Hz.

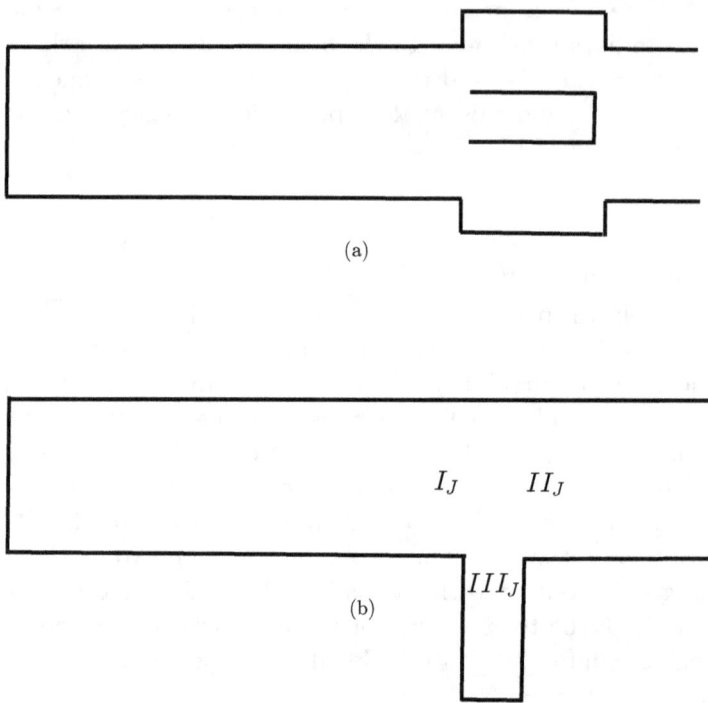

FIGURE 18. Configuration for the closed channel in the tongue anterior

Piriform sinuses

The piriform sinuses are two short cavities, or tubes, that surround the larynx as shown in Figure 19(a). These sinuses open into the vocal tract at the level of the laryngeal vestibule. The physical configuration is abstracted in Figure 19(b).

In the simulation, we take the sinuses to be constant cross-sectional area tubes, so that their combined cross-sectional area is $A^{(III)} = 1.5$ cm^2 and their length is $l_{III} = 1.8$ cm. We take the opening of the piriform sinuses to be 3.5 cm above the vocal folds (Dang & Honda 1997). Further, $A^{(I)} = A^{(II)} = 4$ cm^2, with $l_I = 3.5$ cm and $l_{II} = 13.8$ cm. There are five mode frequencies below and near 4000 Hz: 484 Hz, 1481 Hz, 2500 Hz, 3393 Hz, and 4084 Hz. The lowest frequency for which $Y^{(III_J)eff} = \infty$ and there is no radiation from the mouth is 4806 Hz.

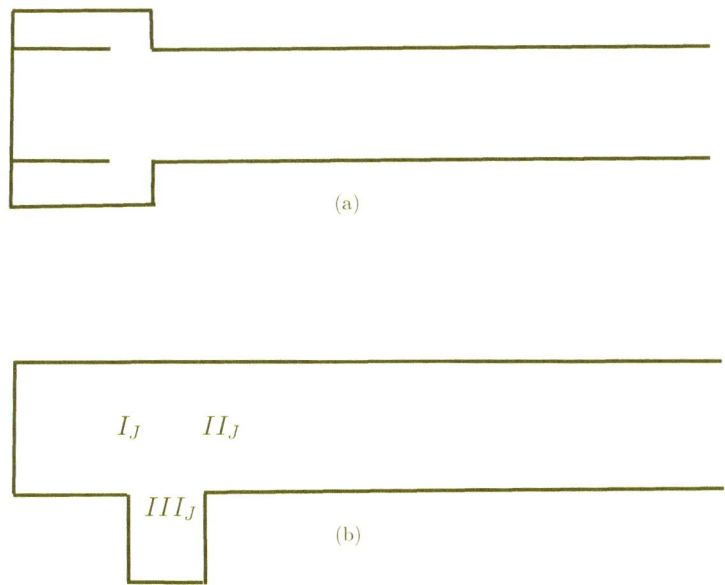

FIGURE 19. Configuration for piriform sinuses

Conclusion

With Chapter 12 we have completed most of our basic considerations of speech acoustics. Of course there are other topics that have not been examined, but we have covered the fundamentals of the acoustics of speech. There is a question that remains however, and that is the large amplitudes of the acoustic waves in the finite-length tube when the piston is producing peak amplitudes of pressure at just less than 1% atmospheric, or about 8.3 cm H_2O. We mentioned these large amplitudes in Chapter 2 and in Chapter 7. In the latter case we estimated the overall sound pressure level (OSPL) at the face of the piston moving with a fundamental frequency and harmonics that decline in amplitude as we expect the glottal source does. That estimate was 141 dB. In the next chapter we consider whether this is too large for the real vocal tract, and examine whether the physics of air near the larynx can provide a better estimate of the power in sound production by this apparent volume velocity source. This requires an outline of a theory of air motion that is not described by the acoustic approximation.

References

Dang, J., Honda, K., & Suzuki, H. (1994). Morphological and acoustical analysis of the nasal and paranasal cavities. *Journal of the Acoustical Society of America*, **96**. p 2088.

Dang, J. & Honda, K. (1997). Acoustic characteristics of the piriform fossa in models and humans. *Journal of the Acoustical Society of America*, **101**. p 456.

Stevens, K.N. (1998). *Acoustic Phonetics*. MIT Press, Cambridge, MA. 187-96, (pp 494-9).

Chapter 13: Fluid Mechanics and Aeroacoustic Sources

Introduction

We begin to explore the descriptions of air motion in the vocal tract in approximations to the mass and momentum conservation laws that differ from the acoustic approximation introduced in Chapter 2. It turns out that the acoustic approximation is not the appropriate approximation for air motion at certain times and in certain regions of the vocal tract. However, air in the vocal tract obeys the same conservation laws no matter which mathematical approximation to these laws that we find appropriate.

First, we provide a word of caution. Physical quantities do not possess measurable acoustic and a non-acoustic components, but are unitary measurable physical quantities. This applies, for instance, to pressure. We choose to describe the variation of pressure in space and time using various mathematical approximations to the conservation laws in order to keep the representations understandable by humans, but pressure is all that we measure.

The air motion in the piston-in-a-tube picture with the acoustic approximation describes voiced speech, and is commonly referred to as the source-filter theory of voiced speech. However, we wish to go beyond source-filter theory to obtain a more complete picture of air motion in the vocal tract during speech. For instance, the boundary at the laryngeal end of the tube is not always closed like the face of a piston, but more like a valve that opens and closes. Only when the valve is closed is the laryngeal boundary a solid, and when it is open it is more like the end of a sub-tube that extends the air channel farther to the left through the trachea and below. In order to obtain a more realistic picture of the voice source and its relation to the air motion in the rest of the vocal tract, we must consider the air flow near the larynx in detail, and, thus, leave behind the piston-in-a-tube picture.

Not only does the idea of a moving piston as the voice source need amending, but there are situations when the picture provided by the acoustic approximation in the supralaryngeal vocal tract needs to change in order to obtain a good approximation to the air flow. These situations are associated with relatively high air flow velocities and small cross-sectional areas. These situations often produce turbulent flow. This results in sources of energy for air motion in the rest of the vocal tract that can be described by the acoustic approximation. One of the defining features of turbulent flow is rotational air motion, which is described below.

Often, a good approximation in the vocal tract is that both an acoustic air field and a non-acoustic motion, and these interact only weakly. For instance, this seems to be the case for air motion above the the larynx during voiced speech. This is discussed below.

Both the voice source and turbulent flow require a more thorough understanding of the mechanics of air than we have so far attained. The first part of this chapter is an introduction to the mechanics of air under an important approximation that differs from the acoustic approximation.

The Euler model

We recall from Chapter 1 that the mechanics of air is generally described in what is considered an Eulerian frame. The Eularian frame is the frame in which space is fixed and the air particles move through the space. We generally do not follow individual fluid particles, which would entail using a different frame of reference, the Lagrangian frame, for each of an infinity of air particles.

The discussion of air motion here ignores the compressibility property of air. This, indeed, is a big assumption, and it appears to be a counterintuitive one to make for air. The reason that we can consider air to be in *incompressible flow* has to do with the size of the regions that we examine. More will be said about this later in the chapter. To consider air flow to be incompressible flow is to approximate the speed of sound c_0 defined in Equation (9) of Chapter 1 to be infinite. The incompressibility approximation is made in order to simplify the development of ideas and equations. Further, we assume that the density of air is uniform throughout the region of space that we consider. This density is denoted ρ_0.

We also neglect viscosity, which means that particles of air interact with one another only through the *pressure stress*. This, along with the incompressible flow approximation, enables a tractable discussion of the mechanics of air in certain regions of the vocal tract. With these approximations, we obtain what Lighthill (1986) terms the *Euler model* for air motion. [Euler model is not Eulerian frame.] Beyond the Euler model, we further neglect variations of force due to gravity, and, as already mentioned, any variation in air density. We also neglect variations in entropy caused by such phenomena as heat conduction.

Steady and quasi-steady motion: streamlines and stream tubes

We consider steady air flow for now. This may seem a little odd given that at the end of these considerations we examine the relation between unsteady motion under the acoustic approximation and the unsteady air motions described under the Euler model. There are two reasons that a study of steady air motion is important. The first is that one needs to understand steady motion before the more complicated unsteady motion can be understood. Secondly, it is very often the case that the *quasi-steady approximation* can be made. In this approximation, air motion in the unsteady case is pictured as a series of instantaneous steady motions. This is much like the fact that a movie with a series of static frames close enough together in time appears to show the viewer continuous motion. In more precise words, the quasi-steady approximation means that the accelerations of air particles at a fixed point in space (i.e. the Eulerian frame) are presumed to have no effect on the physics of flow at proximate times. The quasi-steady approximation is often a good approximation to the study of air motions in the vocal tract.

For steady air flow, each point in space can be associated with an arrow, a *particle velocity vector*, whose direction indicates the direction of flow and whose length indicates the *particle speed*. Unlike our previous discussions, we are now considering motion in three spatial dimensions. [Direction and speed, together, constitute velocity. Particle velocity, which is represented by the particle velocity vectors, is denoted **u**, and the particle speed is $|\mathbf{u}|$. We also use $u = \pm|\mathbf{u}|$ to denote particle velocity, when the motion is one-dimensional and the positive and negative directions are well-defined. Until now, particle

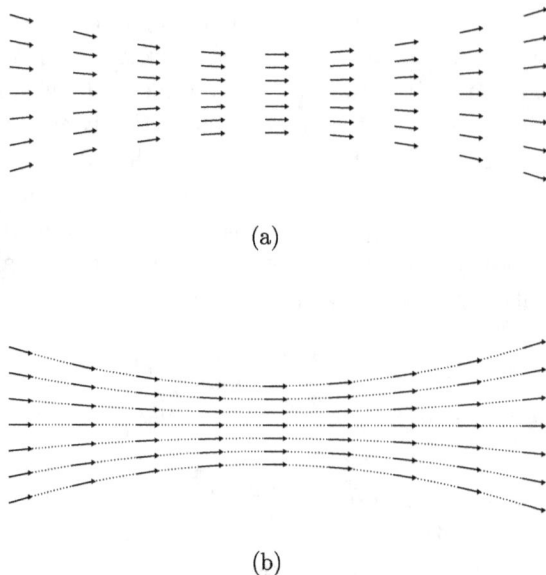

FIGURE 1. Velocity vectors and streamlines

velocity always been along the x-axis.] An example of steady flow is shown in Figure 1(a), where the arrows of velocity are shown at various points in a plane represented by the surface of the paper. Further, we can imagine connecting neighboring points with curves that are indicated by these arrow at the bottom of Figure 1(b). The resulting curves, which are shown in Figure 1(b) in dots, are called streamlines, and for a steady flow, the particles of air follow these streamlines.

We take the geometric construction one step further by combining streamlines to form stream tubes. Consider a simple closed curve, or loop, in the three-dimensional space in which air is flowing, as well as all the streamlines that pass through that curve. Figure 2 shows the curve C_1 with a sample of streamlines passing through C_1. These streamlines form a surface in three-dimensional space that forms the boundary of a stream tube. Because air flows along the streamlines that form the surface boundary of the stream tube in steady flow, air does not enter or leave a stream tube.

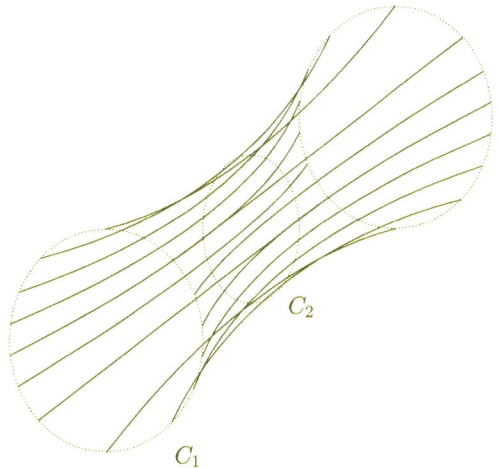

FIGURE 2. Stream tube

Conservation laws for steady flow in the Euler model

With the property that air flows within stream tubes, we come to the connection between the geometric objects we call streamlines and stream tubes and the mass and energy conservation laws for the steady flow in the Eulerian model. Consider a very thin stream tube. Stream tubes can vary in cross-sectional area, and there is a simple relation between stream tube cross-sectional areas and particle velocities within the thin stream tube at those cross-sections. Suppose A_1 is a cross-sectional area of the stream tube cross-section at curve C_1, as shown in Figure 2. Because the stream tube is very thin, we can assume that the particle velocity through A_1, say u_1, is orthogonal to the cross-section and nearly constant over the cross-section. Consider the cross-sectional area at another curve C_2 along the stream tube, A_2, with particle velocity of u_2. When there is no compressibility and uniform density the rate of air volume flow through the two cross-sections must be equal.

$$u_1 A_1 = u_2 A_2 . \tag{1}$$

Equation (1) is a way of expressing the mass conservation for incompressible air of uniform density. Figure 2 shows the situation where $A_1 > A_2$, so that Equation (1) implies that $|u_1| < |u_2|$.

Is there an analogous way of employing streamlines or stream tubes to express energy conservation in steady, incompressible flow of air with uniform density ρ_0? There is, and this is known as the *Bernoulli equation*. In the derivation of the Bernoulli equation, entropy producing mechanisms, such as heat conduction are assumed to be unimportant (Batchelor 1967). In fact, we are already assuming that entropy is a constant of time and space, just as in the acoustic approximation.

$$p + \frac{\rho_0}{2}|\mathbf{u}|^2 = \text{Constant along a streamline} . \qquad (2)$$

With $u = \pm|\mathbf{u}|$ along a streamline Equation (2) can be written as

$$p + \frac{\rho_0}{2}u^2 = \text{Constant along a streamline} . \qquad (3)$$

Equations (2) and (3) are for steady flow.

It is often the case that the constants for large groups of streamlines are identical. In particular, the constant is the same among neighboring streamlines when identical conditions apply to the streamlines in some region of space. In this case the common constant is known as the *pressure head*. If we can assume that all the streamlines in the stream tube in Figure 2 have the same pressure head, then it follows from Equations (1) and (2) that $p_1 > p_2$, if $A_1 > A_2$. As a narrow stream tube constricts, the particle speed increases and the pressure decreases.

Momentum conservation is a slightly more mathematically complicated, even for steady flow. We postpone introducing this conservation law, and further, state it only for a special case later in this chapter.

Rigid body motion

Here we consider some facts about solid body motion that relate to air particle motion. We consider the motion of bodies that are not simply points, but possess some extent. In fact, we first consider solid, rigid bodies where the particles within the body do not change their mutual distances with time.

Extended rigid bodies can move through space in complicated ways. In this development we consider the motion of rigid bodies for very brief durations of time and in a fixed three-dimensional coordinate system with orthogonal axes, x, y, and z. First we focus on the extended rigid body shown in Figure 3. The body can translate through space

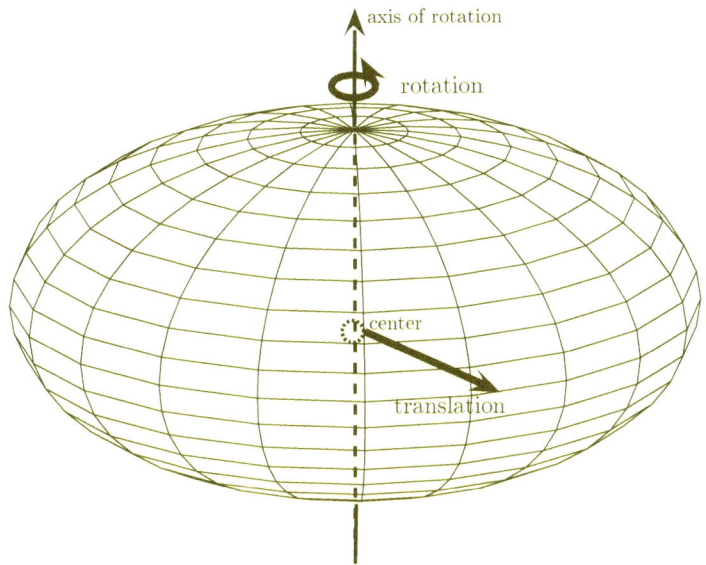

FIGURE 3. Rigid body motion

in a brief instant of time. Because the body is rigid, all the particles in the body translate in the same way. This *translational motion* is indicated in Figure 3 with a bold arrow from the chosen point showing the magnitude and direction of the motion of all the points within the rigid body due to translational motion.

Further, the rigid body can move about any chosen point in the body. A theorem, *Euler's theorem (regarding rigid body motion)* (Goldstein 1950), requires that this motion, in a very brief duration of time, be a rotational motion about a line through the chosen point in the body. Here we show the chosen point to be the center of spheroidal body in Figure 3. This line is known as the *axis-of-rotation*. Figure 3 shows the spheroidal rotating about an axis-of-rotation containing the chosen body point. The axis-of-rotation moves with the translational motion of the body. Note that the points within the solid do not move with the same magnitude and direction due to rotational motion. The magnitude of the movement of a point in the body depends on that point's distance from the axis-of-rotation. The direction of a point's movement depends on its orientation with respect to the axis-of-rotation. If we were to consider more than a brief duration of time, the direction of the axis-of-rotation could change through time as well.

An example of rotational motion is that of a wheel turning on its axle. The axis-of-rotation is the line of symmetry that runs along the center of the wheel's physical axle. The axis translates with the wheel, but otherwise does not move. In the simplest model of earth's movement, the earth can be viewed as a sphere rotating about an axis through the earth's center and translating through space as it revolves around the sun.

In order to proceed we need to assign a direction to each axis-of-rotation. In the case of the wheel, we could consider one side of the wheel to be the inside and the other the outside, and then say that the axis-of-rotation is directed from the inside to the outside of the wheel. In the case of the earth, the axis-of-rotation can be considered to be directed from the southern hemisphere to the northern hemisphere. Further, rotation is quantified in something named the *angular velocity*, Ω. The earth rotates at the rate of $|\Omega| = 2\pi$ radian per day or $|\Omega| = 0.0000157\dot{2}\pi$ radian per second. The angular velocity Ω has a sign that depends on the direction of the orientation of the rotation with respect to the axis-of-rotation. If the rotation is clockwise looking in the direction of the axis-of-rotation, Ω is given a positive sign, and, otherwise it is given a negative sign. The angular velocity is positive in Figure 3. [Looking from the direction of the axis-of-rotation without the arrow toward the end with the arrow, the rotation is clockwise. Looking in the opposite direction, for which Figure 3 lends itself, the rotation is counter-clockwise. We use the former viewpoint to define the sign of Ω.]

Motion of "extended" air particles

We follow Lighthill (1986) in our description of air particle motion in the Euler model. The particle of air is supposed to be extended slightly from that of a mathematical point to a very small sphere. There are three kinds of motion that small extended air particles can experience during a very brief amount of time. Two of the types of motion are those that rigid bodies exhibit: translation and rotation about an axis that runs through the body. As with the rigid body, there is an angular velocity Ω associated with the direction of the axis-of-rotation, which has either a positive or negative value depending on whether it is turning clockwise or not with respect to the direction of the axis. Instead of quantifying the rotation with Ω, we use a quantity called vorticity and denoted ω. [Vorticity should be represented by a vector in three-dimensional space, but in our applications the axis-of-rotation is specified. This means that we know the direction of the axis-of-rotation and ω can be either be negative or positive, depending whether the rotation is clockwise or counter-clockwise with respect to the axis-of-rotation.] In the special case that the extended particle is a small sphere, and the axis-of-rotation goes through the center of the sphere, ω is twice Ω. If a particle of air possesses rotational motion at position $\mathbf{x} = (x, y, z)$ in three-dimensional space at time t, then the vorticity $\omega(\mathbf{x}, t)$ is non-zero. Note that vorticity is defined in the Eulerian frame because it is a function of spatial position $\mathbf{x}$ and time t.

Another kind of motion that the sphere of air can experience in a very brief moment of time is a change from a spherical shape into an ellipsoidal shape, with stretching in certain directions, and shrinking in other directions. This is called *pure straining motion*. Because we are working in the incompressible approximation, stretching in one direction must be compensated for by shrinking in other directions in order for the volume of the small sphere to remain constant. A rigid body cannot experience pure straining motion.

We use the example of what it known as *shear flow* to illustrate that the three types of motion can be present simultaneously, and that the rotational motion is not always obvious until all three components of motion are identified. Shear flow exists when there is a steady flow of air over a smooth, solid surface, as shown in Figure 4(a). The velocity of air at the surface is zero, and we suppose that the particle velocity away from the surface is directed parallel to the surface with particle velocity U. The magnitude of the particle velocity as a function of distance from the plate increases from zero to $|U|$ is an example of shear flow. [Shear flow cannot be generated by only pressure stress, which is the only stress in the Euler model. In the Euler model, the shear flow is added to solid surfaces where there is neighboring air flow. This is often idealized as an infinitely thin sheet of vorticity where there is a jump in velocity from that of the solid to the air outside the infinitely thin *vortex sheet*. These sheets of vorticity are discussed below. The

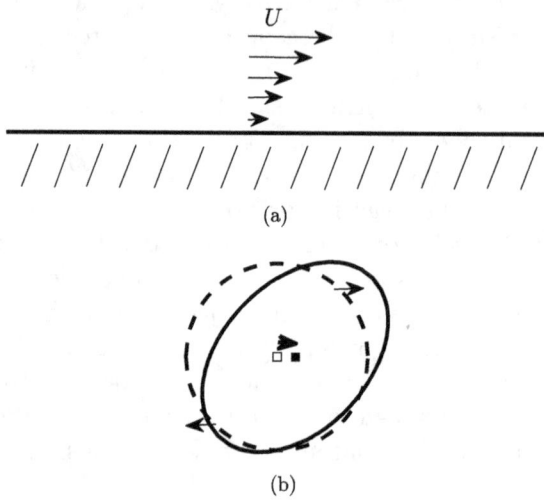

FIGURE 4. Velocity vectors for shear flow

Euler model permits us to consider shear flow; we just don't attribute the cause of the shear flow to viscosity and *viscous stress*.]

Figure 4(a) shows the speed and direction of air particles flowing over a plane solid surface in terms of arrow length and direction, respectively. The thick line represents the plane surface, which extends into and out of the paper. We see that the speed of air particles is a function of distance from the plane solid surface. Also, the spatial variables of interest are in the plane of the paper; nothing varies in the direction orthogonal to the paper.

The motion of a small sphere of air pictured as a circle in the plane of the paper in Figure 4(a) is examined in Figure 4(b). In Figure 4(b), the motions of the sphere are shown with the sphere projected onto the two-dimensional plane of the paper. The translational motion of the sphere is indicated by the thick arrow pointed to the right from the center of the circle (small open square) to the center of the ellipse (small filled square). If this translational motion is subtracted off, it is seen that different parts of the sphere are moving either to the left and the right, at varying speeds. We have shown the movements of the top-most and bottom-most faces of the circle with narrow arrows, when the translational motion is subtracted off. These arrows indicate shear

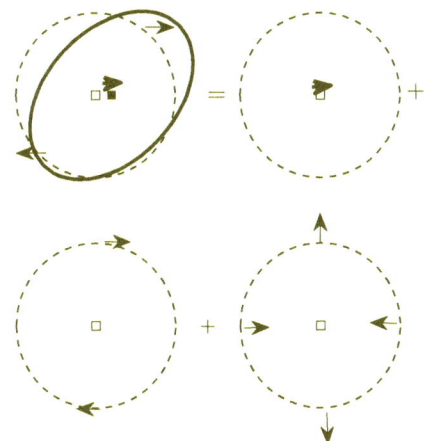

FIGURE 5. Decomposition of the motion of a small fluid sphere in two-dimensions

flow is present, and in a brief duration the circle becomes a rotated ellipse. [In three dimensions, the sphere becomes an ellipsoid.]

Figure 5 is a pictorial equation that states that the total motion of the small sphere is equal to the sum of three more elementary motions. Just as in Figure 4(b), the motions of the sphere are shown with the sphere projected onto the two-dimensional plane of the paper in Figure 5. In this shear flow, the axis-of-rotation goes through the center of the sphere and points out of the plane of the paper, which we will take to be the z-direction. The reader can see that the angular velocity Ω and vorticity ω are negative.

Over very brief durations of time, the extended particle of air experiences the two components of rigid body motion: translation and rotation, as well as a third component: pure straining motion. Thus, the steady motion of air over a solid surface necessarily means that rotational air motion is present. This also is true of unsteady motion, although it is slightly more complicated to show. In the case of unsteady motion, the magnitude and direction of the vorticity (i.e. rotational motion) depends on the magnitude and direction of the flow away from the solid surface. In all cases, the magnitude of the vorticity is directly proportional to the particle speed of the flow near the solid surface $|\mathbf{U}|$.

Two-dimensional air motion

While air motion occurs in three spatial dimensions, often there are motions that vary with, at most, two spatial directions. This is the case with the example of shear flow over a planar surface considered above. The planar surface is considered to lie in a plane into and out of the paper, and there is no variation in the motion in the direction into and out of the paper.

Here we consider other air flows that do not change substantially in a single direction. The only variation in the air flow is in planes perpendicular to that single direction. When this occurs the description of motion can be economized from three to two dimensions, because the direction for which there is no variation can be eliminated from the description. We often take this direction to be the z-direction.

The line vortex

A two-dimensional motion appears when we consider vorticity within cylindrical regions and the vorticity is directed parallel to the axis of the cylinder. We assume that the axis of the cylinder is, in turn, parallel to the z-direction. In other words, we consider a cylindrical region of fluid with circular cross-sectional with radius R and uniform vorticity ω in the $\pm z$-direction, as shown in Figure 6. The left panel of Figure 6 shows the situation looking in the negative y-direction. The cylinder of vorticity is cut-away to reveal arrows showing the local vorticity within the cylinder to be in the positive z-direction. The sign of the vorticity is indicated by the small circles to be positive for clockwise rotation with respect to the z-axis, which is pointed out of the paper.

We might think that this cylinder of vorticity could be conceived as an infinitely long, thin rigid rod spinning about its axis. However, this would be incorrect. Rather, the thin cylinder of vorticity should be conceived as composed of even thinner cylinders of air, each with vorticity ω. In fact, the difference between a rotating solid cylinder, or rod, and a cylindrical region of uniform vorticity is profound. The vorticity in the cylinder of air entails that the air outside the cylindrical region of vorticity move in a way that is consistent with this

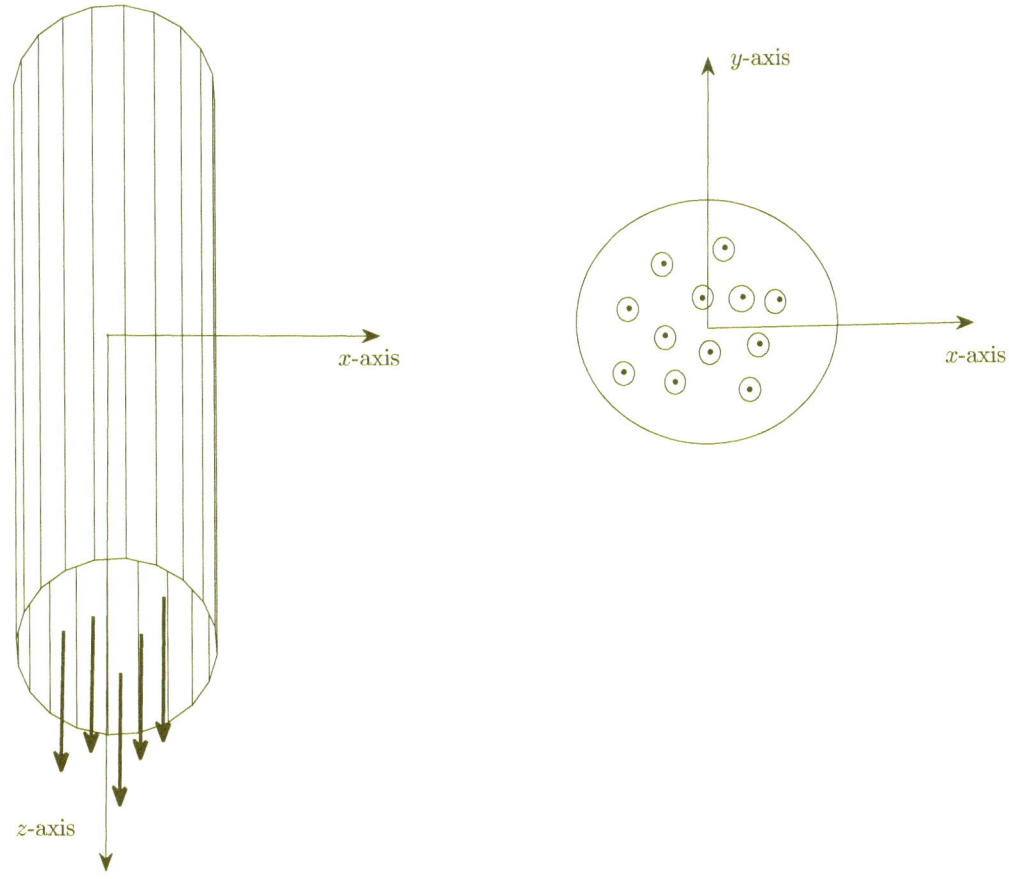

FIGURE 6. Vortex tube

region of rotational air motion provided that the air particle velocity is continuous. This would not be the case for a solid rod in the Euler model where viscosity is neglected. The absence of viscosity permits the velocity of the rod's surface to be discontinuous with that velocity of the air at that surface. Thus, the rod spins without any movement of the surrounding air. One of the distinguishing features of vorticity is that there must be motion of air outside regions of vorticity in the Euler model. Further, the motion of the air outside the region of vorticity is not rotational motion.

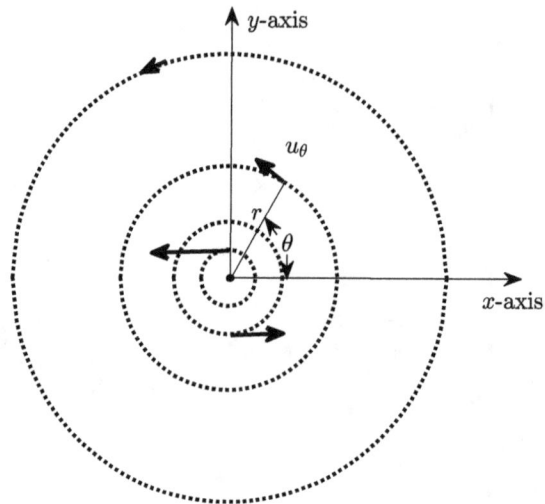

FIGURE 7. Line vortex

What determines the motion outside of these cylindrical regions of vorticity? It is not the vorticity ω alone, but it is the product of the vorticity and the cross-sectional area of the cylinder that contains the vorticity, A_ω. This is equivalent to something called *circulation* Γ, where
$$\Gamma = \omega A_\omega . \tag{4}$$
Note that the sign of Γ is the same as the sign of ω.

We can imagine shrinking the cross-sectional area of the cylinder to zero and increase the magnitude of vorticity to infinity at the same time, so that circulation of the cylinder remains constant. With the radius going to zero, we obtain a line parallel to the z-axis. This is a *line vortex*. In the x-y plane, a line vortex appears as a point.

The line vortex induces the flow shown in Figure 7. This figure shows the streamlines of the flow near a line vortex located at $(x, y) = (0, 0)$. The streamlines, which show the paths followed by air particles in steady flow are circular. However, the flow, or motion, is not rotational anywhere, except at the line vortex itself, where $(x, y) = (0, 0)$. While the air particles translate in circular paths, the particles themselves are not rotating about any axis that runs through the particle. The arrows indicate the direction and magnitude of the particle velocities along various circular paths. Figure 7 shows a particle at the distance r that is at angle θ with respect to the x-axis. The particle speed is denoted $|u_\theta|$, which is constant along circular streamlines. The streamlines have been drawn so that the particle speeds on each streamline decrease by one-half compared to the neighboring, more inward streamline. It can be shown that (Howe 2007)

$$u_\theta = \frac{\Gamma}{2\pi r}, \qquad (5)$$

where r is the distance to the line vortex.

Suppose that there is, in addition to the flow shown in Figure 7, an air particle velocity **U** in the x-y plane imposed near the line vortex. It turns out that the line vortex moves with this same velocity **U**. Further, if we remain in the frame-of-reference of the moving line vortex, the picture in Figure 7 remains valid.

The vortex sheet

Another idealization of vorticity is something called the *vortex sheet*. Suppose we abut cylindrical vortex tubes possessing small radii R each containing uniform vorticity ω with their axes directed parallel to the z-axis, as shown in Figure 8. In the x-y plane, this configuration is shown in Figure 9 by the symbol "X", which means that the vorticity is negative and the air particles are spinning counter-clockwise with respect to the z-axis. In Figure 8, note that the arrows representing the vorticity now point in the negative z-direction. This is indicated in Figure 9 with the use of the X.

The abutting vortex tubes, each have circulation $\Gamma = \omega(\pi R^2)$ This means that the circulation per-unit-length of x

$$\text{circulation per-unit-length} \equiv \mathcal{C} = \frac{\Gamma}{2R} = \frac{\pi R \omega}{2}. \qquad (6)$$

Before we take limits as $R \to 0$ and $\omega \to \pm\infty$, keeping the circulation per-unit-length $\mathcal{C}$ constant, let's note the particle velocity (in the x-direction) at the top and bottom of each vortex tube. At the top of the vortex tube (maximum y-coordinate) and at the bottom of a vortex

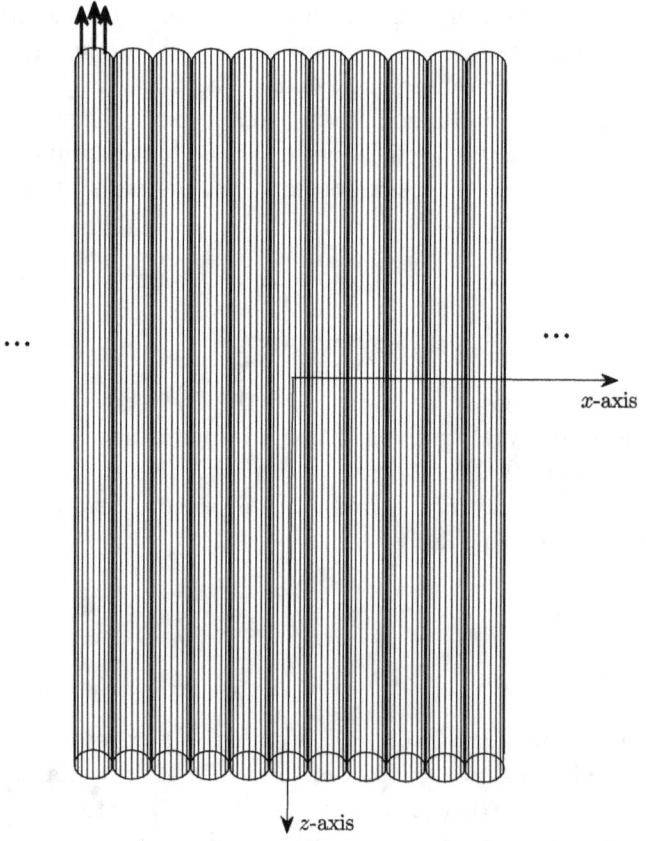

FIGURE 8. Cylindrical vortex tubes abutting one another

tube (minimum y-coordinate) the particle velocity are, respectively,

$$V = -\Omega R = -\frac{\omega}{2}R = -\frac{\mathcal{C}}{\pi} \quad \text{and} \quad -V = \frac{\mathcal{C}}{\pi}, \qquad (7)$$

where V is in the positive x-direction and $-V$ is in the negative x-direction.

With Equation (7), as we create the vortex sheet by letting as $R \to 0$ and $\omega \to \pm\infty$ keeping $\mathcal{C}$ constant, we leave the V and $-V$ unchanged. Thus, the vortex sheet means that there is a jump of particle velocity in the plane of its surface and in the direction orthogonal to the vorticity. If we change or frame of reference to be one traveling at

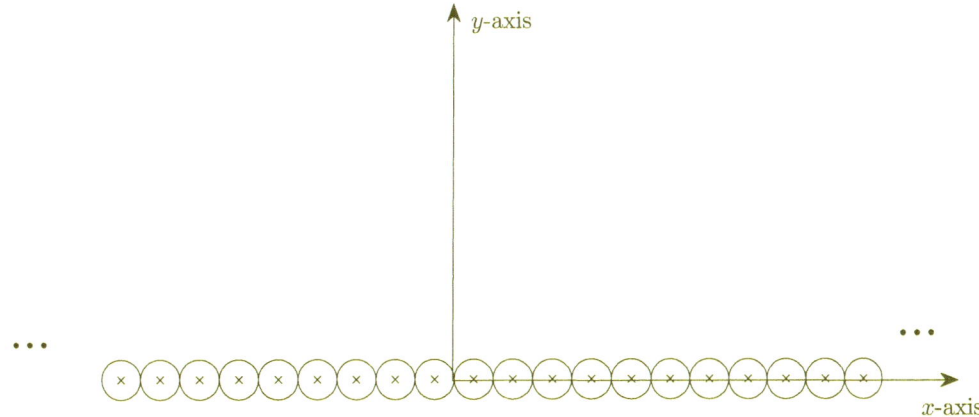

FIGURE 9. Cylindrical vortex tubes abutting one another in two dimensions

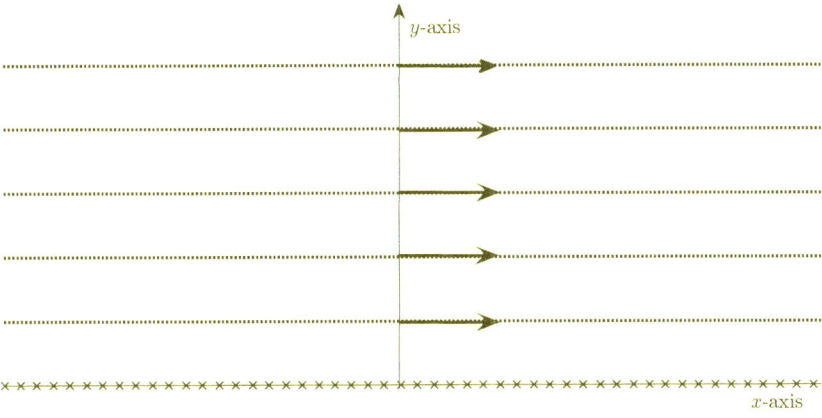

FIGURE 10. Velocity induced by a vortex sheet

constant particle velocity $-V$ along the x-direction, then the vortex sheet represents a situation with zero particle velocity below and particle velocity $U = 2V$ above. It turns out that without any other solid surfaces or vorticity that the particle velocity is uniformly U in the x-direction above the vortex sheet. This is shown in Figure 10.

The vortex sheet in Figure 10 also convects. For the vortex sheet, it is the flow induced by the sheet itself that causes the sheet to convect. It is mathematically consistent to have the sheet convect with signed speed $U/2$ for the case shown in Figure 10. However, there is no substantial change in configuration due to convection, because it has infinite extent.

Momentum conservation

We end this section by stating momentum conservation for the steady Euler model with constant density and no entropy variation. First we need to state that in such a model there is a nonlinear term called the *Lamb vector*, which involves both vorticity and particle velocity. We denote the Lamb vector with the symbol $\mathcal{L} = \mathcal{L}(\omega, \mathbf{u})$. This is the part of the nonlinear term that was neglected in the calculation of particle acceleration under the acoustic approximation in Equation (8) of Chapter 2. Further, $|\mathcal{L}| = 0$ wherever the vorticity $\omega = 0$ or the particle speed $|u| = 0$. Because $\mathcal{L}$ is a vector, it has a direction, and we let the variable s be the coordinate in this direction. Momentum conservation for the specified model can be stated as

$$\frac{\Delta_s(p + \rho_0 u^2/2)}{\Delta s} \approx -\rho_0 |\mathcal{L}| . \qquad (8)$$

where the approximation improves as Δs gets small. In words, Equation (8) says that there is a spatial rate-of-change, or gradient, of pressure head $p + \rho_0 u^2/2$ is opposite to the direction of the Lamb vector and this gradient has magnitude proportional to the magnitude of the Lamb vector. In particular, where there is no vorticity, there is no change in pressure head: only regions separated by vorticity can possess different pressure heads.

We can make this equation a little more understandable if we apply it to the vortex sheet shown in Figure 10. It can be shown that $|\mathcal{L}| = u(y)\omega$, where $u(y)$ is the particle velocity in the x-direction. Also, $\mathcal{L}$ is in the negative y-direction. In this case $\Delta s = -\Delta y$ and $\Delta_s(p + \rho_0 u^2/2) = -\Delta_y(p + \rho_0 u^2/2)$. Momentum conservation, Equation (8), states that traversing this thin region of vorticity from below to above the sheet

$$\frac{\Delta_y(p + \rho_0 u^2/2)}{\Delta y} \approx \rho_0 u(y)\omega . \qquad (9)$$

In fact, we can solve this relation as the approximation to a vortex sheet gets better, that is as $R \to 0$. Quoting the result in the limit as

$R \to 0$, where from Equation (7) we have $\omega R \to 2V = U$, the jump in pressure head above the vortex sheet from the pressure head below the vortex sheet is

$$\left[(p + \rho_0 u^2/2)\right]_y = \frac{\rho_0}{2} U^2 . \qquad (10)$$

where the square brackets $\left[\,\cdot\,\right]_y$ are used to denote jump in the y-direction across the vortex sheet. There is a derivation of this result in Howe (2007). With reference to Figure 10, Equation (10) implies that the pressure p is the same on both sides of the vortex sheet.

Summary of vorticity

Both the line vortex and the vortex sheet are used as mathematical idealizations to model real air flows. The line vortex can be used to model very localized, long cylindrical regions of vorticity, such as a tornado funnel. The vortex sheet can often be used to model shear flows where there are neighboring regions of air moving at different velocities. And, although this involves a solid with adjacent air, it turns out that the shear flow of air along the surface can often be idealized as a vortex sheet.

At this point we note that patterns of vorticity help determine the flow outside of the vorticity. Conversely, irrotational air motion can require that certain patterns of vorticity be present elsewhere. For instance, when there is shear flow we can infer the existence of a vortex sheet in the region of greatest change of air particle velocity.

We have seen that vorticity can move, or convect, according to the surrounding air particle velocity. We have not discussed much about how vorticity comes to exist, or how it may cease to exist. The Euler model does not allow either phenomenon in the body of the air away from solid boundaries. In fact, the only way to get vorticity to cease to exist is to leave behind the Euler model and to allow viscosity. We do not make this step in this book, other than to examine air flow near a solid surface. However, the Euler model does permit vorticity to come into existence at solid surfaces. How vorticity comes to reside in the air away from the solid boundary is a part of the next section.

Dynamics of rotational air motion

Air flow in the vocal tract always involves vorticity, and the acoustic approximation does not allow for such rotational air motion. We explore the use of the Euler model to describe the air flow in regions with considerable rotational air motion. In particular, we explore the Euler model under the quasi-steady approximation as a model for flow in the glottal region and other highly constricted regions of the vocal tract. These considerations lead us to an exploration of rotational air motion and its effects on air flow dynamics. We focus on the forces between the air and solid boundaries in this exploration.

More considerations of length scale

We now examine the conditions under which the approximation called the Euler model is valid. In the process, we relate the Euler model to situations in the vocal tract. First we introduce some notation. Let $\mathcal{F}_{max}$ denote the highest frequency of the air motion that we consider. The air motion may have motions of higher frequency, but we consider only those of frequencies up to $\mathcal{F}_{max}$. This provides a minimum time scale of $T_{min} = 1/\mathcal{F}_{max}$. Further, we let ℓ be the length scale of the region under consideration. So, for instance, for the glottis at the larynx in an adult we can take ℓ to be approximately 1 or 2 cm. Similar length scales can apply at the teeth for sibilant fricatives.

In order for the flow to be approximately incompressible, we need

$$\ell \ll c_0 T_{min} = c_0/\mathcal{F}_{max} = \lambda_{min} , \qquad (11)$$

where c_0 is the speed of sound defined in Equation (9) of Chapter 1, and λ_{min} is the wavelength of sound corresponding to $\mathcal{F}_{max}$. We require that the length scale of the region of interest to be much smaller than the wavelength of sound that corresponds to the highest frequency under consideration. [In the case that $\ell = 1$ cm we can take $\mathcal{F}_{max} \leq 3,500$ Hz. However, often, we are not stringent about frequency restrictions, because the resulting theories can still provide practical answers. Recall that we also neglect phenomena such as heat conduction that could cause density variation.] Restricting frequency from above allows us to neglect variations in density over the region of interest. Also, this means that signals travel from one end of the region to another end in durations much smaller than T_{min}. Compared to acoustic signals traveling over regions on the order of λ_{min}, the time it takes to traverse the region of length scale ℓ is very small. Thus, to say that flow is incompressible is equivalent to approximating the speed of sound to be infinite.

A necessary part of the Euler model is that we can neglect viscosity. Viscosity appears in a term of the equation for momentum conservation before any approximations are made. A factor in this term is a physical constant called the kinematic viscosity ν. For normal atmospheric conditions $\nu \approx 0.23$ cm^2/s. Consider steady flow with the characteristic length scale ℓ and characteristic particle speed $|U|$. For the case of the glottis or a tight tongue constriction, we could consider $\ell \backsim 1$ or 2 cm and $U \backsim 1000$ cm/s. In an argument beyond the scope of this book, the effect of the term containing the kinematic viscosity can be shown to be related to the other terms by a dimensionless number, called *Reynolds number* (Batchelor 1967)

$$\mathcal{R} = \frac{|U|\ell}{\nu} . \qquad (12)$$

The larger $\mathcal{R}$, the less important the viscous term in regions away from solid boundaries. Near solid boundaries viscosity is always important, and this is discussed below. For the values of the quantities provided above, we would have $\mathcal{R} \backsim 5000$ or so. With this magnitude for the Reynolds number we would expect that viscous stress can be neglected in the air flow away from solid boundaries. [Viscous stress is the force per unit area that is provided by viscous friction. Where pressure is a stress that acts in directions orthogonal to mathematical surfaces in the air, viscous stress works tangentially to these surfaces. To neglect viscosity is to neglect viscous stress.]

On the other hand, we must have that the speed $|U|$ is much less than the adiabatic speed of sound c_0. That is, we consider only low *Mach number* flow because we are approximating the density of the air to be the constant rest density ρ_0.

$$M = |U|/c_0 \ll 1 . \qquad (13)$$

It would seem that the Euler model, which neglects viscous stress, is, indeed, a valid model for air motion in the vocal tract. However, even with high Reynolds number flows, viscous stress is always a factor near solid boundaries. In the Euler model for steady flow, the shear flow can be pictured as a vortex sheet that is contiguous with the solid surface,

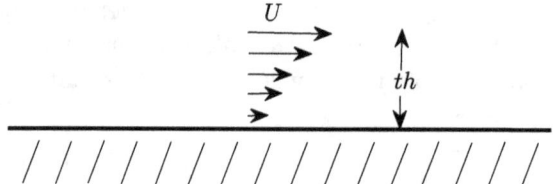

FIGURE 11. Boundary-layer flow

if the shear flow is confined to a thin region next to the solid surface. By thin we mean that the shear flow is such that the speed of the flow goes from zero to the external flow speed $|U|$ in a short distance, th, along a line perpendicular to the sold surface to the external flow, as shown in Figure 11. The region is termed the *boundary layer*. It can be shown that the ratio of the thickness of a steady boundary layer, th to the length scale ℓ is (Lighthill 1986)

$$\frac{th}{\ell} \approx \frac{1}{\sqrt{\mathcal{R}}}. \tag{14}$$

Therefore, a large Reynolds number $\mathcal{R}$ means that the boundary layer is thin compared to the length scale ℓ, and that the steady boundary layer can be pictured as a vortex sheet on the surface of the solid in an Euler model air flow.

We can only give the outline of an argument as to why boundary layer thickness follows Equation (14). There are two processes that move vorticity from one place to another: convection by flow and diffusion by viscosity. Vorticity in the boundary layer is convected in the direction of the flow just outside the boundary layer, just like the vorticity in a vortex sheet in the main flow. Let this be the horizontal direction. Because the external flow is along the expanse of the solid surface, so too the vorticity of the boundary layer is convected along the expanse of the solid surface. Vorticity in the boundary layer also moves away from the solid surface, in the vertical direction, by a diffusion process governed by viscous stress. The ratio of the rate at which vorticity travels due to viscous stress diffusion mechanism to rate it moves by the convection mechanism is also given by th/ℓ. This ratio is given by Equation (14). When viscous stress is neglected in the main flow we use the vortex sheet to represent the boundary layer.

Viscous stress is also neglected in the acoustic approximation. However, unsteady boundary layers do occur at solid surfaces due to nearby motion described by acoustic waves. As we saw in Chapter 11 these boundary layers have a small overall effect on air motion outside of them for all frequencies, other than extremely low frequencies, in the acoustic approximation. The reason for this has to do with the fact that the region of large viscous stress in oscillatory air motion remains very close to any solid boundaries, independent of Reynolds number.

The duct system

For the remainder of the chapter, the vocal tract is pictured as a duct system with rectangular cross-sections. This greatly simplifies matters without losing the important aspects of the discussion. The axis along the duct system is the x-axis, as shown in Figure 12. Orthogonal to the x-axis, and in the plane of the paper is the y-axis. Orthogonal to both the x and y axes is the z-axis directed out of the paper. The depth of all ducts in the z-direction is the same, and this depth is denoted d.

Irrotational flow

Air flow with no vorticity is called irrotational flow. Motion under the acoustic approximation is irrotational, but irrotational motions under the Euler model (or approximation) can also be considered. Again, we consider steady, irrotational air flow.

Figure 13 shows the streamlines below the midline of an idealized glottal section in the midsagittal plane, where the mirror image for the streamlines is assumed above the midline. The situation has been idealized so that there is uniform air flow upstream and downstream of the glottal region. The flow that is shown represents the unique irrotational flow that conforms to the Euler model with uniform air flow specified upstream of the glottal region.

More realistically, we need to add the rotational motion in the boundary layers in the form of vortex sheets along some of the solid surfaces, assuming a high Reynolds number flow. The Reynolds number $\mathcal{R}$ in Equation (12) is computed in terms of the length scale ℓ based on the dimensions of the vocal folds, and U, the particle speed

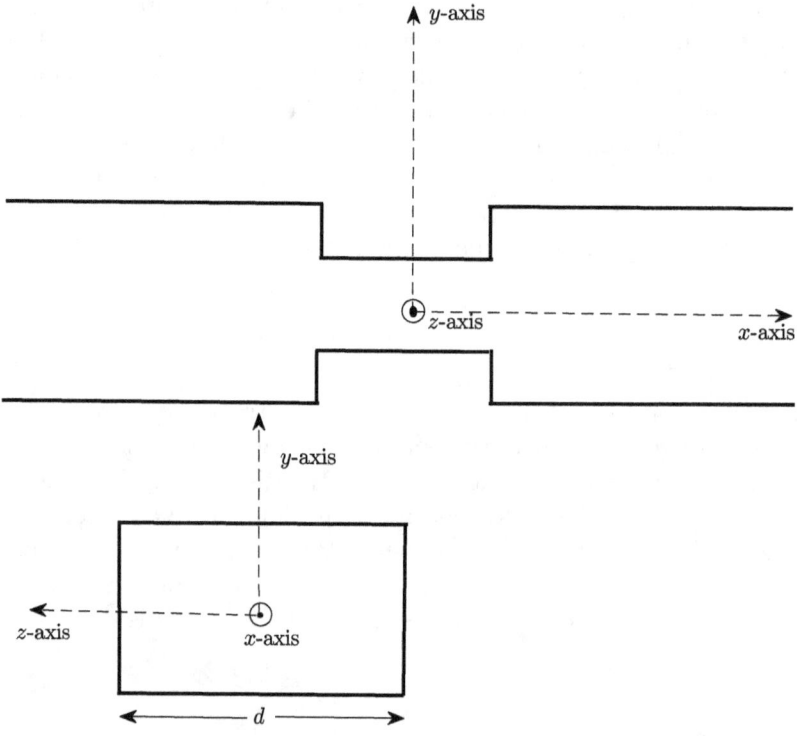

Figure 12. Duct system

through the glottis. U is assumed to be positive here. Note that this particle speed is much greater in the glottal region between the vocal folds than the particle speeds in the tracheal and supra-glottal regions. This is shown in Figure 13 by the fact that the streamlines are so close together in the glottal region compared to the tracheal or supraglottal regions. We have approximately incompressible flow for the length scales of the glottis at sufficiently low frequencies. We also make the quasi-steady approximation, and actually only consider steady situations.

Figure 13 may appear to represent a realistic flow situation. However, the unreality of this flow becomes apparent when we consider the power that is required to move the flow down the duct. Experimentally, at moderate-to-high glottal particle speeds, 500 cm/s to 5000 cm/s,

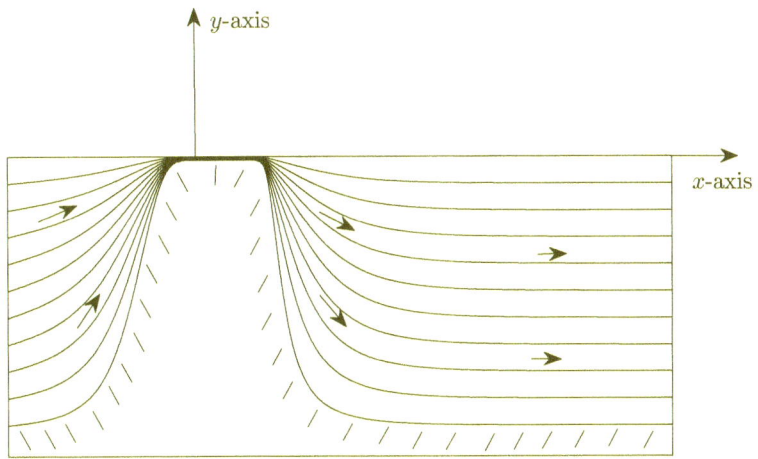

FIGURE 13. Proposed flow over obstruction in a duct

say, we would find that it can take substantial amounts of power to move the air. This expenditure of power, or energy per unit time, is much higher than what would be expected if the power was needed to overcome only viscous friction between the air and the solid boundaries of the duct. There must be another force exerted between the vocal folds and air in the direction of flow along the duct axis. We explore the conservation equations to find this force.

We reconsider mass, momentum, and energy conservation, when steady air flow is in a tube or duct system, as in Lighthill (1986). The conservation equations in Equations (1), (3), and (8) were derived without considering external forces on the air. External forces can be exerted by solid boundaries that contain the air flow of interest. Below, we rewrite the momentum conservation equation, Equation (8), to account for external forces due to solid surfaces.

A configuration of ducting is shown in Figure 14, where there is a straight duct with, possibly, a change in cross-sectional area, that is rather gradual along the axis of the duct. The air is assumed to flow from left to right, and the pressure head $p + \rho_0 u^2/2$ is constant throughout the duct. This is the case, because there is no vorticity anywhere in the duct, except at the duct walls. Because we assume that there is no vorticity in the main flow of the duct, the magnitude

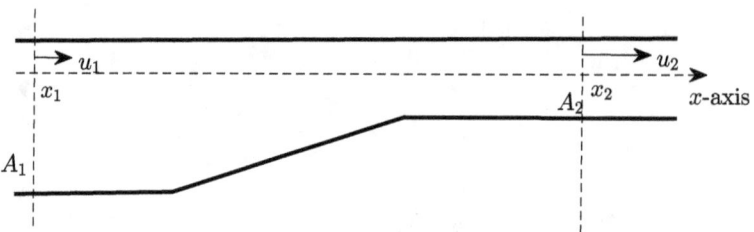

FIGURE 14. Flow in a duct with a gradual change in cross-sectional area

of the Lamb vector $|\mathcal{L}|$ is zero in Equation (8) for this situation, and pressure head must be the same throughout the flow.

We consider the region $\mathcal{B}$ of the duct between the locations x_1 and x_2 on the duct axis, at large distances from the sloping wall, so that the flow is in the x-direction at these two locations. The cross-sectional area at location x_1 is A_1 and the cross-sectional area at location x_2 is A_2. We let u_1 and u_2 be the corresponding air particle velocities, which are directed along the x-axis.

The rate that momentum in the x-direction that enters region $\mathcal{B}$ is $(\rho_0 A_1 u_1)u_1 = \rho_0 A_1 u_1^2$, and the rate that momentum in the x-direction that leaves region $\mathcal{B}$ is $\rho_0 A_2 u_2^2$. The difference of these quantities must equal the net force in the x-direction on the region of air, $\mathcal{B}$. This is a statement of x-direction momentum conservation. The force in the x-direction on the air in the region at x_1 is $p_1 A_1$, and the force in the x-direction on the region at x_2 is $-p_2 A_2$. There is the possibility of a force in the x-direction exerted by the duct on the air, which is denoted F_x. Equating rate of momentum change in the x-direction to the force in the x-direction, we obtain

$$p_1 A_1 - p_2 A_2 + F_x = \rho_0 A_2 u_2^2 - \rho_0 A_1 u_1^2 . \qquad (15)$$

[As discussed above, the Lamb vector is zero.]

Energy conservation for steady internal air flow in the Euler model can also be written. The rate at which pressure at A_1 does work on region $\mathcal{B}$, i.e. the power input to region $\mathcal{B}$ is $p_1 A_1 u_1$. The corresponding quantity at A_2 is $-p_2 A_2 u_2$. This power input affects the rate that kinetic energy leaves region $\mathcal{B}$, which is $(\rho_0 A_2 u_2)u_2^2/2 - (\rho_0 A_1 u_1)u_1^2/2$.

Thus,

$$p_1 A_1 u_1 - p_2 A_2 u_2 = (\rho_0 A_2 u_2)\frac{u_2^2}{2} - (\rho_0 A_1 u_1)\frac{u_1^2}{2}$$
$$= \rho_0 A_2 \frac{u_2^3}{2} - \rho_0 A_1 \frac{u_1^3}{2} \ . \tag{16}$$

If we invoke mass conservation for incompressible flow, Equation (1), then Equation (16) can be rewritten in a way similar to energy conservation in Equation (3)

$$p_1 - p_2 = \frac{\rho_0}{2} u_2^2 - \frac{\rho_0}{2} u_1^2 \ . \tag{17}$$

This is simply the Bernoulli equation.

Equations (15) and (17) can be used to eliminate u_1 and u_2, and to express F_x in terms of the remaining quantities.

$$F_x = -(p_1 A_1 + p_2 A_2)\left(\frac{A_1 - A_2}{A_1 + A_2}\right) \ . \tag{18}$$

In the case of the duct shown in Figure 14, there is a net force on the air in the negative x-direction. This force is exerted on the air by the sloping wall of the duct, and because it is in the direction opposite to the flow, it is termed a *drag force*.

Let's consider the region surrounding the glottis in Figure 13. We place location x_1 somewhere upstream of the glottis and location x_2 somewhere downstream of the glottis. We wish to find the force exerted by the vocal folds on the air in the direction of the duct axis, which is in the x-direction. If we assume that the downstream side of the duct opens to the atmosphere, then we can take $p_2 = 0$. When $A_1 = A_2$, Equation (12) says that the drag force is zero. We know, empirically, that this is not the case. Further viscous friction is not nearly large enough to account for the observed magnitude of F_x. The ingredient that is missing is vorticity in the air flow.

We experience drag forces in our running, driving a car, flying in an aircraft, or moving in a boat. In these cases, a solid object is moving through air or water, instead of air or water moving across a solid object. The situation of the glottis and vocal folds is more like that of a building in a strong wind, where the air moves and the solid is relatively stationary. In order to have drag forces in these examples, vorticity needs to be present in the body of air flow away from solid boundaries. Vorticity "fixes" the problem of no drag force for flow in a vocal tract duct when $A_1 = A_2$.

Flow separation

In the Euler model, vorticity cannot be generated in air alone but only at the surface of a solid. Further, vorticity at solid surfaces has very limited influence on the flow outside the boundary layer. The reason for this is that the solid surface has the effect of canceling the long range influence of the rotational motion in the boundary layer, which remains thin due to high flow speeds. [To prove this would require a more in-depth course in fluid mechanics.] Thus, vorticity must enter the air flow away from the surfaces in order for it to have a substantial influence on the air flow.

So how does vorticity end up in the air away from solid surfaces? *Flow separation* is a mechanism by which the vorticity of the boundary layer can enter the main flow of air. We include viscous stress in the discussion, so we put aside the Euler model briefly. The resulting description is, perhaps, the most complicated one encountered in this chapter. We follow Lighthill (1986) to explain how vorticity leaves the surface of the solid into the air flow away from the solid. We assume that the boundary layer is thin because the Reynolds number $\mathcal{R}$ is large. This also means that vortex diffusion by viscous stress is much weaker than vortex convection, so that vorticity moves a much greater distance along the surface than it moves away from the surface by diffusion [see Equation (14)].

We consider Figure 13 again. The streamlines get very close together going from upstream of the glottis into the glottis. This means that the corresponding stream tubes are getting much narrower in cross-sectional area, because the width of the channel d in the dimension into the paper does not change along its length across the paper. We know that this means that the particle speeds within the stream tubes increases going into the glottis. The opposite occurs leaving the glottis: the particle speeds within the stream tubes decrease going from the glottis to the downstream side of the glottis.

With the flow going from left-to-right, the vorticity in the boundary layer is pointed in the negative z-direction. Figure 15 is a close-up

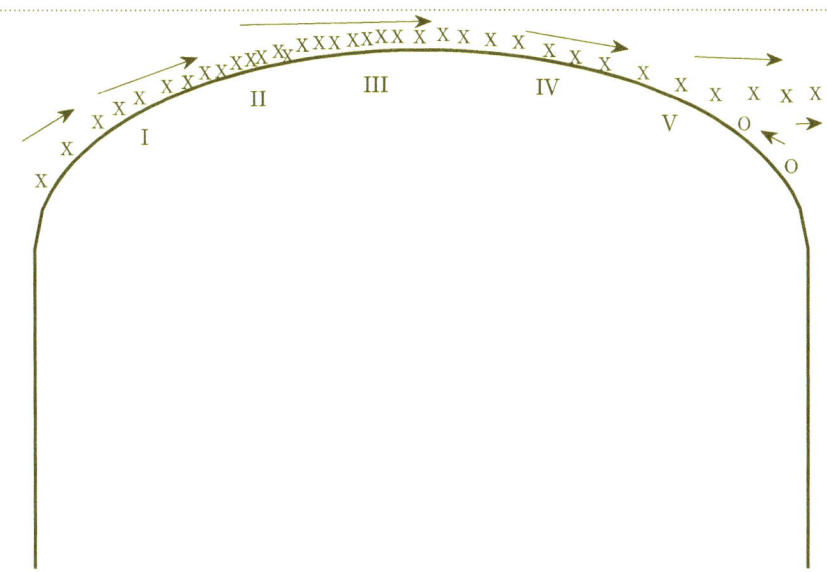

FIGURE 15. Flow separation from an obstruction in a duct

of the glottal duct without the streamlines. The dashed line is the centerline of the duct. Consider a stream tube that is adjacent to the boundary layer vortex sheet and the solid surface. The air in the stream tube is moving at a higher speed at point II than upstream at point I. This means that there is more negative vorticity in the boundary layer at II than at I. [Negative vorticity is indicated by X's in Figure 15. The density of the X's indicates the amount of negative vorticity.] Thus, even with negative vorticity convecting from I to II, this only goes to help supply the extra necessary negative vorticity at II. There is no problem in this scenario with the boundary layer remaining at the surface of the solid.

However, let's examine the air flow leaving the glottis, say from point III to point IV. The quantity of the negative vorticity at IV is less than the negative vorticity at III. But vorticity at III is convected to IV. This means that there is an excess of negative vorticity at IV, unless there is positive vorticity production at IV. As long as the change in the flow speed in the stream tube just outside the boundary layer is not too large, enough negative vorticity is diffused away before reaching

IV, and there does not need to be positive vorticity production. Thus, the flow can stay "attached" to the solid boundary as long as there is not too much slowing of the flow. The fact that the vortex sheet stays attached to the surface after a small amount of deceleration of flow is confirmed experimentally for a vibration model of vocal folds (Šidlof, Doaré, Cadot, Chaigne 2011). However, this cannot be the case for position V because viscous stress diffusion cannot keep up with convection. So at V, there must be positive vorticity production, and a reversed flow outside of any boundary layer appears between IV and V, and the boundary layer separates from the solid surface and enters the flow away from the solid. [The positive vorticity is indicated by the small open circles in Figure 15.] This separation of the boundary layer vortex sheet results in the vortex sheet entering the air flow away from the solid surface. This process of flow separation is sometimes referred to as *vortex shedding*.

Now that we have made a physical argument for how a vortex sheet can enter flow away from solid surfaces, we return to the Euler model. We simply stipulate that the boundary layer vortex sheet separates into the main flow at a certain location on the surface of the solid.

Drag force on air near the vocal folds

Suppose that the trachea and supraglottal ducts have equal cross-sectional area, A, and height $2h$, so that $A = 2hd$. The minimum distance between the vocal folds 2ζ, as shown in Figure 16. We set $A_g = 2\zeta d$ as the minimum area of the glottis. For simplicity, symmetry across the midline is assumed.

A *jet* is formed between the vortex sheets that result because of flow separation from the two vocal folds. A jet is a region of relatively high particle speed surrounded by a region, or regions, of low particle speed. Initially, we take the air particle speed surrounding the glottal jet to be zero. The jet can have a cross-sectional area, A_σ, that is different from the minimum cross-sectional area of the glottis A_g. Let σ be the ratio of these areas.

$$\sigma = \frac{A_\sigma}{A_g}. \tag{19}$$

This means that the distance between the two vortex sheets is $2\sigma\zeta$ [Figure 16].

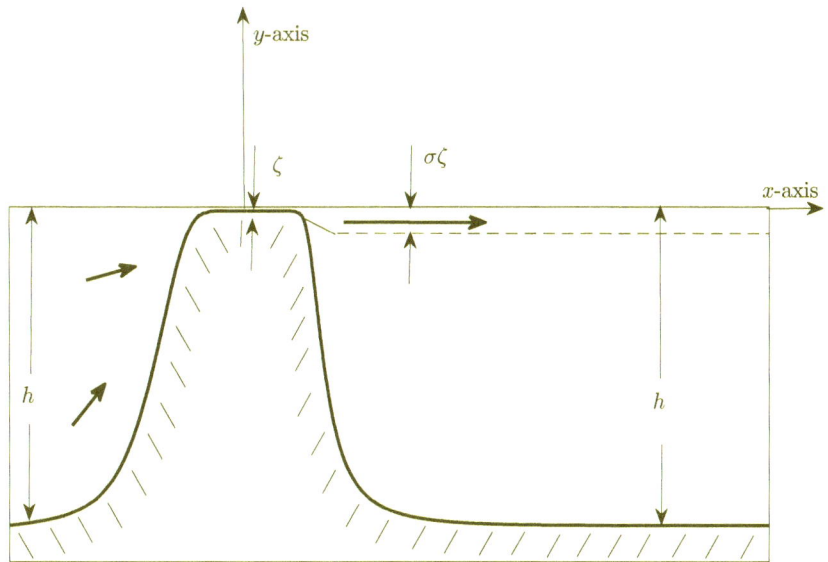

FIGURE 16. Flow over an obstruction in a duct with separation and vortex sheet

The effect of the vortex sheets that bound the jet is to greatly narrow the effective cross-sectional area for air flow downstream of boundary layer separation. Therefore, instead of using $A_2 = A$ in Equation (13), we have $A_2 = A_\sigma$, where A_σ is the cross-sectional area of the jet. Also, we let $p_1 = p_T$ be the tracheal pressure just upstream of the glottis. For now, we assume that $p_T > 0$.

$$F_x \approx -p_T A \left(\frac{A - A_\sigma}{A + A_\sigma} \right). \tag{20}$$

For $A_\sigma \ll A$

$$F_x \approx -p_T A, \tag{21}$$

and the vocal folds "feel" the full force, $-F_x$ of the tracheal pressure p_T. In this approximation, the pressure on the upstream side of the vocal folds is close to p_T above atmospheric pressure, and on the downstream side of the vocal folds the pressure above atmospheric pressure is assumed to be zero.

The vortex sheets that bound the jet indicate that a fundamental change occurs in the flow from the irrotational situation. It does this in such a way as to have the pressure on the downstream faces of the vocal folds be essentially atmospheric pressure, while the upstream face experiences the higher tracheal pressure. Pressure head is different between the region behind the vocal folds and the region in front of the vocal folds.

From Equation (20), it can be seen that the drag force F_x is a function of time when A_σ is a function of time. Of course this is the case if the vocal folds open and close in their vibration. It is shown below that the resulting time-variable drag force creates a sink for the energy of irrotational air motion, which goes into creating energy for vorticity. On the other hand, the fluctuating drag force can be considered to be an acoustic source.

Sometimes shed vorticity is random in nature, particularly when it is part of a turbulent flow. In this case the drag force also possesses a random time-variable component. This is the scenario that is provided in sibilant frication, where a turbulent jet strikes the teeth and the shed vorticity is a part of the resulting turbulent flow.

Other views of the relation between vorticity and forces between solids and air

We have provided an elementary view of how vortex shedding can lead to a drag force between air and obstructions in a duct. It would require a, more or less, complete course in fluid mechanics to understand the relation between vorticity and the forces between solids and air in greater generality. The references listed below, particularly Batchelor (1967), Lighthill (1986), and Howe (2003, 2007), can help a mathematically inclined and interested reader understand this relation.

We content ourselves with some basic statements instead of a thorough course. Vorticity is one way to provide impulse to air. As discussed in Chapter 4, impulse has the dimensions of momentum, and thus, vorticity adds momentum to air outside of the regions of vorticity. In fact, we have seen in the examples of a line vortex and a vortex sheet above that there is momentum outside the region of vorticity that is completely the result of the vorticity. The Lamb vector in Equation (8) expresses the fact that vorticity adds or subtracts momentum to or from the rest of the flow. The time rate-of-change of momentum is force. It turns out that the time rate-of-change of *vortex impulse* can be the dominant part of the force between the air and surrounding solids.

In many situations, a non-zero time rate-of-change of the vortex impulse exists due to changes in either the location of vorticity or the amount of vorticity in the air flow. We have seen that vortex sheets enter the air flows because of flow separation. This process has a beginning when the flow starts, so that the vortex sheet grows as time increases from the start of the flow. This is a situation where the amount of vorticity increases with time, and, thus, there is a time rate-of-change of vortex impulse. Further, a line vortex convected by irrotational flow over a fence is as an example of vortex impulse that changes with time. Here the amount of vorticity remains constant, but the location of the vorticity changes with time, which gives a time rate-of-change of impulse. The result of this changing impulse is a fluctuating force between the air and the fence that acts as an acoustic source (McGowan & Howe 2006).

A slightly different view of the relation between vorticity and forces between solids and air is obtained in a theorem proved by Howe (1995, 2003, 2007). The theorem relates the Lamb vector $\mathcal{L}$ in Equation (4) to the flow that *would* occur around a solid surface if there was no vorticity. The Lamb vector resembles what is termed the *Coriolis force* in classical mechanics for rotating rigid bodies (Goldstein 1950). In the relation derived by Howe, it appears as though the Coriolis-like force associated with the vorticity acts on the irrotational flow that *would* occur if vorticity was not present. Consider the flow after the separation point, which would have flowed down the leeward side of the vocal fold as in Figure 13, but is redirected to flow downstream near the separation point as in Figure 16. The flow is accelerated both in terms of magnitude of particle speed, because it is restricted to a thin jet, and in terms of directional change. This is instantiated as a force between the air and the solid surface according to Howe's theorem.

Aeroacoustic sources in speech

The connection between vortex shedding and drag force on an obstruction has just been examined. Drag force is intimately connected with the vorticity that enters the flow due to flow separation. In this section we examine the particular ways that unsteady drag forces go into creating the voice source and sibilant fricatives.

The modern study of sound generated by fluid flow, including air flow, was initiated with the publication of Lighthill's 1952 paper on jet noise. This paper examines the sound generated by turbulent jets when solid boundaries are not present. Also, this paper introduces the idea of the acoustic analogy, which has become to be known as *Lighthill's acoustic analogy*. The acoustic analogy is the idea that regions of unsteady vorticity can be sources, or sinks, of acoustic wave motion for other spatial regions where the acoustic approximation is appropriate. Further, the regions of space dominated by unsteady vorticity can very often be described mathematically with an Euler model. The field that grew out of Lighthill's work is known as *aeroacoustics*.

There has been a very large body of both empirical and theoretical work done in aeroacoustics since 1952, particularly as it pertains to noise generation by aircraft and naval vessels. The concepts and mathematical tools that have been based on Lighthill's acoustic analogy have come to be applied to many sound generation problems, including speech production. Further information on the nature of acoustic sources can be found in the first chapter of Lighthill's *Waves in Fluids* (Lighthill 1978). This book combines prose and mathematics in its description of acoustic sources, but it does not discuss aerodynamic sources in general. For those interested in a more in-depth account of aeroacoustics, Howe's *Theory of Vortex Sound* (Howe 2003) serves as a mathematical introduction to the subject.

Matching

We have seen that physical quantities are best described by different approximations in different regions of the vocal tract. These physical quantities, and their mathematical representations, must agree where the regions come together. That is, the mathematical solutions for the physical quantities from different regions must be matched where the regions adjoin and overlap by adjusting constants that appear in mathematical representations of the physical quantities. This process is called *matching*. Thus, the pressure fluctuations or air particle velocity fluctuations that occur near the glottis, which are best described by the Euler model, must agree with the pressure or air particle velocity fluctuations, which are best described by the acoustic approximation in a region where both approximations are valid. The mathematical technique of matching solutions is a procedure that is not well understood in general, even by some practitioners of the procedure. However, the matching done in recent analyses involving speech are relatively elementary and valid (e.g. Howe & McGowan 2011a).

The reason that we need to match solutions from different regions is because we are ultimately solving a boundary value problem from the lungs up to the mouth opening, and into the atmosphere. We necessarily traverse regions of air where different approximations are appropriate. When source-filter theory was discussed in Chapter 7, the air motion is described by the acoustic approximation from the piston boundary to the mouth. No matching procedure needed to be employed. This is not the case when aerodynamics is considered in more detail. Matching is a mathematical technique that permits us to implement Lighthill's acoustic analogy in many cases.

The quasi-steady approximation and statistical stationarity

The quasi-steady approximation is often made in studies of the physics of speech. Under the quasi-steady approximation for the motion of air, unsteady situations are pictured as a series of steady ones, where air particle acceleration is neglected. In this approximation, terms in the conservation equations with time rate-of-change of air particle velocity or volume velocity are neglected. Therefore, at each instant of time, we consider the air flow to approximate a steady air flow under the same conditions. The quasi-steady approximation is not necessarily appropriate for glottal volume velocity during voicing, so we eventually include a time derivative term for an equation for glottal volume velocity below.

Turbulent flow can also provide for acoustic sources, particularly when it is incident on a solid obstacle. The two defining properties of turbulence is the presence of vorticity and of a statistically random, unsteady flow. Therefore, the quasi-steady approximation is not appropriate when turbulence is a major factor in the air flow. Instead, we assume statistical stationarity for the air flow. This means that expected values of physical quantities, such as particle velocity and pressure, as well as products of these physical quantities, do not change with time.

Glottal volume velocity without vorticity

We go back to consider the case of irrotational flow through and near the glottis shown in Figure 13 with the cross-sectional areas of the trachea and supraglottal tract equal to A. There is no drag force on the vocal folds in the irrotational flow case according to Equation (18).

We imagine that the lungs contract very quickly, so that air flow starts in the lower bronchi and move up into the upper bronchi. There are high pressure fronts moving through the upper bronchi at close to the adiabatic speed of sound c_0 because the motion of air is approximately governed by the acoustic approximation. These fronts eventually enter the trachea providing a pressure above atmospheric of, say, p_I. We imagine that the wave front in the trachea is very abrupt in time and space.

The details of the air flow through the glottis are examined first in a steady situation. Thus, we consider the situation when the glottis is open to some degree, and we can derive such quantities as pressure, air volume velocity, and particle velocity. Based on the small length scale of the glottis we use the Euler model with constant density for air flow. When, eventually, we do allow for vocal fold movement, we do not explain the mechanism by which the the vocal folds begin to open and close, or oscillate, as a result of the airflow through the glottis. We simply assume that they do oscillate at a frequency on the order of 100 Hz.

When the over-pressure p_I impinges on the glottis, air flow through the glottis begins. Let u_g be the air particle velocity at the location of minimum glottal area A_g. Mass conservation given in Equation (1) means that volume velocity Q is the same at the glottal end of the trachea, in the glottis, and just downstream of the glottis, where

$$Q = u_g A_g . \tag{22}$$

Not only is the Euler model valid near the larynx, but also the acoustic approximation can be made just upstream and downstream of the glottis because the flow is irrotational. We denote this pressure

p in the supraglottal tract. The corresponding particle velocity is $u = p/\rho_0 c_0$. Just downstream of the glottis, where the Euler model is also valid, we must have

$$u = \frac{Q}{A}, \quad (23)$$

where A is both the tracheal and supraglottal tract cross-sectional area. It is very convenient to work in a region where both the Euler model for incompressible glottal flow and the acoustic approximation are valid.

There is reflection and transmission at the glottal entrance associated with volume velocity Q. Continuity of pressure and volume velocity would give a pressure in the trachea just below the glottis of

$$p_T = 2p_I - \rho_0 c_0 u = 2p_I - p . \quad (24)$$

Therefore $-p$ is the pressure of the reflected wave traveling away from the glottis in the trachea when p is the pressure of the transmitted wave traveling away from the glottis in the supraglottal tract. Further, using the acoustic approximation we obtain

$$p = \rho_0 c_0 u . \quad (25)$$

By continuity of pressure through the glottis in the steady situation, we have

$$p = p_T . \quad (26)$$

Combining Equations (24), (25), and (26) produces

$$\rho_0 c_0 u = p_I = p_T . \quad (27)$$

Equation (27) just expresses the known relation between particle velocity and pressure in the incident wave from the lungs, which is the only source of energy in this scenario. Of course, multiplying both sides of Equation (27) by twice the duct cross-sectional area A gives

$$2\rho_0 c_0 Q = 2p_I A . \quad (28)$$

We use this later.

The acoustic relation holds between pressure and particle velocity when there is no drag force, with pressure produced by the lungs suddenly rising from zero to p_I, and then remaining constant. However, we would not hear much in the way of sound. We would only hear a pop as the overpressure from the lungs reaches our ears, again neglecting reflection in the supralaryngeal vocal tract. There would be no time variation in the overpressure after the wavefront passes the listener, and, thus, no sound perceived after the pop.

Therefore, we allow time variation in the glottal width 2ζ so that $\zeta = \zeta(t)$ in order to obtain continuous sound from the glottal region. The quasi-steady approximation is not made here, so we find the mass of air that accelerates in the glottal region. In other words, continuity of pressure may no longer be a good assumption when the glottal aperture is varying in area. We must add a lumped mass element to account for the large differences between A_g and A. Suppose that length L_g is such that $L_g A_g$ is the volume of air within the glottis. It can be expected that $L_g \approx 1$ cm. As an approximation for this purpose, the vocal folds are taken to be a thin shutter that has the same depth as the duct, that is d, but has opening width $\zeta(t)$. In the case that $\zeta(t) \ll d$, the length of the lumped mass element is

$$\ell_g = \frac{4\zeta}{\pi} \ln\left(\frac{2h}{\pi\zeta}\right). \tag{29}$$

where $2h$ is the width of the tracheal and supraglottal ducts (Morse & Ingard 1986, Howe & McGowan 2011b).

Because $\zeta = \zeta(t)$ is a function of time, $\ell_g = \ell_g(t)$ is a function of time. Viscous effects not included in the Euler model should be taken into account for the very smallest openings. Viscosity does mitigate the effect of the lumped mass element at very narrow openings, but we ignore this effect here. Equation (28) should be modified to account for time rate-of-change of particle velocity in the glottis

$$\rho_0(L_g + \ell_g)A_g \left(\frac{\Delta_t u_g}{\Delta t}\right) + 2\rho_0 c_0 Q = 2p_I A ,$$

or $\tag{30}$

$$\rho_0(L_g + \ell_g) \left(\frac{\Delta_t Q}{\Delta t}\right) + 2\rho_0 c_0 Q \approx 2p_I A ,$$

where Equation (22) has been invoked. The term $-\rho_0(L_g + \ell_g)u_g(\Delta_t A_g/\Delta t)$ has been neglected in terms of sound production (Howe & McGowan 2012). This completes our examination of the situation without drag force on the vocal folds.

An argument very similar to the one just concluded can be made in the presence of vortex shedding. Instead of both the acoustic approximation and the Euler model being simultaneously valid in the glottal region, we assume that only the Euler model is valid when there is vortex shedding. The region where both the acoustic approximation and the Euler model are valid is moved farther away from the glottal region, but the matching arguments are made with the same result in the acoustic field above the glottis. The vorticity, which decays into turbulence above the glottis, is ignored in this matching. Here, we simply add the effect of drag force due to vortex shedding to the Equation (30) in the next section, supposing that the matching has been accomplished.

Voice source basics

In the discussion leading to Equation (20) and (21), we showed that there is a substantial drag force on the vocal folds due to vortex shedding when the folds are open. We now reference Figure 16 for the following discussion. We note that the change in pressure head in traversing the vortex sheet on the supraglottal side of the glottis from bottom to top is $(\rho_0/2)u_\sigma^2$, because both the jet and the region of air below the jet are at the same pressure p. This difference in pressure head follows from Equation (10).

Recall from Equation (19) that the ratio of the cross-sectional area where the vortex sheet separates from the vocal fold to the minimum glottal area is σ. In our considerations of separation we have $\sigma > 1$. Also, by mass conservation

$$Q = u_g A_g = u_\sigma A_\sigma . \tag{31}$$

where u_σ is the particle velocity in the jet at the place where flow separation occurs.

It turns out that of the drag force on the air is approximately $-(\rho_0/2)u_\sigma^2 A$. This is the argument. The pressure in the trachea at the glottis is p_T. By continuity of pressure head under the quasi-steady approximation $p_T = (\rho_0/2)u_\sigma^2 + p$. [We have not given up on the quasi-steady approximation, except in the equation for time evolution of glottal volume velocity, as in Equation (30)]. Thus, the difference in pressure between the tracheal side of the vocal folds and the supralaryngeal side of the vocal folds is $\Delta p = p_T - p = (\rho_0/2)u_\sigma^2$. For small A_σ, i.e. $A_\sigma \ll A$, the drag force on the air is $F_x \approx -\Delta p A = -(\rho_0/2)u_\sigma^2 A$. Thus, we associate the term $-(\rho_0/2)u_\sigma^2 A$ with the fluctuating drag force.

The voice source involves a fluctuating drag force that is the result of vortex shedding. This drag force is a sink of irrotational air motion into rotational air motion. Later we use this fact to amend our estimates of the amplitudes of the piston in the finite-length tube that can be used as a first-approximation to the voice source in source-filter theory. On the other hand, this same fluctuating drag force term serves as an acoustic source. The details behind this assertion is the subject of the remainder of the present section.

Now we go back and incorporate the drag force due to vorticity into Equation (25). A simple version of this representation comes from Equation (30) with the addition of drag force.

$$\rho_0(L_g + \ell_g)\left(\frac{\Delta_t Q}{\Delta t}\right) + 2\rho_0 c_0 Q = 2p_I A - \frac{\rho_0}{2}u_\sigma^2 A \;,$$

or

$$\rho_0(L_g + \ell_g)\left(\frac{\Delta_t Q}{\Delta t}\right) + 2\rho_0 c_0 Q = 2p_I A - \frac{\rho_0}{2}\left(\frac{h}{\sigma \zeta}\right)^2 u^2 A \;, \qquad (32)$$

or

$$\rho_0(L_g + \ell_g)\left(\frac{\Delta_t Q}{\Delta t}\right) + 2\rho_0 c_0 Q = 2p_I A - \frac{\rho_0}{2}\left(\frac{h}{\sigma \zeta}\right)^2 \frac{Q^2}{A} \;.$$

Because the drag force term is subtracted from the positive $2p_I A$, the drag force is a sink of irrotational air motion into vorticity. The fluctuating drag force is, at the same time, an acoustic source; a fluctuating force is a pressure source. In other words, the drag force term acts as a source of energy for acoustic wave motion according the formal mathematical representation (Howe & McGowan 2011b). The pressure waves that travel down the supraglottal tract are π out of phase with those that travel down the subglottal tract, when they both leave the glottal region. That is, pressure generated on the tracheal side of the glottis is $-p = -\rho_0 c_0 u = -\rho_0 c_0 Q/A$. This is the negative of the pressure on the supraglottal side of the glottis p. This is typical of fluctuating forces when the forces are in a region that has small length scales compared to the wavelength of sound emitted. The pressure source is a *dipole source*, because there are waves of opposite phase traveling in either direction.

In the source-filter theory, on the other hand, the glottal volume velocity Q on the supraglottal side of the glottis is considered to be the source. This is a volume velocity source, which is a *monopole source* for the supraglottal vocal tract. We are not concerned with the subglottal tract in this theory. The glottal volume velocity Q corresponds to the volume velocity of the piston of Chapter 7 in the source-filter picture of vocal tract acoustics. In the present chapter, we have removed the volume velocity from being the source quantity to both the pressure generated by the lungs p_I and the time-varying drag force on the vocal folds as the source quantities.

The final relation in Equation (32) can be written

$$\rho_0(L_g + \ell_g)\left(\frac{\Delta_t Q}{\Delta t}\right) + 2\rho_0 c_0 Q + \frac{\rho_0}{2}\left(\frac{h}{\sigma\zeta}\right)^2 \frac{Q^2}{A} = 2p_I A . \qquad (33)$$

Equation (33) can be designated a *Fant equation* (Howe and McGowan 2011b), because it is a modification of the equation derived for the glottal jet by Fant in his monograph (Fant 1960). The effect of both the first and third terms on the left-hand-side of Equation (33), the acceleration and drag force terms, respectively is to "sculpt" the volume velocity Q into a time varying waveform. This is why we can consider the fluctuating drag force to be a source of sound: without it, and ignoring the effect of the acceleration term, we would hear only the passing of the pressure wave front from the lungs. With the drag force term we obtain a time-varying acoustic wave that is heard as sound.

The exact form that the volume velocity takes depends on how the glottal width at the place of flow separation behaves. We do not pursue vibration mechanisms in this chapter, but a preliminary model for vocal vibration is offered in Chapter 15. However, we can imagine ζ to be mathematically specified, and find the corresponding Q by solving Equation (33) numerically. This was done by turning Equation (33) into a differential equation as $\Delta t \to 0$, and then using a fourth-order Runge-Kutta algorithm with $d = 1$ cm, $h = 1$ cm, so that $A = 2hd = 2$ cm^2. The vocal folds oscillate at $\mathcal{F} = 120$ Hz, the pressure from the lungs $p_I = 10,000$ dyne/cm$^2 \approx 10$ cm H$_2$O, and $\sigma = 1.0$ here.

Figure 17 shows a specified function glottal area $A_g(t) = 2\zeta(t)d$ (Figure 17(a)), the resulting glottal volume velocity, $Q(t)$ (Figure

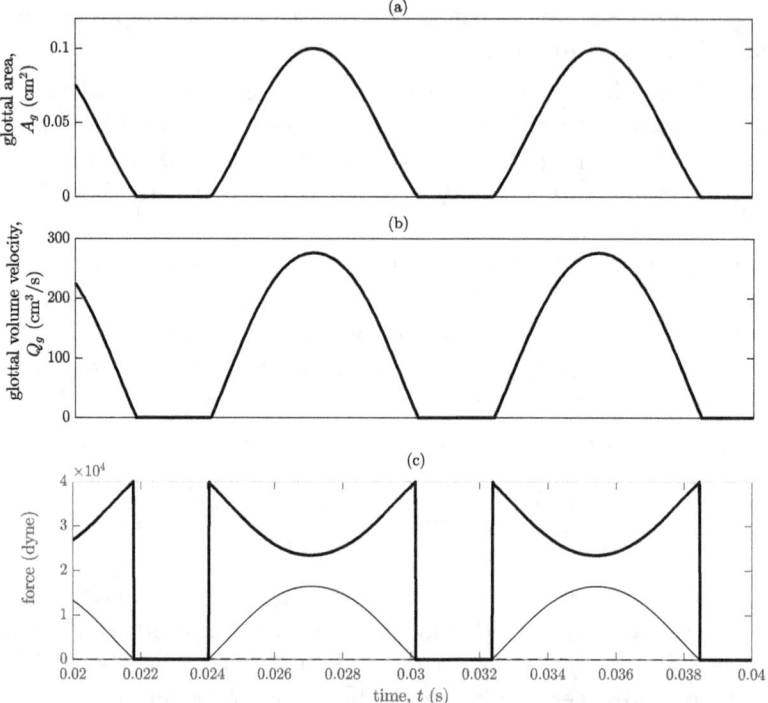

FIGURE 17. Glottal parameters during voicing: (a) glottal area A_g, (b) volume velocity Q, and (c) force terms. The drag force (third term on the left-hand-side of equation (33)) in bold black, the acceleration term (first term on the left-hand-side of equation (33)) in thin, dashed black, the acoustic propagation term (second term on the left-hand-side of equation (33)) in thin dark, and the constant right-hand side pressure term in thin gray at 40000 dyne.

17(b)), and the four terms from Equation (33) in Figure 17(c). The glottal width $\zeta(t)$ is specified as a truncated, raised sinusoid. There is no noticeable skewing in resulting volume velocity $Q(t)$, because the acceleration term is so small, as seen in by the thin, dashed black line that is near zero level in Figure 17(c). The quasi-steady approximation for glottal volume velocity appears to hold here. This situation can change with the addition of interaction with the acoustics of the subglottal and supraglottal tracts. The drag force term, which is the third term on the left-hand-side of Equation (33), is represented by the thick, solid line. The second term on the left-hand-side of Equation (33) is the acoustic propagation term and the drag force term, repsctively. The sum of the second and third terms nearly has the value of the pressure term on the right-hand-side of Eaution (33).

Energy considerations for the voice source

The vortex shedding, drag force term and the acoustic propagation term, the third and second terms on the left-hand-side in Equation (33) are the major terms to balance the pressure source term from the lungs on the right-hand-side. Estimates of the power that goes into acoustics can also be made.

According to Equation (33) the acoustic wave that results for the subglottal tract and the supraglottal tract have the same magnitude in terms of particle velocity $u = Q/A$ and pressure $p = \rho_0 c_0 u$. The power generated in planes perpendicular to the tract axes to produce acoustic wave motion is

$$\mathcal{P}_{acoustics} = 2puA = 2\rho_0 c_0 \left(\frac{Q^2}{A}\right). \tag{34}$$

With the vorticity entering the flow from the vocal folds, there is energy that is going from irrotational air motion to rotational air motion in the form of a vortex sheet. The kinetic energy contained within the vortex sheet per unit length times the rate it is being convect in the x-direction provides the power input to vorticity, $\mathcal{P}_{vorticity}$. We refer to Equations (6) and (7) and Figure 16 for the definition of quantities. The convection velocity of the vortex sheet is $u_\sigma/2$. Unsteady airfoil theory can be used to find the kinetic energy per unit length in the vortex sheet (Howe 2007).

$$\text{kinetic energy per unit length of vortex sheet} = \rho_0 d \frac{|\mathcal{C}|^2}{8\kappa}, \tag{35}$$

where κ is the hydrodynamic wavenumber, $\kappa = 2\pi \mathcal{F}/(u_\sigma/2)$. $\mathcal{F}$ is the frequency of vocal fold oscillation. From Equation (7) with $u_\sigma = 2V$

$$\mathcal{C} = -\frac{\pi}{2} u_\sigma. \tag{36}$$

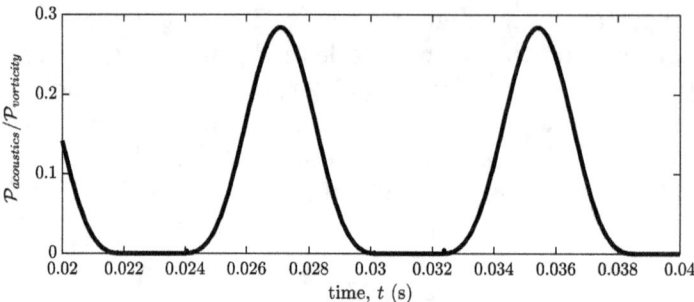

FIGURE 18. Ratio of acoustic power to vortex sheet power for the situation in Figure 17

With Equations (35) and (36), the power that is expended into the two vortex sheets is

$$\begin{aligned}\mathcal{P}_{vorticity} &= \frac{\rho_0 d\pi}{128\mathcal{F}} u_\sigma^4 \\ &= \frac{\rho_0 d\pi}{128\mathcal{F}} \left(\frac{Q}{A_\sigma}\right)^4 \\ &= \frac{\pi}{2048} \frac{\rho_0 Q^4}{\mathcal{F} d^3 (\sigma\zeta)^4} \,.\end{aligned} \qquad (37)$$

The ratio of the acoustic power in Equation (34) to the vortex sheet power in Equation (37) is

$$\begin{aligned}\frac{\mathcal{P}_{acoustics}}{\mathcal{P}_{vorticity}} &= \frac{2\rho_0 c_0 \left(\frac{Q^2}{A}\right)}{\frac{\pi}{2048} \frac{\rho_0 Q^4}{\mathcal{F} d^3 (\sigma\zeta)^4}} \\ &= \frac{4096}{\pi} \frac{c_0 \mathcal{F} d^3 \sigma^4}{A} \frac{\zeta^4}{Q^2} \,.\end{aligned} \qquad (38)$$

Using the same parameters that were used to generate Figure 17, Figure 18 is the plot of the ratio of powers given in Equation (38). The acoustic power is at most one-quarter of the power in the production of the vortex sheet. In a single cycle only 14% of the air flow energy goes into acoustics, and the rest goes into rotational air motion. This means that only 7% of the airflow energy goes into the energy of acoustic wave motion in the supraglottal tract, because about one-half of the energy of acoustic wave motion propagates into the subglottal region.

The rotational motion in the vortex sheet from the vocal folds eventually broadens into a turbulent jet in the vocal tract because the vortex sheet is unstable. We can expect that much of a measured velocity field would include this random rotational air motion, whose relation to pressure fluctuations would not be according to the acoustic approximation. This observation may help to clarify the results of researchers who have measured air motion either in vocal tracts or models of vocal tracts (e.g. Teager & Teager 1983a,b; Barney, Shadle, & Davies 1999; Shadle, Barney, & Davies 1999).

Theory predicts that the portion of the air flow field that is governed by the acoustic approximation interacts with the turbulent field very weakly (Howe 1998). Further, it can be expected that the turbulence and its associated vorticity in the vocal tract is damped, even as it is convected toward the lips. We do not have an estimate of the ratio of fluctuating volume velocity at the lips that results from convection of vorticity to the fluctuation volume velocity at the lips the arrives by acoustic propagation. The former travels at a speed of the mean air flow, which can vary greatly from below 500 cm/s to 5000 cm/s. On the other hand, acoustic disturbances propagate at speeds of up to 34,100 cm/s, or up to to 70 faster than the convected vorticity travels from glottis to lips. This may mean that the volume velocity at the lips due to vorticity fluctuations are substantially damped compared to the volume velocity fluctuations due to acoustic propagation, because vorticity spends a longer time in the vocal tract. This is a research question: how much of the fluctuating volume velocity at the lips is the result of vorticity fluctuation generated at the glottis at a much earlier time than the volume velocity fluctuation due to acoustic propagation (Shadle, Barney, & Davies 1999)?

Figure 17 shows that the peak piston volume velocity $Q_{pst} = 275$ cm^3/s corresponds to $p_I = 10,000$ dyne/cm$^2 \approx 10$ cm H$_2$O. We would expect a peak volume velocity of about $Q_{pst} = 230$ cm^3/s for $p_I = 8.3$ cm H$_2$O. This is less than what would be the case if there was no loss of energy into rotational motion for which $Q_{pst} = 400$ cm^3/s with a tube cross-sectional area of 2 cm^2. The power input at the piston face is shown in Figure 19, where we use the same parameters used to plot Figure 5 of Chapter 7 for the current simulations except now the peak piston volume velocity Q_{pst} is reduced from 400 cm^3/s to 230 cm^3/s. The OASPL at the piston face for $Q_{pst} = 230$ cm^3/s is 136 dB. This

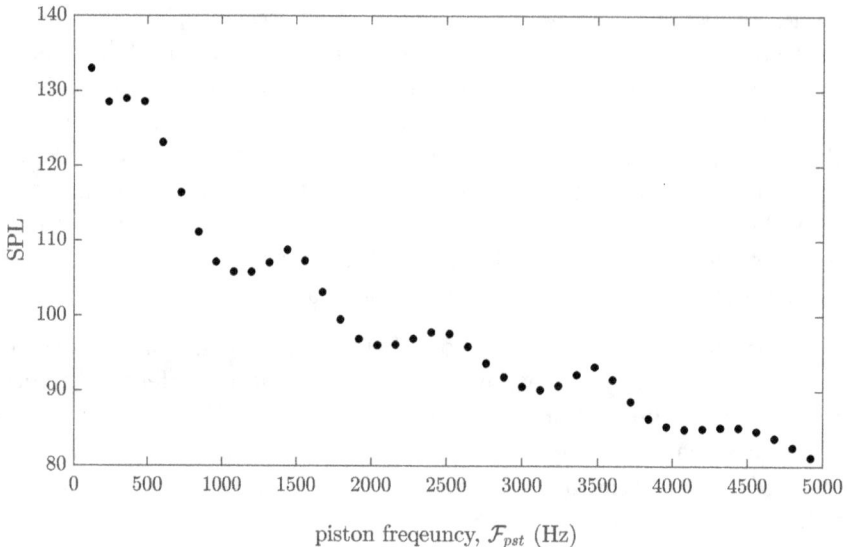

FIGURE 19. SPL at the piston face

can be compared to the OASPL found in Chapter 7 of 141 dB for $\mathcal{Q}_{pst} = 400$ cm^3/s.

The acoustics for a listener one meter from the lips

We use Equation (4) of Chapter 11 with $K_s = 2$ to estimate what the sound pressure level, or SPL, one meter from the lips ($+\ell_{rad}$) in the case that the amplitude of the piston is $\mathcal{Q}_{pst} = 230$ cm^3/s. [Here, we ignore any contribution to fluctuating volume velocity due to convected vorticity.] This equation is written

$$p_a(r,t) = \rho_0 \frac{\dot{Q}(x=L', t-r/c_0)}{2\pi r} , \qquad (39)$$

where $L' = L + \ell_{rad}$. We use an approximation to the Green's function in Equation (13) of Chapter 7 to find $\dot{Q}(x = L', t - r/c_0) \approx \Delta Q(x = L', t - r/c_0)/\Delta t$. Neglecting the terms involving $\bar{p}_m^{space}(x)$ as small, and replacing L with L', and setting $p_m^{space}(x) \approx \cos(k_m x)$, as in Equation

(14) of Chapter 7 gives

$$\mathcal{G}_Q^p(x, y = 0, t - \tau) \approx -\frac{2\rho_0 c_0^2}{AL'}\mathcal{H}(t - \tau) \times \sum_{m=1}^{\infty} e^{-\gamma_m(t-\tau)} \frac{\sin(\hat{\omega}_m(t - \tau))}{\hat{\omega}_m} \cos(k_m x) ,\quad (40)$$

where the mode wavenumbers have been modified to take account of the longer L'.

According to Equation (10) of Chapter 2

$$\frac{\Delta_t Q(x,t)}{\Delta t} = -\frac{A}{\rho_0}\frac{\Delta_x p}{\Delta x} . \quad (41)$$

Applying Equation (41) to Equation (40) results in

$$\mathcal{G}_Q^Q(x, y = 0, t - \tau) \approx -\frac{2c_0}{L'}\mathcal{H}(t - \tau) \times \sum_{m=1}^{\infty} e^{-\gamma_m(t-\tau)} \frac{\cos(\hat{\omega}_m(t - \tau))}{\hat{\omega}_m} \sin(k_m x) . \quad (42)$$

This is the Green's function for $Q(x,t)$ in the finite-length tube for a piston (volume velocity) source at $y = 0$. Evaluating this at $x = L'$ provides

$$\mathcal{G}_Q^Q(x = L', y = 0, t - \tau) \approx \frac{2c_0}{L'}\mathcal{H}(t - \tau) \sum_{m=1}^{\infty} (-1)^m e^{-\gamma_m(t-\tau)} \frac{\cos(\hat{\omega}_m(t - \tau))}{\hat{\omega}_m} . \quad (43)$$

As for Equation (16) we take the source to be $Q_{pst}(t) = \mathcal{H}(t)\mathcal{Q}_{pst}\sin(\omega_{pst}t)$, which is convolved with the Green's function in Equation (42). Neglecting transients, the result is

$$Q(x = L', t) \approx \frac{2c_0 \mathcal{Q}_{pst}\mathcal{H}(t)}{L'} \times \sum_{m=1}^{\infty}(-1)^m \frac{\sqrt{\omega_{pst}^2 - \gamma_m^2}}{\sqrt{(\omega_{pst}^2 - \omega_m^2)^2 + 4(\gamma_m^2\omega_{pst}^2)}} \sin(\omega_{pst}t + \theta_m)$$

$$\text{where } \theta_m = \arctan\left(\frac{\omega_{pst}[(\omega_m^2 - \omega_{pst}^2) - 2\gamma_m^2]}{\gamma_m[\omega_{pst}^2 + \omega_m^2]}\right) . \quad (44)$$

Using the definition of θ_m in Equation (44) and finding $\dot{Q}(x = L', t) \approx \Delta_t Q(x = L', t)/\Delta t$, and

$$\dot{Q}(x = L', t) \approx \frac{2c_0 \mathcal{Q}_{pst} \mathcal{H}(t)}{L'} \sum_{m=1}^{\infty} (-1)^m \frac{\omega_{pst}}{(\omega_{pst}^2 - \omega_m^2)^2 + 4(\gamma_m^2 \omega_{pst}^2)} \times \\ \left[\gamma_m (\omega_m^2 + \omega_{pst}^2) \cos(\omega_{pst} t) - \omega_{pst} ((\omega_m^2 - \omega_{pst}^2) - 2\gamma_m^2) \sin(\omega_{pst} t) \right]. \quad (45)$$

Combining Equations (39) and (45) we obtain the fluctuating pressure at a distance r from the lips.

$$p_a(r) \approx \frac{\rho_0 c_0 \mathcal{Q}_{pst} \mathcal{H}(t)}{\pi r L'} \sum_{m=1}^{\infty} (-1)^m \frac{\omega_{pst}}{(\omega_{pst}^2 - \omega_m^2)^2 + 4(\gamma_m^2 \omega_{pst}^2)} \times \\ \left[\gamma_m (\omega_m^2 + \omega_{pst}^2) \cos(\omega_{pst} t) - \omega_{pst} ((\omega_m^2 - \omega_{pst}^2) - 2\gamma_m^2) \sin(\omega_{pst} t) \right]. \quad (46)$$

Equations (19) and (20) of Chapter 7 with Equation (46) can be used to calculate SPL at a distance 1 m = 100 cm from the lips. The parameters used to generate Figure 19 for SPL at the piston, provide SPL levels shown in Figure 20 at 100 cm from the lips. The OASPL at 100 cm from the lips is 65 dB.

Source-tract interaction

We have not included the effect of reflected acoustic waves in Equation (33). The reflected acoustic waves impinging on the glottis, from both the subglottal and supraglottal vocal tracts, are the result of the acoustic waves that propagate from the glottis at an earlier time, say τ, where $\tau < t$. Let p_{sub} be the pressure that impinges on the glottis due to reflections in the subglottal vocal tract, and p_{sup} be the pressure that impinges on the glottis due to reflections in the supraglottal vocal tract. The following should be the case.

$$\rho_0 (L_g + \ell_g(t)) \left(\frac{\Delta_t Q(t)}{\Delta t} \right) + 2\rho_0 c_0 Q(t) + \frac{\rho_0}{2} \left(\frac{h}{\sigma \zeta(t)} \right)^2 \frac{Q(t)^2}{A} \\ = (2p_I + p_{sub} - p_{sup}) A, \quad (47)$$

where $p_{sub} = p_{sub}(Q(\tau))$ and $p_{sup} = p_{sup}(Q(\tau))$, for $\tau \leq t$.

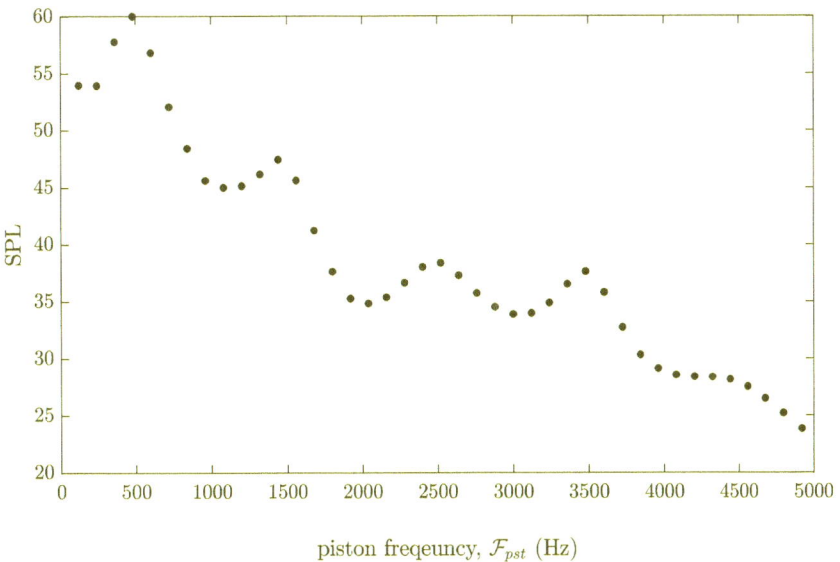

FIGURE 20. SPL at 1 m from the lips

Expressions for p_{sub} and p_{sup} in terms of $Q(\tau)$ with $\tau < t$ can be found (McGowan & Howe 2011).

Sources for turbulent jet flow and three-dimensional acoustic propagation

The topic of frication in speech brings us close to Lighthill's original work on jet noise (Lighthill 1952). When a jet is essentially in open air, without solid boundaries close to its flow beyond its nozzle, it is possible to derive expressions for pressure that propagates acoustically in terms of properties of the jet flow by using scaling arguments. We also consider what occurs when a small solid object is placed in the jet flow.

We exhibit the famous eighth power law of Lighthill for a jet in open air. For details see Howe (2003). Assume that an observer is at a great distance from a jet. First, we assume that there are regions in the jet, loosely called *eddies*, for which the flow properties are correlated. The characteristic length scale of these eddies is denoted l. We also assume that the characteristic velocity of the turbulence, U, provides a low Mach number flow, i.e. $M = U/c_0 \ll 1$. For a jet, we take U to be the mean velocity of the jet. A low Mach number in the jet allows us to neglect density variations in the turbulent jet, and, thus, treat the flow as incompressible. Further, the characteristic frequency, $\mathcal{F}$, of velocity and pressure fluctuation of an eddy is $\mathcal{F} \sim U/l$. So the characteristic wavelength of sound, λ, is much larger than the length scale of the eddies, i.e. $\lambda = c_0/\mathcal{F} \sim l/M \gg l$. Further the differences in acoustic travel times from different regions of an eddy is small as long as the observer is many wavelengths from the eddy. We are making the far-field approximation.

With these approximations, the acoustic power, $\mathcal{P}_{eddy}$, radiated by a single eddy out to the mathematical sphere containing the observer and centered at the eddy is

$$\mathcal{P}_{eddy} \sim \rho_0 l^2 U^3 M^5 . \tag{48}$$

[The derivation of this scaling involves the concept of *turbulent stress*, and this is beyond the scope of this book.] If V_0 denotes the total volume of turbulence, then the total acoustic power, $P_{jet\ noise}$, from V_0/l^3 uncorrelated eddies scales as

$$\mathcal{P}_{jet\ noise} \sim \rho_0 \frac{V_0}{l} U^3 M^5 = \frac{\rho_0}{c_0^5} \frac{V_0}{l} U^8 . \tag{49}$$

The acoustic power radiated into the air increases as the eighth power of the characteristic velocity of the turbulence.

Perhaps as important as the result in Equation (49) is an estimate of the scaling of efficiency with which the kinetic energy of the the jet is converted into the energy of acoustic wave motion. An estimate of the power needed to sustain the jet is

$$\mathcal{P}_{jet} \sim \rho_0 \frac{V_0}{l} U^3 . \tag{50}$$

Equations (49) and (50) gives the scaling

$$\frac{\mathcal{P}_{jet\ noise}}{\mathcal{P}_{jet}} \sim M^5 . \tag{51}$$

For $M \ll 1$, M^5 can be a very small number. This is a good thing because the great majority of increases of aircraft jet energy with jet speed goes to moving the plane and not into noise production. For speech, $U = 3500$ cm/s is a substantial jet velocity during frication, for which $M \approx 0.1$.

If we put a small, rigid object in the jet we can create some more noise. We again follow Howe (2003) in this discussion. Let's say that an area A_w of the surface of the rigid object is wetted by the jet turbulence. We suppose that $\lambda = c_0/\mathcal{F} \sim \sqrt{A_w}/M \gg \sqrt{A_w}$, so that the flow is approximately incompressible around the object. It can be shown that the power of the sound emitted by this object through the sphere containing the observer many wavelengths away from the object scales as

$$\mathcal{P}_{object\ in\ jet} \sim \rho_0 A_w U^3 M^3 \ . \tag{52}$$

Comparing Equations (49) and (52), the small object in a jet enhances the efficiency with which acoustic power is radiated over that of the jet alone by a factor of $1/M^2$. This can be a very large factor when $M \ll 1$.

This greater efficiency can be explained physically. In the case of the jet without solid boundaries, the turbulence cannot support any net fluctuating force in the air alone. Thus, the predominate sources of sound in such a turbulent field are turbulent stresses, which can be decomposed into two opposing fluctuating forces that are very close together. Because these forces are in opposite directions, they nearly cancel each other as acoustic sources, except for the fact that there are small differences in the amount of time that it takes the signal from each fluctuating force to reach an observer many wavelengths away from the region of stress. When a small object is placed in the turbulent jet, that object can provide a fluctuating force on the air. There is a large drag force component to this fluctuating force. In fact, this situation is analogous to the time-varying drag force on the vocal folds, except that we are dealing with random vorticity in a turbulent jet flow.

Sources due to turbulent flow in the vocal tract

We consider turbulence in a tube or duct infinitely extended along its axis in both directions. If turbulence occupies a volume V_0, then we can use the reasoning that provides the scalings for power in Equations (49) and (52) for plane wave acoustics in the duct. It turns out that the turbulence is a more efficient acoustic source both with and without a small object in its midst. Without a small object we would have,

$$\mathcal{P}_{jet\ noise} \sim \rho_0 A \frac{V_0}{l^3} U^3 M^3 = \frac{\rho_0 A}{c_0^3} \frac{V_0}{l^3} U^6 \ . \qquad (53)$$

with A as the cross-sectional area of the duct. The power of the propagated jet noise in the duct scales as the characteristic velocity to the sixth power, U^6, as opposed to U^8 for three-dimensional propagation.

If there is a small object within the turbulence that has an area A_w wetted by the turbulence, then

$$\mathcal{P}_{object\ in\ jet} \sim \rho_0 \frac{A_w^3}{l^2 A} U^3 M \ . \qquad (54)$$

The object in the jet exhibits more acoustic efficiency as a function of Mach number, M, than the turbulence alone for the same reasons as for radiation into three-dimensional space. The small object as an obstruction to turbulent flow is what we call a pressure source in speech production.

It is difficult to find generalities pertaining to how turbulent sources behave in the vocal tract. The reason for this is that the sources are necessarily coupled with the resulting acoustic field, because of the resonance properties in the vocal tract. [Also, the nature of turbulence in a duct or tube is likely to be different than if it exists without solid boundaries to contain it.] Each vocal tract geometry must be considered on its own. However, one generality that can be offered: when an object is placed in a turbulent flow so that it substantially obstructs that flow in the general direction of the vocal tract axis, there is a relatively efficient source due to the fluctuating drag force. The reason for this relative efficiently is the same as that for the jet in three-dimensional space: without a solid boundary a net fluctuating force cannot be supported.

As an example of when significant obstruction to turbulent air flow in the vocal tract has been worked out (Howe & McGowan 2005). The major result is that the pressure spectrum of the turbulence at the surface of the teeth during sibilance is significantly amplified by the resulting fluctuating drag force between the teeth and the air. The resulting predictions are parameterized by measurable quantities and predict sound pressure levels on an absolute scale. The fluctuating drag force between the air and teeth is the acoustic source for sibilance.

References

Barney, A., Shadle, C. H., & Davies, P. O. A. L. (1999). Fluid flow in a dynamic mechanical model of the vocal folds and tract. I. Measurements and theory. *Journal of the Acoustical Society of America*, **105**. p 444.

Batchelor, G. K. (1967). *An Introduction to Fluid Dynamics*. Cambridge University Press, Cambridge, England. (pp 156 - 162, 211-212, 517-520).

Fant, G. (1960). *Acoustic Theory of Speech Production*. Mouton, The Hague. (pp 265-268).

Goldstein, H. (1950). *Classical Mechanics*. Addison-Wesley Publishing Company, Reading, MA. (Chapter 4).

Howe, M.S. (1995). "On the force and moment exerted on a body in an incompressible fluid, with application to rigid bodies and bubbles at high and low Reynolds number." *Quarterly Journal of Mechanics and Applied Mathematics*, **48**. p 401.

Howe, M.S. (1998). *Acoustics of Fluid-Structure Interaction*. Cambridge University Press, Cambridge, England. (pp 138-143).

Howe, M.S. (2003). *Theory of Vortex Sound*. Cambridge University Press, Cambridge, England.

Howe, M.S. & McGowan, R.S. (2005). "Aeroacoustics of [s]." *Proceedings of the Royal Society A*, **461**. p 1005.

Howe, M.S. (2007). *Hydrodynamics and Sound*. Cambridge University Press, Cambridge, England. (pp 109, 201, 218-219, 253-273).

Howe, M.S. & McGowan, R.S. (2011a). "Production of sound by unsteady throttling of flow into a resonant cavity, with application to voiced speech." *Journal of Fluid Mechanics*, **672**. p 428.

Howe, M.S. & McGowan, R.S. (2011b). "On the generalized Fant equation." *Journal of Sound and Vibration*, **330**. p 3123.

Howe, M.S. & McGowan, R.S. (2012). "On the role of glottis-interior sources in the production of voiced sound." *Journal of the Acoustical Society of America*, **131**. p 1391.

Lighthill, M.J. (1952). "On sound generated aerodynamically. Part I: General Theory." *Proceedings of the Royal Society of London*, **A211**. p 564.

Lighthill, J. (1978). *Waves in Fluids.* Cambridge University Press, Cambridge, England. (Chapter 2).

Lighthill, J. (1986). *An Informal Introduction to Theoretical Fluid Mechanics.* Clarendon Press, Oxford, England. (chapters 1 through 5, pp 200-228).

McGowan, R.S. & Howe, M.S. (2006). "Compact Green's functions extend the acoustic theory of speech production." *Journal of Phonetics*, **35**. p 259.

McGowan, R.S. & Howe, M.S. (2011). "Source-tract interaction with prescribed vocal fold motion." *Journal of the Acoustical Society of America*, **131**. p 2999.

Morse, P.M. & Ingard, K.U. (1986). *Theoretical Acoustics.* Princeton University Press, Princeton, NJ. (pp 483-488).

Shadle, C.H., Barney, A.,& Davies P. O. A. L. (1999). Fluid flow in a dynamic model of the vocal folds and tract. II. Implications for speech production studies. *Journal of the Acoustical Society of America*, 105. p 456.

Šidlof, P., Doaré, O., Cadot, O., & Chaigne, A. (2011). Measurement of flow separation in a human vocal folds model. *Experiments in Fluids*, **51**. p 123.

Teager, H.M. & Teager, S.H. (1983a). The effects of separated air flow on vocalization. In D. M. Bless and J. H. Abbs, editors, *Vocal Fold Physiology.* College Hill Press, San Diego, CA. (pp 124-143).

Teager, H.M. & Teager, S.H. (1983b). Active fluid dynamic voice production models, there is a unicorn in the garden. In I. Titze and R. Scherer, editors, *Vocal Fold Physiology.* Denver Center for the Performing Arts, Denver CO. (pp 387-401).

Chapter 14: Scaling, Curvature, and Speech Development

Introduction

The knowledge that has been gained in the previous chapters is now applied to understand certain aspects of speech development in very young children in the age range from 12 months to 48 months. This is a time of great anatomical change in the vocal tract that continues to be substantial until about the eighth year of life. (e.g. Vorperian, Wang, Chung, Schimek, Durtschi, Kent, Ziegert, & Gentry 2009; Denny & McGowan 2013). In this chapter we argue that the anatomy of a young child's vocal tract can make it very difficult, or impossible, to produce speech sounds in an adult-like manner. While we expect children to have energy in higher frequency bands of their magnitude spectra than adults, we argue that it is difficult, or impossible, for very young children to produce adult-like sounds, even accounting for a uniform scaling of frequency based on vocal tract length.

We review two kinds of problems that could occur for very young children. The first has to do with tongue surface *curvature* and our hypothesis that children cannot make curvatures whose magnitudes are greater than those of adults in certain extreme articulatory scenarios. The second type of problem is that normal mode, or formant, frequencies scale differently than does the level of sound produced in turbulent flow. This makes it impossible for a very young child to simultaneously produce a magnitude spectrum that is simply a scaled adult magnitude spectrum for certain sounds that involve both acoustic propagation and noise production.

The first problem is illustrated in a longitudinal acoustic study of [ɹ] production, where it is shown that stressed syllable-initial [ɹ] for young Americans growing up in a rhotic dialect area was the last of all the rhotic sounds to develop for a small group of children (McGowan, Nittrouer, & Manning 2004). For adults, stressed syllable-initial [ɹ] is the strongest [ɹ] in comparison with [ɹ] in other contexts, including medial and syllable-final [ɹ]. Here, the term strong is used in the sense that the first three normal mode frequencies, or formant frequencies, attain very low values. In particular, the third formant frequency, denoted F3 here instead of $\mathcal{F}_3$, is very low for strong [ɹ]. We believe that the stressed, syllable-initial [ɹ] is extreme, in the sense that tongue curvatures and spatial changes in curvature involved in its production are large.

The first problem could also explain another observed phenomenon: children do not produce sibilant fricative magnitude spectra that are simply high frequency versions of adult sibilants. Sibilant fricatives are speech segments that are difficult for children to master (e.g. McGowan & Nittrouer 1988). We believe that tongue curvature explanations offered here can serve as unifying explanations for why both sibilant fricatives and strong [ɹ] are some of the later segments to develop into adult-like productions.

The second problem is illustrated in the oft reported fronting of velar releases in the speech production of young children. We show here that this is due to the fact that formant frequencies scale inversely with vocal tract length, while the intensity of sound that is produced at the release of a velar or alveolar consonant depends on other geometric and aerodynamic parameters that affect acoustic source properties. We believe that there are reasons to expect actual fronting, as well as perceptual errors by adults in judging place-of-articulation made by adult transcribers.

Scaling and curvature

Scaling

Vocal tract length scaling is employed when comparing a child's and an adult male's vocal tracts. [When speaking about hypotheses and simulations, we speak of a typical child and a typical male adult. Only when we are discussing previous results do we refer to groups, or populations, of children and adults.] The scale factor, $\mathcal{S}$, is the ratio of the adult's vocal tract length L_{adult} to the child's vocal tract length L_{child}.

$$\mathcal{S} = \frac{L_{adult}}{L_{child}}. \qquad (1)$$

It will often be applied to the child's vocal tract to make it easier to compare shapes of the tracts. Note that $\mathcal{S} > 1$, and, typically $\mathcal{S} < 2$. We use $\mathcal{S} = 3/2$ in examples. The child's lengths can be scaled up using $\mathcal{S}$ for direct comparison between child and adult area function shapes. Thus, if L_{child} is a length for the child, the scaled length is

$$L_{scaled} \equiv \mathcal{S} L_{child} = L_{adult} , \qquad (2)$$

which follows from Equation (1).

From Chapter 5 onward, we have discovered that the normal mode wavenumbers k_m are directly proportional to the inverse of vocal tract length, as long as the ratios of cross-sectional areas of sub-tube sections remain the same. That is, as long as the area function has the same shape when the x-axis is scaled according to ratios in vocal tract length, then k_m is directly proportional to vocal tract length. We can call this property of area functions equivalence of area function with vocal tract length changes. Equivalence does not require that the area functions be the same, because the areas may not have the same values. All that is required is that the ratios among pairs of cross-sectional areas be the same, so that equivalent area functions have the same shape when their x-axes are scaled to the same length. Because $\mathcal{F}_m = c_0 k_m/(2\pi)$, the normal mode frequencies $\mathcal{F}_m$ are also proportional to the inverse of vocal tract length, when area functions are equivalent. Also, we use the notation F1, F2, and F3 for $\mathcal{F}_m, m = 1, 2, 3$, respectively, in the present chapter and refer to formant frequencies.

There are factors other than formant frequencies that affect whether a child's speech magnitude spectrum resembles that of an adult. Aerodynamic sources, as discussed in Chapter 13, also play a role in the spectral properties of speech. We point to acoustic power generated by turbulence striking the teeth as our prime example. In particular we refer to Equation (54) of Chapter 13 as an example of an important power level that can change the perception of a speech sound. In particular, power level, in conjunction with noise duration, helps to distinguish sibilant from non-sibilant frication. Acoustic power production does not scale as simply as formant frequencies do with vocal tract length.

Criterion for the child's speech to be adult-like

We state the following strong criterion in order to say that a child produces adult-like speech. The scaled child's vocal tract, as expressed in Equation (2), should be able to produce a speech sound that matches

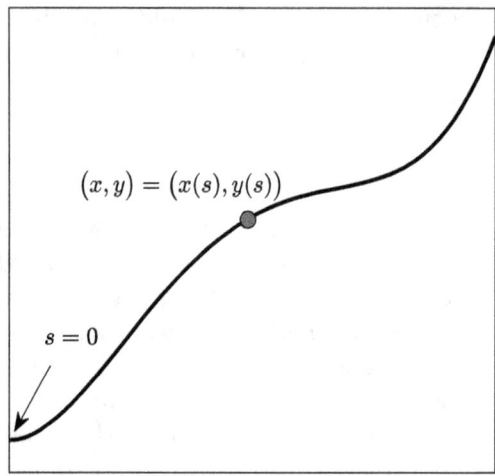

FIGURE 1. A two-dimensional curve

that of the adult in the frequency domain in order to say that the child can produce adult-like speech. At the very least, the formant frequencies of the scaled child's vocal tract and the adult vocal tract should match. This requires that the child have an area function that is a equivalent to the adult's area function: the child's area function becomes has the same shape as the adult's area function when the scale factor S is used to multiply the x-values of the child's area function. While matching of formant frequencies is a necessary condition for the child to produce adult-like speech, it is not sufficient. The properties of acoustic sources are not preserved between equivalent area functions.

Curvature

We characterize surface shapes using a property called curvature. In particular, we examine tongue surfaces in the midsagittal or coronal planes, as well as outer vocal tract wall surfaces in these planes. Figure 1 shows the outline of a surface, which is a curve when viewed in the midsagittal or coronal plane. The points of the curve are specified by Cartesian coordinates, (x, y). In fact, the points on the curve can be considered to be functions of a parameter s that increases as we move along the curve. s can be the particular parameter of the length of the curve from its start at $s = 0$, and we will assume this from here on. Thus, we can write $(x, y) = (x(s), y(s))$. We want to say something about how rapidly the curve bends at any point. To do this we use the concept of curvature, which can be defined geometrically using circles.

Suppose that $(x(s_1), y(s_1)) = (x_1, y_1)$ for some $s_1 > 0$ is a point on the curve. We can find a tangent line to the curve at (x_1, y_1), as shown in Figure 2. The perpendicular to the tangent at (x_1, y_1) can also be drawn, and circles that share the tangent to the curve at (x_1, y_1) can also be considered. The centers of these circles with different radii have their centers along the perpendicular line. There is a particular circle with radius r_1 that approximates the curve the best, and this is termed the *radius-of-curvature*. The curvature at (x_1, y_1) $\mathcal{K}_1$ is

$$\mathcal{K}_1 = \pm \frac{1}{r_1} . \tag{3}$$

Therefore, we define curvature $\mathcal{K}_1$ at (x_1, y_1) to be positive or negative the reciprocal of the radius-of-curvature. The sign of the curvature is decided by convention. Here we take curvatures to be positive if the approximating circle is "outside" the curve. They are negative if the approximating circle is "inside" the curve. [Of course, we need to define what outside and inside correspond to, but we will not be too concerned with this.]

We can define

$$\dot{\mathcal{K}} \equiv \frac{\Delta_s \mathcal{K}}{\Delta s} , \tag{4}$$

where $\Delta s \to 0$. The is the rate-of-change of curvature with respect to curve length. [As opposed to the previous uses where the "dot" notation was associated with time rate-of-change, it now denotes a particular spatial rate-of-change.] $|\dot{\mathcal{K}}|$ is a measure of how rapidly curvature changes along the curve.

Experimentally, we can only approximate this quantity, because we cannot measure points that are infinitely close together. We can, however, consider curvatures at two distinct points, $(x(s_1), y(s_1)) = (x_1, y_1)$ and $(x(s_2), y(s_2)) = (x_2, y_2)$, with $\mathcal{K}_1 = -1/r_1$ and $\mathcal{K}_2 = 1/r_2$. For two points on the curve, (x_1, y_1) and (x_2, y_2), $|s_2 - s_1|$ is the distance along the curve between the points (x_1, y_1) and (x_2, y_2). The *average spatial rate-of-change of curvature* is denoted

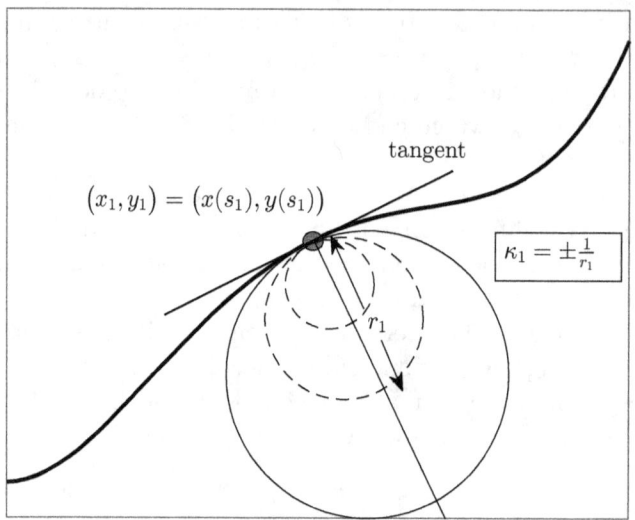

FIGURE 2. Curvature at a point on a curve

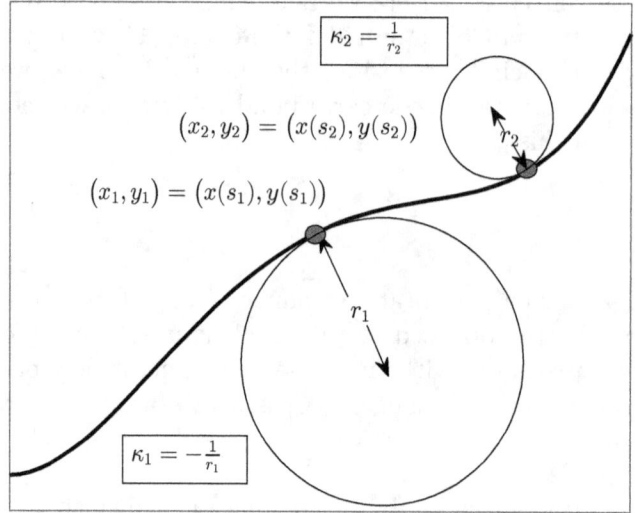

FIGURE 3. Curvature at two points on a curve

$\langle\dot{\mathcal{K}}\rangle$. Average spatial rate-of-change of curvature is defined

$$\langle\dot{\mathcal{K}}\rangle \equiv \frac{\mathcal{K}_2 - \mathcal{K}_1}{s_2 - s_1}. \tag{5}$$

[As opposed to the previous uses where $\langle \cdot \rangle$ notation was used to denote time average, it now denotes a particular spatial average.] We often refer to $\langle\dot{\mathcal{K}}\rangle$ as spatial rate-of-change of curvature, without the term "average". For instance, if $r_1 = 2$ cm and $r_2 = 4$ cm, then $\mathcal{K}_1 = -0.5$ cm^{-1} and $\mathcal{K}_2 = 0.25$ cm^{-1}. With these numerical values, $\mathcal{K}_2 - \mathcal{K}_1 = 0.75$ cm^{-1}. For $s_2 - s_1 = 3$ cm, we have $\langle\dot{\mathcal{K}}\rangle = 0.25$ cm^{-2}.

Because curvature $\mathcal{K}$ has units of inverse length (e.g. cm^{-1}) and $\langle\dot{\mathcal{K}}\rangle$ has units of inverse length squared (e.g. cm^{-2}), the corresponding scaled quantities are given by

$$\mathcal{K}_{scaled} = \frac{\mathcal{K}_{child}}{\mathcal{S}} \quad \text{and} \quad \langle\dot{\mathcal{K}}\rangle_{scaled} = \frac{\langle\dot{\mathcal{K}}\rangle_{child}}{\mathcal{S}^2}. \tag{6}$$

Thus, for $\mathcal{S} > 1$, curvature and spatial rate-of-change of curvature are scaled down. Further, the percentage differences between scaled and unscaled is greater for $\langle\dot{\mathcal{K}}\rangle$ than for $\mathcal{K}$. Using the the same values that were used after Equation (5), we obtain $\mathcal{K}_{1\ scaled} = -0.333$ cm^{-1}, $\mathcal{K}_{2\ scaled} = 0.167$ cm^{-1}, and $\langle\dot{\mathcal{K}}\rangle_{scaled} = 0.111$ cm^{-2} when $\mathcal{S} = 3/2$.

Hypotheses regarding tongue surface curvature

We now state two plausible hypotheses regarding tongue surface curvature and spatial rates of change of curvature for young children in either the midsagittal or coronal planes. These hypotheses are in terms of maximum values of these quantities.

The hypotheses are that the maximum absolute value of scaled curvature and scaled spatial rate-of-change of curvature for a young child is less than the same quantities for the adult. Symbolically,

$$\max|\mathcal{K}_{adult}| \geq \max|\mathcal{K}_{scaled}| = \frac{1}{\mathcal{S}}\max|\mathcal{K}_{child}|,$$
$$\text{and} \tag{7}$$
$$\max|\langle\dot{\mathcal{K}}\rangle_{adult}| \geq \max|\langle\dot{\mathcal{K}}\rangle_{scaled}| = \frac{1}{\mathcal{S}^2}\max|\langle\dot{\mathcal{K}}\rangle_{child}|,$$

where max means maximum value. Note that the first inequality can be stated in terms of radius-of-curvature: the minimum scaled radius-of-curvature of the child is greater than the minimum radius-of-curvature of the adult.

Even if the child's maximum scaled values in Equation (7) are less than those of the adult's, his or her maximum unscaled values could be greater than those of the adult. For instance, the child may be able to produce a maximum unscaled curvature of $|\mathcal{K}_{child}| = 0.6$ cm^{-1}, which corresponds to the child being able to produce a minimum unscaled radius-of-curvature of $r = 1.67$ cm. The adult could have a maximum curvature of $|\mathcal{K}_{adult}| = 0.5$ cm^{-1}, which corresponds to a minimum radius-of-curvature of 2 cm. With a scaling factor of $\mathcal{S} = 3/2$, the child's scaled curvature is $|\mathcal{K}_{scaled}| = 0.4$ cm^{-1} < 0.5 cm^{-1} $= |\mathcal{K}_{adult}| < 0.6$ cm^{-1} $= |\mathcal{K}_{child}|$.

What are the reasons to make the hypothesis expressed in Equation (7)? We can only appeal to intuition here by claiming that making highly curved tongues and tongues where curvature changes rapidly is difficult. This is particularly true for the tongue when it is unbraced by the palate during speech production. By difficult, we mean that the motor units need to be highly spatially differentiated and simultaneously controlled. Because the young child learning to speak does not possess the experience that the adult speaker does, we expect that the hypothesis expressed in Equation (7) is true. That is, we expect that the child cannot produce articulations more difficult, even allowing for length scale differences. However, we do not have physiological evidence in terms of, say, the independence of control of motor units in the tongues of the child versus the adult (Denny & McGowan 2012).

If the curvature hypotheses are true, then young children cannot produce strong [ɹ]

Acoustic simulations

We refer to the simulation of the production of [ɹ] in Figures 9 and 11 in Chapter 12. These figures show schematic area functions for the production of [ɹ]. We argued that we could strengthen [ɹ] production in Figure 9 of Chapter 12 by narrowing the area function in the pharyngeal region, as shown in Figure 11 of Chapter 12. The first two formant frequencies are virtually the same for the two area functions, but the area function corresponding to Figure 11 of Chapter 12 has a much lower third formant frequency F3 than the corresponding to Figure 9 of Chapter 12. Therefore, adding a pharyngeal narrowing to the two other constrictions at the lips and palate acoustically strengthens the [ɹ] by lowering F3.

Area functions and vocal tract anatomy

Before proceeding, we need to relate area functions to the vocal tract as viewed in the mid-sagittal plane. We are most interested in comparing the adult with the child in the region behind the palatal constriction. For this purpose, we make the crude assumption that the anatomical features in the dimension lateral to the mid-sagittal plane of the child's scaled vocal tract are approximately the same as that of the adult's. This means that differences between the child and the adult in distance from the upper tongue surface to the outer vocal tract surface of hard and soft palates and rear pharyngeal wall are proportional to the vocal tract cross-sectional areas. Distances from the upper tongue surface to the outer vocal tract surface are generally measured by the length of line segments between the structures, which are perpendicular to the mid-line of the vocal tract.

Are there particular anatomical, or geometric, factors that make the production of a strong [ɹ] particularly challenging for young children? We believe that there are factors that make the production of strong [ɹ] difficult for children 18-months of age, say. Some factors making strong [ɹ] production difficult for very young children have been discussed previously (Denny & McGowan 2012). We review those factors and add additional factors here. The relevant factors in comparing adult and young children's articulation of [ɹ] are:
1) the orientation of the axes of the oral and pharyngeal cavities,
2) the relative axial lengths of the oral and pharyngeal cavities,
3) palate doming, [Young palates are less "domed" and flatter in sagittal planes than adult palates in these plane (Hiki & Ito, 1986).]
4) the size of adenoidal tissue, and
5) the orientations of external tongue muscles.

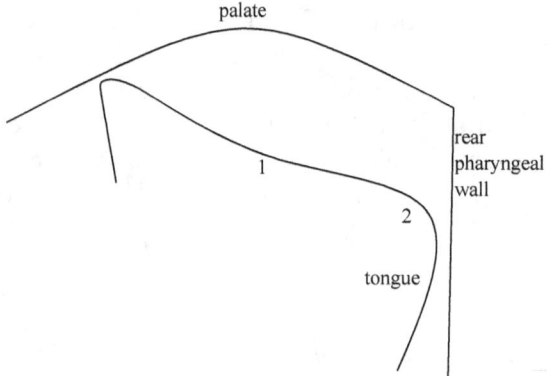

FIGURE 4. The adult's mid-saggital vocal tract for for strong [ɹ]

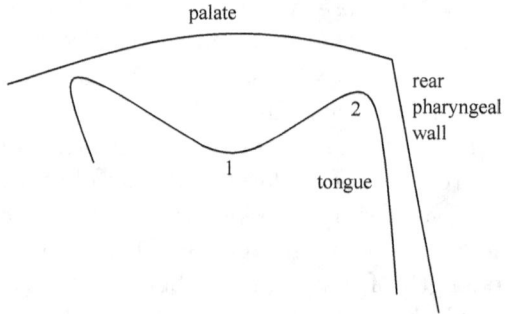

FIGURE 5. The scaled child's mid-saggital vocal tract vocal tract for strong [ɹ]

We argue that factors 2, 3, and 5 combine to require more extreme tongue surface curvature and average spatial rate-of-change of curvature for the young child in order to attain the changes in area function curvature to produce an adult-like strong [ɹ]. Figure 4 shows mid-sagittal configurations for strong [ɹ] production for the adult, and Figure 5 shows the vocal tract of the scaled child. As with the area function, the child's vocal tract has been scaled by a factor that is equal to the ratio of the length of the adult vocal tract to the length of the child's vocal tract S in Figure 5. We show a tongue tip up [ɹ] production, but the argument below also applies to bunched tongue [ɹ] production.

The length of the child's pharyngeal cavity compared to its oral cavity is less than the same ratio of lengths for the adult (factor 2). Thus, for the scaled child to attain the same area function as the adult, he or she must have pharyngeal narrowing higher in the pharynx than the adult does. With the palate less domed in the midsagittal plane (factor 3), the palatal constriction is lower with respect to the pharyngeal narrowing than for the adult. This means that the scaled child makes a palatal constriction that is lower and a pharyngeal constriction that is higher than the adult possesses. As a consequence, the two constrictions' heights are more nearly the same for the scaled child compared to the adult. This means that the maximum curvature of the tongue between the palatal constriction and pharyngeal narrowing, shown to be at point 1 of the tongue surface outline in the midsagittal plane in Figures 4 and 5, needs to be greater for the scaled child than for the adult. Factor 3 actually adds to this effect again, because the relative flatness of the palate for the scaled child means that the tongue must make up for the the lack of curvature in the palate.

This could also mean that the average rate of curvature change in the tongue from point 1 to the pharyngeal narrowing at point 2 in Figures 4 and 5 could be greater for the scaled child than for the adult. Factor 1 actually works to reduce the required maximum curvature of the tongue in the region of the pharyngeal narrowing for the scaled child. On the other hand, the already noted tendency of the palatal constriction and pharyngeal narrowing to have more nearly equal vocal tract heights for the scaled child than for the adult, works to toward making the the maximum tongue curvature in the pharyngeal region larger for the scaled child than for the adult. Factors 2 nd 3 should overwhelm factor 1 though. This would make the curvature at point 2 large for the scaled child, along with the maximum average spatial rate-of-change of curvature between points 1 and 2.

Factors 4 pertains to the hypothesis for children older than three years. In fact, the growth of adenoidal tissue, factor 4, is probably not a factor in the very youngest of children learning to speak, say, at 18 months.

Regarding factor 5, we consider two sets of opposing muscles that are external, opposing muscles used to position the tongue. One set is the anterior genioglossus muscle and the styloglossis muscle and the other set is the posterior genioglossus muscle and hyoglossus muscle. These two sets of opposing muscles are more orthogonal in the direction in which they act for the adult than for the young child. These muscles provide for less vertical motion in the young child than for the adult. Thus, they would be less effective for raising the tongue narrowing at point 2 for the child.

It seems that the hypotheses expressed in Equation (7) are violated if the child produces a strong [ɹ]. This does not prove the hypotheses of course. However, we explore another phenomenon that is common in children's speech production that could be explained by the hypotheses.

Sibilant fricatives and tongue curvature

Various acoustic properties of sibilant fricatives spoken by American children, aged 3 to 7 years, and by American adults have been examined (McGowan & Nittrouer 1988). The amplitudes of the second formants were relatively high for the children compared to the adults during sibilant production. Because the great majority of the second formant energy is behind the tongue constriction for sibilant fricatives, it was argued that the children's tongue constrictions are generally not as tight during sibilant production as those for adults. In the present notation, we expect the scaled constriction cross-sectional area of the scaled child A_{scaled} to be greater than the constriction cross-sectional area of the adult A_{adult}.

$$A_{scaled} = \mathcal{S}^2 A_{child} > A_{adult} . \tag{8}$$

Because $\mathcal{S} > 1$, it is possible for the inequality in Equation (8) to be true and still have $A_{child} < A_{adult}$. We cannot speak to that possibility at the present time. We argue below that the empirical relation expressed in Equation (8) is the result of the hypotheses expressed in Equation (7). This cannot be shown with mathematical certainty, but a plausibility argument is presented.

We are not certain about the tongue surface curvatures near the tongue tip during sibilant production, because the constriction is formed using three dimensional deformation of the tongue near the tip, and can be affected by individual variation in palate morphology. However, we can consider aspects of tongue surface shape in a coronal plane near the tip. The doming of the child's palate in coronal planes

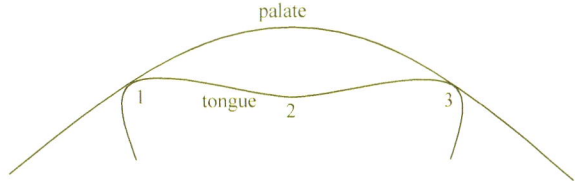

FIGURE 6. The adult's coronal section near the tongue tip for a sibilant fricative

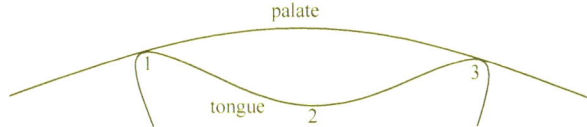

FIGURE 7. The scaled child's coronal section near the tongue tip for a sibilant fricative

near the sibilant constriction is not as great as that of the adult's (Hiki & Itoh 1986).

Again, we consider the child's vocal tract scaled up by the factor S. It can be imagined that the adult's tongue and palate in a coronal plane near the maximum constriction would look like the sketch shown in Figure 6. Let's suppose that the inequality in Equation (8) does not hold. Thus, we suppose that the scaled child has $A_{scaled} = A_{adult}$ instead. The doming of the palate in the coronal plane for the adult means that the palate has a large curvature. This, in turn, means that the tongue surface can have very little curvature in the coronal plane between the sides of the tongue at points 1 and 3, and still form a constriction of small area, A_{adult}. We suppose the maximum curvature between these points occurs at point 2. [We show the tongue braced in this plane, but this is not necessarily the case at the end of the tongue tip.]

What would be the case if the child were to attempt to attain $A_{scaled} = A_{adult}$? Figure 7 shows that the tongue needs to form most of three sides of the constriction channel, because the curvature of the child's palate the coronal plane has a small magnitude. This requires a large-magnitude tongue surface curvature in the coronal plane between the braced sides of the tongue at points 1 and 3.

The large curvatures, and, possibly, rapid average spatial rates-of-change in curvature would be in contradiction to the relations expressed in Equation (7) for a coronal plane near the tip of the tongue. Here the tongue is braced on the sides for both the adult and the child, so we could expect that we could attain larger values for these quantities than for unbraced configurations. However, there must be limitations, even in the braced conditions.

The other possibility for the scaled child to attain $A_{scaled} = A_{adult}$ is to have his or her tongue braced further apart laterally. This would mean that the curvature at (x_2, y_2) would not need to be large. However, this would mean that for the unscaled child, that the tongue surface would be very close to the palate. The air pressure reduction in the fast air flow could be enough to suck the tongue tip to the palate, thus producing a stop. The child would avoid this, perhaps. In conclusion, if our hypotheses in Equation (7) are assumed to be true, then the inequality in Equation (8) must hold.

Difficulties in velar and alveolar stop releases

Fronting of velar consonants is a well-attested phenomenon in young children's speech. We believe that this phenomenon, at least for the youngest children, is partly the result of a scaling mismatch between the physics of acoustic propagation and the physics of the turbulent acoustic source at the teeth during stop release. We have seen that formant frequencies, scale according to the ratio of adult vocal tract length to child vocal tract length, or $\mathcal{S}$, if the child produces is to produce adult-like speech sounds. However, the relation between turbulent acoustic sources of children and adults possess more complicated scaling properties. For instance, if the length of a finite-length tube is scaled down by a factor of two, we would expect the formant frequencies to increase by a factor of two for an equivalent area function. This is not true of turbulent acoustic source properties. In fact, we examine an instance where the predominant turbulent acoustic source property, intensity in higher frequency bands, depends on the unscaled distance from the tongue tip constriction to the teeth. We do not expect that $\mathcal{S}$ has any effect on the intensity in the higher-frequency bands. What constitutes a higher-frequency band, on the other hand, does depend on $\mathcal{S}$.

Here we concentrate on stressed syllable-initial voiceless velar and alveolar stops, which are aspirated in American English. First, we consider some empirical measures of the noise after the release bursts for both some adults and children. We then argue that a young child's intended velar turbulent acoustic source patterns do not resemble those of the adult, because the child has the impossible task of simultaneously matching the formant frequency pattern at release. The result is that the child produces an intended velar release with ambiguous acoustic cues for adult listeners.

Another aspect of velar fronting is what has been termed "undifferentiated tongue gesture", where the tongue covers a large area of the palate during closure. Indeed, tongue contact with the palate is a function of tongue surface curvature. We do not address this aspect of velar fronting in this book. However, the reader can imagine that limitations on tongue surface curvature can lead to relatively large areas of palate covered by the tongue during stop closure.

Turbulent acoustic source properties for velar and alveolar release

We performed a small study of voiceless, aspirated velar and alveolar stop release noise spectra as they evolve in time with one adult male subject and one adult female subject. We used a multi-scale spectral analysis, where the bandwidth of the analysis increases with frequency, while the time resolution also increases with frequency (Stockwell, Manshinha, & Lowe 1996). This analysis shares this property with wavelet analysis, but it is more easily interpretable in terms of standard Fourier analysis than are wavelets.

The results for the two adults producing velar and alveolar releases in two different vowel contexts are shown in Figures 8 and 9. Figure 8 shows the analysis of the male speaker and female speaker for both velar [k] and alveolar [t] releases into the vowel [a]. Figure 9 shows the analysis of the male speaker and female speaker for both velar [k] and alveolar [t] releases into the vowel [i]. These figures show the amplitude of different octave bands for the time interval from about 5 ms before and 15 ms after the release of the consonant. The center frequencies of the bands are given in the legends of these figures.

In both vowel contexts the amplitude of the band centered at 4.4 kHz, represented by the thick gray line, remains high after the alveolar

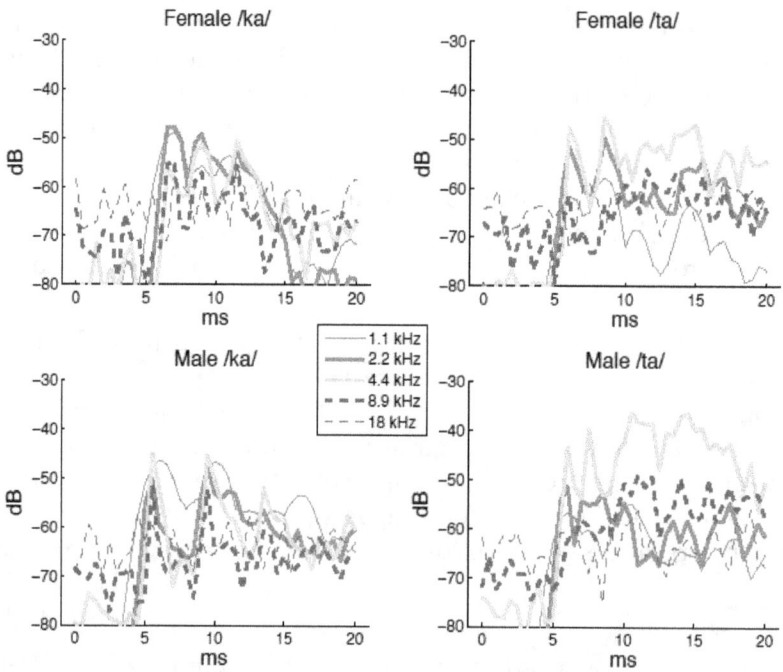

FIGURE 8. Octave band analysis of one male and one female speaker producing [ka] and [ta]

burst, but not after the velar burst. For the high, front vowel context, [i], the next higher band centered at 8.9 kHz, represented by thick dashed lines, also shows the same difference in behavior according to place-of-articulation. [The female speaker does show somewhat elevated energy in the band centered at 4.4kHz after the velar [k] release with the [i] vowel, but the band centered at 8.9 kHz does not show a sustained high energy after release. Further, the amplitude of her 4.4 kHz centered band does not remain high during the entire 15 ms interval after the burst, indicating that multiple releases of the [k] may have occurred. We cannot discount the possibility that the [k] release was somewhat fronted for the female speaker in this front vowel context.]

These results are, perhaps, not the expected results. After all, it is known that the velar release is slower than the alveolar in general. We interpret the results as follows. The sustained strong 4.4 kHz and 8.9

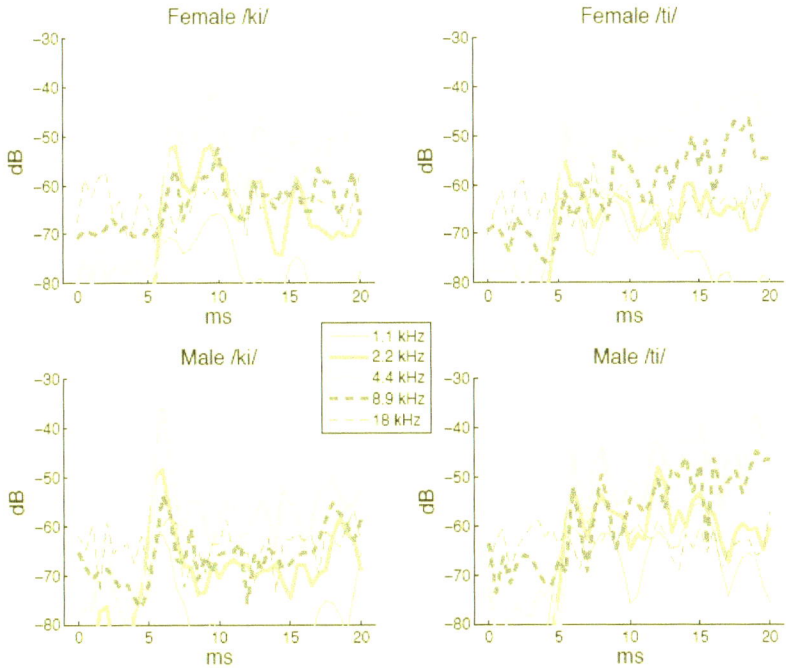

FIGURE 9. Octave band analysis of one male and one female speaker producing [ki] and [ti]

kHz centered bands after alveolar release is due to the the strong air flow from the glottis during aspiration being directed toward the teeth by the tongue blade as the tongue tip moves from the palate. This creates a sustained strong turbulent acoustic source at the teeth for a relatively long time after an alveolar stop release. Also, we hypothesize that the high intensity high frequency energy is generated at the teeth, during velar release. However, a directing of air flow toward the teeth is not sustained after velar release, so the high amplitudes in the 4.4 kHz and 8.9 kHz centered bands are relatively brief. Overall, we hypothesize that a major acoustic cue in place-of-articulation for a stop is the turbulent acoustic source's duration of high-frequency energy.

What about the same analysis performed on the speech of a young child? Figures 10 and 11 show the same analysis applied to alveolar releases of a 30 month-old child who was playing with an adult. Figure 10 shows some purported alveolar releases, while Figure 11 shows

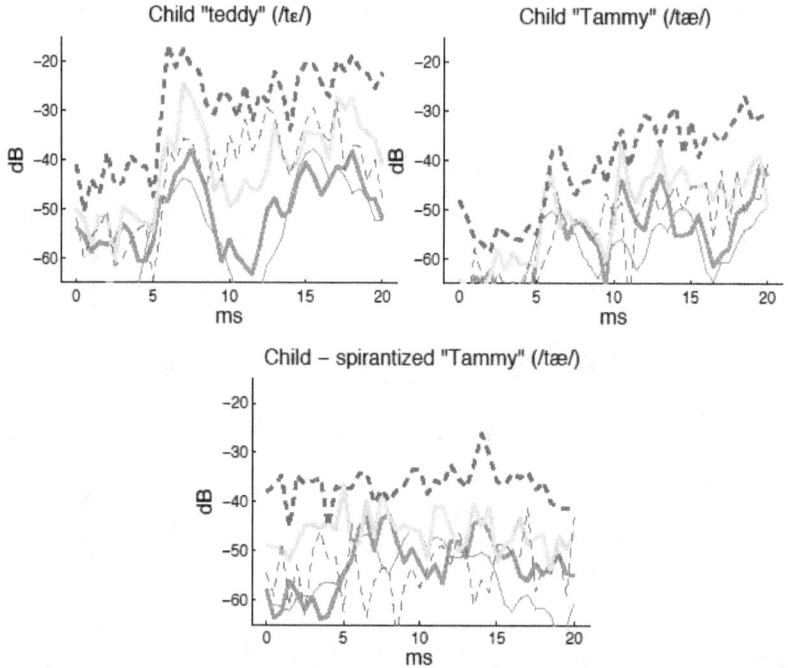

FIGURE 10. Octave band analysis of 30 month-old speaker producing purported [t] releases Center frequencies are given in the legends of Figures 8 and 9.

purported velar releases. Because of the shorter vocal tract, it is the band with center frequency of 8.9 kHz that has the highest sustained amplitude for the alveolar releases in Figure 10. [The turbulent acoustic sources are "filtered" according to formant frequencies, which scale inversely with vocal tract length.]

A comparison of Figures 10 and 11 indicate that there is not a substantial difference in the way that the octave band centered at 8.9 kHz exhibits a sustained high energy for this child for both alveolar and velar releases. They indicate that this young child's intended velar releases have post-release turbulent acoustic source characteristics that similar to those of alveolar releases. The spectrum at the bottom Figure 11 is probably a case of phonemic substitution: of /t/ for /k/, which may just indicate some confusion in the phonemic distinction for this young speaker.

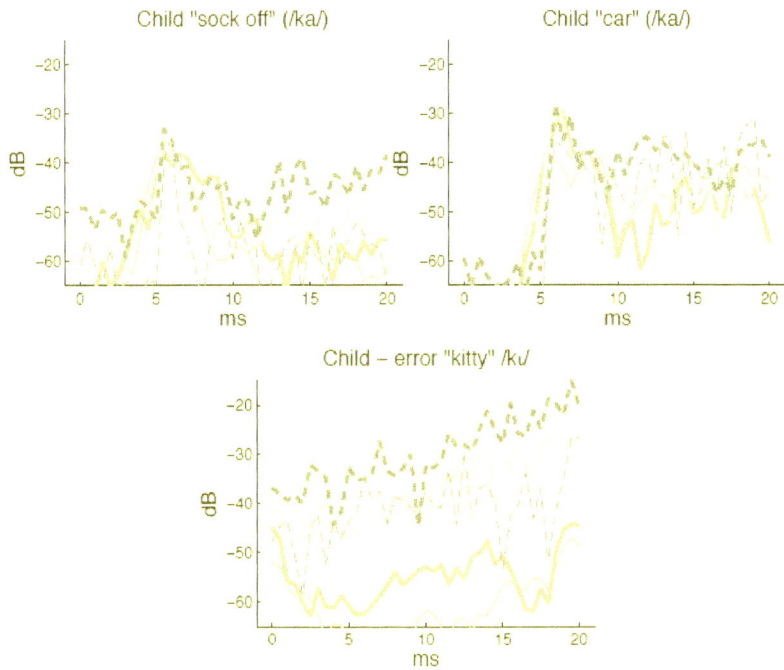

FIGURE 11. Octave band analysis of 30 month-old speaker producing purported [k] releases. Center frequencies are given in the legends of Figures 8 and 9.

A reasonable explanation for these magnitude spectra can be provided, even if we assume that the child is releasing their purported velar consonants at their velum. We consider the way that the power of the turbulent acoustic source at the teeth scales according to physical parameters, such as jet velocity. Instead of using Equation (54) of Chapter 13 to explore the scaling, we use a simpler empirical result that has been theoretically verified (Howe & McGowan 2005). [The expression in Chapter 13 contains parameters that can actually depend on one another, such as jet velocity U and turbulence correlation length l, so this is not the preferred expression for exploring the scaling of noise power for a single speaker.] OASPL has been measured outside the lips during sibilance, and it was found that (Hixon, Minifie, & Tait 1967; Badin 1989)

$$\text{OASPL} \sim (\Delta p_0)^{2.2} . \tag{9}$$

where Δp_0 is the time average pressure difference from behind the teeth to in front of the teeth. This is known as *intra-oral pressure*. Recall that OASPL, or overall SPL, is the total sound pressure level over all frequencies, as defined in Equation (20) of Chapter 7. The symbol $\sim$ means "scales as". We also have

$$\Delta p_0 \sim \frac{\rho_0}{2} U^2 . \qquad (10)$$

where U is specifically the jet velocity at the teeth. So that

$$\text{OASPL} \sim U^{4.4} . \qquad (11)$$

Equation (11) gives a simple scaling for intensity for the listener, but what does the jet velocity U at the teeth depend on? It depends on the subglottal, or tracheal, pressure p_T, the area of the tongue constriction, and the distance of the tongue constriction to the teeth, d_{ct}. If we assume that place-of-articulation has no affect on p_T, then the OASPL during sibilance depends only on d_{ct} for the continuum of tongue constriction areas that occur during stop release.

The child has a shorter distance between the velum and alveolar ridge than does the adult. This means that there is less of difference in the jet velocity U, through its dependence on d_{ct} between the two places of articulation for the child than for the adult. Further, d_{ct} of the child is less than that of the adult for velar stop releases, meaning that children's velar releases are more like alveolar releases, rather than the other way around.

We have argued our point based on sibilant OASPL, but because most of this energy is in the high-frequency octave band (centered at between 4.4 kHz and 8.9 kHz for two adults and centered at 8.9 kHz for a 30 month-old child), we believe that the argument holds for this octave as well. Further, this phenomenon may account for some transcribers describing fronted, when, perhaps, there is no velar stop tongue fronting.

Here we have a spectral property that does not scale with vocal tract length: sustained high intensity, high frequency noise production for velar and alveolar stop release. Another acoustic cue for place-of-articulation are the formant frequency trajectories from the release of tongue constriction. We explore the reason that the child cannot produce a magnitude spectrum that is a simply a constant scaled version of that of the adult. From the arguments above, the child would

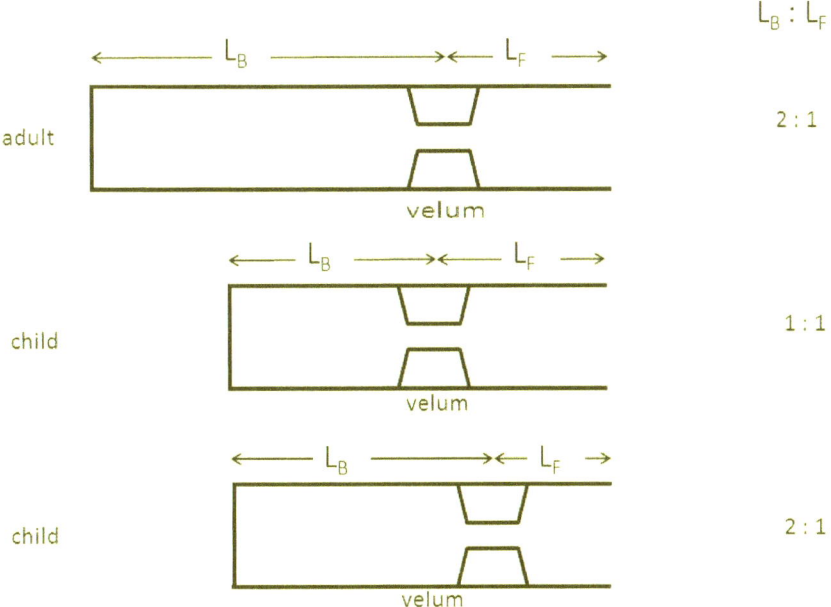

FIGURE 12. A child must front his or her tongue to attain an F2-F3 pinch

need to move the velar place-of-articulation farther back in the vocal tract to obtain a noise source upon release like that of the adult velar release. We now argue that to obtain the adult-like formant frequency pattern for velar place-of-articulation, the child would, indeed, need to front the tongue, which makes the velar release noise source even more alveolar-like.

One of the expected features of the formant frequency trajectories at [k] release is what has been termed the "F2-F3" pinch (Stevens 1998), where the second and third formant frequencies appear to emerge in close proximity at the time of the adult's velar release. The explanation for this pinch is that the constriction at the velum produces a rear sub-tube that is approximately twice the length of the front sub-tube. The rear sub-tube is approximately a half-wave resonator and the front tube a quarter-wave resonator, and, thus, they possess nearly identical resonant frequencies ascribed to F2 and F3.

Figure 12 illustrates this phenomenon, with L_B and L_F denoting the rear sub-tube length and front sub-tube length. The ratio of pharyngeal cavity length to oral cavity length is less for the young child compared to the adult. This means that the child needs to make his or her constriction farther forward with respect to his or her teeth compared to the adult in order to produce the pinch in F2 and F3 at the time of release of the intended velar. The top panel shows the case of velar release for the adult, and the middle panel for a velar release for the young child, accounting for the relatively small pharynx length-to-mouth length ratio in young children. In order that the young child attain the F2-F3 pinch it is necessary to move the tongue place-of-articulation forward along the hard palate.

Not only is the velum closer to the teeth for the child, but the child has a reason to move the intended [k] constriction even closer to the teeth. It is plausible that the child's [k] release occurs at a location that has a comparable distance to the teeth that the adult has for his or her [t] release. This diminished distance has substantial consequences for the other acoustic cue: the spectral properties of the turbulent acoustic source as a function of time. Roughly, moving a tongue constriction closer to the teeth would result in stronger sibilance.

We conclude that what adults hear as a fronted velar release can, indeed, be fronted. However, while we hear a turbulent acoustic source indicating a frontal release, we could find, with further investigation, that the formant trajectories contain the F2-F3 pinch for an intended velar.

Conclusion

We have seen two examples of late-developing segments, [ɹ] and sibilants, for English speaking children that could be the result of anatomical mismatches and hypotheses regarding a child's ability to manipulate the curvature of the tongue surface. We believe that this warrants empirical investigation.

It turns out that turbulent acoustic sources do not scale the same way as acoustic propagation. This makes it impossible for the scaled child to produce both the turbulent acoustic source cues and the formant frequency trajectory cues that the adult can produce.

References

Badin, P. (1989). Acoustics of voiceless fricatives: production theory and data. *Transmission Laboratory of KTH STL-QPSR*, **3/1989**, p 35.

Denny, M. & McGowan, R. S. (2012). Implications of peripheral muscular and anatomical development for the acquisition of lingual control of speech production: a review. *Folia Phoniatrica et Logopaedia*, **64**, p 105.

Denny, M. & McGowan, R. S. (2013). Sagittal area of the vocal tract in young female children. *Folia Phoniatrica et Logopaedia*, **64**, p 297.

Hiki, S. & Ito, H. (1986). Influence of palate shape on lingual articulation. *Speech Communication*, **5**, p 141.

Hixon, T. J., Minifie, F. D. & Tait, C. A. (1967). Correlates of turbulent noise production for speech. *Journal Speech and Hearing Research*, **10**, p 133.

Howe, M. S. and McGowan, R.S. (2005). "Aeroacoustics of [s]." *Proceedings of the Royal Society A*, **461**, p 1005.

McGowan, R.S. & Nittrouer, S. (1988). Differences in fricative production between children and adults: evidence from an acoustic analysis of /ʃ/ and /s/. *Journal of the Acoustical Society of America*, **83**, p 229.

McGowan, R.S., Nittrouer, S., & Manning, C. J. (2004). Development of [ɹ] in young, Midwestern, American children. *Journal of the Acoustical Society of America*, **115**, p 871.

Stevens, K.N. (1998). *Acoustic Phonetics*. MIT Press, Cambridge, MA. (pp 365-7).

Stockwell, R.G., Manshinha, L., & Lowe, R.P. (1996). Localization of the complex spectrum: the S transform. *IEEE Trans. Sig. Proc.*, **44**, p 998.

Vorperian, H.K., Wang, S., Chung, M.K., Schimek, E.M., Durtschi, B.B., Kent, R.D., Ziegert,A.J. & Gentry, L.R. (2009). Anatomical development of the vocal tract: an imaging study. *Journal of the Acoustical Society of America*, **125**, p 1666.

Chapter 15: A Layered Structure Model for Vocal Fold Vibration: First Results

Introduction

We present a mathematical layered structure model for vocal fold vibration that is based on the cover-body conception of the morphology of the vocal folds. The purpose of this model is to provide an understanding for solid-air interaction that can be related to the physiology and morphology of the vocal folds. Further, we wish to start simply so that the physics can be understood, but at the same time, allow enough conceptual room in the model so that more physical details can be added later. The portions of the theory dealing with the tissue are linear, which may limit the model to small amplitude vibration. Certainly the non-linearities of collision need to added to obtain a more complete model of vocal fold oscillation.

The starting point for the layered structure model is Hirano's picture of layered tissues (Hirano 1974, 1988). Hirano describes the set of tissue layers of a vocal fold of an adult human cut in a coronal plane through the middle portion of its anterior-posterior length. There is a cover layer, which consists of sublayers. From glottal midline laterally they are: epithelium, Reinke's space, the intermediate layer, and the deep layer. The latter two layers constitute the vocal ligament, and this ligament appears to be unique to humans. Lateral to the cover layers is the body of the folds, which is the vocalis muscle. The two basic layers of cover and body have become the basis for some descriptions of vocal fold mechanics (e.g. Fujimura 1981). In these descriptions it is the absolute and relative elasticities of the cover and the body that determine many features of vocal fold oscillation.

As a first approximation to the mechanical properties of this layered structure, the model here is presumed not to vary in the anterior-posterior dimension, so that its geometry and dynamics can be modeled in a coronal plane, similar to other models, such as the two-mass model. Also, in this initial model of the vocal folds, the vocal ligament is neglected, so that the model is more applicable to non-humans, such as canines. Hirano emphasized the importance of relating the histology of the layers to their mechanical properties. For instance, he described Reinke's space as a "...loose fibrous component. It is pliable and extremely important to the singing voice." (Hirano 1988, p 52). Thus, in the present model, the outer, epithelial layer is assumed to possess the elasticity of the cover layer, and it is known here as the elastic sheet and presumed infinitely thin. Reinke's space is viewed as an incompressible fluid material and is known simply as the incompressible fluid layer here. The elastic sheet and incompressible fluid layers are considered together when the dynamical equations are derived. The vocalis, or elastic body, is viewed as a sprung mass in the the coronal plane. The air in the glottis is a layer that is referred to as the air channel.

We say a few words regarding the two-mass model (Ishizaka & Matsudaira 1968; Ishizaka & Flanagan 1972) before continuing onto the present model. The two-mass model is ingenious in demonstrating the importance of the relative phasing among different regions of the cover layer. However, there are two short-comings to this model that we seek to remedy. The first is that the description of the air flow is too crude for a good understanding of the role of the air layer in the vibration. The second is that it is difficult to relate the mechanical parameters of the two-mass model to the measurable physio-mechanical properties of vocal folds. It is hoped that what is presented here is a substantial step in remedying these short-comings.

The static base configuration

The static model is shown in Figure 1. We describe the base condition when there is no motion in any of the layers. It is two-dimensional, so that only a planar section in the coronal plane is shown. A rectangular block of mass m has dimensions l_g in the x-direction and d into the paper. The block is attached to a massive wall with a spring with spring constant k and damping ratio $\mathcal{R}$. [The face of the block closest to the air channel is referred to here as the close face.]

A thin elastic sheet is placed over the close face of the block and attached at the ends $x = -l_g/2$ and $x = l_g/2$, and it has a rest length in the x-direction of l_g. This elastic sheet is impervious to fluid flow. A volume of incompressible fluid V_f is injected between the elastic sheet and the block in such a way that the block bows so that a fluid layer of

Acoustics of Speech Production

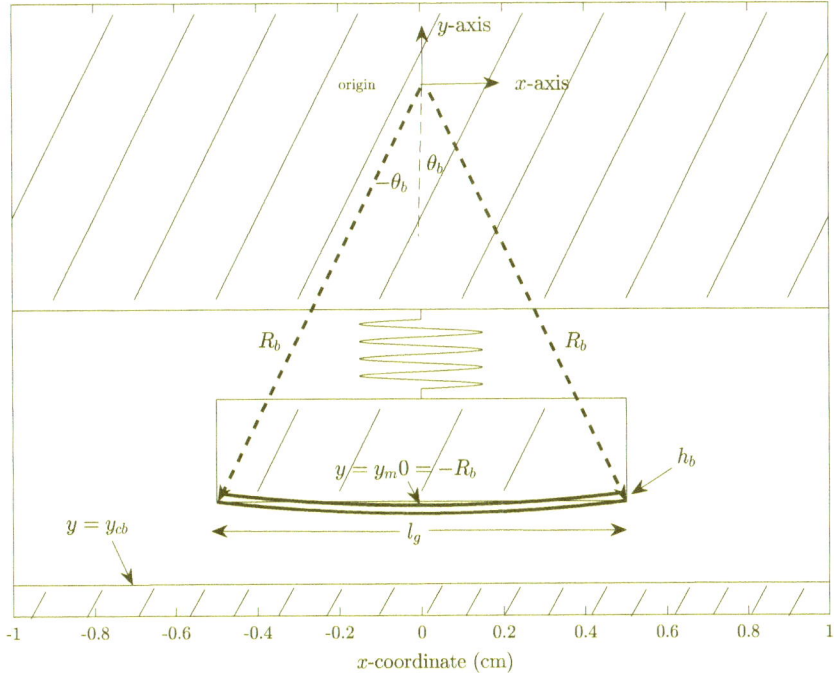

FIGURE 1. The static configuration. The glottal channel width is exaggerated

constant depth h_b forms on the close face. The density of the fluid is denoted ρ_f. When there is no motion, the injection of incompressible fluid causes the elastic sheet to stretch into a sector of a circle of radius R_b between angles $-\theta_b$ and θ_b. The sheet has a tension constant T, which is the constant of proportionality between the pressure difference across the sheet and the curvature of the sheet per unit length in the direction into the page. The close face of the block surface is bowed into the sector of a circle of radius $R_b - h_b$.

The values of R_b, θ_b, and h_b are found in two steps In the first step, no bowing of the close face is permitted, as shown in Figure 2, and the values of R_b and θ_b are found for for this configuration. The two equations that are solved simultaneously for R_b and θ_b are

$$A_f \equiv \frac{V_f}{d} = R_b^2[\theta_b - \sin(\theta_b)\cos(\theta_b)], \tag{1}$$

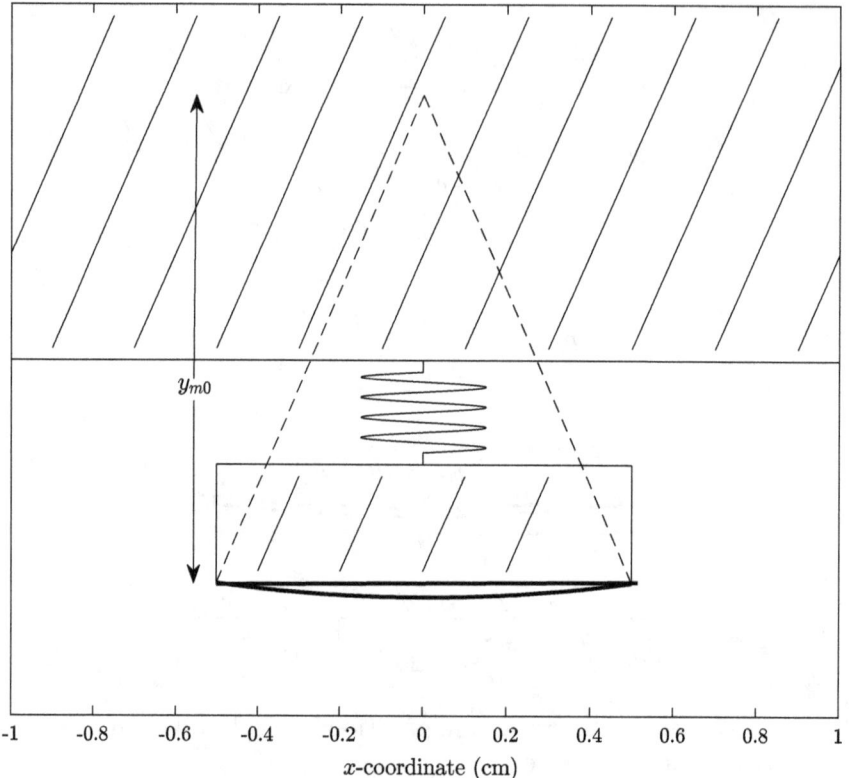

FIGURE 2. The static configuration before bowing. The glottal channel width is exaggerated

and
$$l_g = 2R_b \sin(\theta_b) \,, \qquad (2)$$
where d is the length of the folds orthogonal to the x-y plane.

The second step in defining the incompressible fluid layer is to use these values and to allow bowing of the close face so that a fluid layer of constant depth (in the radial direction) is formed. This incompressible fluid channel has depth h_b, as shown in Figure 1. This means that h_b satisfies the quadratic equation
$$A_f = \theta_b \big(R_b^2 - (R_b - h_b)^2 \big) \,. \qquad (3)$$

The pressure in the incompressible fluid layer above atmospheric is denoted p_{fb}. In the base, static configuration its value is given

$$p_{fb} = \frac{T}{R_b}. \tag{4}$$

All pressures are referenced to atmospheric pressure, so that the air pressure at the elastic sheet surface is $p_a = 0$ in this base condition.

In this initial model, the motions are symmetric about the plane that includes the air channel mid-line and extends orthogonally into and out of the page in Figure 1. A solid wall below the sprung mass represents this mid-channel plane. The mid-channel plane is at a distance 0.1 cm below the unbowed close face of the block, or mass. [The y-coordinate of the unbowed closed face is taken to be the same as the y coordinate of the ends of the elastic sheet in this work.] The y-coordinate of the mid-channel plane is given by $y_{cb} = y_{m0} - 0.1$. It is supposed that there is a weightless boot surrounding the region containing the spring between the block and the massive upper wall, so that when air does flow, it must pass between the elastic sheet and the mid-channel plane. This boot is not drawn in Figure 1.

The following parameters are held constant throughout this chapter.

$$l_g = 1.0 \text{ cm}, \quad d = 1.0 \text{ cm}, \quad y_{m0} = -2.0594 \text{ cm}, \tag{5}$$

$$\rho_f = 1.0 \text{ g/cm}^3, \quad V_f = 0.04 \text{ cm}^3.$$

Values obtained using these parameters are

$$R_b = 2.1192 \text{ cm}, \quad \theta_b = 0.2382 \text{ rad} = 0.076\pi \text{ rad}, \tag{6}$$

$$h_b = 0.040 \text{ cm}, \quad y_{cb} = -2.1594 \text{ cm}.$$

Dynamics

The dynamics of the model are described by a coupled set of equations describing the dynamics of the four layers: the air channel (air in the glottis), the elastic sheet (epithelium), the incompressible fluid (Reinke's space), and the mass-spring system (vocalis). The dynamics of the elastic sheet layer and the incompressible fluid layer dynamics are considered together.

Mass-spring system for the vocalis

The mass-spring system is assumed to be linear. The mass, or block, moves in the y-direction, and its position is denoted by y_m, which can be taken as the y-coordinate of the block's unbowed close face. The equation of motion is

$$\frac{d^2 y_m}{dt^2} + 2\mathcal{R}\omega_0 \frac{dy_m}{dt} + \omega_0^2(y_m - y_{m0}) = \frac{\mathfrak{F}'_f}{m}. \qquad (7)$$

where t is time, y_{m0} is the rest position of the mass, $\omega_0^2 = k/m$, and $\mathfrak{F}'_f$ is the fluctuating force due to the fluctuation in the pressure of the incompressible fluid p_f at the surface of the block. (Any steady net force due to steady pressure in the incompressible fluid is assumed to be accounted for in the term $-\omega_0^2 y_{m0}$ on the left-hand-side.)

The values of the parameters pertaining to Equation (7) are

$$m = 2.0 \text{ g}, \quad k = 1.14 \times 10^6 \text{ g/s}^2, \text{ and } \mathcal{R} = 0.02. \qquad (8)$$

This gives $\omega_0 = 755$ rad/s, which corresponds to a natural frequency $\mathcal{F}_0 = \omega_0/2\pi \approx 120$ Hz. The fluctuating force $\mathfrak{F}'_f$ is a function of the dynamics of the other layers.

Elastic sheet with an underlying incompressible fluid

A polar coordinate system is employed to write the dynamical equations for elastic sheet and incompressible fluid layers. This coordinate system moves with the mass, so that if the change in mass position is Δy_m, then the origin of this polar coordinate system is moved from the fixed origin in Figure 1 in the y-direction by the same increment, Δy_m. When the system of equations for all the layers are solved simultaneously, care is taken to reference the resulting dynamics of the elastic sheet and incompressible fluid to the fixed coordinate system. In the moving polar coordinate system, the radial direction is denoted by the variable r, and the angular direction is denoted by the variable θ. θ varies between $-\theta_b$ and θ_b. A vector that is pointing in the $-y$ direction is in the $\theta = 0$ direction, and rotations counter-clockwise looking into the paper are in the positive θ direction.

Radial displacements of the elastic sheet ζ from the circular sector of the base condition in the moving polar coordinate system are considered. $\zeta = \zeta(\theta, t)$, and the position of the elastic sheet surface at angle θ is given by $r_s(\theta, t) = R_b + \zeta$. In a polar coordinate system, the curvature of the elastic sheet, $\mathcal{K}(\theta, t)$, in the coronal plane is given by

$$\mathcal{K}(\theta, t) = \frac{2(\partial \zeta/\partial \theta)^2 - r_s \partial^2 \zeta/\partial \theta^2 + r_s^2}{\left(r_s^2 + (\partial \zeta/\partial \theta)^2\right)^{3/2}}. \tag{9}$$

This follows from the definition of curvature as the magnitude of the second derivative with respect to curve length of the mapping of from curve length to the Cartesian two-space (do Cormo 1976).

It is assumed that ζ/R_b and its derivatives with respect to θ are small compared to unity. Expanding Equation (9) to first order in ζ/R_b and its derivatives provides the approximation

$$\mathcal{K}(\theta, t) \approx \frac{1}{R_b} - \frac{1}{R_b^2}\left(\zeta + \frac{\partial^2 \zeta}{\partial \theta^2}\right). \tag{10}$$

$p_a = p_a(\theta, t)$ denotes the pressure of the air (above atmospheric) at the surface of the elastic sheet. Thus,

$$\left. p_f \right|_s - p_a = T\left(\frac{1}{R_b} - \frac{1}{R_b^2}\left(\zeta + \frac{\partial^2 \zeta}{\partial \theta^2}\right)\right), \tag{11}$$

where $\left. p_f \right|_s = \left. p_f \right|_s (\theta, t)$ is the pressure of the incompressible fluid at the elastic surface. With p_f' defined as the pressure due to non-zero $\zeta + \partial^2 \zeta/\partial \theta^2$, so that $p_f = p_{fb} + p_f'$, it follows from Equations (4) and (11) that

$$\left. p_f' \right|_s = -\frac{T}{R_b^2}\left(\zeta + \frac{\partial^2 \zeta}{\partial \theta^2}\right) + p_a, \tag{12}$$

where $p_a = p_a(\theta, t)$ is the fluctuating air pressure at the elastic sheet surface. [A prime superscript is not used because the static, base condition is $p_a = 0$ with no air flow.]

We follow Howe (2007) in deriving an equation of motion for surface waves on a water-like, incompressible fluid. In that derivation, the incompressible fluid can be considered to be shallow on the scale of the wavelength of the surface waves and gravity provides the restoring force for surface disturbances. In the present model, the restoring force of gravity is replaced by the tension in the elastic sheet. If the linearized momentum equation in the radial direction for the incompressible fluid layer is integrated from some radius r, with $R_b - h_b < r < R_b + \zeta$, to $R_b + \zeta$, then

$$\rho_f \int_r^{R_b+\zeta} \frac{\partial u_r}{\partial t} dr = p_f(r,\theta,t) - p_f(R_b+\zeta,\theta,t)$$

$$\equiv p_f(r,\theta,t) - p_f\bigg|_s (\theta,t) \quad (13)$$

$$= p_f(r,\theta,t) + \frac{T}{R_b^2}\left(\zeta + \frac{\partial^2 \zeta}{\partial \theta^2}\right) - \frac{T}{R_b} - p_a(\theta,t) ,$$

where u_r is the fluid particle velocity in the radial direction. The final equality in Equation (13) follows from Equation (12). The left-hand-side scales as $\rho_f \omega^2 \zeta h_b$, where ω is a characteristic (circular) frequency for the disturbance. The term on the right-hand-side that involve ζ and $\partial^2 \zeta/\partial \theta^2$ scales as $T(\zeta/R_b^2 + \zeta/\lambda^2)$, where λ is the characteristic wavelength for the disturbance on the elastic sheet. In order for the rate-of-change of radial momentum, the left-hand-side of Equation (13) be negligible, in relation to term involving ζ and $\partial^2\zeta/\partial\theta^2$ in Equation (13). Thus, we can make the shallow fluid approximation if

$$1 \ll \frac{T}{\rho_f h_b \omega^2}\left(\frac{1}{R_b^2} + \frac{1}{\lambda^2}\right) . \quad (14)$$

It will be seen below that λ consistently scales as l_g. Further, $\omega = \omega_0$, the natural (circular) frequency of the mass-spring system, and $R_b \geq l_g$. It is assumed that $T > 10^4$ dyne/cm. Using the values given in Equations (5) and (6), it is seen that the shallow fluid approximation is valid in the situation considered here, so that the rate-of-change of radial momentum in the incompressible fluid layer can be neglected. Therefore, p_f does not depend on r.

$$p_f = p_f(\theta,t) . \quad (15)$$

The linearized momentum equation for the incompressible fluid is written in the angular dimension

$$\frac{\partial u_\theta}{\partial t} = -\frac{1}{\rho_f R_b}\frac{\partial p_f}{\partial \theta} . \quad (16)$$

Thus, in the linear approximation, the fluid particle velocity in the angular direction u_θ is not a function of r, so that $u_\theta = u_\theta(\theta,t)$.

By mass conservation the divergence of the incompressible fluid velocity is zero, so that its integral through the fluid layer is also zero

$$0 = \int_{R_b-h_b}^{R_b+\zeta} \left(\frac{\partial u_r}{\partial r} + \frac{1}{r}\frac{\partial u_\theta}{\partial \theta} \right) dr \qquad (17)$$

$$\approx \frac{\partial \zeta}{\partial t} + \frac{h_b}{R_b}\frac{\partial u_\theta}{\partial \theta} ,$$

where the kinematic relation $u_r = \partial \zeta/\partial t$ at $r = R_b + \zeta$ and $u_r = 0$ at $r = R_b - h_b$ in our local coordinate system. The approximations $h_b/R_b, \zeta/R_b \ll 1$ are used in Equation (17). Taking the derivative of Equation (16) with respect to $R_b\theta$, and the time derivative of Equation (17), and using Equations (4) and (11) to write p_f in terms of ζ and p_a produces

$$\frac{\partial^2 \zeta}{\partial t^2} + \frac{Th_b}{\rho_0 R_b^4}\left(\frac{\partial^2 \zeta}{\partial \theta^2} + \frac{\partial^4 \zeta}{\partial \theta^4} \right) = \frac{h_b}{\rho_0 R_b^2}\frac{\partial^2 p_a}{\partial \theta^2} ,$$

or (18)

$$\frac{\partial^2 \zeta}{\partial t^2} + \omega_T^2\left(\frac{\partial^2 \zeta}{\partial \theta^2} + \frac{\partial^4 \zeta}{\partial \theta^4} \right) = \frac{h_b}{\rho_0 R_b^2}\frac{\partial^2 p_a}{\partial \theta^2} ,$$

where $\omega_T = \sqrt{\dfrac{Th_b}{\rho_0 R_b^4}}$.

Equation (18) is the equation of motion for the elastic sheet and incompressible fluid that is used here.

The constraints that are applied along with Equation (18) are

$$\zeta(\theta_b, t) = 0 ,$$
$$\zeta(-\theta_b, t) = 0 , \qquad (19)$$
$$\frac{\partial^2 \zeta(\theta_b, t)}{\partial \theta^2} = 0 ,$$
$$\int_{-\theta_b}^{\theta_b} \zeta(\theta, t) d\theta = 0 .$$

The first two constraints state that there is no movement of the elastic sheet at its endpoints. The third constraint says that there is not extra restoring force due to the perturbation ζ at the downstream boundary. The fourth constraint states that there is no volume added to the underlying fluid due to the perturbation, ζ.

Solutions to the homogeneous equation for ζ

Set $\zeta = \hat{\zeta} e^{i\omega t}$. In this case of harmonic motion, the homogeneous form of Equation (18) can be written

$$\frac{d^4\hat{\zeta}}{d\theta^4} + \frac{d^2\hat{\zeta}}{d\theta^2} - \left(\frac{\omega}{\omega_T}\right)^2 \hat{\zeta} = 0 . \qquad (20)$$

The constraints are those in Equation (19) expressed in the frequency domain.

Let

$$\delta = \frac{\omega}{\omega_T} \quad \text{and} \quad \gamma = \sqrt{\frac{1 + \sqrt{1 + 4\delta^2}}{2}} . \qquad (21)$$

From Equation (21) relationship between γ and $\delta^2 = (\omega/\omega_T)^2$ is

$$\delta^2 = \left(\frac{\omega}{\omega_T}\right)^2 = \frac{(2\gamma^2 - 1)^2 - 1}{4} . \qquad (22)$$

The characteristic equation that is associated with Equation (20) has roots $i\gamma$, $-i\gamma$, $\sqrt{\gamma^2 - 1}$, and $-\sqrt{\gamma^2 - 1}$. Four linearly independent solutions to Equation (20) can be written

$$\cos(\gamma\theta), \; \sin(\gamma\theta), \; \sinh(\sqrt{\gamma^2 - 1}\,\theta), \; \text{and} \; \cosh(\sqrt{\gamma^2 - 1}\,\theta) . \qquad (23)$$

These solutions can be combined to produce linearly independent solutions that satisfy the constraints in Equation (19). One group of these solutions can be written

$$G_m(\theta) = \sin(k_m \theta) , \; \text{where} \; k_m = \frac{(m\pi)}{\theta_b}, m \; \text{an integer} . \qquad (24)$$

The k_m are the values of γ that are required to fit the conditions in Equation (19). Let ω_{Gm} be the circular frequency that corresponds to k_m. Figure 3 indicates the locations of the values of k_m on the γ-axis with open circles. From Equation (22) it follows that

$$\left(\frac{\omega_{Gm}}{\omega_T}\right)^2 = \frac{(2k_m^2 - 1)^2 - 1}{4} ,$$

or $\qquad (25)$

Acoustics of Speech Production

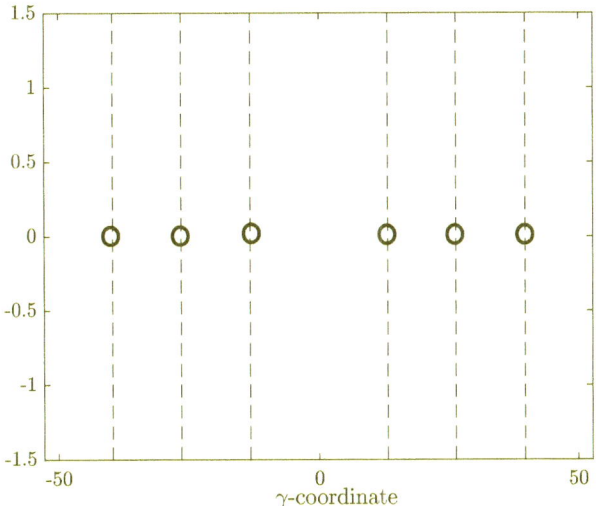

FIGURE 3. Allowed non-zero values of γ, or k_m, for functions G_m.

$$\omega_{Gm} = \omega_T \frac{\sqrt{(2k_m^2 - 1)^2 - 1}}{2}.$$

Another group of solutions can be written

$$F_n(\theta) = \frac{\sinh\left(\sqrt{\kappa_n^2 - 1}\,(\theta - \theta_b)\right)}{\sinh\left(2\sqrt{\kappa_n^2 - 1}\,\theta_b\right)} - \frac{\sin\left(\kappa_n(\theta - \theta_b)\right)}{\sin\left(2\kappa_n\theta_b\right)},$$

where (26)

$$\sqrt{\kappa_n^2 - 1}\,\tan\left(\kappa_n\,\theta_b\right) = \kappa_n \tanh\left(\sqrt{\kappa_n^2 - 1}\,\theta_b\right), \quad n \text{ an integer.}$$

Figure 4 shows the locations of the values of κ_n on the γ-axis with open square boxes. The κ_n are the values of γ that are required to fit the conditions in Equation (19). Let ω_{Fn} be the circular frequency that corresponds to κ_n. As in Equation (25) for ω_{Gm}, the following Equation (27) holds for ω_{Fn}.

$$\left(\frac{\omega_{Fn}}{\omega_T}\right)^2 = \frac{(2\kappa_n^2 - 1)^2 - 1}{4},$$

or (27)

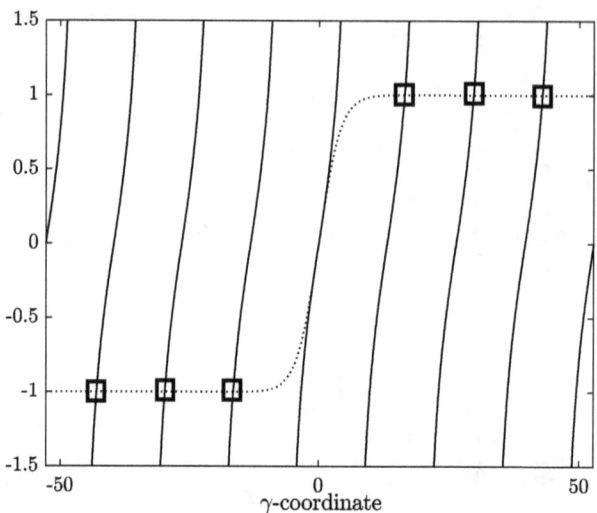

FIGURE 4. Allowed non-zero values of γ, or κ_n, for functions F_n.

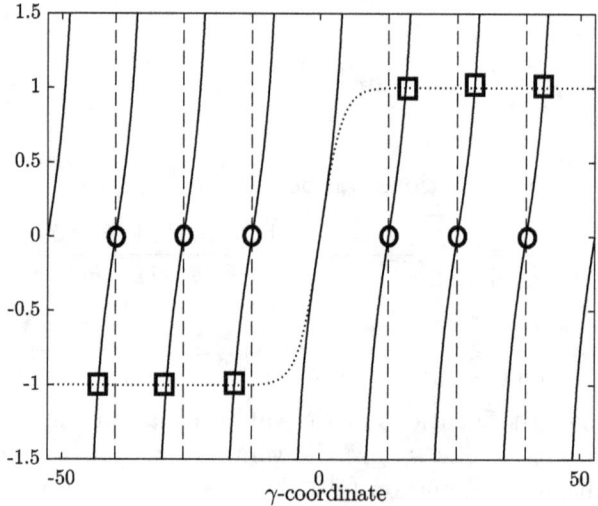

FIGURE 5. Allowed non-zero values of k_m and κ_n.

$$\omega_{F_n} = \omega_T \frac{\sqrt{(2\kappa_n^2 - 1)^2 - 1}}{2} .$$

Figure 5 shows the values of k_m and κ_n together. We need only take positive values of k_m and κ_n, as the negative values do not result in different eigenvalues with linearly independent functions. In this chapter, eigenfunctions are often referred to as modes or components.

Solution to the non-homogeneous equation for ζ

In this section, solutions to Equation (18) with the constraints expressed in Equation (19) are written using an expansion in terms of a set of orthogonal eigenfunctions derived from the eigenfunctions in Equations (24) and (26). [Transients are neglected in this analysis.] The theory of self-adjoint differential operators can be applied to our problem because the operator $d^4/d\theta^4 + d^2/d\theta^2$ is self-adjoint (Ince 1956). According to Courant & Hilbert (1953), there is a set of orthonormal eigenfunctions, $\hat{\zeta}_n$, with eigenvalues, λ_n, $n = 1, 2, 3, ...$ that also satisfy any imposed homogeneous boundary conditions. That is, there is a set of orthonormal functions $\hat{\zeta}_n$ such that

$$\frac{d^4\hat{\zeta}_n}{d\theta^4} + \frac{d^2\hat{\zeta}_n}{d\theta^2} - \lambda_n\hat{\zeta}_n = 0 ,$$
with $\lambda_n = (\omega_n/\omega_T)^2$, and where , \hfill (28)

$$\hat{\zeta}_n(-\theta_b) = 0 \quad \hat{\zeta}_n(\theta_b) = 0 \quad \frac{d^2\hat{\zeta}_n}{d\theta^2}(\theta_b) = 0 .$$

Further, this set of orthonormal eigenfunctions is complete in the sense that any sufficiently smooth function $g(\theta, \omega)$ that satisfies Equation (28) can be written as an absolutely and uniformly converging sum of the $\hat{\zeta}_n$ eigenfunctions.

$$g(\theta, \omega) = \sum_{n=1}^{\infty} \hat{c}_n(\omega)\hat{\zeta}_n(\theta) . \tag{29}$$

We extend this theorem, without proof, to the case of the constraints expressed in Equation (19), transformed from the time domain to the frequency domain. [An outline of a proof is as follows. The constraints in Equation (19) contain three homogeneous boundary conditions. The final constraint states that the solution must be orthogonal to the

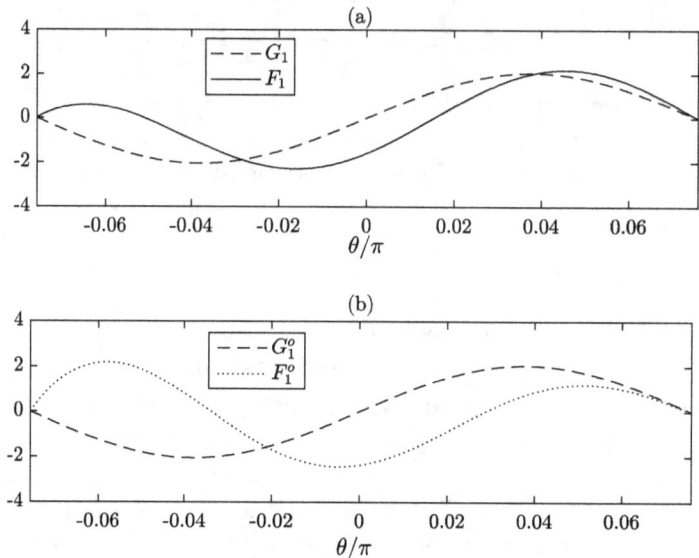

FIGURE 6. Functions G_1 and F_1 before and after orthogonalization process

constant function. The functions that satisfy the first three constraints, as well as the fourth, form a closed subspace of the function space that satisfy the first three constraints.]

In order to generate an appropriates set of orthonormal set of eigenfunctions, the Gram-Schmidt orthogonalization process can be employed (Courant & Hilbert 1953). First, the eigenfunctions G_m and F_n are ordered so that their eigenvalues form an increasing sequence. It can be seen from Figure 5 that this ordering is $G_1, F_1, G_2, F_2, \ldots$ with corresponding eigenvalues $(\omega_{G1}/\omega_T)^2$, $(\omega_{F1}/\omega_T)^2$, $(\omega_{G2}/\omega_T)^2$, $(\omega_{F2}/\omega_T)^2$, ... These eigenfunctions can all be normalized by their magnitudes, so we will assume that this has been done without changing notation. Figure 6(a) shows normalized G_1 and F_1. The first two orthonormal eigenfunctions are

$$G_1^o = G_1 \tag{30}$$

$$F_1^o = \frac{F_1 - \langle F_1, G_1^o \rangle G_1^o}{\|F_1 - \langle F_1, G_1^o \rangle G_1^o\|}.$$

$\Big[$The usual definitions of inner product, $\langle \cdot \rangle$, and norm, $\| \cdot \|$, apply: $\langle f, g \rangle = \int_{-\theta_b}^{\theta_b} f(\theta)g(\theta)d\theta$ and $\|f\| = \sqrt{\langle f, f \rangle}.$ $\Big]$ These orthonormal functions are shown in Figure 6(b). The orthogonalization process can proceed indefinitely through the ordered set of eigenfunctions. Note that the eigenvalue of G_1^o is $(\omega_{G1}/\omega_T)^2$ and the eigenvalue of F_1^o is $(\omega_{F1}/\omega_T)^2 - \langle F_1, G_1^o \rangle(\omega_{G1}/\omega_T)^2$.

With Equation (30) providing the first two orthonormal eigenfunctions, a truncated Equation (29) produces the following first approximation to a solution, $\zeta(\theta, t)$, to Equation (18) under constraints (19) in the time domain.

$$\zeta(\theta, t) \approx C_G(t)G_1^o(\theta) + C_F(t)F_1^o(\theta). \tag{31}$$

In this approximation, Equation (18) becomes

$$\left(G_1^o(\theta)\frac{d^2 C_G(t)}{dt^2} + F_1^o(\theta)\frac{d^2 C_G(t)}{dt^2} \right) + \\ \left(C_G(t)\left(\frac{d^4 G_1^o(\theta)}{d\theta^4} + \frac{d^2 G_1^o(\theta)}{d\theta^2} \right) + \right. \\ \left. C_F(t)\left(\frac{d^4 F_1^o(\theta)}{d\theta^4} + \frac{d^2 F_1^o(\theta)}{d\theta^2} \right) \right) = \frac{h_b}{\rho_0 R_b^2}\frac{\partial^2 p_a}{\partial \theta^2}. \tag{32}$$

Using the fact that G_1^o and F_1^o are eigenfunctions, Equation (32) can be written

$$G_1^o(\theta)\frac{d^2 C_G(t)}{dt^2} + F_1^o(\theta)\frac{d^2 C_G(t)}{dt^2} + \omega_{G1}^2 C_G(t)G_1^o + \\ (\omega_{F1}^2 - \langle F_1, G_1^o \rangle \omega_{G1}^2)C_F(t)F_1^o \\ = \frac{h_b}{\rho_0 R_b^2}\frac{\partial^2 p_a}{\partial \theta^2}. \tag{33}$$

Recall that $\langle \cdot \rangle$ denotes inner product in function space here and not time average.

Multiplying Equation (33) by G_1^o and integrating with respect to θ, and then doing the same with F_1^o as the multiplicative factor, results in two ordinary differential equations for $C_G(t)$ and $C_F(t)$.

$$\frac{d^2 C_G(t)}{dt^2} + \omega_{G1}^2 C_G(t) = \frac{h_b}{\rho_0 R_b^2} \left\langle \frac{\partial^2 p_a}{\partial \theta^2}, G_1^o \right\rangle,$$

and (34)

$$\frac{d^2 C_F(t)}{dt^2} + \left(\omega_{F1}^2 - \langle F_1, G_1^o \rangle \omega_{G1}^2\right) C_F(t) = \frac{h_b}{\rho_0 R_b^2} \left\langle \frac{\partial^2 p_a}{\partial \theta^2}, F_1^o \right\rangle.$$

The terms on the right-hand sides are the projections of the forcing function $(h_b/(\rho_0 R_b^2))\partial^2 p_a/\partial \theta^2$ onto the eigenfunctions, or modes, G_1^o and F_1^o.

Fluctuating force on the mass-spring system

Equation (12) and the fact that pressure in the incompressible fluid p_f is independent of radial position permits the calculation of pressure at the bowed closed face of the mass-spring system. The force on the mass in the y direction, $\mathfrak{F}'_f$, that appears in Equation (7) is then,

$$\mathfrak{F}'_f = \left((R_b - h_b) \int_{-\theta_b}^{\theta_b} p'_f(\theta) \cos(\theta) d\theta \right.$$

$$\left. + h_b \sin(\theta_b)\left(p_f'(-\theta_b) + p'_f(\theta_b)\right) \right) d,$$

(35)

where p'_f is given by Equation (12).

The air layer

The final layer to be discussed is the air layer between the mid-channel plane and the elastic sheet. The air pressure at the surface of the elastic sheet, $p_a(\theta, t)$ needs to be determined so that Equation (34) can be solved and Equation (35) evaluated. It is supposed that the air flow in the channel is potential and separates from the elastic sheet at the position where the sheet and the mid-channel plane are closest.[4] The air upstream and outside of the resulting jet is at rest.

---

[4]This is not quite correct, as was shown in Chapter 13. The separation should be a little downstream of the position where the elastic sheet is closest to the mid-channel

The disturbances within the air channel between the plane and elastic sheet are characterized by frequency $\mathcal{F}_0$, which has a corresponding wavelength of sound that is large compared to the length scale of the air channel l_g. Thus, the air flow is approximately incompressible. Further, the quasi-steady approximation is made, and the Bernoulli equation [Equation (3) of Chapter 13] can be invoked upstream of the separation point, as well as in the jet downstream of the separation point.

$$p_a + \frac{\rho_a}{2} u_a^2 = P_{head}(t) \ . \tag{36}$$

where $p_a = p_a(x, y, t)$ is air (static) pressure and $u_a = u_a(x, y, t)$ is the air speed in the channel. $P_{head}(t)$ denotes the pressure head for the air upstream of the separation point. We are again assuming no dependence in the dimension into the page.

Upstream of the point of flow separation, velocity potential lines in the x-y plane were approximated by circular sections between the plane and the elastic sheet. The equal potential lines intersect these surfaces in their normal directions, because there is no air flow across either surface. For an epithelium that is extremely bent with a negative curvature, this can be a poor approximation, but this is not expected because the base configuration has positive curvature and the perturbations to this curvature made by ζ are not extreme.

It is possible to write expressions for the surface of the elastic sheet in the fixed coordinate system. The position of a point (x_s, y_s) on the surface of the elastic sheet is given by

$$x_s = (R_b + \zeta)\sin(\theta) \quad \text{and} \quad y_s = -(R_b + \zeta)\cos(\theta) + \Delta y_m \ . \tag{37}$$

[Recall that $\Delta y_m = y_m - y_{m0}$.] At each time t the tangent to the elastic sheet makes an angle α with respect to the x-axis, with

$$\tan(\alpha) = \frac{\frac{\partial y_s}{\partial \theta}}{\frac{\partial x_s}{\partial \theta}} \ . \tag{38}$$

The normal to the surface of the elastic sheet makes an angle $\alpha + \pi/2$ with the x axis, while the mid-channel plane has its normal directed $\pi/2$ to the x-axis. Thus, the circular sector that we seek turns through α radians. Let D be the perpendicular distance from (x_s, y_s) to the

---

plane. However, we believe that this first approximation does not affect the main results here.

plane so that
$$D = \Delta y_m - (R_b + \zeta)\cos(\theta) - y_{cb}. \qquad (39)$$
Let s_p denote the arc length of one of the circular sectors. If α is zero, then the radius of the circular sector is infinite, and the equal potential line is straight, and the arc length $s_p = D$. Otherwise, the radius of the circular sector r_p is
$$r_p = \frac{D}{\sin(\alpha)} = \frac{\Delta y_m - (R_b + \zeta)\cos(\theta) - y_{cb}}{\sin(\alpha)}. \qquad (40)$$
The sector arc length is
$$s_p = \alpha r_p = \left(\Delta y_m - (R_b + \zeta)\cos(\theta) - y_{cb}\right)\frac{\alpha}{\sin(\alpha)}, \qquad (41)$$
which also is valid for $\alpha = 0$.[5]

---

[5]Another piece of the air flow picture through the glottis during voicing, or phonation, is the variablity in the discharge coefficient of the glottal jet during each cycle of vocal fold oscillation (Park & Mongeau 2007). Howe & McGowan (2010) took these data as indicating that the glottal jet contraction ratio σ changes through the glottal cycle. These authors assumed that the jet contraction ratio changed with changes in the separation point from infinitely thin vocal folds in such away that the separation occurs at the downstream end of the vocal folds near their tips as the vocal folds open and on the upstream side of the vocal folds near their tips as the vocal fold close.
This picture has important consequences on the *suction forces* on the tips of the folds, which are also known as *Coanda forces*. This force is missed when a quasi-one dimensional model for air flow is used, such as in the two-mass model of Ishizaka and Matsudaira (1968). These force must exist in order for the fluid particles follow a curvlinear path around the vertical extremes of the vocal folds, The curvlinear paths require that streamlines of equal flux are very close to one another near the tips, and, by the Bernoulli equation [Equation (36) with spatially and time constant pressure head], and the pressures become very small around the tip. When the folds are opening the tips of the folds experience atmospheric pressure [neglecting the transmitted wave], and when the golds are closing they experience the suction forces. In the present model some details two-dimensional air flow are modeled. However, the changes in separation point location is always at the point of minimum glottal width– the local changes of this separation point according to Howe and McGowan (2010) are not modeled here. If these local changes were accounted for the separation point would be just upstream or at the glottal minimum during the opening phase (i.e. the downward movement of the block), and the location would be downstream of the during the closing phase the closing phase (i.e. the upward movement of the block). This effect should enhance the ability of the air flow to input energy into the vocal fold vibration with a suction force.

The Bernoulli relation expressed in Equation (36), is invoked to find p_a along each equal-potential line in terms of $P_{head}(t)$ and $Q(t)$. $p_a(\theta, t)$ equals the p_a along the potential line that intersects the elastic sheet at θ. We make the quasi-steady approximation in writing an equation for the volume flow through the air channel, $Q(t)$. Using Equation (33) of Chapter 13 under the quasi-steady approximation results in

$$\left(\frac{h}{\sigma \xi}\right)^2 \frac{1}{2A} Q^2(t) + 2c_0 Q(t) \approx \frac{2Ap_I}{\rho_0} , \qquad (42)$$

where $2p_I$ is the subglottal pressure established with a traveling wave from the lungs with pressure p_I long enough in the past that transient effects have decayed. σ is the jet contraction ratio and ξ is the minimum distance between the elastic sheet and the fixed plane.[6] The subglottal and supraglottal ducts have rectangular cross-section of height $2h$ and cross-sectional area $A = 2hd$.[7] It follows from Equations (23), (24), and (25) of Chapter 13 that the pressure head upstream of separation is

$$P_{head}(t) = 2p_I - \frac{\rho_0 c_0}{A} Q(t) . \qquad (43)$$

Upstream of the separation point, the air pressure on the surface of the epithelium p_a is found from the Bernoulli equation in Equation (36). Downstream of the separation point $p_a = \rho_0 c_0 Q/A$.

The feedback due to resonances in the supra-glottal and sub-glottal tracts is neglected here. The parameters that appear in the equations for air flow were assigned values

$$\sigma = 1.0, \quad h = 2.0 \text{ cm} , \qquad (44)$$

$$\text{and} \quad p_I = 4000 \text{ dyne/cm}^2 \approx 0.004 \text{ atmosphere} .$$

These assignments were used in the simulations presented below. Other parameter values assigned in Equations (5), (6), and (8) were also used for these simulations.

---

[6] A notational change has been made here from that of Chapter 13, so that ξ denotes the minimum distance from the elastic sheet to the mid-channel plane in the present Chapter.

[7] We assumed that the cross-sectional areas of the subglottal and supraglottal ducts are equal In the derivation of these equations, but this is not critical to the overall argument.

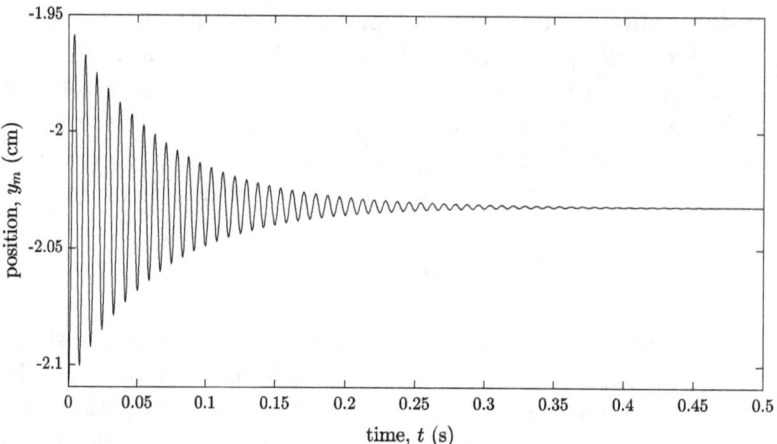

FIGURE 7. Mass position when there is no interaction with the incompressible fluid layer

Simulations

In all of the following simulations, the initial $\Delta y_m = -0.05$ cm. The situation where the elastic sheet-incompressible fluid dynamics are "turned off" is illustrated first. In this case $\zeta(\theta, t) = 0$ for all θ and all time t, which implies that $\mathfrak{F}'_f = 0$. Figure 7 shows the y-coordinate of the block $y_m(t)$ versus time. The motion is transitory: it is exponentially damped, and it has substantially died away by $t = 0.4$ s.

Presently we permit full coupling between the fluid layer and the mass-spring system. The following simulations were done with an initial $\zeta = 0.01 G_1^o$, and epithelial tension $T = 2 \times 10^5$ dyne/cm. The natural frequency for G_1^o associated with these values for sheet tension, are $\omega_{G1}/(2\pi) \approx 194$ Hz. For F_1^o the natural frequency is $\omega_{F1}/(2\pi) \approx 254$ Hz. The natural frequency of the block-spring system is 120 Hz.

Figure 8 shows y_m versus time for a duration of 6 s. After the transient has died there appears to be steady motion, as shown in Figure 8. There is net energy input to the mass-spring system from the other layers, as shown in Figure 9. Its peak-to-peak amplitude of y_m is about 0.007 cm at 5 s into the oscillation, which is a very small amplitude, as shown in Figure 10(a). This corresponds to a 1.2

Acoustics of Speech Production

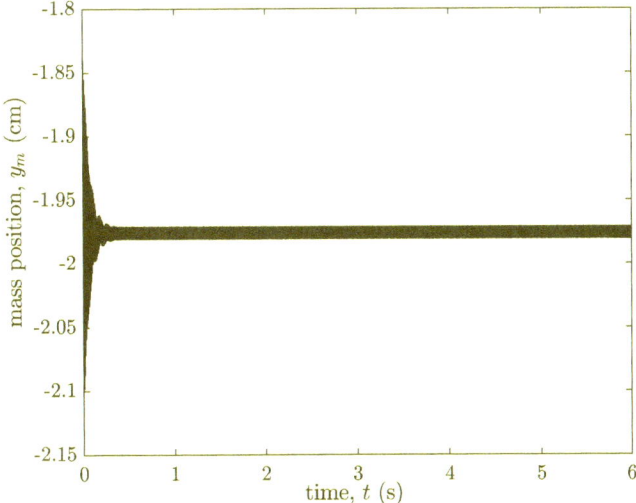

FIGURE 8. Mass position for $T = 2 \times 10^5$ dyne/cm.

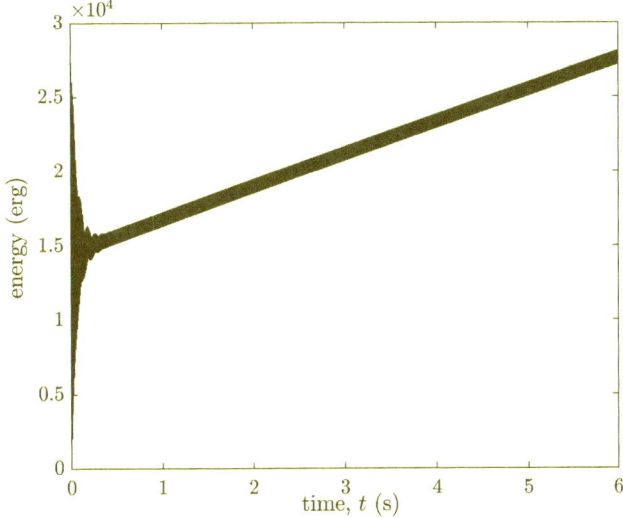

FIGURE 9. Energy input to the mass-spring system for $T = 2 \times 10^5$ dyne/cm.

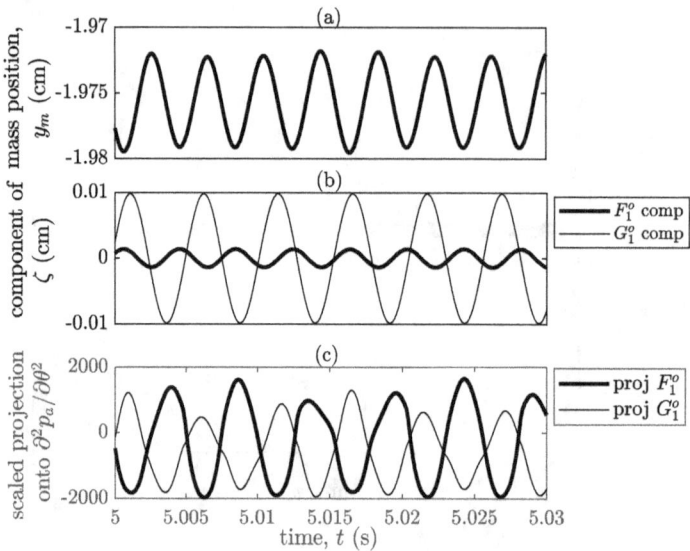

FIGURE 10. Details the simulations shown in Figures 8 and 9, a) mass position versus time, b) components of ζ, and c) scaled projections of $\partial^2 p_a/\partial \theta^2$.

cm³/s peak-to-peak variation in air volume velocity and a variation of about 31 cm/s in maximum air channel particle velocity peak-to-peak amplitude. The energy input over the first 6 s is very small for $T = 2 \times 10^5$ dyne/cm. Figure 9 shows that the total energy input into the mass-spring system, or work done on the mass-spring system is about 27,000 erg. If the 15,000 erg of energy that is input during the transient phase is discounted, then the energy input during the steady phase is 12,000 erg.

Figure 10(a) shows that the position of the mass y_m versus time t, and while exhibiting oscillatory motion, it also possesses some irregularity. This is the result of beating among the three natural frequencies involved: 120 Hz for the mass-spring system, and 194 Hz and 254 Hz for the system of the elastic sheet and incompressible fluid. Figure 10(b) shows the relative phase between G_1^o component and the F_1^o component is continuously changing because of the frequency differences between these components. Figure 10(c) shows that the continuously changing relative phase between the G_1^o and F_1^o

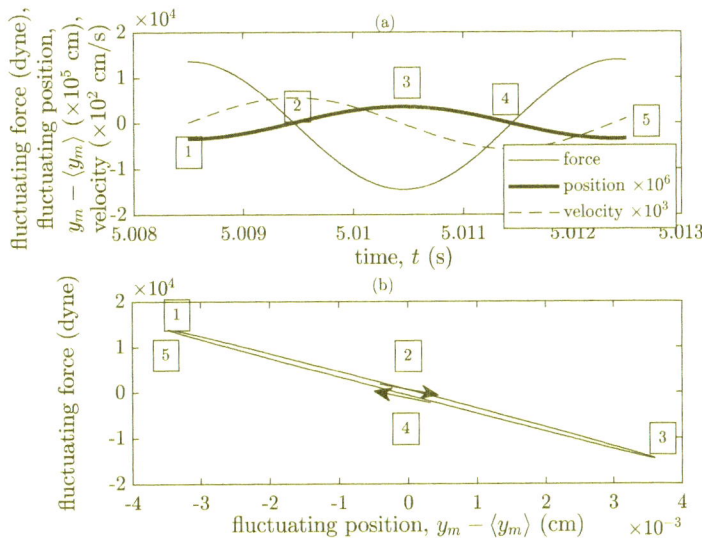

FIGURE 11. Details for $T = 2 \times 10^5$ dyne/cm: a) force on the mass-spring $\mathfrak{F}'_f$ position y_m and velocity of the mass-spring, b) force on the mass-spring $\mathfrak{F}'_f$ versus position y_m.

components results in a complex pattern of projections onto $\partial^2 p_a / \partial \theta^2$ in Figure 10(c). These projections are calculated by integrating the product of G_1^o or F_1^o and $\partial^2 p_a / \partial \theta^2$ from $-\theta_b$ to θ_b, and they appear on the right-hand-sides in Equation (34).

In order to understand the details of the mechanism by which energy is transferred from the air to the mass-spring system, the time interval from 5.0085 s to 5.0125 s is examined. Figure 11(a) shows the fluctuating force on the mass-spring system, the fluctuating mass position, and its velocity as a function of time, with five particular times indicated.[8] These times correspond to two minima, two zeros, and a maximum for mass positions y_m as a function of time. These same times are also indicated on Figure 11(b) and Figure 12(a), 12(b), where arrows represent the direction of increasing time. The plot of fluctuating force versus fluctuating mass position, $y_m(t)$, is shown in Figure 11(b). The integral of fluctuating force against the differential

---

[8]Fluctuating here means that the time average has been removed.

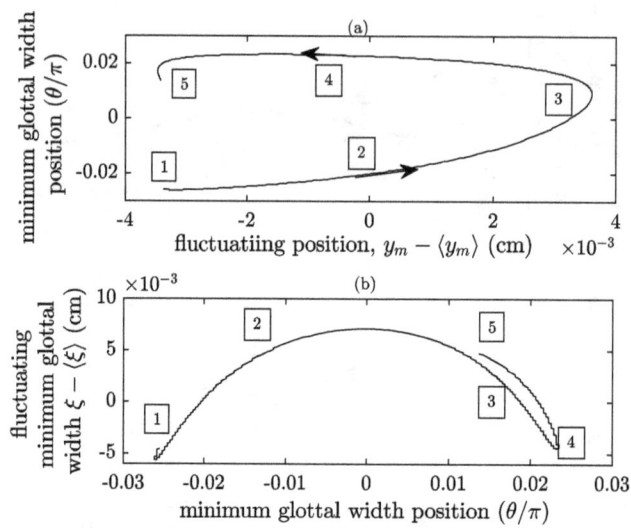

FIGURE 12. Details for $T = 2 \times 10^5$ dyne/cm: a) position of minimum glottal width versus position of the mass-spring y_m, and b) mimimum glottal width versus position of minimum glottal width

of position over the is positive, so there is net work done on the mass-spring system [≈ 1.9 erg] over the corresponding time interval.

The situation here is that of a lightly damped mass-spring system being driven by a force that is oscillating at approximately twice the natural frequency of the mass-spring system [approximately 250 Hz versus 120 Hz]. Figure 11(a) shows the expected phase relation between the fluctuating force and the mass velocity: the mass velocity lags the force by nearly $\pi/2$ in phase [Figure 2 in Chapter 7]. The mass-spring system being driven well away from its natural frequency means that power input by the fluctuating force is minimal.

Figure 12(a) is a plot that relates the location of the minimum width of the air channel as a function of θ versus fluctuating mass position. Figure 12(b) shows the fluctuating air channel minimum width versus its fluctuating position.

The movement of the separation point could enable maximum power from the fluctuating mass into the mass-spring system under a certain scenario. In this secenario the separation point would be farther downstream [i.e. with larger θ] during the upward movement of the mass [moving from negative to positive $y_m - \langle y_m \rangle$, or from time 1 to 3 in Figure 12(a)] than on its downward movement [moving from positive to negative $y_m - \langle y_m \rangle$, from time 3 to 5 in Figure 12(a)]. This would mean that more of the elastic sheet surface area would be exposed to the static pressure in Equation (36) behind the separation on the upward movement than on the downward movement. These static pressures should be greater than the static pressure downstream of the separation point. Less surface area would be exposed to the generally greater air pressure in the glottal jet during the upward movement than during the downward movement of the mass. However, this is not the case in Figure 12(a). In fact, the opposite occurs here. Also, Figure 12(b) is ambiguous as to the effect of minimum glottal width in energy exchange here.

Robustness of vibrations

The oscillations shown in the above simulations are small. It is shown presently that this may be due to physical factors that have been neglected in the model so far. One such factor is what could be called an entry condition for the air channel. For humans and other animals, the cross-sectional area change from the trachea to the glottal region is rapid, yet continuous. This change depends on the exact position of the vocal folds, and could produce large time-varying second derivatives of air pressure with respect to θ, which is the function of interest for the projections of the forcing functions in Equation (32). The entry conditions should affect F_o^1 more than G_1^o, because the former function possesses a non-zero second derivative with respect to θ at the entrance to the glottis ($\theta = -\theta_b$), while the latter function does not (Equations 24 and 25).

Therefore, a simulation was run with an additional $\partial^2 p_a / \partial \theta^2 = 10^3 \sin(\omega_{F1} t)$ dyne/cm^2 term imposed for $-\theta_b < \theta < -\theta_b + 0.1$ to see whether this plausible extra entry condition could make the phenomena observed above more robust. We still have $T = 2 \times 10^5$ dyne/cm in these simulations. Figures 13(a) and 14(a) show a greater amplitude in y_m oscillation in the case with the extra entry pressure second derivative imposed compared to the case when it is not imposed in Figure 8(b). The peak-to-peak amplitude is approximately 0.03 cm

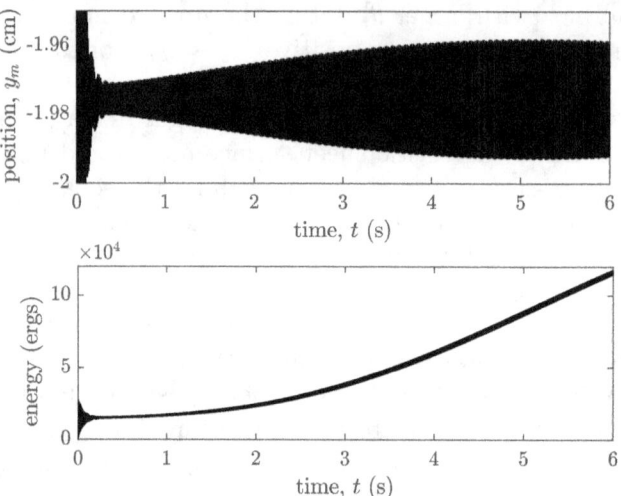

FIGURE 13. Mass position, y_m, and energy input for enhanced inlet air pressure amplitude

in the former figure compared to 0.007 cm in the latter figure. The amount of work that is done on the mass-spring system, discounting the work in the transient phase, in the case of the extra entry second derivative of pressure is about 10^4 erg in 6 s, as shown in Figure 13(b), compared to under 1200 erg without the extra entry pressure second derivative, as shown in Figure 9(b).

Figure 14 shows a reason for this more robust vibration. Figure 14(b) indicates that the amplitude of the F_1^o component in the ζ function is now about 0.6 that of the G_1^o component. Without the entry pressure second derivative this factor had been 0.1 in Figure 10(b). This could be due to the fact that the projection of both F_1^o and G_1^o onto the second derivative of air pressure are both larger with the added entry condition than without it as shown in Figures 14(c) and 10(c).

Figure 15 shows plots analogous to those in Figure 11, and Figure 16 are analogous to those in Figure 12, over the time interval 5.0085 s to 5.0125 s. The curve of fluctuating force plotted against fluctuating mass position, encloses a substantially greater area in Figure 15(b) [6.4 erg] than it does in Figure 11(b) [1.9 erg], meaning that substantially more work is done on the mass-spring system in the former case than in the latter. This is consistent with the greater energy input over 6

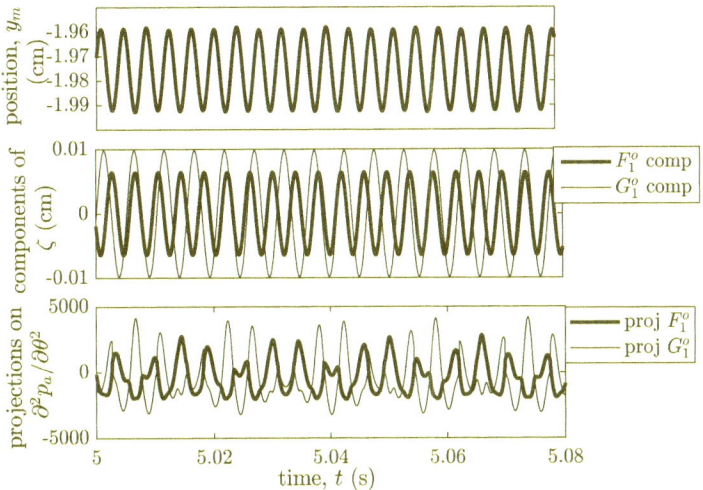

FIGURE 14. Details for oscillation in Figure 13. a) mass position y_m, b) components of ζ, and c) scaled projections of $\partial^2 p/\partial\theta^2$.

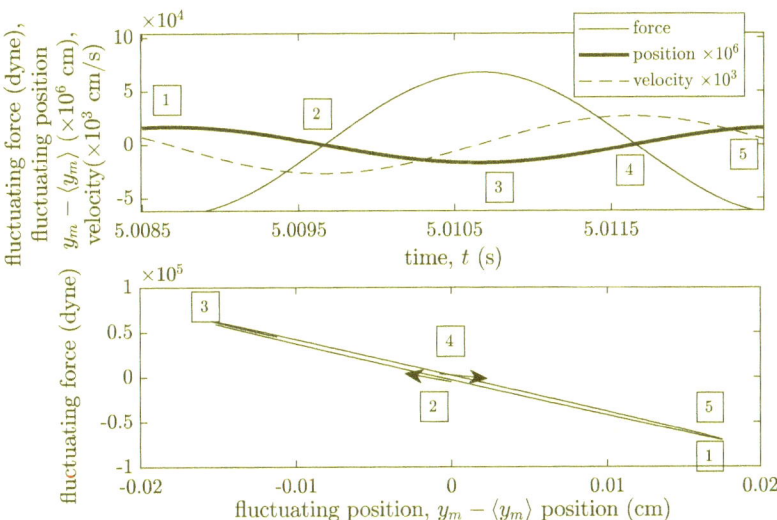

FIGURE 15. Details for oscillation in Figure 14. a) force on mass, mass position y_m, and mass velocity, and b) force on mass versus mass position.

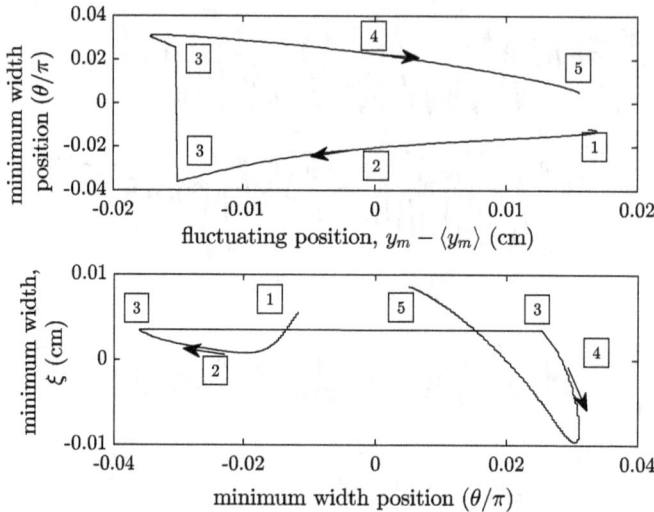

FIGURE 16. More details for oscillation in Figure 14. a) position of channel minimum width versus mass position y_m. and b) minimum channel width versus position

s duration shown in Figure 13 with the extra input condition than in Figure 9 without the extra input condition.

One cause of this greater energy input for the system with the extra entry condition is the fact that its separation point is farther downstream during the upward movement of the mass-spring system than during its downward movement, as seen in Figure 16(a). This is opposite to what Figure 12(a) shows for the system without the extra entry condition.

Finally, because the vibrations of the system with the extra second derivatives of pressure at the glottal entrance are so much larger than without this condition, a series of frames of the entire system is shown with the extra entry conditions in Figures 17 through 19. The 21 frames shown in these two figures are over the interval from 5.00805 s to 5.0125 s in the condition with the enhanced entry condition. The heavy horizontal line represents the mid-channel plane, the heavy wavy line is the elastic sheet surface, the light line of a circular sector is the interface between the incompressible fluid and the mass, or block. The vertical line from this interface and the fixed, horizontal dashed line

Acoustics of Speech Production

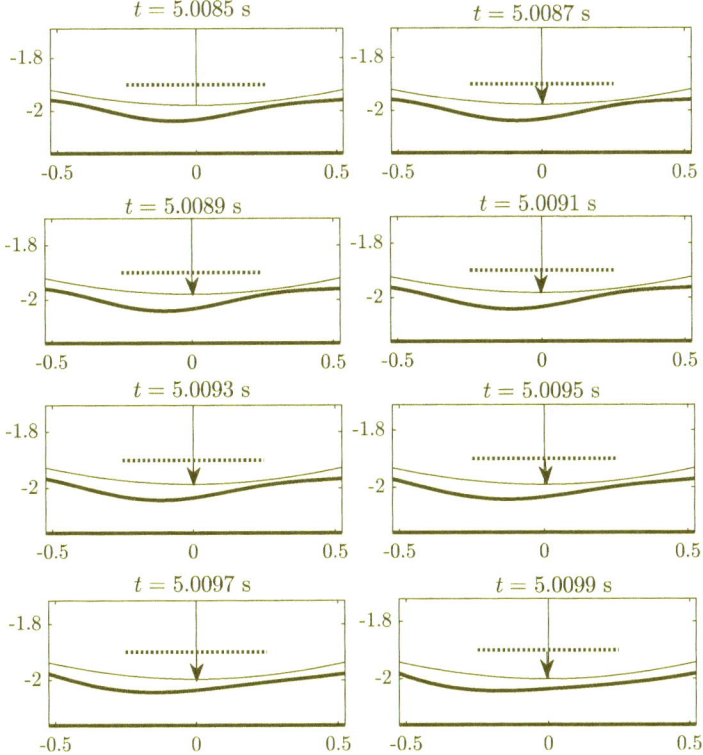

FIGURE 17. Motion of the model over one cycle

is intended to help the reader to visualize the vertical position of the mass. The arrow on the vertical line indicates the direction of the movement of the mass.

Extensions to larger vibration amplitudes with vocal fold collision

The model and simulations that have been presented do not have large amplitude vibrations that could involve collision of the vocal folds. Thus, this is only a preliminary examination of the model in terms of phonation. It is certainly possible to extend the current model to allow for collisions between the vocal folds. Also, this model should

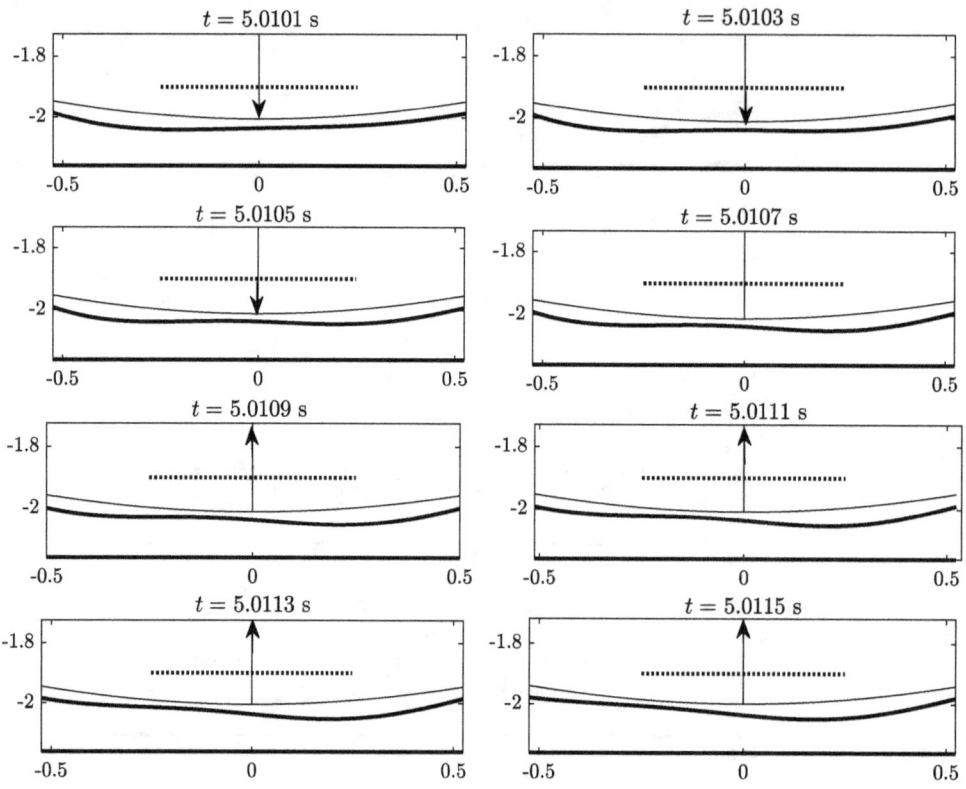

FIGURE 18. Motion of the model over one cycle, continued

be compared with the results of the empirical eigenfunction theory of vocal fold vibration (Berry, Herzel, Titze, & Krischer 1994).

There appears to be even more to the behavior of the epithelium when collision is present than a mucosal wave. In the considerations of Howe & McGowan (2013), based on data from Berry, Montequin, & Tayama, N. (2001), the maximum elevation of the epithelium remains located relatively far downstream in the glottis while the folds move apart, and then this maximum elevation moves rapidly upstream just before the folds begin to move together, and, then, the maximum elevation moves relatively slowly downstream as the vocal folds move close together. The simulation with the extra entry condition appears to model this behavior.

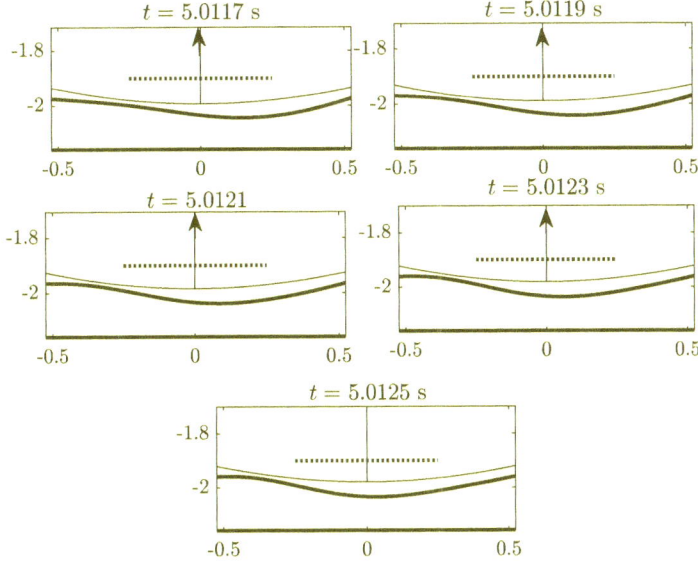

FIGURE 19. Motion of the model over one cycle, continued

Further refinement of this model would be to include the effect of the Coanda force as the folds move together. This was discussed in a previous footnote in the present chapter. During the closing phase of the vocal fold oscillation, the separation point would be forward of the point of the epithelium closest to the center-line of the glottis, and the Coanda force would be part of the total force acting to close the folds. There would be no Coanda force during the opening phase with the separation point occurring very close to the point on the epithelium closest to the center-line.

Conclusion

A specific mathematical model for vocal fold vibration based on the cover-body conception of the layered structure of the vocal folds has been outlined. The model folds oscillate for a range of elastic sheet tensions for one mass-spring stiffness. One mechanism that allows air to do work on the mass-spring system involves the hysteresis in the location of the separation point and the position of the mass.

There is much that should be done with the model. Simulations need to be run with more values for the natural (circular) frequency of the elastic sheet-shallow incompressible fluid system. Also, the effects of a variable depth incompressible fluid layer depth should be explored. It is certainly possible to extend the model to the situation of colliding vocal folds.

This work is part of a project to build a relatively simple mathematical model where there are clear correspondences to the physiology of the vocal folds, and whose physics can be understood. For instance, a non-zero bending stiffness for the epithelium may need to be added. It is also a model for which more details of the mechanics of the vocal folds can be added, such as more mechanical and constitutive properties of the Reinke's space material, so that their effects can be understood in the context of major vibration mechanisms.

References

Berry, D.A., Herzel, H., Titze, I., & Krischer, K. (1994). Interpretation of biomechanical simulations of normal and chaotic vocal fold oscillations with empirical eigenfunctions. *Journal of the Acoustical Society of America*, **95**, p 3395.

Berry, D.A., Montequin, D.W., & Tayama, N. (2001). High-speed digital imaging of the medial surface of the vocal folds, *Journal of the Acoustical Society of America*, **110**, p 2539.

do Cormo, M.P. (1976). *Differential Geometry of Curves and Surfaces*. Prentice-Hall, Inc., Englewood Cliffs, NJ (pp 1-25).

Courant, R. & Hilbert, D. (1953). *Methods of Mathematical Physics, Volume I*. Interscience Publishers, Inc., New York. (p 50; pp 358-360).

Fujimura, O. (1981). Body-cover theory of the vocal fold and its phonetic implications. In *Vocal Fold Physiology* K. N. Stevens and M. Hirano (Eds.) (University of Tokyo Press, Tokyo Japan). pp 271-81.

Hirano, M. (1974). Morphological structure of the vocal cord as a vibrator and its variations, *Folia Phoniatrica* **26**, p 89.

Hirano, M. (1988). Vocal mechanisms in singing: Laryngological and phoniatric aspects, *Journal of Voice*, **2**, p 51.

Howe, M.S. (2007). *Hydrodynamics and Sound.* Cambridge University Press, Cambridge, England. (pp 299-300).

Howe, M.S. & McGowan, R.S. (2010). On the single-mass model of the vocal folds. *Fluid Dynamics Research*, **42**, p 1.

Howe, M.S. & McGowan, R.S. (2011). On the generalized Fant equation, *Journal of Sound and Vibration*, **330**, p 3123.

Howe, M.S. & McGowan, R.S. (2013). Aerodynamic sound of a body in arbitrary, deformable motion, with application to phonation, *Journal of Sound and Vibration*, **332**, p 3909.

Ince, E.L. (1956). *Ordinary Differential Equations.* Dover Publications, Inc., New York. (p. 128).

Ishizaka, K. & Flanagan, J.L. (1972). Synthesis of voiced sounds from a two-mass model of the vocal cords. *The Bell System Technical Journal*, **51**, p 1233.

Ishizaka, K. & Matsudaira, M. (1968). Analysis of the vibration of the vocal cords. *Journal of the Acoustical Society of Japan*, **24**, p 311.

McGowan, R.S. & Howe, M.S. (2010). Comments on single-mass models of vocal fold vibration, *Journal of the Acoustical Society of America*, **127**, EL215-EL221.

Park, J.B. and Mongeau, L. (2007). Instantaneous orifice coefficient of a physical, driven model of the human larynx. *Journal of the Acoustical Society of Japan*, **121**, p 422.

Index

acceleration,
    of a point mass, definition of, 6
    of gravity, 8
    particle, 31

acoustic perturbation theory, 164, 167-77, 254-7, 272-84

acoustic quantities,
    first-order, 170
    second-order, 170

acoustic delay time, 49, 309

adiabatic,
    process, 13, 172-4, 321
    invariant, 173

adiabatic speed of sound, 12, 42

admittance,
    acoustic, 226
    effective, 233, 244-51
    radiation, 290-1, 306-7, 310-2
    reactive, 307
    resistive, 307
    tube, 236

adjoint problem, 183

aeroacoustics, 391-410

air particle, 13-5, 25, 366-70

air particle velocity, see velocity, particle

angular velocity, 366-7

anti-node, 135-6, 161, 208

approximation,
    acoustic, 25, 27-32, 61, 131, 170, 314, 320-1
    far-field, 307-8
    low-frequency, 234, 289-93, 306-8, 312
    quasi-steady, 36-4, 377, 393, 453, 455

area function, 268, 275-7, 421-4

atmospheric conditions, see normal atmospheric conditions

axis-of-rotation, 365-9

beat frequency, see frequency, beat

beating, 109, 121

Bernoulli equation, 364, 384, 453

boundary condition, 47, 62-9, 73-8, 137-9, 142-6, 148-9, 183-4, 203, 240, 252-3, 268, 270, 445

boundary layer, 379-80, 385-8
    acoustic momentum, 320-4
    acoustic thermal, 320-4

c-g-s units, 2

cavity affiliation, 284-5

cm H_2O, 11

chain rule, 59

circular function, 17-20, 69-72, 86-7, 132-3, 149-151, 197, 228

circulation, 371-3

Coanda force, 454, 467

complex
  conjugate, 222-3
  exponential function, 220-3
  number, 215-20
  plane, 215-8, 220-2
  variable, 132, 212-3, 215-30

compression, 43-9, 52-7, 61-9

conservation,
  energy, 27, 51, 363-4
  mass, 27, 29-30, 180, 182, 186, 314-5, 363, 445
  momentum, 27, 30-2, 138, 182, 186, 262-4, 315, 383-4

continuum, 13, 26, 30, 443

convolution sum, 105-6, 210
  double, 185, 187, 201, 210

convolution theorem, 212

Coriolis force, 391

Courant condition, 35

curvature, 413, 416-19, 439
  average spatial rate-of-change, 417-19
  spatial rate-of-change, see curvature, average spatial rate-of-change

damped sinusoid, 113, 117, 200

damping, 93, 111-30, 152, 154, 193-212, 225, 228, 305-6
  boundary layer,
    acoustic momentum, 305, 320-4
    acoustic thermal, 305, 320-4
  friction, 109, 111-23, 194
  radiation, 305-12
  wall vibration, 305, 313-9

damping constant, 112, 129, 305-6, 325-9

damping ratio, 114-7, 196, 198-9

decibel, definitions of, 56

density, 9
  perturbation, 12, 27, 33
  rest, 12, 27, 29-30

difference,
  centered first, 4
  centered second, 6

differential equation, 34-5, 399

displacement, 16

displacement position, see displacement

dissipation, 69, 111, 120

distensibility, 23, 315

distributed system, 149-50, 167, 169

dyne, 8

eddy, 407-8

effective length, 203, 291

Ehrenfest's theorem, 170, 172-4

eigenfunction, 139-46, 150, 241, 446-7, 449-51, 463, 465, 469
   semi-, 233, 239, 267-8
      left, 145, 183-4, 187, 240, 268
      right, 145, 183-4, 187, 240

eigenvalue, 140-2, 241, 268, 272, 325, 446-7,

energy,
   heat, 52, 111-2
   kinetic, 1, 19, 50
   potential, 1, 19, 50

energy dissipation, see dissipation,

energy density,
   kinetic, 50
   potential, 51
   total, 51

entropy, 12-3, 25, 27, 172, 320-1, 361

equation of state, 12, 27

erg, 50

Euler model, 360-91

Euler's theorem (for complex exponentials), 220-1, 230

Euler's theorem (regarding rigid body motion), 365

evanescent wave, 276-7

expansion, see rarefaction,

exponential damping, see damping,

extrapolation in time, see time-stepping,

Fant equation, 399

flanged opening, 62, 290, 306-12

flow,
   incompressible, 360, 363, 382, 384
   irrotational, 381-5
   shear, 367-9, 379
   turbulent, 189, 360, 407-10

flow separation, 386-91, 468

force, 7-8
   external, 1, 96-126, 194-9
   drag, 385, 388-91, 468

friction, 111-23
gravitational, 8, 11
restoring, 1, 15-6, 21-3, 50, 58, 443
sinusoidal, 196

formant synthesis, 210

Fourier analysis, 92, 156, 427

Fourier sum, 149

Fourier synthesis, 149

Fourier transform, 89, 228
    inverse, 89

fourth-order Runge-Kutta algorithm, 35, 399

frame(-of-reference),
    Eulerian, 14-5, 26, 30-1, 37, 47
    Lagrangian, 14-5, 31, 37-8, 47

frequency 17, 71
  beat, 109
  circular, 17, 71
    normal mode, 140, 199-200
      reduced, 200, 326
    forcing, 106, 116, 194
    natural, 17, 71, 93, 126, 194
      perturbed, 128-9
    reduced, 112
    resonance, 199, 203
  forcing, 106-29, 193, 196, 198, 201
  formant, 141, 210, 413-6, 433
  normal mode, 140-3, 149, 164-165, 167-91, 203-204, 239-44, 252-4, 272, 275, 283, 291-2, 294, 296-302
    reduced, 200, 203,
  mode, 319
    single path, 351-6
  natural, 17, 22, 71, 94, 104, 149, 196, 199, 288
    reduced, 112
  resonance, 199, 203

frequency domain, 86-91, 100, 183, 211-2, 229, 416, 446

friction, 12, 20, 30-1, 69, 111-7
    coefficient of, 112
    viscous, 319-21

gradient, 30-4, 40, 138-9, 143
    pressure, 263-4, 266

Gram-Schmidt orthogonalization, 450

Green's function, 129, 164, 167, 177-90, 199-201, 206-9, 329

half-wave resonator, 354, 433

heat conduction, 12, 320-4

Heisenberg's uncertainty principle, 91

Helmholtz resonator, 20-3, 150, 168-9, 333-47

Hilbert transform, 328

Hooke's law, 15, 21, 93

Howe's theorem, 391

imaginary number, 215
  part, 219

impedance
  acoustic, 225-6
  characteristic, 42, 227
  radiation, 290-1, 310-1

impulse, 96-105, 390-1
  unit, 97-9
  vortex, 390-1

impulse response function, 93, 96-105,
  113, 127, 129, 167

initial condition, 16, 94-116, 139,
  154-6, 162-4, 180, 183, 273

initial value problem, 94-7, 105-7,
  146-8, 154-64 179, 183

intensity, 53-7, 81-6, 124-8, 151-2

interference,
  constructive, 102-3,
  destructive, 103, 181

intra-oral pressure, see pressure,
  intr-oral

jet, 389-91, 397, 407-10, 424-34

jetting, 305-6, 312, 327

kinematic quantities, see properties,
  kinematic

Kramer-Kronig relations, 328

Lamb vector, 375-6, 390-1

line vortex, 370-3

Lighthill's acoustic analogy, 391-3

linear equation, 42

linear operator, 24, 150

linear superposition, see superposition,

linearization, 27-9, 42, 45

lumped mass element, 258, 261-5,
  286-302, 333-5, 396

lumped spring element, 333, 335-7

lumped system, 149-50

m-k-s units, 54

Mach number, 379
  acoustic, 45-6

mass, 2
  point, 2-3, 14

mass-spring system,
  friction damped 112-23
    forced, 118-9, 121-4, 194-9, 313
    linear, 93-130

simple, 15-7, 22-3, 50, 71

matching, 392-3

method of successive approximation, 170

mode, 201

momentum, 7, 96-8

motion,
   equation of, 16, 18, 93-4, 100-1, 105-6, 115, 190, 442, 445
   irrotational (air), 25, 27, 381-5, 393-6
   Newton's second law of, 7-8, 16, 30-1, 93
   Newton's third law of, 8
   one-dimensional, 25, 30, 36, 259,
   periodic, 17, 69, 72, 78, 84, 119, 172
   pure straining motion, 367, 369
   rotational (air), 25, 312, 320-1, 360, 365-7, 369, 371-2, 377, 382, 386, 397, 401-3
   simple harmonic, 17, 22, 72, 113
      damped, 113
   translational, 50, 364-5, 368-9
   wave, 23, 74, 318
      acoustic, 42, 40-50, 61, 69, 73, 78, 91, 131-2, 209-10
      dispersive, 318
      plane, 56, 233-4 321

multiplicative inverse, 221-3

Navier-Stokes equation, 30

nasal, 349-55

node, 135, 208

nonlinear steepening, 37-9, 45, 56

normal atmospheric conditions, 2, 9-10, 13, 20, 27, 321

normal direction, 9-10

normal mode, 150-64, 271-86

normal functions, see normal mode

no-slip condition, 320

normalized Gaussian, 90-91

one-dimensional propagation, see motion, wave, plane

orthogonal functions, see normal functions,

oscillation, 19, 72-3, 95, 113, 126, 401

particle velocity vector, 361

period, 17, 71

periodic system, see motion, periodic

phase, 71, 134, 194-99, 202, 219, 224, 272-4, 398
   initial, 71, 87, 95-6, 114, 134, 139, 148, 154-5, 180

relative, 193, 196-7, 226

phase speed, 315, 317-8

piriform sinus, 356

plane wave propagation, *see* motion, wave, plane

Plemelj relations, *see* Kramer-Kronig relations,

power, 52-6, 107-11, 118-24, 197-9, 201-8, 227-8, 288, 401-3, 407-9, 431

pressure, 2, 9-10
  atmospheric, 9-11, 21, 27, 44-5
  intra-oral, 432
  perturbation, 12, 27, 33, 44-7
  rest, 12, 27, 33, 173

pressure head, 364, 376, 453, 455

properties,
  thermodynamic, 8-13, 25-6, 226
  kinematic, 7-13, 33, 37, 42, 226

pulse,
  particle velocity, 35-7, 48, 61-9
  pressure, 33-9, 48, 61, 61-9

quarter-wave resonator, 138-47, 433

radiation, 203, 305-12, 352-3

radius-of-curvature, 417, 420

rarefaction, 43-5, 47-8, 51, 54-5, 62-63, 66, 69, 73

real part, 219

reflection coefficient, 237-9

retarded time, 49, 77, 309

Reynolds number, 379-80, 382

scaling, 407-10, 413-34

secant line, 4, 71

self-adjoint equation, 183, 449

self-adjoint problem, 183

side branch, 333, 347-56

signed amplitude, 243-7, 267, 272-3, 276-82, 296-300

sink,
  acoustic, 50, 65-6, 131, 148, 152, 154, 164, 167

sinusoid, *see* circular function

sloshing behavior, 203, 264

sound pressure level (SPL), 204-6
  overall (OASPL), 204-6, 431-2

source,
  acoustic, 49-50, 63, 65-6, 77, 131, 148-9, 152, 154, 164, 167, 178-90,

390-403, 407-10, 424-34
dipole, 398
mass, 179
monopole, 398
pressure, 178, 185-6, 188-9,
  206-208, 211, 398, 401, 410
volume velocity, 178-179, 188-189,
  199, 206-9, 398

source-filter theory, 130, 210-11, 397-8

spectrum,
  amplitude, see magnitude,
  continuous, 89-91, 100, 201
  discrete, 87-9, 148, 201
  magnitude, 88-91, 100, 148, 184

speed,
  of a point mass, 6
  particle, 361, 364

spherical spreading, 309

spring constant, 15, 17, 21-23, 33, 93,
  168-9, 313

spring stiffness, see spring constant,

statistical stationarity, 78, 393

steadiness, 78-86, 111, 118, 126, 131,
  190

steady conditions, see steadiness,

steady acoustic radiation pressure,
  170-6, 254-7, 277-88

steady phase, 118-126, 193, 209-10

steady state phase, see steady phase

stiffness, see spring constant

Stokes layer, see boundary layer,
  acoustic momentum

stress,
  normal, 9, 11
  pressure, 360
  viscous, 367-8, 386,
  turbulent 408-9

streamline, 258-62, 361-2

stream tube, 259-60, 362-3

superposition, 43-4, 57, 74, 94, 131,
  137

temperature, 2, 8, 11-3, 52, 321

thermal diffusivity, 321

thermodynamic quantities, see
  properties, thermodynamic

time domain, 86-91, 100, 184, 211,
  229, 449-50

time stepping, 35

transmission coefficient, 237, 239

transfer function, 211

transient, 117-26, 193-4, 202, 210

transient phase, *see* transient

transient response, see transient,

turbulence, see flow, turbulent

U-tube manometer, 10-1

velocity,
  of a point mass, 3-4
  particle, 26
  volume, 27
    unit, 179-82

vortex impulse, *see* impulse, vortex

vortex shedding, *see* flow separation,

vortex sheet, 367, 373-7, 379-80, 386-90, 401-2

vorticity, 321, 330, 367, 369-71, 390-3, 401-2

viscosity, 361, 379-380
  kinematic, 321, 378

wall vibration, 312-19

Watt, 54

wave,
  incident, 137, 224, 235-9
  reflected, 62-9, 73-8, 137, 235-9
  standing, 131-64, 239, 467
  transmitted, 131-2, 134-7, 146-9, 235-9
  traveling, 131-2, 224, 227, 273, 467-8

wave equation, 39-40, 42-3, 45, 57-60, 135-7, 139, 148-9, 182-3, 185-7, 208-9, 230, 236, 240, 265, 314-5

wavelength, 133-4, 140

wavenumber, 133-4
  normal mode, 131-67, 325-6
  mode, 323, 325

work, 173-4, 197, 458-64

weight, 8, 11,

zero-crossing, 176, 284-6

www.ingramcontent.com/pod-product-compliance
Lightning Source LLC
Chambersburg PA
CBHW080847020526
44118CB00037B/2248